HEAT PIPES and SOLID SORPTION TRANSFORMATIONS

Fundamentals and Practical Applications

HEAT PIPES and SOLID SORPTION TRANSFORMATIONS

Fundamentals and Practical Applications

EDITED BY
L.L. Vasiliev • Sadik Kakaç

CRC Press
Taylor & Francis Group
Boca Raton London New York

CRC Press is an imprint of the
Taylor & Francis Group, an **informa** business

CRC Press
Taylor & Francis Group
6000 Broken Sound Parkway NW, Suite 300
Boca Raton, FL 33487-2742

First issued in paperback 2017

ISBN 13: 978-1-138-07737-9 (pbk)
ISBN 13: 978-1-4665-6414-5 (hbk)

Library of Congress Cataloging-in-Publication Data

Heat pipes and solid sorption transformations : fundamentals and practical applications / editors, L.L. Vasiliev, Sadik Kakaç.
 pages cm
 Includes bibliographical references and index.
 ISBN 978-1-4665-6414-5 (hardback)
 1. Heat pipes. I. Vasiliev, L. L. (Leonard Leonidovich) II. Kakaç, S. (Sadik)

TJ264.H425 2013
621.402'5--dc23   2013004219

Visit the Taylor & Francis Web site at
http://www.taylorandfrancis.com

and the CRC Press Web site at
http://www.crcpress.com

Contents

Preface.. vii

Editors... xi

Contributors... xvii

1 A Review of Modelling Approaches to Heat and Mass Transfers
   in Porous Wicks 1
   Sassi Ben Nasrallah and Marc Prat

2 Thermally Powered Adsorption Cooling: Recent Trends and
   Applications... 29
   B. B. Saha and I. I. El-Sharkawy

3 Optimisation of Adsorption Dynamics in Adsorptive Heat
   Transformers: Experiment and Modelling... 63
   Yuri I. Aristov

4 Mechanisms of Intensive Heat Transfer for Different
   Modes of Boiling 109
   Victor V. Yagov

5 A Review of Practical Applications of Heat Pipes and
   Innovative Application of Opportunities for Global Warming 145
   M. Mochizuki, A. Akbarzadeh and T. Nguyen

6 Heat Pipes and Thermosyphons for Thermal Management of
   Solid Sorption Machines and Fuel Cells 213
   L. L. Vasiliev and L. L. Vasiliev Jr.

7 Modelling of Heat and Mass Transfer in Sorption and
   Chemisorption Heat Converters and Their Optimisation................ 259
   L. L. Vasiliev, O. S. Rabinovich, N. V. Pavlyukevich and M. Yu. Liakh

8 Sorption Systems with Heat Pipe Thermal Management for
   Hydrogenous Gas Storage and Transportation 283
   L. L. Vasiliev and L. E. Kanonchik

9 Fundamental Questions of Closed Two-Phase Thermosyphons 319
   M. K. Bezrodny

10 **Thermal Control Systems with Variable Conductance Heat Pipes for Space Application: Theory and Practice** 357
    V. M. Baturkin

11 **Thermosyphon Technology for Industrial Applications** 411
    Marcia B. H. Mantelli

12 **Fluid Flow and Heat Transfer with Phase Change in Minichannels and Microchannels** 465
    V. V. Kuznetsov and S. A. Safonov

Index 497

Preface

This book covers the state of the art of adsorption research and technologies for relevant applications based on the use of the efficient heat transfer devices—heat pipes and two-phase thermosyphons—with the objectives of energy efficiency and sustainability. The severities of energy crisis and environmental problems have been calling for rapid developments in Freon-free air conditioning and heat pump technologies, the heat exchangers, which are considered as the components of prime importance. The concerns of energy consumption and environmental pollution urge researchers to work on the development of clean energy and the utilization of waste energy. From this viewpoint, interest in fuel cells and thermally activated (heat pipe heat exchangers) adsorption systems using natural refrigerants and/or alternative to hydrofluorocarbon-based refrigerants has increased significantly. The quest to accomplish a safe and comfortable environment has always been one of the main preoccupations of the sustainability of human life. Accordingly, during the past few decades, research aimed at the development of thermally powered adsorption cooling technologies has intensified. They offer two main benefits: (1) reduction in energy consumption and (2) adoption of environmentally benign adsorbent/refrigerant pairs, without compromising the desired level of comfort conditions.

The efficiency of new power sources (co-generation, tri-generation systems, fuel cells, photovoltaic systems) can be increased with the help of heat pipe heat exchangers, solid sorption heat pumps, refrigerators, accumulators of heat and cold, heat transformers, and fuel gas (natural gas and hydrogen) storage systems. Low-temperature power systems are generally significantly less expensive to build than high-temperature ones. Since a major barrier to acceptance, this is a main concern for fuel cell technology, in general. All these arguments are considered as the fundament to launch this book. Finally, the heat pipe thermal control of the spacecraft is also considered in detail in some chapters of the book.

The analysis of transport phenomena (including heating and cooling) in porous media using heat pipes and thermosyphons technology is considered in detail in Chapters 1, 5, 6, 10, and 11 of this book. Against the fact that technology of two-phase cooling systems such as heat pipes is well developed and known, some aspects of the fundamental phenomena are still not analysed in detail. The latest development in the modelling of heat and mass transfers in porous wicks in relation to the study of conventional heat pipes, loop heat pipes, and two-phase thermosyphons are is welcome. The main numerical approaches are analysed, and the advantages and drawbacks of direct methods, semi-direct methods (pore network models), and mean-field classical methods (Darcy's scale) are considered and discussed.

The global warming crisis is real and it is our responsibility to put forth an effort to reduce and prevent further damage to the global environment. One of the efficient tools to do this is heat pipe and thermosyphons application in different branches of industry. Chapter 5 contains a detail review of numerous successful heat pipe applications ranging from computer electronics to renewable energy. Heat pipes, vapour chambers, and thermosyphons have emerged as the most significant technology and cost-effective thermal solution owing to their excellent heat transfer capabilities, high efficiency, and structure simplicity.

Chapter 6 deals with some aspects of the heat pipe–based thermal control of fuel cells, solid sorption transformers, and electronic components and air-condition devices in a renewable energy context. Firstly, it briefly relates heat pipe design for fuel cells application, adsorption cooling and heating (including solar cooling), snow melting, and ground heating and cooling.

Finally, it outlines different heat pipe designs based on nanofluid and nanocoating application. The combination of fuel cell and solid sorption transformer is also discussed in this chapter based on heat pipe technology application.

The basic principles of adsorption cooling in terms of adsorption equilibrium, kinetics, and heat of adsorption are discussed in Chapters 2, 3, 7, and 8. Some thermally activated adsorption cooling cycles are overviewed. These systems have the following advantages: firstly, exploiting renewable energy or waste heat of temperature below 100°C, and secondly, using very low electricity for the circulation of heat transfer fluids (hot, cold, and chilled water). Finally, a three-bed dual evaporator-type advanced adsorption cooling cum desalination cycle is shown. The evaporators work at two different pressure levels and produce simultaneously cooling effects and potable water from a single low-grade energy input. Silica/water-based direct contact condensation and evaporation cycles can significantly improve heat and mass transfer. Advances in understanding and trends in studying the adsorption dynamics (Chapter 3) are the subject of prime interest as the booming progress in the materials science offers a huge choice of novel porous solids, which may be used for adsorption transformation of low-temperature heat. An overview of original and literature data on several classes of materials potentially promising for some very important applications, namely, metalaluminophosphates metal–organic frameworks, ordered porous solids, and various composites a useful for further its application in modern designs of adsorption coolers and heat pumps using heat pipe thermal management to increase the efficiency of sorption cycles. A mathematical model of coupled heat and mass transfer during the isobaric adsorption process has been used for detailed dynamical analysis and for extraction of the heat and mass transfer coefficients. The most important findings and general regularities that have been revealed for systematic studying of the adsorption dynamics of water, methanol, and ammonia in adsorption heat transformer systems are summarised, illustrated, and discussed.

The fundamental mechanisms of intensive heat transfer, which are observed not only in nucleate boiling but also in transition and film boiling (Chapter 4), are typical not only for heat pipes and thermosyphons but also for other two-phase heat transfer devices, including the subcooled liquid boiling. The availability of three-phase boundaries is considered as a peculiarity of nuclear boiling, which distinguishes it from other modes of convective heat transfer. It is important to note that the only mechanism that can explain the extremely high heat transfer intensity of boiling is evaporation in the vicinity of interlines, that is, the boundaries of contact of three phases. Chapter 7 covers some complex models of sorption and chemisorptions heat pumps and refrigerators. These models have been elaborated in Chapters 2, 3, 7, and 8 on the basis of the present state of the art in the field of transport phenomena in porous media. The models describe thermodynamics and kinetics of sorption processes as well as transport of single- and two-phase sorbate fluid in adsorbers and its phase transitions (condensation and evaporation). A set of computer programs for numerical modelling and optimization of sorption heat converters has been developed with the use of elaborated models.

The programs allow investigation of various heat conversion devices designed as a combination of one or several reactors with condensers and evaporators. The programs have the options for varying reactor parameters, describing sorption processes and heat and mass transport phenomena in it, temperatures of heat carrier thermal reservoirs, resistance of control valves connecting the reactors, and time of their opening and closing. As a result of modelling, the programs help to calculate main characteristics of thermal energy conversion such as the coefficient of performance, the specific heat production (specific cold production), and the maximal temperature difference between the working fluid and adsorbent during the process. Besides, the time and space distribution of temperatures and concentrations over the reactors is available in tables and graphical representation. The concept of flow boiling and flow condensation heat transfer in microchannels is analysed in detail in Chapter 12, which guarantees the novel cooling strategies in the areas requiring successful thermal management. The influence of capillary forces, disjoining pressure, wall roughness, and vapour shear stress on configuration of the microscale interface and the mathematical model of the annular flow in a rectangular microchannel is considered. The stability of meniscus in the channel corners and establishment of limit radius of meniscus curvature at high vapour velocities seem to be the limiting factors for the accumulation of liquid in the corners. A pressure drop model was proposed for rectangular microchannel which accounts the capillary pressure using statistical parameters of a gas–liquid flow. Some features of evaporating and condensing heat transfer in a small-sized channel at low heat fluxes are considered for closed and open microchannels. Experimental data and numerical modelling of evaporation and condensation inside the rectangular channel show the significant effect of capillary forces on film surface curvature and heat transfer.

We are very grateful to all the authors, Sassi Ben Nasrallah, Marc Prat, B.B. Saha, I.I. El-Sharkawy, Yuri I. Aristov, Victor V. Yagov, Masataka Mochizuki, Aliakbar Akbarzadeh, Thang Nguyen, O.S. Rabinovich, N.V. Pavlyukevich, M.Yu. Liakh, Larisa E. Kanonchik, Mikhail K. Bezrodny, V.M. Baturkin, Marcia Barbosa Henriques Mantelli, Vladimir V. Kuznetsov, and S.A. Safonov, who provided the substance of this book and its full success.

During the preparation of this book, many colleagues at the A.V. Luikov Heat and Mass Transfer Institute were involved in the fruitful collaboration. We are thankful to the members of the Porous Media Laboratory and A.V. Luikov Heat and Mass Transfer Institute. Our heartful thanks are extended to Drs. A.A. Antukh, Grakovich, A.G. Kulakov, D.A. Mishkinis, M.I. Rabetsky, and A.S. Zhuravlyov.

We also thank Laurie Schlags, project coordinator, Editorial Project Development, Jonathan W. Plant, executive editor, and other individuals at CRC Press for their close collaboration and guidance during the preparation of this book.

L. L. Vasiliev
Porous Media Laboratory, A.V. Luikov Heat and
Mass Transfer Institute
Minsk, Belarus

Sadik Kakaç
Department of Mechanical Engineering,
TOBB University of Economics and Technology,
Ankara, Turkey

Editors

L. L. Vasiliev is the president of the New Independent States (NIS) Scientific Association 'Heat Pipes' and a general leading scientist of Porous Media Laboratory, A.V. Luikov Heat and Mass Transfer Institute, National Academy of Sciences, Belarus. He was born on 3 January 1937 in Simferopol, Crimea, the USSR.

He started his scientific career with studies of thermal properties (thermal conductivity, heat capacity and thermal diffusivity) of solid materials at cryogenic temperatures and developed a new non-stationary method of their measurements (1960–1964) at A.V. Luikov Heat and Mass Transfer Institute under the supervision of Professor Alexis Luikov. He received his first doctoral degree (Candidate of Sciences) in 1964 in Minsk with the thesis 'Thermal Properties of Solid Materials at Cryogenic Temperatures'. In 1972, he got his second doctoral degree (Doctor of Science) in thermal physics from A.V. Luikov Heat and Mass Transfer Institute with the thesis presentation 'Heat and Mass Transfer in Capillary Porous Media Based on Heat Pipe Phenomena'. In 1978, he was elected as a professor of mechanical engineering in Belarusian Academy of Sciences. From 1967 to 1986, he was the head of Cryogenic Laboratory of A.V. Luikov Heat and Mass Transfer Institute. During the same period, he gave lectures on two-phase heat and mass transfer in capillary porous media in Belarusian State University and Belarusian Polytechnic University. In 1986, Cryogenic Laboratory of A.V. Luikov Heat and Mass Transfer Institute was renamed as Porous Media Laboratory, and Professor Vasiliev remains as the head of this laboratory.

Twenty-seven doctor/engineers can call him a supervisor; three were habilitated under his supervision. Three of his disciples are now the head of the laboratory in A.V. Luikov Heat and Mass Transfer Institute.

Professor Vasiliev was frequently invited to visit as a guest professor or leading scientist in scientific institutions all over the world. In 1967, he had his stage at the Aerothermodynamics Laboratory in Meudon, France, and spent 10 months there. During 1969–1982 period, he was regularly invited as visiting scientist to the Institute of Thermomechanics, Academy of Sciences of the Czech Republic, Prague, and to the State Institute of Machinery, Behovize, Czechoslovakia. Many times he was invited as a visiting professor to some institutions in Poland, Bulgaria, and Baltic countries.

He contributed his knowledge and scientific experience to many national and international scientific associations and committees.

From 1967 to 1975, he was a member of the editorial board of *Journal of Engineering Physics and Thermophysics*—National Academy of Sciences of Belarus (translated to English by Springer Inc.). Since 1981, he has been a member of advisory board of *Journal of Applied Thermal Engineering* (in the past, *Journal of Heat Recovery Systems*), Elsevier. Since 1997 he has been an associate editor of *International Journal of Thermal Sciences*, Elsevier; in January 2006, he was elected as a member of editorial board of *International Journal of Low Carbon Technologies*, Manchester University Press.

In 1976, he was elected as the member of the International Scientific Committee of the organizing conference on Heat Pipes. In 1991, he was elected as the president of the NIS Scientific Association Heat Pipes and enjoying his position till now. He was the editor of the Proceedings of the Seventh International Heat Pipe Conference (two volumes) (Begell House Inc., 1993), co-editor of *Microscale Heat Transfer—Fundamentals and Applications* (Springer, 2005), and co-editor of *Mini-Micro Fuel Cells—Fundamental and Applications* (Springer, 2008).

In 1993, he was the initiator of the series of international conferences 'Minsk Seminars—Heat Pipes, Heat Pumps, Refrigerators, Power Sources' and the editor of the eight Proceedings of these Conferences held in Minsk. The last one, 'Heat Pipes, Heat Pumps, Refrigerators, Power Sources', was held on 12–15 September 2011.

He was elected as scientific committee member of international conferences in Russia, Belarus, China, India, Poland, Canada, France, England, Brazil, Korea, Australia, and Italy several times.

Now he is a member of the Scientific Council of the International Centre for Heat and Mass Transfer, which demonstrates his esteem and respect in the scientific world.

In recognition of his national and international scientific reputation, Professor Vasiliev obtained the Arcot Ramachandran Professorship in Madras Institute of Technology, India, in 2005, and numerous prestigious awards such as Belarusian State Award (1981), USSR Council of Ministers Award (1983), and Czechoslovak Silver Medal (1986). He was awarded one gold, two silver, and one bronze medals of USSR Exhibitions on Advances of Industrial and Agricultural Products in Moscow, USSR. In 1988, he was awarded the silver medal of USSR Exhibitions for his book with co-authors *Hydrodynamic and Heat Transfer in Porous Components of Aircraft Design* (Mashinostroenie, 1988) in Moscow.

In 2005, he was given the Luikov Award and Medal of National Academy of Science of Belarus. In 2011, he was given the George Grover Distinguish Scientist Award and Medal as an appreciation for his outstanding contribution to the development of the science and technology of heat pipes (Committee on International Heat Pipe Conferences).

He authored or co-authored about 340 publications in journals and conference proceedings. He is the author and co-author of 14 books, has 218 Soviet author certificates on invention, and owns 12 patents.

His main fields of activity were and still are heat and mass transfer in capillary porous media with phase change, thermophysical properties of solid materials at low temperatures, heat pipes, heat pumps, heat exchangers, solid sorption machines, solar collectors, energy recovery systems, thermal and gas storage systems, electronic components cooling, and mini/microscale heat transfer with phase change.

In addition to the scientific excellence he demonstrated throughout his career, we want to emphasize the important role he played during his guidance for 45 years in the Porous Media Laboratory, A.V. Luikov Heat and Mass Transfer Institute, National Academy of Sciences. The practical importance of the efficiency of Porous Media Laboratory is the fact that his colleagues were awarded 34 medals for different merits in their scientific activity.

Sadık Kakaç was born on 1 October 1932 in Çorum, Turkey. He received his degree of Dip.-Ing (1955) from the Department of Mechanical Engineering at the Technical University of Istanbul (ITU).

Professor Kakaç joined the chair of Heat Technique as a research/teaching assistant (1955–1958) in the Department of Mechanical Engineering at ITU, and then he went to Massachusetts Institute of Technology (MIT) under a scholarship to study nuclear engineering; he received his MS in mechanical engineering in 1959 and his MS in nuclear engineering in 1960, both from MIT. Receiving a scholarship from the UNESCO in 1963, he went to UK, and in 1965, he received his PhD from the Victoria University of Manchester, UK.

He started his academic life in the Department of Mechanical Engineering at the Middle East Technical University (METU) (1960) as an instructor. He then became associate professor (1965) and professor (1970); then he also served different governmental and research positions; he was elected as a member of the Turkish Scientific and Technological Research Council (1972–1980) and appointed as the secretary general of the Turkish Atomic Energy Commission (1978–1980), representing Turkey in a number of scientific endeavours abroad as a member of NATO Science Committee (1979–1980), the OECD NEA Steering Committee (1978–1980), the Cento Scientific Coordinating Board (1972–1974). He served as the chairman of the Department of Mechanical Engineering at METU (1976–1978), and then he was invited as a visiting professor to the Department of Mechanical Engineering at the University of Miami (1980–1982), and in 1982, he was appointed as a full-time professor of the Department of Mechanical Engineering at the University of Miami with tenure. He served as the chairman of the department during 1990–1998.

He has maintained his high quality of research for more than 50 years. He has authored or co-authored, publishing over 200 scientific papers on transient and steady-state laminar, turbulent forced convection, two-phase flow instabilities, fuel-cells modelling, micro heat transfer with sleep flow, and recently, he has been concentrated in research on convective heat transfer enhancement with nanofluids in single-phase and two-phase flow conditions. He performed very valuable contributions in all these areas, and he has been known as one of the most recognized scientists in the field of heat transfer.

He is highly respected by his peers. He has received many international recognition: Alexander von Humboldt Senior Distinguished U.S. Scientist Award for his outstanding contributions on heat transfer and two-phase flow was bestowed to him in 1989; Science Award by the Association of Turkish–American Scientists in 1994; American Society of Mechanical Engineers (ASME) Heat Transfer Memorial Award in 1997. Because of his contributions to research and education through his research work and books, he received distinguished service awards from the Middle East Technical University in 1998 and from the Turkish Scientific and Technical Research Council in 2000. He has received a 'lifetime achievement' award from the TOBB University of Economics and Technology, Ankara (2010) and iNEER Leader Award for his contribution to International Network for Engineering Education and Research (2012).

He received the Doctor Honoris Causa from the University of Ovidius, Romania (1998), the University of Reims, France (1999), and Odessa State Academy of Refrigeration (2007). He is a member of the Turkish Academy of Sciences (1999), and a foreign member of the Academy of Sciences of the Republic of Bashkortostan of Russian Federation (1998), and the Brazilian Academy of Sciences (2012). He is an honorary professor of Shanghai Institute of Electrical Power (1986), Xi'an Jiaotong University, China (1988), and Gandhi Institute of Technology and Management College of Engineering, India (1992); he is a fellow member of ASME.

He is a member of the scientific council and executive committee, and fellow of the International Centre of Heat and Mass Transfer.

He has organized and directed well-known NATO Advanced Study Institutes on various topics of thermal and fluid sciences for the last 35 years bringing the eminent scientists as lecturers and the young scientists together from NATO and non-NATO countries for exchange of ideas, educating young scientists, and dissemination of information on emerging topics providing very valuable service to the heat transfer community.

He has been frequently invited as lecturer and speaker by various institutions in the United States, Europe, China, Malaysia, Singapore, Brazil, and other countries; he organized short courses on the thermal design of heat exchangers in Taiwan, Singapore, Thailand, France, and Turkey.

He has been serving as an editor in editorial board of major journals on heat and mass transfer, and energy. He is the author or co-author of very popular textbooks of *Convective Heat Transfer* (CRC Press, 1995), *Thermal Design of Heat Exchangers* (CRC Press, 1997), and *Heat Conduction* (CRC Press, 2008); he edited 14 volumes in the field of thermal sciences and heat exchanger fundamentals and design, including *Handbook of Single-Phase Convective Heat Transfer* (John Wiley, 1987) and *Boilers, Evaporators, and Condensers* (John Wiley, 1991), which all became permanent reference books in the field.

Contributors

Aliakbar Akbarzadeh
Japan Thermal Technology Division
Fujikura Ltd.
Tokyo, Japan

Yuri I. Aristov
Boreskov Institute of Catalysis
Russian Academy of Sciences
Novosibirsk, Russia

V.M. Baturkin
National Technical University
  of Ukraine "Kyiv Polytechnic
  Institute"
Kyiv, Ukraine

and

Institute of Space Systems
Bremen, Germany

Sassi Ben Nasrallah
Laboratoire d'Etudes des Systèmes
  Thermiques et Energétiques
Ecole Nationale d'Ingénieurs de
  Monastir
Université de Monastir
Monastir, Tunisia

Mikhail K. Bezrodny
Heat-and-Power Engineering
  Department
National Technical University
  of Ukraine "Kyiv Polytechnic
  Institute"
Kyiv, Ukraine

I.I. El-Sharkawy
Mechanical Power Engineering
  Department
Faculty of Engineering
Mansoura University
El-Mansoura, Egypt

Larisa E. Kanonchik
A.V. Luikov Heat and Mass Transfer
  Institute
National Academy of Science of
  Belarus
Minsk, Belarus

Vladimir V. Kuznetsov
Laboratory of Multiphase Systems
Kutateladze Institute of
  Thermophysics
Siberian Branch of Russian
  Academy of Sciences
Novosibirsk, Russia

M. Yu. Liakh
A.V. Luikov Heat and Mass Transfer
  Institute
National Academy of Science of
  Belarus
Minsk, Belarus

Marcia B. H. Mantelli
Mechanical Engineering
  Department
Heat Pipe Laboratory
Federal University of Santa
  Catarina
Florianópolis, Santa Catarina,
  Brazil

Masataka Mochizuki
Japan Thermal Technology Division
Fujikura Ltd.
Tokyo, Japan

Thang Nguyen
Japan Thermal Technology Division
Fujikura Ltd.
Tokyo, Japan

N.V. Pavlyukevich
A.V. Luikov Heat and Mass Transfer
    Institute
National Academy of Science of
    Belarus
Minsk, Belarus

Marc Prat
Institut de Mécanique des Fluides
    de Toulouse
Université de Toulouse
Toulouse, France

O.S. Rabinovich
A.V. Luikov Heat and Mass Transfer
    Institute
National Academy of Science of
    Belarus
Minsk, Belarus

S.A. Safonov
Kutateladze Institute of
    Thermophysics
Siberian Branch of Russian
    Academy of Sciences
Novosibirsk, Russia

B.B. Saha
Mechanical Engineering Department,
    Faculty of Engineering
International Institute for Carbon-
    Neutral Energy Research (I2CNER)
Kyushu University
Fukuoka, Japan

L. L. Vasiliev
A.V. Luikov Heat and Mass Transfer
    Institute
National Academy of Science of
    Belarus
Minsk, Belarus

Leonard L. Vasiliev, Jr.
A.V. Luikov Heat and Mass Transfer
    Institute
National Academy of Science of
    Belarus
Minsk, Belarus

Victor V. Yagov
National Research University
    Moscow Power Engineering
    Institute
Moscow, Russia

1

A Review of Modelling Approaches to Heat and Mass Transfers in Porous Wicks

Sassi Ben Nasrallah and Marc Prat

CONTENTS

1.1   Introduction..1
1.2   Digital Porous Media...3
1.3   Direct Approaches..4
1.4   Pore-Network Approach..6
1.5   Models Based on the Continuum Approach 14
       1.5.1   Introduction..14
       1.5.2   Non-Equilibrium Models 16
       1.5.3   Local Thermal Equilibrium Models..18
       1.5.4   Local Thermal Equilibrium Models with Explicit Tracking
                of Macroscopic Interfaces 18
       1.5.5   Multi-Phase Mixture Model..19
       1.5.6   Application to Evaporators...21
1.6   Conclusion 23
Nomenclature..24
References..25

1.1 Introduction

Although the technology of two-phase cooling systems, such as conventional heat pipes or loop heat pipes (LHPs) or capillary pumped loops (CPLs), is well developed and of widespread use (Peterson 1994, Faghri 1995), many aspects of it are not well understood (Smirnov 2010). Furthermore, the improved design of such systems still often requires costly series of trial and error experimental campaigns. It is, therefore, desirable to develop numerical tools aiming at both improving our understanding of systems and helping the engineer in designing improved systems with a limited number of experimental tests. The increasing demand for miniaturised devices is also a factor for developing advanced numerical simulations. Developing numerical simulations can be interesting at various scales: at the scale of the system (a whole LHP, e.g. as in Launay et al. [2007] and Kaya et al. [2008]), a component of the system (e.g. the evaporator or the condenser of an LHP), a subelement of a component (the porous wick of a heat pipe or an LHP) or even at smaller scales (e.g. the cross section of a single groove of a heat pipe; see Stephan and Busse 1992). In this chapter, we focus on the numerical simulation tools at the porous wick scale. The porous wick is a key component of the evaporator in an LHP and is often encountered in conventional heat pipes as well as in many other devices (thermal spreaders, etc.). The coupled heat and mass transfers with liquid–vapour phase change occurring in an LHP evaporator porous wick are encountered in many other domains of engineering, such as nuclear engineering (Lipinski 1984, Fichot et al., 2006), geothermal engineering (Woods 1999), and drying (Perre et al., 1993). Although our interest is mainly cooling devices, the models presented in this chapter are clearly of broader interest.

As sketched in Figure 1.1, simulations and the associated models can be developed at various scales. The microscopic approaches refer to *ab initio* simulations, molecular dynamics and associated theoretical approaches, such as the density functional theory. These approaches are useful for the fundamental study of the phenomena at the scale of a single pore (typically of nanometric size) over a very small time scale. To the best of our knowledge, this type of approach has not yet been applied in relation with the study of transfer in the porous wick of two-phase cooling devices, and therefore is not discussed in this chapter. The next scale is the pore scale (also referred to sometimes as 'mesoscale'), in which direct approaches can be developed. The governing equations are solved at the pore scale over digital porous structures. On account of computational costs, the direct simulations are generally limited to domains containing only a few pores. By contrast, the pore-network simulations allow much larger domains containing up to 10^6 pores, but with important simplifications in the modelling of transport phenomena between two adjacent pores. The last scale considered

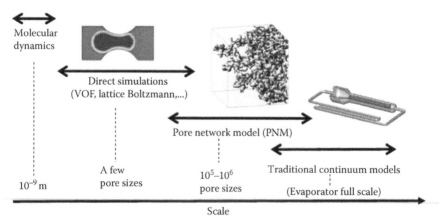

FIGURE 1.1
Various methods for the numerical simulation of transport phenomena in porous media.

in this chapter is Darcy's scale. This scale corresponds to the most classical approach, in which the porous medium is viewed as a fictitious continuum or a set of interacting fictitious continua.

In this chapter, we give an overview of these various approaches (with the exception of microscopic approaches), with a special focus on pore-network and continuum approaches.

1.2 Digital Porous Media

Before presenting the various simulation approaches, it is worth mentioning the concept of digital porous media. Digital porous media refer to numerical porous structures. These numerical porous structures can be used as input data for the approaches described in Sections 1.3 through 1.5, especially with regard to the direct and the pore-network approaches. Three-dimensional (3D) numerical porous structures can be obtained, for example, from direct imaging of real porous media using micro x-ray computerised tomography (Coker et al., 1996, Manke et al., 2011), nuclear magnetic resonance imaging (Baldwin et al., 1996) or focused ion beam scanning electron microscopy (Karwacki et al., 2011). As an example, Figure 1.2 shows the reconstructed

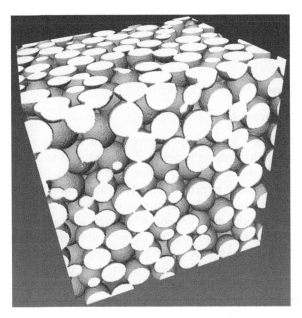

FIGURE 1.2
Digital random packing of spherical particles obtained from a real random packing using micro x-ray computerised tomography at IFPEN. (From Horgue, P., *Ph.D. dissertation*, University of Toulouse, 2012.)

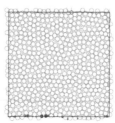

FIGURE 1.3
Example of numerically generated porous matrix.

image of a real random packing of spherical particles obtained using micro x-ray computerised tomography (Horgue 2012).

A popular alternative is to generate the microstructures numerically. This can be done starting from some statistical properties of a real porous structure (Liang et al., 1998) or purely numerical (Torquato 2002), a typical example being the numerical generation of a random packing of spherical particles (Soppe 1990).

Figure 1.3 shows a simple example in which the porous matrix is formed by a two-dimensional (2D) array of randomly distributed discs.

Once a numerical microstructure is obtained, the next step is to compute the transport phenomena of interest over this microstructure, that is, in the pore space (fluid flows, for instance) together possibly with the transport in the solid phase (in the case of thermal transfer, for instance).

1.3 Direct Approaches

In direct approaches, the transport problem of interest is directly solved at the pore scales. Increasingly popular examples in this category are the Lattice Boltzmann method (LBM) (Chen and Doolen 1998) and the direct simulations with diffuse interface models (Anderson et al., 1998). Whereas the latter are based on the discretisation of the Navier–Stokes equations, the former treats a fluid as an ensemble of artificial particles. Although the LBM method is perhaps more popular in the porous media literature (Olson and Rothman 1997, Porter et al., 2009), notably because complex geometry can be handled relatively simply with this method, diffuse interface models are also well adapted to direct simulations in microstructures of complex shapes. It is therefore anticipated that these methods will also be frequently used in future works. As an example, Figure 1.4 shows the simulation of an immiscible two-phase flow in a 2D array of discs performed using the so-called volume of fluid (VOF) method (Hirt and Nichols 1981), implemented in the commercial code Fluent™ (Horgue 2012).

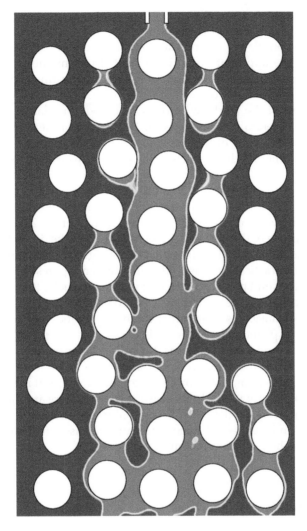

FIGURE 1.4
Example of direct simulation of two-phase flow. Liquid (in light grey) is injected from the top and flows through the random array of discs (in white) under the action of gravity and viscous effects. Note the gas (in dark grey) entrainment induced by this gravity-driven liquid flow. (From Horgue, P., *Ph.D. dissertation*, University of Toulouse, 2012.)

Although these methods seem very attractive since they allow studying the process at the pore scale without any further modelling or simplifications, they have serious computational limitations. As discussed in more depth in Joekar-Niasar et al. (2010), the direct methods are presently too memory and computational-time demanding for simulating transport phenomena over a representative elementary volume (REV) of a microstructure. The REV size being itself generally small compared to the size of a porous domain in most engineering applications, it is clear that the direct simulation of transport phenomena over the whole wick of an LHP evaporator is completely out of reach for the moment. This also explains why, despite the apparent attractiveness of direct methods, little results of practical interest have been obtained so far using this type of method (as far as two-phase flows or even more complex processes are concerned). Although most of the works with direct methods in relation with porous media can be essentially categorised as illustrative or methodological, it is surmised that this type of method could be advantageously used in conjunction with other methods, such as the pore-network methods described in Section 1.4.

As a result, it is not surprising that this type of method has practically not been used so far for the study of transfer phenomena in evaporator porous wick. The only work of which we are aware (Xuan et al., 2011) can be seen as a good example of the computational limitation of the direct method. The computational domain is 2D and of unrealistic high porosity, much too small compared with practical sizes as well as in number of pores considered (only a few pores), the solid phase is formed of isolated particles (the solid phase is thus non-percolating), the wick is fully liquid saturated and so on.

A partial conclusion is that direct methods can be very useful for the careful study of phenomena at the scale of one or a few pores, so as to provide guidance or reference solutions, for instance, for modelling adapted to larger scales. When the objective is the simulation at the scale of an REV or a whole porous domain, direct methods are of no practical interest because of computational limitations. In this case, it is much more appropriate to rely on the pore network models (PNMs) or continuum models described now. Naturally, it can be expected that significant increases in computer performances might change this situation in the future.

1.4 Pore-Network Approach

Initially developed mostly in relation with oil recovery problems, pore scale models have been successfully applied to many other domains, including the study of heat and mass transfers with vaporisation in the porous wick of LHPs (Figus et al., 1999, Prat 2010, Louriou and Prat 2012). As given in Figure 1.5, PNMs are based on the representation of pore space in terms of a network of pores (or sites) connected by throats (or bonds). The 'pores' correspond roughly to the larger voids whereas the throats connecting the pores correspond to the constrictions of the pore space. As discussed in Prat (2010), the pore network can be constructed from the digitally reconstructed porous structure (see Section 1.3). This leads to 'morphological' or

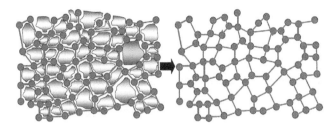

FIGURE 1.5
Schematic of a pore network. The pore space is represented as a network of sites (~ intersections of the pore space) connected by bonds (corresponding to the constrictions of the pore space).

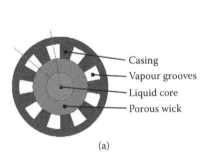

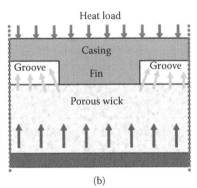

(a)

(b)

FIGURE 1.6
(a) Cross section of a capillary cylindrical evaporator (b) two-dimensional unit cell of a capillary evaporator.

'unstructured' pore networks. Although using morphological pore networks is certainly the most attractive approach, using simpler pore networks can be very instructive.

In fact, morphological pore networks have not yet been used in relation with LHP-related problems. The available studies (Prat 2010 and references therein) are based on simple 2D square networks. Computations are easier than for a morphological network, and it is still possible to incorporate information from the 'real' microstructure such as throat and pore size distributions. Also, to understand some effects, a simple network is generally sufficient. However, it can be anticipated that 3D morphological networks will be used in future works. Accordingly with the above description, only simple networks are considered in what follows.

Since a recent review is available (Prat 2010), only a short overview of the technique and main results is proposed in this chapter. As presented in many previous studies, a typical LHP consists of an evaporator, a reservoir (usually called a compensation chamber), vapour and liquid transport lines and a condenser. As illustrated in Figure 1.6, the evaporator consists of a liquid-passage core, vapour-evacuation grooves, an outer casing and a capillary porous wick. Heat applied to the outer casing leads to the vaporisation of the liquid inside the wick or at the wick/vapour groove interface. The produced vapour is collected in the vapour grooves and flows through the vapour transport line towards the condenser. The menisci formed at the wick/grooves interface or inside the wick adjust themselves to establish a capillary suction that balances the total pressure drop in the device. Pore network (PN) simulations are developed for analysing the transfers within the evaporator, which clearly represents a crucial component of an LHP.

Figure 1.6 shows a sketch of the cross section of a cylindrical evaporator. Although real systems are often more complicated and may include, for example, a secondary wick structure and a bayonet, only the simpler situation depicted in Figure 1.6 has been studied so far using PNM. Owing to the spatial periodicity of the evaporator structure, computations are restricted to a unit cell of the structure, such as the one shown in Figure 1.6. PNMs have been used to study steady-state regimes (Figus et al., 1999, Coquard 2006, Coquard et al., 2007), as well as transient regimes (Prat 2010, Louriou and Prat 2012). For the sake of brevity, we consider some results only for the steady-state regimes in this chapter. The main results obtained for transient regimes are summarised at the end of this section.

As discussed, for example, in Figus et al. (1999) and sketched in Figures 1.6b and 1.7, two main regimes are distinguished regarding the phase distribution within the wick. For low-to-moderate heat loads (Figure 1.6b), the wick is assumed to be fully saturated (this situation is referred to as the all-liquid wick) and vaporisation takes place at the wick/groove interface. Above a certain heat load, bubble nucleation occurs within the wick, leading to the formation of an internal vaporisation front, as illustrated in Figure 1.7. The

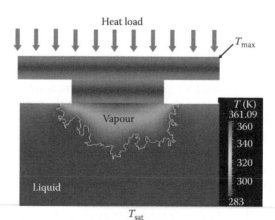

FIGURE 1.7
Steady-state regime with well-developed vapour pocket.

situation with the vapour pocket is referred to as the vapour–liquid wick. To describe the transition from the all-liquid wick to the vapour–liquid wick, the simplest approach is to assume that the transition occurs when a given boiling incipient superheat value is reached somewhere under the fin, which is the hottest region within the wick. As reported in Prat (2010), visualisation experiments in a model porous medium have confirmed the development of the vapour pocket inside the porous medium.

The PNM summarised in what follows is the one developed in Coquard (2006). Compared to a previous model presented in Figus et al. (1999), several new features were introduced: the conduction heat transfer in the external casing is computed, convection heat transfer is taken into account within the wick in addition to heat conduction and phase change, variations in density in the vapour phase with pressure and temperature are taken into account assuming an ideal gas behaviour and heat loss due to the vapour convective flow in the vapour grooves is taken into account. To facilitate the parametric studies, the model developed in Coquard (2006) also presents some significant differences in terms of model formulation. In the classical PNM, the disorder of the porous structure is taken into account through the random distribution of pore volumes and throat sizes. The disorder affects the transport since the throat conductance (see Equation 1.2) varies randomly in the hydraulic network and the local thermal conductance of a bond in the thermal network varies randomly as well (Figus et al., 1999). The randomness of the wick microstructure also affects the capillary effects since the capillary invasion threshold associated with a throat (of radius r) of the hydraulic network is defined as (assuming a perfectly wetting liquid)

$$p_{cth} = \frac{2\sigma}{r} \tag{1.1}$$

This local capillary pressure, Equation 1.1, represents the pressure difference across a meniscus in a throat that must be overcome for the throat to be invaded. Since r is a random variable, p_{cth} varies randomly in the hydraulic network.

In this classical conception of PNMs, the pore/throat size distribution is imposed and the permeability of the network must be computed afterwards (Prat 2010). Similarly, the effective thermal conductivity is computed once a local thermal conductance is assigned to each bond of the thermal network. This is not convenient in the prospect of a parametric study on the effect of the wick macroscopic properties (permeability, effective thermal conductivity, etc.). It is much more convenient to impose macroscopic transport properties directly as input parameters.

Since the pore size distribution in an evaporator wick is expected to be rather narrow, a simple solution is to neglect the effect of disorder on the transport while considering its effect on capillarity. (We know from previous studies that the shape of the liquid–vapour front and associated capillary fingering directly depend on the capillary effects associated with porous medium disorder.) Such a PNM in which the disorder of the porous medium is fully taken into account regarding capillarity but considered in an average sense regarding the computation of the transport phenomena is called a mixed PNM. For example, if the desired permeability of the porous medium is k, the hydraulic conductance of each bond in the square network is given by $g = k\,e$, where e is the 2D network thickness. Similarly, the thermal conductance of a bond of the thermal network is specified as a function of the desired thermal effective conductivity, which typically takes two values depending on the fluid, liquid or vapour, occupying the pore space. An additional advantage of the mixed PNM is that the nodes of the thermal and hydraulic networks representing the wick are the same. Hence, contrary to a classical PNM (Figus et al., 1999), it is not necessary to use a finer network for the thermal network.

Under these circumstances and under the assumptions of a steady-state process, homogeneous and isotropic capillary structure, negligible gravitational and radiative effects and local thermal equilibrium between the porous structure and the working fluid, the problem to be solved within the framework of the mixed PNM can be expressed as follows (for the most complicated case, i.e. in the presence of the vapour pocket, the all-liquid wick situation is a simpler subcase):

Heat conduction in casing:

$$\nabla \cdot (\lambda_a \nabla T) = 0 \tag{1.2}$$

Heat transfer and flow in porous wick:

$$(\rho C_p)_i \mathbf{u}_i \cdot \nabla T = \nabla \cdot (\lambda_{\text{effi}} \nabla T) \tag{1.3}$$

$$\nabla \cdot \rho_i \mathbf{u}_i = 0 \tag{1.4}$$

$$\mathbf{u}_i = -\frac{k}{\mu_i} \nabla p_i \tag{1.5}$$

with $i = \ell$ in the wick liquid region (where $\rho_\ell = \mathrm{cst}$) and $i = v$ in the wick vapour region (where $\rho_v = p_v M / RT$). The Clausius–Clapeyron relationship:[2]

$$p_s(T_{\text{sat}}) = p_{\text{ref}} \exp\left(-\frac{h_{vl}M}{R}\left[\frac{1}{T_{\text{sat}}} - \frac{1}{T_{\text{ref}}}\right]\right) \tag{1.6}$$

Continuity and energy conservation conditions at the vapour/liquid interface within the wick:

$$T_\ell = T_v = T_{\text{sat}}(p_v) \tag{1.7}$$

$$\rho_v \mathbf{u}_v \cdot n = \rho_\ell \mathbf{u}_\ell \cdot n \tag{1.8}$$

$$\left[-\lambda_{\mathrm{effl}} \nabla T\right] \cdot n = \left[-\lambda_{\mathrm{effv}} \cdot \nabla T\right] \cdot n + h_{vl} \dot{m} \tag{1.9}$$

Boundary condition and energy conservation conditions at the wick/casing interface:

$$\nabla p \cdot n = 0 \tag{1.10}$$

$$\left[-\lambda_{\mathrm{effi}} \nabla T\right] \cdot n = \left[-\lambda_a \cdot \nabla T\right] \cdot n \quad i = l \, or \, v \tag{1.11}$$

Boundary conditions at the porous wick/groove interface:

$$-\lambda_{\mathrm{effl}} \nabla T \cdot n = h_{vl} \dot{m} \quad \text{or} \quad \lambda_{\mathrm{effl}} \nabla T \cdot n = h_c (T - T_{\mathrm{gr}}) \tag{1.12}$$

where the convection heat transfer coefficient h_c is specified as in Kaya and Goldak (2006).

Boundary condition at the casing/groove interface:

$$\lambda_a \nabla T \cdot n = h_c (T - T_{\mathrm{gr}}) \tag{1.13}$$

Boundary condition at the entrance of the wick (wick/liquid–passage core interface):

$$T = T_{\mathrm{sat}} - \Delta T_o, p = p_s (T_{\mathrm{sat}}) \tag{1.14}$$

Boundary condition at the casing external boundary (heat flux):

$$\lambda_a \nabla T \cdot n = Q \tag{1.15}$$

Spatial periodicity conditions are imposed on the lateral sides of the computational domain.

The method of solution is similar to that presented in Figus et al. (1999). Once discretised on the network, Equations 1.2 through 1.15 lead to systems of equations quite similar to the ones obtained using standard finite difference or finite volume techniques, the mesh in the wick being formed by the sites of the pore network. Note that computational nodes are also placed in the metallic casing so as to compute the heat transfer within the casing and the fin.

Hereafter, we summarise the procedure when a vapour pocket exists within the wick. The all-liquid wick situation is easier to compute. Initially, the first row of pores under the fin is assumed to be occupied by vapour (experimental visualisations have revealed a quick vapour invasion along the fin). Equations 1.2 through 1.15 are solved using an iterative method as described in Coquard (2006). This gives, in particular, the pressure field in the vapour and liquid regions within the wick. Then we compute the pressure jump $\delta p = p_v - p_\ell$ over each meniscus present within the wick or at the wick/groove interface. Each meniscus for which the computed pressure jump is greater than its capillary pressure threshold (see Equation 1.1), that is, $\delta p > (2\sigma/r)$, is moved into the adjacent pore. This gives a new phase (liquid/vapour) distribution within the wick. Equations 1.2 through 1.15 are then solved again until convergence for this new phase distribution and stability of menisci is tested as explained before, possibly leading to new pore invasions. This procedure is then repeated until the criterion of meniscus stability, that is $\delta p \leq (2\sigma/r)$, is fulfilled for each meniscus present in the wick.

The interesting feature of the model is that it permits us to track explicitly the position of the liquid/vapour interface within the wick. This is illustrated in Figure 1.7, which shows a typical example of phase distribution and temperature field obtained with this model. More details (mesh size, computational domain size, fluid properties, etc.) on this computation can be found in Coquard et al. (2007). Note the irregular (fractal) internal vaporisation front typical of this type of invasion process controlled by the competition between viscous and capillary effects (Prat and Bouleux 1999).

As exemplified in Coquard et al. (2007), this model is well-adapted to explore the impact of the wick transport properties on evaporator performance. To illustrate this, the casing overheat is defined as

$$\Delta T_{max} = T_{max} - T_{sat} \tag{1.16}$$

where T_{max} is the maximum temperature of the casing and T_{sat} the saturation temperature. The temperature of the casing cannot exceed a certain temperature, and this defines a casing overheat operating limit.

The heat flux balance over the computational domain can be expressed as follows:

$$Q = Q_p + Q_c + Q_l + Q_v + Q_{evap} \tag{1.17}$$

where Q is the applied heat load, Q_p the parasitic heat flux (which is the heat flux lost by conduction at the entrance of the wick), Q_c the heat flux lost by convection in the groove, Q_l the heat flux to heat up the liquid, Q_v the heat flux to heat up the vapour and Q_{evap} the heat flux used to vaporise the liquid. Of special interest is the parasitic heat flux, which can possibly contribute to heat up the compensation chamber and therefore modify the operating thermodynamic condition in the loop. It is therefore desirable to limit the parasitic heat flux.

Figure 1.8a shows the variation in the casing overheat as a function of the applied heat load for different values of the thermal conductivity λ_m of the porous matrix. It shows that a high thermal conductivity leads to a much greater casing overheat operating limit compared with a wick of low thermal conductivity. However, as illustrated in Figure 1.8b, a high

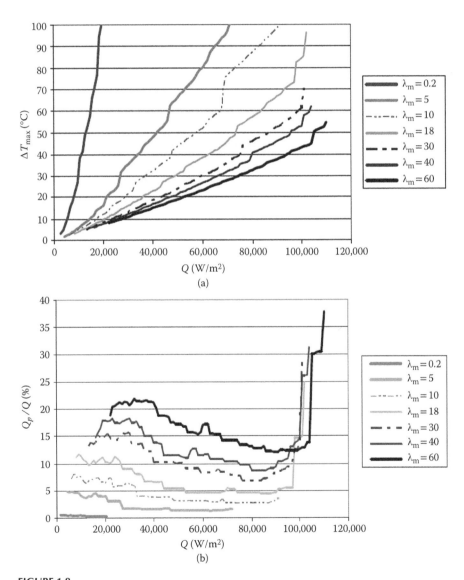

FIGURE 1.8
(a) Evolution of casing overheat as a function of applied heat load for different values of porous matrix thermal conductivity (b) evolution of parasitic flux as a function of applied heat load for different values of porous matrix thermal conductivity.

conductivity increases the parasitic heat flux. Together with other results concerning the capillary operating limit (not shown here), this suggests that a two-layer wick should lead to improved performance. The layer in contact with the casing should be of relatively high thermal conductivity (so as to limit the casing overheat) whereas the layer in contact with the liquid artery should be of lower thermal conductivity (so as to limit the parasitic heat losses) and with smaller pores (so as to avoid the vapour breakthrough across the wick). As shown in Coquard et al. (2007), this two-layer wick leads to better performance, at least according to this type of simulation.

As mentioned earlier, transient regimes have been studied as well (Louriou 2010, Prat 2010). The numerous results obtained include the identification of a vapour pocket pressurisation scenario leading to possible premature vapour breakthrough across the wick, oscillation phenomenon due to a succession of liquid reinvasions/vapour breakthroughs in the groove, hysteretic effects induced by non-monotonous changes in the applied heat load and a heat pipe effect within the wick (Louriou and Prat 2012).

1.5 Models Based on the Continuum Approach

1.5.1 Introduction

Within the framework of the continuum approach to porous media, the porous material is seen as a fictitious continuum medium or as a set of several interacting continua in which three phases – namely the liquid (l), solid (s) and vapour (v) – can be present. The porous structure is assumed to be rigid and inert. The heat and mass transfer governing equations in this system can be, in principle, obtained using an upscaling technique (Whitaker 1999), allowing one to obtain the governing equations of interest from those expressed at the pore scale through a series of mathematical operations. A popular approach in this context relies on the concept of REV. Such an elementary or averaging volume V is associated with every point in space. Then, the macroscopic transport equations are obtained by averaging the pore scale equations over this volume and by using closure hypotheses (Whitaker 1999). For some variable p_β associated with the β-phase, the pressure in the β-phase for example, the method consists of defining the corresponding variable at the scale of the fictitious continuum as $\langle p_\beta \rangle^\beta$ with

$$\langle p_\beta \rangle^\beta = \frac{1}{V_\beta} \int_{V_\beta} p_\beta \, dV \qquad (1.18)$$

where V_β represents the volume of the β-phase contained within the averaging volume V. However, the upscaling techniques are difficult to use in the presence of moving menisci at the pore scales. As a result, it is more exact to consider the equations presented in this section as mostly empirical and phenomenological.

The macroscopic equations governing the simultaneous heat and mass transfer in the different regions are as follows:

In the single-phase regions:

Mass balance equations:

$$\frac{\partial \varepsilon \rho_l}{\partial t} + \nabla \cdot (\rho_l \mathbf{u}_l) = 0 \tag{1.19}$$

$$\frac{\partial \varepsilon \rho_v}{\partial t} + \nabla \cdot (\rho_v \mathbf{u}_v) = 0 \tag{1.20}$$

where ρ_l and ρ_v are the liquid and vapour densities, $\mathbf{u}_l$ and $\mathbf{u}_v$ are the average liquid and vapour flow velocities and ε is the porosity.

Momentum equations:

$$\mathbf{u}_l = -\frac{k}{\mu_l}[\nabla p_l - \rho_l \mathbf{g}] \tag{1.21}$$

$$\mathbf{u}_v = -\frac{k}{\mu_v}[\nabla p_v - \rho_v \mathbf{g}] \tag{1.22}$$

where k is the absolute permeability, μ_l and μ_v are the liquid and vapour viscosities, $\mathbf{g}$ is the gravity vector and p_l and p_v are the liquid and vapour pressures.

In the two-phase regions:

Mass balance equations:

Assuming that the liquid and vapour densities do not vary significantly within the averaging volume, the macroscopic mass balance equations are expressed as

$$\frac{\partial \varepsilon s \rho_l}{\partial t} + \nabla \cdot (\rho_l \mathbf{u}_l) = \dot{m} \tag{1.23}$$

$$\frac{\partial \varepsilon (1-s) \rho_v}{\partial t} + \nabla \cdot (\rho_v \mathbf{u}_v) = -\dot{m} \tag{1.24}$$

where s is the liquid saturation ($s = \varepsilon_l/\varepsilon$) and $\dot{m}$ is the vaporisation rate.

Momentum equations:
Using the generalised Darcy laws, the liquid and vapour flow velocities $\mathbf{u}_l$ and $\mathbf{u}_v$ are given by

$$\mathbf{u}_l = -\frac{kk_{rl}}{\mu_l}\left[\nabla p_l - \rho_l g\right] \tag{1.25}$$

$$\mathbf{u}_v = -\frac{kk_{rv}}{\mu_v}\left[\nabla p_v - \rho_v g\right] \tag{1.26}$$

where k_{rl} and k_{rv} are the relative permeabilities of liquid and vapour.
Capillary pressure:
The macroscopic capillary pressure, which results from the presence of curved menisci within the averaging volume, is given by

$$p_c = p_v - p_l \tag{1.27}$$

The capillary pressure depends on the pore geometry, fluid physical properties and phase saturation. For a given porous medium and a given fluid, the capillary pressure and the relative permeabilities are generally assumed to depend only on saturation s.

Energy equations: The macroscopic description of heat transfer in a porous medium subject to a two-phase flow with phase change is often investigated using a single-temperature equation model. This model is based on the local thermal equilibrium assumption that means that the macroscopic (averaged) temperatures of the three phases are close enough so that a single temperature suffices to describe the heat transport process. The concept of thermal equilibrium involves the volume-averaged temperatures and should therefore not be confused with the classical assumption of local thermodynamic equilibrium. The local thermal equilibrium is no longer valid when the particles or pores are not small enough, when the phase thermal properties differ widely, when convective transport is important and when there is a significant internal heat generation (Duval et al., 2004). Under these conditions, separate transport equations for each phase are required leading to non-equilibrium models.

1.5.2 Non-Equilibrium Models

On the basis of a steady-state closure at the microscopic scale, a comprehensive generalised three-temperature model taking into account the vaporisation in the porous medium was presented in Duval et al. (2004). When the non-traditional convective terms are negligible and the three macroscopic temperature gradients are sufficiently close to each other, the thermal energy equations of this model can be expressed in the two-phase region as follows:

In the vapour phase:

$$\rho_v \varepsilon (1-s) C_{pv} \frac{\partial T_v}{\partial t} + \rho_v C_{pv} \mathbf{u}_v \cdot \nabla T_v - \dot{m} C_{pv}(T_{sat} - T_v)$$

$$= \nabla \cdot \left(\lambda_{effv} \nabla T_v \right) - H_{vv}(T_v - T_{sat}) - H_{vl}(T_l - T_{sat}) - H_{vs}(T_s - T_{sat}) \qquad (1.28)$$

In the liquid phase:

$$\rho_l \varepsilon s C_{pl} \frac{\partial T_l}{\partial t} + \rho_l C_{pl} \mathbf{u}_l \cdot \nabla T_l + \dot{m} C_{pl}(T_{sat} - T_l)$$

$$= \nabla \cdot \left(\lambda_{effl} \nabla T_l \right) - H_{ll}(T_l - T_{sat}) - H_{lv}(T_v - T_{sat}) - H_{ls}(T_s - T_{sat}) \qquad (1.29)$$

In the solid phase:

$$\rho_l (1-\varepsilon) C_{ps} \frac{\partial T_s}{\partial t} = \nabla \cdot \left(\lambda_{effs} \nabla T_s \right) - H_{ss}(T_s - T_{sat}) - H_{sv}(T_v - T_{sat}) - H_{sl}(T_s - T_{sat})$$

$$(1.30)$$

where T_s, T_l and T_v are the volume-averaged temperatures of the solid, liquid and vapour phases, respectively. T_{sat} is the saturation temperature. C_{ps}, C_{pl} and C_{pv} are the heat capacity of the solid, liquid and vapour, respectively. λ_{effs}, λ_{effl} and λ_{effv} are the effective thermal conductivity of the solid, liquid and vapour, respectively. $H_{\alpha\beta}$ (with α = v, l, s and β = v, l, s) is the effective heat transfer coefficient.

Mass rate of vaporisation: When the volume fraction gradients and the pseudo-convective contributions are negligible, the mass rate of vaporisation (Duval et al., 2004) is given by

$$\dot{m} h_{vl} = H_l (T_l - T_{sat}) + H_v (T_v - T_{sat}) - H_s (T_s - T_{sat}) \qquad (1.31)$$

where $h_{vl} = h_{vsat} - h_{lsat}$ represents the heat of vaporisation.

In the single-phase regions, there is no phase change and the thermal energy equations for the two-temperature model can be expressed as follows:

In the vapour region:

$$\rho_v \varepsilon C_{pv} \frac{\partial T_v}{\partial t} + \rho_v C_{pv} \mathbf{u}_v \cdot \nabla T_v = \nabla \cdot \left(\lambda_{effv} \nabla T_v \right) - H_{vs}(T_v - T_s) \qquad (1.32)$$

$$\rho_s (1-\varepsilon) C_{ps} \frac{\partial T_s}{\partial t} = \nabla \cdot \left(\lambda_{effs} \nabla T_s \right) - H_{vs}(T_s - T_v) \qquad (1.33)$$

In the liquid region:

$$\rho_l \varepsilon C_{pl} \frac{\partial T_l}{\partial t} + \rho_l C_{pl} \mathbf{u}_l \cdot \nabla T_l = \nabla \cdot \left(\lambda_{effl} \nabla T_l \right) - H_{ls}(T_l - T_s) \qquad (1.34)$$

$$\rho_s(1-\varepsilon)C_{ps}\frac{\partial T_s}{\partial t} = \nabla \cdot \left(\lambda_{\text{effs}}\nabla T_s\right) - H_{ls}\left(T_s - T_l\right) \qquad (1.35)$$

As mentioned before, the use of thermal non-equilibrium models is desirable when the conditions are such that a significant difference can exist between the averaged temperatures associated with each phase. In the case of the two-phase cooling devices, it is widely considered that the local thermal equilibrium assumption represents a valid approximation. In fact, it is even often assumed that the liquid–vapour–wick region is isothermal without much loss of accuracy. This is especially true for the single-component liquid–vapour system because the local thermodynamic equilibrium assumption (again not to be confused with the local thermal equilibrium assumption) implies that the liquid–vapour region temperature in the wick is nearly equal to equilibrium saturation temperature T_{sat}.

1.5.3 Local Thermal Equilibrium Models

When local thermal equilibrium is assumed to be valid, the single phase-wick regions can be described by a single spatial average temperature T and the following approximation can be used: $T_s \approx T_l \approx T$ in the liquid–wick region; $T_s \approx T_l \approx T$ in the vapour–wick region. The macroscopic thermal energy equations in the vapour–wick and liquid–wick regions are given by

$$\left(\rho_v\varepsilon C_{pv} + (1-\varepsilon)\rho_s\varepsilon C_{ps}\right)\frac{\partial T}{\partial t} + \rho_v C_{pv}\mathbf{u}_v \cdot \nabla T = \nabla \cdot \left(\lambda_{\text{effv}}\nabla T\right) \qquad (1.36)$$

$$\left(\rho_l\varepsilon C_{pl} + (1-\varepsilon)\rho_s\varepsilon C_{ps}\right)\frac{\partial T}{\partial t} + \rho_l C_{pl}\mathbf{u}_l.\nabla T = \nabla \cdot \left(\lambda_{\text{effl}}\nabla T\right) \qquad (1.37)$$

where λ_{effv} and λ_{effl} are the effective thermal conductivity of the vapour–wick and the liquid–wick regions, respectively. The modelling of the liquid–vapour region is discussed in Sections 1.5.4 and 1.5.5.

1.5.4 Local Thermal Equilibrium Models with Explicit Tracking of Macroscopic Interfaces

The governing equations are coupled by internal boundary conditions at moving and irregular interfaces separating the various regions. The locations of theses interfaces are not known *a priori* and depend on the coupled flow and heat and mass transfer in each region. A numerical procedure for such a multi-region problem needs to explicitly track the moving interface, thus calling for complex coordinate mapping or numerical remeshing. For a boiling problem in porous media with two regions (liquid–wick and liquid–vapour–wick regions), Ramesh et al. (1990a,b) use a coordinate transformation so as to deal with an immobilised interface. The transformation serves two purposes: first, it fixes the domain in the transformed space; and, second, it explicitly introduces the non-linearity due to the moving interface into the equations. However, the resulting equations are naturally more complicated than the original equations, leading to serious numerical difficulties.

1.5.5 Multi-Phase Mixture Model

To overcome these difficulties, Wang and Beckermann (1993a,b) proposed, when the thermal equilibrium is valid, a two-phase mixture model in which the liquid and vapour phases are regarded as constituents of a multi-phase mixture. This model has several advantages. First, significantly fewer governing equations are to be solved. Second, the model leads to a single domain formulation (there is no need to distinguish explicitly the various regions: liquid, vapour, liquid–vapour). Hence there is no need to track the interfaces explicitly. The set of conservation equations of heat, mass and momentum for such a mixture is deduced from the macroscopic governing equations of each phase in each region (Wang and Beckermann 1993a,b, 1994, Wang 1997, Najjari and Ben Nasrallah 2002, 2003, 2005). By assuming that radiative heat transfer is negligible the mixture governing equations can be written as follows:

Continuity equation for the two-phase mixture:

$$\varepsilon \frac{\partial \rho}{\partial t} + \nabla \cdot (\rho \mathbf{u}) = 0 \tag{1.38}$$

where ρ and $\mathbf{u}$ are the density and velocity of the mixture given by

$$\rho \mathbf{u} = \rho_l \mathbf{u}_l + \rho_v \mathbf{u}_v \tag{1.39}$$

$$\rho = \rho_l s + \rho_v (1 - s) \tag{1.40}$$

Momentum equation for the two-phase mixture:

$$\mathbf{u} = -\frac{k}{\mu} (\nabla p - \rho_k g) \tag{1.41}$$

where $\rho_k = \rho_l [1 - \beta_l (T - T_0)] \lambda_l + \rho_v [1 - \beta_v (T - T_{sat})] \lambda_v$,

$$\mu = \frac{[\rho_l s + \rho_v (1 - s)]}{(k_{rl} / v_l) + (k_{rv} / v_v)}; \quad v = \frac{\mu}{\rho}, \, v_1 = \frac{\mu_1}{\rho_1}; \quad \lambda_1 = \frac{v}{v_1} k_{rl}$$

$$\nabla p = \lambda_l \nabla p_l + (1 - \lambda_l) \nabla p_v \tag{1.42}$$

By replacing $\mathbf{u}$ by its expression in Equation (1.43), the following pressure equation is obtained:

$$\varepsilon \frac{\partial \rho}{\partial t} + \frac{k}{v} \nabla^2 p - \nabla p \cdot \nabla \left(\frac{k}{v} \right) + \nabla \left(\frac{k}{v} \rho_k g \right) = 0 \tag{1.43}$$

Energy equation: A unified form of the energy conservation equations for solid, liquid and vapour phases is given by the volumetric enthalpy H equation:

$$\Omega \frac{\partial H}{\partial t} + \nabla(\gamma_h u H) = \nabla\left(\frac{\Gamma_h}{\rho}\nabla H\right) + \nabla\left(f(s)\frac{k\Delta\rho h_{vl}}{v_v}g\right) \tag{1.44}$$

where $H = \rho(h - 2h_{vsat})$ and $\rho h = \rho_l s h_l + \rho_v(1-s)h_v$. h_l and h_v are the enthalpies of the liquid and vapour phases; they are related to temperature by the relationships

$$h_s = c_s T + h_s^0 \tag{1.45}$$

$$h_l = c_l T \tag{1.46}$$

$$h_v = c_v T + \left[(c_l - c_v)T_{sat} + h_{fg}\right] \tag{1.47}$$

whereas the coefficients Ω, γ_h, Γ_h/ρ and $f(s)$ are given by

$$\Omega = \varepsilon + \rho_s c_s(1-\varepsilon)\frac{dT}{dH}; \quad \gamma_h = \frac{[s\rho_l + \rho_v(1-s)][h_{vsat}(1+\lambda_1) - h_{lsat}\lambda_1]}{(2h_{vsat} - h_{lsat})s\rho_l + \rho_v(1-s)h_{vsat}},$$

$$\frac{\Gamma_h}{\rho} = \frac{\rho_l h_{fg}}{\rho_l h_{fg} + (\rho_l - \rho_v)h_{vsat}}D + \lambda_{eff}\frac{dT}{dH},$$

$$D = \frac{\sqrt{\varepsilon k}\sigma}{\mu_l}\frac{k_{rl}k_{rv}}{(v_v/v_l)k_{rl} + k_{rv}}(-J'(s)); f(s) = \frac{k_{rl}k_{rv}/v_l}{k_{rl}/v_l + k_{rv}/v_v}; \Delta\rho = \rho_l - \rho_v$$

The temperature and liquid saturation can be deduced from the volumetric enthalpy H by

$$T = \begin{cases} \dfrac{H + 2\rho_l h_{vsat}}{\rho_l c_l} & H \le -\rho_l(2h_{vsat} - h_{lsat}) \\[2mm] T_{sat} & -\rho_l(2h_{vsat} - h_{lsat}) < H \le -\rho_v h_{vsat} \\[2mm] T_{sat} + \dfrac{H + \rho_v h_{vsat}}{\rho_v c_v} & -\rho_v h_{vsat} < H \end{cases} \tag{1.48}$$

$$s = \begin{cases} 1 & H \le -\rho_l(2h_{vsat} - h_{lsat}) \\[2mm] -\dfrac{H + \rho_v h_{vsat}}{\rho_l h_{fg} + (\rho_l - \rho_v)h_{vsat}} & -\rho_l(2h_{vsat} - h_{lsat}) < H \le -\rho_v h_{vsat} \\[2mm] 0 & -\rho_v h_{vsat} < H \end{cases} \tag{1.49}$$

The expression of dT/dH is different in each zone:

In the liquid zone: $\left(H \leq -\rho_1\left(2h_{\text{vsat}} - h_{\text{lsat}}\right)\right)$, $\dfrac{dT}{dH} = \dfrac{1}{\rho_1 c_1}$

In the vapour zone: $\left(-\rho_v h_{\text{vsat}} < H\right)$, $\dfrac{dT}{dH} = \dfrac{1}{\rho_v c_v}$

In the two-phase zone: $\left(-\rho_1(2h_{\text{vsat}} - h_{\text{lsat}}) < H \leq -\rho_v h_{\text{vsat}}\right)$, $\dfrac{dT}{dH} = 0$

The velocities of individual phases can also be computed as follows:

$$\rho_1 u_1 = \lambda_1 \rho u + j \tag{1.50}$$

$$\rho_v u_v = (1 - \lambda_1)\rho u - j \tag{1.51}$$

where j is a mass diffusion flux:

$$j = -\rho_1 D \nabla(s) + f(s)\frac{k\Delta\rho}{\nu_v}g \tag{1.52}$$

The capabilities of the two-phase mixture model have notably been shown through application to boiling with thermal convection in a porous layer heated from below (Wang et al., 1994, Najjari and Ben Nasrallah 2005), to boiling with thermal mixed convection in an inclined porous layer discretely heated (Najjari and Ben Nasrallah 2003) and to pressure-driven boiling flow adjacent to a vertical heated plate inside a porous medium (Najjari and Ben Nasrallah 2002).

1.5.6 Application to Evaporators

There have been several works based on the continuum approach to porous media devoted to the study of transfers within a subregion of a capillary evaporator, such as the one shown in Figure 1.6b. In Cao and Faghri (1994a,b) solutions were developed assuming a completely liquid-saturated wick. A qualitative discussion of the boiling limit in a capillary structure was provided. A conclusion was that reasonably accurate results can be obtained by a 2D model especially when the vapour velocities are small for certain working fluids such as Freon-11 and ammonia. The vapour pocket within the wick was first considered in Demidov and Yatsenko (1994) and then in Figus et al. (1999). Among other things, these works show that the size of the vapour pocket increases with the applied heat load. The comparison with

PN simulations reported in Figus et al. (1999) indicates that the continuum model used in Figus et al. (1999) and Demidov and Yatsenko (1994) was consistent only with an extremely narrow pore size distribution (so as to have a sharp transition between the liquid zone and the vapour zone). In other terms, the existence of a two-phase zone within the wick was not taken into account. A similar assumption was made in Kaya and Goldak (2006), where the numerical simulations help explain the robustness of LHPs to the boiling limit.

A two-dimensional numerical model based on the two-phase mixture model (see Section 1.5.5) was used in Yan and Ochterbeck (2003) to study the behaviour of a cylindrical CPL evaporator under steady-state operation. In contrast with the simulations reported in Figus et al. (1999), Demidov and Yatsenko (1994) and Kaya and Goldak (2006), the approach leads to significant variations in the saturation within the wick and thus to the existence of an important two-phase liquid–vapour region. The effects of heat load, liquid sub-cooling and effective thermal conductivity of the wick structure on the evaporator performance were studied.

The sharp interface assumption (no two-phase zone within the wick) was again made in a series of similar papers by the same group (Wan et al., 2007, 2009, 2001). As in Figus et al. (1999) or Demidov and Yatsenko (1994), remeshing is needed to track the liquid–vapour interface within the wick. One interesting feature of the model was to use a single set of equations, based on the so-called Brinkman–Darcy–Forchheimer model, within the porous structure and the vapour grooves. Also, the 2D computational domain contains several grooves and this permits us to investigate the evolution of the vapour–liquid interface within the wick over a significantly larger domain than in previous works.

Three-dimensional numerical models have also been proposed (Li and Peterson 2011, Chernysheva and Maydanik 2012). The model presented in Li and Peterson (2011) is, however, restricted to a saturated wick (no vapour pocket) whereas that presented in Chernysheva and Maydanik (2012) is apparently essentially based on the solution of the thermal energy equation without the consideration of partially invaded wick (the wick is either locally saturated or fully dry). Also, convective heat transfer within the wick is ignored. A nice aspect of the model presented in Chernysheva and Maydanik (2012) is the scale of investigation, namely the whole evaporator, as opposed to the small subregion (as shown in Figures 1.6b and 1.9) generally considered in most of the other studies.

It should be pointed out that only steady-state regimes have been numerically simulated in the works mentioned in this section. This is in contrast with the PN simulations, which have been also used to investigate some aspects of transient regimes (Prat 2010, Louriou and Prat 2012).

In summary, this review of the literature on the application of continuum models to the study of heat and mass transfers in capillary evaporators reveals serious and interesting drawbacks in the development of numerical

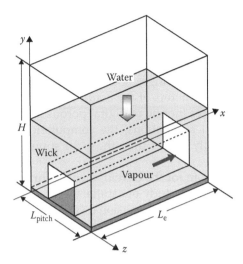

FIGURE 1.9
Three-dimensional computational domain considered in Li and Peterson (2011).

simulations. First, the often-made assumption of sharp interface between the liquid zone and the vapour zone within the wick (no two-phase zone) is questionable. This assumption is neither consistent with the PN simulations nor consistent with the simulations based on the mixture model. Three-dimensional simulations are scarce and based on fairly restrictive assumptions (e.g. saturated wick). Simulation of transient regimes is an open area. This is somewhat surprising owing to the problems often encountered and often not well understood during the start up of LHP. Comprehensive transient 3D models including a proper description of the two-phase zone within the wick with adequate coupling with the transfers in the other parts of the evaporator (vapour grooves, liquid channel, compensation chamber, metallic casing, etc.) are yet to be developed.

1.6 Conclusion

As in other fields involving transport phenomena in porous media, simulations of heat and mass transfers in porous wicks can be performed according to a variety of techniques. These include the direct simulations at the pore scale, PNMs and continuum (Darcy's scale) models. Direct simulations are still little developed in relation with capillary evaporators owing to their very high computational cost. Pore network simulations can be used even in the absence of length scale separation, that is when the traditional continuum approaches are likely to lead to poor results. Furthermore, the relevant transport equations are not solved directly in the pore space and this permits considering systems containing a relatively large number of pores, a distinguishing advantage compared to the direct simulations. Pore network simulations are particularly well adapted to simulate transfers in thin systems (e.g. porous coatings). Also, pore network simulations can be used to evaluate the parameters of the continuum models from 3D images of porous microstructures (notion of digital porous media). Although much less than for the direct simulations, the computational costs for the PN simulations are generally higher than those for simulations based on continuum models.

As a result, there is room for the development of further macroscopic simulations in relation with the study of capillary evaporators or other devices involving porous wicks. Three-dimensional simulations including a proper modelling of two-phase regions are still an interesting open challenge for steady-state as well as for transient regimes.

Nomenclature

C_p	Specific heat capacity (J/(kg-K))	T_{ref}		Reference temperature (K)
G	Gravitational acceleration (m/s²)	T_{sat}		Saturation temperature (K)
h_{vl}	Latent heat of vaporisation (J/kg)	$\mathbf{u}_i$		Filtration velocity (m/s)
H	Effective heat transfer coefficient (W/m³/K)	V		Averaging volume (m³)
k	Permeability (m²)	$V\beta$		Volume of β-phase (V, m³)
k_r	Relative permeability	**Greek letters**		
M	Molecular mass (kg/mol)	ε		Porosity
$\dot{m}$	Mass flux (kg/m²/s)	λ		Thermal conductivity (W/(m-K))
n	Unit normal vector	M		Dynamic viscosity, Pl
p	Pressure (Pa)	$(\rho C_p)_i$		Thermal conductivity
p_c	Capillary pressure (Pa)	ρ_i		Fluid density (kg/m³)
p_{ref}	Reference pressure (Pa)	Σ		Surface tension (N/m)
p_{sat}	Saturation pressure (Pa)	**Subscripts and abbreviations**		
Q	Heat flux (W/m²)	eff		Effective
r	Pore size (m)	ℓ, l		Liquid
R	Universal gas constant (J/mol/K)	S		Solid
r	Throat radius (m)	Sat		Saturation
s	Saturation	V		Vapour
t	Time (s)			
T	Temperature (K,°C)			
T_{max}	Casing maximum temperature (K)			

References

Anderson, D.M., G.B. McFadden and A.A. Wheeler. 1998. Diffuse-interface methods in fluid mechanics. *Annu. Rev. Fluid Mech.* 30: 139–165.

Baldwin, C.A., A.J. Sederman, M.D. Mantle, P. Alexander and L.F.J. Gladden. 1996. Determination and characterisation of the structure of a pore space from 3D volume images. *J. Colloid Interface Sci.* 181: 79–92.

Cao, Y. and A. Faghri. 1994a. Conjugate analysis of a flat-plate type evaporator for capillary pumped loops with three-dimensional vapor flow in the groove. *Int. J. Heat Mass Tran.* 37 (3): 401–409.

Cao, Y. and A. Faghri. 1994b. Analytical solutions of flow and heat transfer in a porous structure with partial heating and evaporation on the upper surface. *Int. J. Heat Mass Tran.* 37 (10): 1525–1533.

Chen, S. and G.D. Doolen. 1998. Lattice Boltzmann method for fluid flows. *Annu. Rev. Fluid Mech.* 30: 329–364.

Chernysheva, M.A. and Y.F. Maydanik. 2012. 3D model of heat and mass transfer in flat evaporator of a cooper-water loop heat pipe. *Appl. Therm. Eng.* 33–34: 124–134.

Coker, D.A., S. Torquato and J.H. Dunsmuir. 1996. Morphology and physical properties of Fontainebleau sandstone via a tomographic analysis. *J. Geophys. Res. B Solid Earth* 101: 17497–17506.

Coquard, T. 2006. Transferts couplés de masse et de chaleur dans un élément d'évaporateur capillaire [Coupled heat and mass transfers in a unit cell of a capillary evaporator (in French)]. *Ph.D. dissertation*, INP Toulouse.

Coquard, T., M. Prat, A. Larue de Tournemine and C. Figus. 2007. Pore-network models as a tool for the analysis of heat and mass transfer with phase change in the capillary structure of loop heat pipe. *Proceedings of the 14th International Heat Pipe Conference, Florianópolis*.

Demidov, A.S. and E.S. Yatsenko. 1994. Investigation of heat and mass transfer in the evaporation zone of a heat pipe operating by the 'inverted meniscus' principle. *Int. J. Heat Mass Tran.* 37 (14): 2155–2163.

Duval, F., F. Fichot and M. Quintard. 2004. A local nonequilibrium model for two-phase flow with phase change in porous media. *Int. J. Heat Mass Tran.* 47: 613–639.

Faghri, A. 1995. *Heat pipe science and technology*, Taylor & Francis, Washington.

Fichot, F., F. Duval, N. Trégourès, C. Béchaud and M. Quintard. 2006. The impact of thermal non-equilibrium and large-scale 2D/3D effects on debris bed reflooding and coolability. *Nucl. Eng. Des.* 236 (19–21): 2144–2163.

Figus, C., Y. Le Bray, S. Bories and M. Prat. 1999. Heat and mass transfer with phase change in a porous structure partially heated. Continuum model and pore network simulations. *Int. J. Heat Mass Tran.* 42: 2257–2569.

Hirt, C.W. and B.D. Nichols. 1981. Volume-of-fluid method for the dynamics of free boundaries. *J. Comput. Phys.* 39: 201–225.

Horgue, P. 2012. Modélisation multi-échelle d'un écoulement gaz-liquide à travers un lit de particules [Multi-scale modelling of liquid-gas flow in fixed beds (in French)]. *Ph.D. dissertation*, University of Toulouse .

Joekar-Niasar, V., S.M. Hassanizadeh and H.K. Dahle. 2010. Non-equilibrium effects in capillarity and interfacial area in two-phase flow: dynamic pore-network modelling. *J. Fluid. Mech.* 655: 38–71.

Karwacki, L., D.A.M. de Winter, L.R. Aramburo, M.N. Lebbink and J.A. Post. 2011. Architecture-dependent distribution of mesopores in steamed zeolite crystals as visualized by FIB-SEM tomography. *Angew. Chem. Int.* 50: 1294–1298.

Kaya, T. and J. Goldak. 2006. Numerical analysis of heat and mass transfer in the capillary structure of a loop heat pipe, *Int. J. Heat Mass Tran.* 49: 3211–3220.

Kaya, T., R. Pérez, C. Gregory and A. Torres. 2008. Numerical simulation of transient operation of loop heat pipes. *Appl. Therm. Eng.* 28: 967–974.

Launay, S., V. Platel, S. Dutour and J.L. Joly. 2007. Transient modeling of LHP for the oscillating behaviour study. *J. Thermophys. Heat Trans.* 21: 487–495.

Liang, Z.R., C.P. Fernandes, F.S. Magnani and P.C. Philippi. 1998. A reconstruction technique for three-dimensional porous media using image analysis and Fourier transforms. *J. Petrol. Sci. Eng.* 21 (3–4): 273–283.

Li, J. and G.P. Peterson. 2011. 3D heat analysis in a loop heat pipe evaporator with a fully saturated wick. *Int. J. Heat Mass Tran.* 54: 564–574.

Lipinski, R.J. 1984. A coolability model for post-accident nuclear reactor debris. *Nucl. Technol.* 65: 53–66.

Louriou, C. 2010. Modélisation instationnaire des transferts de masse et de chaleur au sein des évaporateurs capillaires [Modelling of transient heat and mass transfers in capillary evaporators (in French)]. *Ph.D. dissertation*, University of Toulouse.

Louriou, C. and M. Prat. 2012. Experimental and numerical pore network study of bubble growth by vaporisation in a porous medium heated laterally. *Int. J. Therm. Sci.* 52: 8–21.

Manke, I., H. Markotter, C. Totzke, N. Kardjilov, R. Grothausmann and M. Dawson. 2011. Investigation of energy-relevant materials with synchrotron X-rays and neutrons. *Adv. Eng. Mater.* 13: 712–729.

Najjari, M. and S. Ben Nasrallah. 2002. Étude de l'ébullition en convection mixte dans une couche poreuse verticale. [Numerical study of boiling with mixed convection in a vertical porous layer]. *Int. J. Therm. Sci.* 41 (10): 913–925.

Najjari, M. and S. Ben Nasrallah. 2003. Numerical study of boiling in an inclined porous media. *J. Porous Media.* 6: 71–81.

Najjari, M. and S. Ben Nasrallah. 2005. Numerical study of the effects of geometric dimensions on liquid–vapor phase change and free convection in a rectangular porous cavity. *J. Porous Media.* 8: 1–12.

Olson, J.F. and D.H. Rothman. 1997. Two-fluid flow in sedimentary rock: simulation, transport and complexity. *J. Fluid Mech.* 341: 343–370.

Perre, P., M. Moser and M. Martin. 1993. Advances in transport phenomena during convective drying with superheated steam and moist air. *Int. J. Heat Mass Tran.* 36 (11): 2725–2746.

Peterson, G.P. 1994. *An introduction to heat pipes, modeling, testing and applications*, Wiley Interscience, New York.

Porter, M.L., M.G. Schaap and D. Wildenschild. 2009. Lattice-Boltzmann simulation of the capillary pressure–saturation–interfacial area relationship for porous media. *Adv. Water Resour.* 32 (11): 1632–1640.

Prat, M. 2010. Application of pore network models for the analysis of heat and mass transfer with phase change in the porous wick of loop heat pipes. *Heat Pipe Sci. Technol.* 1 (2): 129–149.

Prat, M. and F. Bouleux. 1999. Drying of capillary porous media with stabilized front in two dimensions. *Phys. Rev. E* 60: 5647–5656.

Ramesh, P.S. and K.E. Torrance. 1990a. Stability of boiling in porous media. *Int. J. Heat Mass Tran.* 33 (9): 1895–1908.

Ramesh, P.S. and K.E. Torrance. 1990b. Numerical algorithm for problems involving boiling and natural convection in porous materials. *Numer. Heat Tran. B* 17: 1–24.

Smirnov, H. 2010. *Transport phenomena in capillary porous structures and heat pipes*, CRC Press, Boca Raton, FL.

Soppe, W. 1990. Computer simulation of random packings of hard spheres. *Powder Technol.* 62: 189–196.

Stephan, P.C. and C.A. Busse. 1992. Analysis of heat transfer coefficient of grooved heat pipe evaporator walls. *Int. J. Heat Mass Tran.* 35: 383–391

Torquato, S. 2002. *Random heterogeneous materials: microstructure and macroscopic properties*. Springer, New York.

Wan, M., J. Liu, J.H. Wan, Z.K. Tu and W. Liu. 2011. An overall numerical investigation on heat and mass transfer for miniature flat plate capillary pumped loop evaporator. *Thermochim. Acta.* 518: 82–88.

Wan, M., W. Liu, Z. Zheng and A. Nakayam. 2007. Heat transfer with phase change in an evaporator of miniature flat plate capillary pumped loop. *J. Therm. Sci.* 16 (3): 254–263.

Wan, M., W. Liu, Z.K. Tu and A. Nakayama. 2009. Conjugate numerical analysis of flow and heat transfer with phase change in a miniature flat plate CPL evaporator, *Int. J. Heat Mass Trans.* 52: 422–430.

Wang, C.Y. 1997. A fixed-grid numerical algorithm for two-phase flow and heat transfer in porous media. *Numer. Heat Tran. B.* 31: 85–105.

Wang, C.Y. and C. Beckermann. 1993a. A two-phase mixture model of liquid-gas flow and heat transfer in capillary porous media—I: Formulation. *Int. J. Heat Mass Tran.* 36: 2747–2758.

Wang, C.Y. and C. Beckermann. 1993b. A two-phase mixture model of liquid-gas flow and heat transfer in capillary porous media—II: Application to pressure driven boiling flow adjacent to a vertical heated plate. *Int. J. Heat Mass Tran.* 36: 2759–2768.

Wang, C.Y., C. Beckermann and C. Fan. 1994. Numerical study of boiling and natural convection in capillary porous media using the two-phase mixture model. *Numer. Heat Tran. A* 26: 375–398.

Whitaker, S. 1999. *The method of volume averaging*, Kluwer, Dordrecht.

Woods, A.W. 1999. Liquid and vapor flow in superheated rock. *Ann. Rev. Fluid Mech.* 31: 171–199.

Xuan, Y., K. Zhao and Q. Li. 2011. Investigation on heat and mass transfer in a evaporator of a capillary-pumped loop with the lattice Boltzmann method: pore scale simulation. *Transp. Porous Med.* 89: 337–355.

Yan, Y.H. and J.M. Ochterbeck. 2003. Numerical investigation of steady state operation of capillary pumping loop evaporator. *J. Electron. Pack.* 125: 251–260.

2

Thermally Powered Adsorption Cooling:
Recent Trends and Applications

B. B. Saha and I. I. El-Sharkawy

CONTENTS

2.1 Introduction 30
2.2 Adsorbent/Refrigerant Pairs for Adsorption Cooling and
     Desalination Applications 30
     2.2.1 Silica Gel/Water Pair 30
     2.2.2 Zeolite/Water Pair 31
     2.2.3 Zeolite Composites/Water Pair 31
     2.2.4 Activated Carbon/Methanol Pair 31
     2.2.5 Activated Carbon/Ammonia Pair 32
     2.2.6 Monolithic Carbon/Ammonia Pair 32
     2.2.7 Activated Carbon Fibre/Methanol Pair 33
     2.2.8 Activated Carbon Fibre/Ammonia Pair 33
     2.2.9 Activated Carbon Fibre/Alcohol Pairs 33
2.3 Two-Bed Adsorption Cooling Cycles 33
     2.3.1 Working Principles 33
     2.3.2 Mathematical Modelling 36
            2.3.2.1 Adsorption Isotherms 36
            2.3.2.2 Adsorption Kinetics 38
            2.3.2.3 Adsorption and Desorption Energy Balance 38
            2.3.2.4 Condenser Energy Balance 39
            2.3.2.5 Evaporator Energy Balance 40
            2.3.2.6 Mass Balance 40
            2.3.2.7 System Performance 41
2.4 Thermally Powered Advanced Adsorption Cooling Cycles 41
2.5 Advanced Adsorption Cooling cum Desalination Cycle 48
     2.5.1 Working Principle of the AACD Cycle 48
     2.5.2 Mathematical Modelling of the AACD Cycle 50
     2.5.3 Performance Results of the AACD Cycle 52
2.6 Conclusions 55
Acknowledgements 55
Nomenclature 56
References 57

2.1 Introduction

The quest for a safe and comfortable environment has always been one of the main preoccupations of the sustainability of human life. Accordingly, research aimed at the development of thermally powered adsorption cooling technologies has intensified in the past few decades. These technologies offer double benefits of reductions in energy consumption, peak electrical demand in tandem with adoption of environmentally benign adsorbent/refrigerant pairs without compromising the desired level of comfort conditions. Alternative adsorption cooling technologies are being developed, which can be applied to buildings [1–4]. These systems are relatively simple to construct as they have no major moving parts. In addition, there is only marginal electricity usage, which might be needed for the pumping of heat transfer fluids. The heat source temperature can be as low as 50°C if a multistage regeneration scheme is implemented [5,6]. However, since the system is driven by low-temperature waste heat, the coefficient of performance (COP) of thermally activated adsorption systems is normally poor [7]. A recent study shows that the cooling capacity of the two-stage silica gel/water refrigeration cycle can be improved significantly when a reheat scheme is used [8]. In the first part of this chapter, several advanced thermally activated adsorption cooling cycles are overviewed. Finally, a three-bed dual evaporator–type advanced adsorption cooling cum desalination (AACD) cycle is introduced in which the evaporators work at two different pressure levels and produce cooling effects and potable water from saline or brackish water. The performance results of the AACD cycle are presented.

2.2 Adsorbent/Refrigerant Pairs for Adsorption Cooling and Desalination Applications

2.2.1 Silica Gel/Water Pair

In the 1980s, silica gel/water systems were investigated mainly in Japan. This pair got attention when Sakoda and Suzuki [9] proposed a transient simulation model of a solar-powered adsorption cooling cycle. The authors [10] also analysed solar-driven adsorption refrigeration system. They reported that a solar COP of 0.2 can be achieved by using a solar collector with dimensions of $500 \times 500 \times 50$ mm and 1 kg silica gel. In 1995, Saha et al. [1] analytically investigated the effects of operating conditions on the cooling capacity and COP of a silica gel/water-based adsorption chiller. The same system was analysed experimentally by Boelman et al. [2]. A lumped transient model and a distributed transient model of a two-bed adsorption cooling system were investigated by Chua et al. [11,12]. Saha et al. [13,14] developed and analysed various multi-stage adsorption cooling cycles using silica gel/water as the adsorbent/refrigerant pair.

2.2.2 Zeolite/Water Pair

Nearly four decades back, Tchernev [15] investigated a zeolite/water system for solar air-conditioning and refrigeration applications. Alefeld et al. [16] used the same pair for heat pump and heat storage systems. Tatler and Erdem-Senatalar [17] and Tatler et al. [18] proposed an arrangement involving zeolite synthesised on metal wire gauzes to eliminate the limitations caused by insufficient heat transfer within the solar collector. Two types of zeolites – namely zeolite 4A and zeolite 13X – were investigated by the authors. Zhang [19] presented a prototype of an adsorption cooling system driven by waste heat from a diesel engine, using a zeolite/water pair. It is reported that the proposed system can achieve a specific cooling power (SCP) and a COP of 25.7 W/kg and 0.38 respectively. Sward et al. [20] presented a model for a thermal wave adsorption heat pump cycle using NaX zeolite/water as the adsorbent/adsorbate pair. The proposed model is used to examine the cycle performance. It is reported that the COP was as high as 1.2 at the heat source, condenser and evaporator temperatures of 120°C, 30°C, and –15°C respectively. A different cycle pattern-periodic reversal forced convective cycle has been studied theoretically by Lai [21] using zeolite 13X/water as the adsorbent/refrigerant pair. Wang et al. [22] presented a novel design of an adsorption air conditioner system that can supply 8°C–12°C chilled water. The system consists of an adsorber, a condenser and an evaporator; all the system components are housed in the same adsorption/desorption chamber. The predicted cooling capacity and COP are 5 kW and 0.25, respectively.

2.2.3 Zeolite Composites/Water Pair

Composite adsorbents made from high conductivity carbon (graphite) and metallic foam with zeolite have been investigated by Guilleminot et al. [23]. The compositions were typically 65% zeolite + 35% metallic foam, and 70% zeolite + 30% graphite. The proposed adsorbents have high thermal conductivity, and hence the system performance was improved. Pons et al. [24] experimentally investigated the performance of a regenerative thermal wave cycle by using an innovative composite adsorbent composed of zeolite and expanded natural graphite. It is reported that the cooling COP improved to 0.9. However, the specific cooling capacity was only 35 W/kg.

2.2.4 Activated Carbon/Methanol Pair

Activated carbon (AC)/ethanol systems have been investigated mainly in Europe since 1980. Pons and Guilleminot [25] studied an AC/methanol system for ice production using renewable energy. Douss and Meunier [26] investigated the possibility of using active carbon of type AC135 along with methanol as a working fluid for refrigeration application. It is reported that for an evaporating temperature lift equal to 25°C, the experimental COP of the intermediate cycle reached 0.5 with regeneration temperature difference (the difference between regeneration and ambient temperatures) equal to 65°C. Critoph [27] studied the possibility of using activated charcoal with different types of refrigerants working at sub-atmospheric and supra-atmospheric pressures. Refrigerants that are sub-atmospheric include methanol, acetonitrile, methyl amine and NO_2. However, ammonia, formaldehyde and SO_2 represent the supra-atmospheric group in the study. Douss and Meunier [28] proposed and analysed a cascading adsorption cycle in which an active carbon/methanol cycle is topped by a zeolite/water cycle. The cycle COP exceeds 1 with a driving heat source of 250°C. Anyanwu et al. [29], Anyanwu and Ezekwe [30] and Wang et al. [31] also conducted studies on various adsorption systems using the same adsorbent/adsorbate pair. El-Sharkawy et al. [32] investigated the isothermal characteristics of methanol on two specimens of ACs: Maxsorb III and Tsurumi activated charcoal. The Dubinin–Radushkevich (D-R) equation is used to correlate the adsorption isotherms of the two assorted adsorbent/adsorbate pairs. Experimental measurements show the superiority of Maxsorb III/methanol pair over the Tsurumi activated charcoal/methanol pair.

2.2.5 Activated Carbon/Ammonia Pair

The possibility of using an AC/ammonia pair for adsorption cooling applications has been investigated by Critoph [27]. He also proposed a novel cycle in which adsorbent heat transfer can be enhanced by forced convection [33]. The adsorption cycle is modelled for AC/ammonia, and results show that a power density of 1–3 kW/kg of adsorbent can be achieved. The heating COP of the adsorption cycle was 1.3 when heat was pumped from 0°C to 50°C. Critoph [34] also used the same pair to build a 10 kW air conditioner. Miles and Shelton [35] experimentally tested a two-bed system using AC/ammonia as the adsorbent/adsorbate pair. An analytical methodology for the prediction of system performance and optimisation has been presented. They reported that through the use of a thermal wave regeneration concept the cycle efficiency can be significantly increased.

2.2.6 Monolithic Carbon/Ammonia Pair

Tamainot-Telto and Critoph [36,37] presented a laboratory prototype of an adsorption refrigeration system using a monolithic carbon/ammonia pair. Critoph and Metcalf [38] suggested a conceptual model of an adsorption refrigeration system based on a plate-type sorption generator using the same adsorbent/adsorbate pair. The results showed that the system could reach a COP of about 0.3 and a specific cooling capacity value above 2 kW/kg of adsorbent.

2.2.7 Activated Carbon Fibre/Methanol Pair

Wang et al. [39] studied the adsorption characteristics of activated carbon fibre (ACF)/methanol and AC/methanol pairs. They considered polyacrylonitrile (PAN)-based ACFs, which are made of organic materials. The authors reported that the adsorption/desorption time for ACF/methanol pair versus AC/methanol pair is about 1/5:1/10. Based on the experimental data, analysis of ideal cycle of ACF/methanol and AC/methanol has been investigated. It was reported that the COP of the ACF/methanol pair is about 10%–20% higher than that of the AC/methanol pair.

2.2.8 Activated Carbon Fibre/Ammonia Pair

A solid sorption heat pump based on ACF/ammonia has been developed and tested by Vasiliev et al. [40]. Results showed that with solar collector efficiency being near 0.7, it is possible to obtain a high solar COP near 0.3. Vasiliev [41] also proposed a solar-powered solid sorption refrigerator using ACF-based adsorbents, 'Busofit', that are saturated with different salts, such as $CaCl_2$, $BaCl_2$ and $NiCl_2$. Ammonia is used as a refrigerant for adsorption onto Busofit.

2.2.9 Activated Carbon Fibre/Alcohol Pairs

Kumita et al. [42] measured the adsorption equilibrium of methanol and ethanol onto ACF and granular active carbon at two adsorption temperatures, 30°C and 50°C, respectively. Adsorption isotherms are predicted from adsorption equilibrium curves. It is reported that ACF has a high adsorption capacity for both ethanol and methanol.

El-Sharkawy et al. [43,44] presented an experimental apparatus to measure adsorption isotherms of ethanol onto ACFs of types (A-20) and (A-15) for adsorption cooling applications. Adsorption kinetics of the ACF (A-20)/ethanol pair has also been experimentally investigated by the same authors [45,46].

2.3 Two-Bed Adsorption Cooling Cycles

2.3.1 Working Principles

The schematic diagram of the two-bed adsorption cooling cycle is shown in Figure 2.1a. The cycle consists of four heat exchangers: an evaporator, a condenser and a pair of adsorber/desorber heat exchangers (sorption elements). The cycle has four modes: A, B, C and D. In mode A, as shown in Figure 2.1a, refrigerant valves V1 and V3 are opened, whereas valves V2 and V4 are closed [47,48]. In this mode, evaporator and SE1 are in adsorption process and condenser and SE2 are in desorption process. In the adsorption-evaporation process that takes place at pressure P^{evap}, the refrigerant in the evaporator is evaporated at an evaporation temperature, T^{evap} and seized heat, Q^{evap}, from the chilled water. The evaporated vapour is adsorbed by adsorbent, at which cooling water removes the adsorption heat, Q^{ads}. The desorption-condensation process takes place at pressure P^{cond}. The desorber (SE2) is heated up to the desorber temperature T^{des}, which is provided by the driving heat source, Q^{des}. The resulting refrigerant vapour is cooled down to a temperature T^{cond}

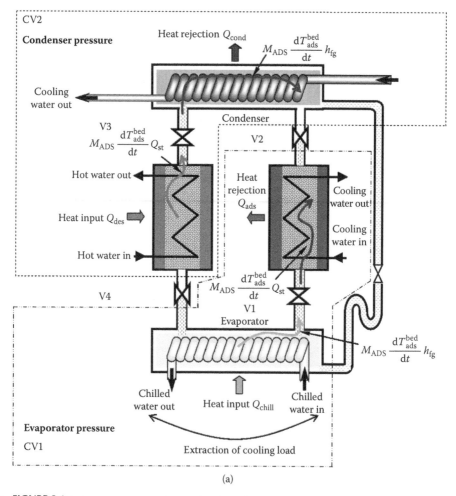

(a)

FIGURE 2.1
Schematic diagram of (a) the two-bed adsorption cooling system in mode A and (b) the two-bed adsorption cooling system in mode C. (From Saha et al. *Int. J. Refrig.*, 30(1), 86–95, 96–102, 2007.)

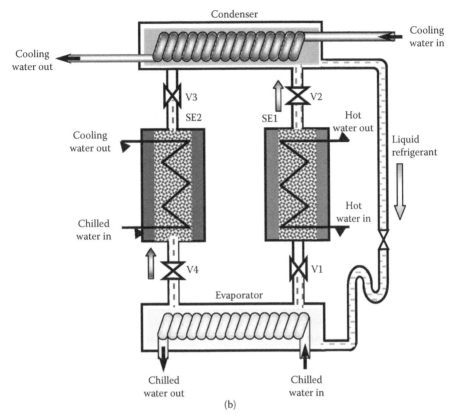

FIGURE 2.1 (*Continued*)

in the condenser by the cooling water, which removes the condensation heat, Q^{cond}. When the refrigerant concentrations in the adsorber and the desorber are at or near their equilibrium levels, the cycle is continued by changing into a short-duration mode named the preheating or pre-cooling mode (mode B), where all refrigerant valves remain closed. In mode B, heat transfer fluid (hot water and cooling water) flows are redirected to ensure the preheating and pre-cooling processes of SE1 and SE2 respectively. When the pressures of desorber and adsorber are nearly equal to those of condenser and evaporator respectively, then the valve between SE2 and evaporator and the refrigerant valve between SE1 and condenser are opened to initiate adsorption and desorption processes respectively. This mode is denoted as mode C. In mode C, valves V2 and V4 are opened, whereas valves V1 and V3 are closed. In this mode, condenser and SE1 are in desorption process and evaporator and SE2 are in adsorption process. As can be noticed from Figure 2.1b and Table 2.1, in mode C the operations of all four refrigerant valves and the processes of

TABLE 2.1

Operation Schedule of Refrigerant Valves and Sorption Elements in a
Two-Bed Adsorption Chiller

	Mode A	Mode B	Mode C	Mode D
Valve 1	Open	Closed	Closed	Closed
Valve 2	Closed	Closed	Open	Closed
Valve 3	Open	Closed	Closed	Closed
Valve 4	Closed	Closed	Open	Closed
SE1	Cooling water	Hot water	Hot water	Cooling water
SE2	Hot water	Cooling water	Cooling water	Hot water

SE1 and SE2 are exactly opposite to those of mode A. After the completion of
mode C, preheating of SE2 and pre-cooling of SE1 start, which are opposite
of mode B and known as mode D. After mode D, the adsorption cooling system returns to mode A. Table 2.1 shows the state of the refrigerant valves and
the choice of hot or cooling water application.

The heat flow diagrams of the adsorption-evaporation and desorption-
condensation processes are shown by CV1 and CV2 respectively, in
Figure 2.1a. In the adsorption-evaporation process, a sensible cooling heat
source (Q_{ads}) is necessary to reject the adsorption latent heat (Q_{ads}^{latent}), which
is developed due to the evaporation of refrigerant vapour in the evaporator
by the extraction of cooling load (Q_{chill}). The heat source (Q_{des}) is also needed
to drive the latent heat (Q_{des}^{latent}) from the bed in desorption mode such that
this latent heat could be extracted through the condenser by a sensible heat.
In the condenser, some latent heat is stored as a warm condensate, which
refluxes back to the evaporator via a capillary U-bend tube and maintains
the pressure difference between the condenser and the evaporator. This
heat depends on the amount of warm condensate, specific heat capacity of
the refrigerant and temperature difference between the condenser and the
evaporator.

2.3.2 Mathematical Modelling

2.3.2.1 Adsorption Isotherms

A sizable number of adsorption isotherm models are used to present the
adsorption capacity of various types of adsorbent/refrigerant pairs. For
instance, Saha et al. [1] presented a modified form of the Freundlich equation
(S-B-K equation) to estimate the adsorption uptake of RD silica gel/water
pair at the equilibrium conditions given by Equation 2.1:

$$q^* = A(T_s)\left[\frac{P_s(T_w)}{P_s(T_s)}\right]^{B(T_s)}$$

(2.1)

where

$$A(T_s) = A_0 + A_1 T_s + A_2 T_s^2 + A_3 T_s^3 \tag{2.2}$$

$$B(T_s) = B_0 + B_1 T_s + B_2 T_s^2 + B_3 T_s^3 \tag{2.3}$$

The Tóth equation [49] is used to predict the adsorption equilibrium of the CaCl$_2$-in-silica gel/water pair [50]:

$$q^* = \frac{q_m K_0 \times \exp[q_{st}/(R \times T)] \times P}{\left\{ 1 + [K_0 \times \exp(Q_{st}/(R \times T)) \times P]^{t_1} \right\}^{(1/t_1)}} \tag{2.4}$$

In the aforementioned equations, q^* is the adsorption uptake at equilibrium conditions. $A(T_s)$ and $B(T_s)$ are temperature-dependent functions defined by Equations 2.2 and 2.3 respectively. $P_s(T_w)$ is the saturation pressure at vapour temperature, and $P_s(T_s)$ presents the saturation pressure at adsorbent temperature. Q_{st} stands for the isosteric heat of adsorption, q_m is the monolayer capacity, K_0 is a pre-exponential constant and t_1 is a dimensionless constant. The numerical values of constants in Equations 2.1 through 2.4 are given in Table 2.2.

The D-R equation is used to estimate the equilibrium uptake of the ACF (A-20)/ethanol pair [47]. The numerical values of q_0 and D were evaluated experimentally by El-Sharkawy et al. [43], and were found to be 0.797 kg/kg and 1.716×10^{-6} 1/K^2 respectively.

TABLE 2.2

Numerical Values of Parameters Used in Equations 2.1 through 2.4

Parameter	Value
A_0	−6.5314
A_1	0.072452
A_2	−0.23951 × 10^{-3}
A_3	0.25493 × 10^{-6}
B_0	−15.587
B_1	0.15915
B_2	−0.50612 × 10^{-3}
B_3	0.5329 × 10^{-6}
K_0	2 × 10^{-12}
q_{st}	2760
w_m	0.8
t_1	1.1

$$q^* = q_0 \exp\left\{-D\left[T\ln\left(\frac{P_s}{P}\right)\right]^2\right\}$$

(2.5)

2.3.2.2 Adsorption Kinetics

The well-known linear driving force (LDF) model (Equation 2.6) is commonly used to estimate the adsorption rate of spherical adsorbent/refrigerant pairs:

$$\frac{\partial q}{\partial t} = k_s a_v \left(q^* - q\right)$$

(2.6)

where $k_s a_v$ is the overall mass transfer coefficient of the adsorbent/refrigerant pair and can be written as

$$k_s a_v = \frac{F_0 D_s}{R_p^2}$$

(2.7)

where R_p is the adsorbent particle radius and F_0 is a constant. The numerical value of F_0 is found to be 15 for spherical adsorbent shapes. Surface diffusivity, D_s, can be expressed by the Arrhenius equation, as shown here:

$$D_s = D_{s0} \exp\left(-\frac{E_a}{RT}\right)$$

(2.8)

In Equation 2.8, D_{s0} is a pre-exponential constant and E_a is the activation energy. For a silica gel/water pair, the numerical values of D_{s0} and E_a are, respectively, 2.54×10^{-4} m²/s and 4.2×10^4 J/mol. El-Sharkawy et al. [45] studied the adsorption kinetics of an ACF (A-20)/ethanol pair using a thermogravimetric analyzer from which the diffusion time constant and, consequently, the overall mass transfer coefficient were evaluated. It was found that the numerical value of F_0 is 11, considering the cylindrical shape of ACF. The numerical values of D_{s0} and E_a were estimated by Saha et al. [46] and found to be 1.8×10^{-12} m²/s and 306.7×10^3 J/kg respectively.

2.3.2.3 Adsorption and Desorption Energy Balance

Using the lumped approach for the adsorption bed, which comprises the adsorbent and the heat exchanger fins and tubes, the energy balance equation is given by

$$\left(Mc_p\right)_{eff}^{bed} \frac{dT_i^{bed}}{dt} + \left[\dot{m}c_p\right]_{i-phase} \frac{dT_i^{bed}}{dt} = \varphi M_{ADS}\left(\frac{dq_i^{bed}}{dt}\right)(Q_{st}) - (\dot{m}c_p)_j\left(T_{j,o} - T_{j,in}\right)$$

(2.9)

where the flag φ is 0 during switching and 1 during adsorption/desorption cycle operation, *i* indicates the adsorption/desorption bed and *j* defines the cooling/heating source. Q_{st} presents the isosteric heat of adsorption. The left-hand side of the adsorber/desorber energy balance equation (Equation 2.9) provides the rate of change of internal energy due to the thermal mass of adsorbent, the refrigerant, as well as the adsorber/desorber heat exchanger during adsorption and desorption. The first term on the right-hand side of Equation 2.9 represents the release of adsorption heat during the adsorption process or the input of desorption heat during the desorption operation. The second term on the right-hand side indicates the total amount of heat released to the cooling water upon adsorption or provided by the hot water for desorption. For a small temperature difference across the cooling/heating fluid such as water, the outlet temperature of the source is sufficiently accurate to be modelled by the log mean temperature difference (LMTD) method; it is given by

$$T_{j,o} = T_i^{bed} + \left(T_{j,in} - T_i^{bed}\right)\exp\left[\frac{-(UA)_i^{bed}}{(\dot{m}c_p)_j}\right] \tag{2.10}$$

Hence, Equation 2.10 expresses the importance of heat transfer parameters: heat transfer area, A^{bed}, and overall heat transfer coefficient, U^{bed}.

2.3.2.4 Condenser Energy Balance

The condenser is a water-cooled shell-tube heat exchanger that liquefies the vapour refrigerant emanating from the desorber during desorption, delivering the condensed liquid to the evaporator via a U tube. The mathematical modelling of the condenser is coupled with the thermodynamic states of desorption beds and the pressure difference between the condenser and the evaporator. The energy balance equation for the condenser can be expressed as follows:

$$(Mc_p)_{eff}^{cond}\frac{dT^{cond}}{dt} = \varphi\left(h_{fg}M_{ADS}\frac{dq_{des}^{bed}}{dt}\right) - (\dot{m}c_p)_w\left(T_{w,o} - T_{w,in}\right) \tag{2.11}$$

Hence, the left-hand side of Equation 2.11 represents the rate of change of internal energy heat required by the metallic parts of heat exchanger tubes due to temperature variations in the condenser. On the right-hand side, the first term gives the latent heat of vaporisation due to the amount of refrigerant desorbed from the desorption bed and the amount of heat that the liquid condensate carries away when it leaves the condenser for the evaporator, and finally the last term represents the total amount of heat released to the cooling water. In the current design, the condenser heat exchanging tubes have been chosen to be corrugated, with the enhanced surface inevitably retaining a thin film of the condensate, which ensures that the condenser and the desorber are always maintained at the refrigerant's saturated vapour pressure. Using the LMTD approach, we can write the outlet temperature of the condenser heat exchanging tube as follows:

$$T_{w,o} = T^{cond} + \left(T_{w,in} - T_{ads}^{bed}\right)\exp\left[\dfrac{-(UA)^{cond}}{(\dot{m}c_p)_w}\right] \qquad (2.12)$$

2.3.2.5 Evaporator Energy Balance

The overall energy balance of the evaporator is governed mainly by the heat and mass interactions between the evaporator and the adsorption bed. The heat released from the condenser is also taken into account. The energy balance of the evaporator can be expressed as given by Equation 2.13:

$$\left[(Mc_p)_{eff}^{evap}\right]\dfrac{dT^{evap}}{dt} = -\varphi h_{fg}\left(M_{ADS}\dfrac{dw_{ads}^{bed}}{dt}\right) - (\dot{m}c_p)_{chill}\left(T_{chill,o} - T_{chill,in}\right) \quad (2.13)$$

The left-hand side of Equation 2.13 represents the change in internal energy due to the sensible heat of the liquid refrigerant and the metal of heat exchanger tubes in the evaporator. On the right-hand side, the first term gives the latent heat of evaporation for the amount of refrigerant adsorbed; the second term shows the cooling capacity of the evaporator. The chilled water outlet temperature is given as follows:

$$T_{chill,o} = T^{evap} + \left(T_{chill,in} - T^{evap}\right)\exp\left[\dfrac{-(UA)^{evap}}{(\dot{m}c_p)_w}\right] \qquad (2.14)$$

2.3.2.6 Mass Balance

The overall mass balance of a refrigerant is governed by the mass interactions during (1) the switching interval and (2) the adsorption/desorption operation period. The rate of change of mass of refrigerant is equal to the amount of refrigerant adsorbed during adsorption as well as that desorbed during desorption. Neglecting the gaseous phase, the mass balance equation for the cooling system is expressed by Equation 2.15:

$$\dfrac{dM_{ref}}{dt} = -M_{ADS}\left[\dfrac{dq_{ads}^{bed}}{dt} + \dfrac{dq_{des}^{bed}}{dt}\right] \qquad (2.15)$$

2.3.2.7 System Performance

As the present system aims at using low-temperature waste heat sources, the end user may be more concerned with instantaneous cooling power and the concept of cycle-averaged cooling power, Q_{chill}^{cycle}, and a COP that is based on cycle-averaged quantities could be a useful yardstick for the designer, and a convenient way for the manufacturer to describe their product. In this context, Q_{chill}^{cycle} and COP are defined by Equations 2.16 and 2.17 respectively:

$$Q_{chill}^{cycle} = \frac{\int_0^{t_{cycle}} \left\{ (\dot{m}c_P)_{chill} \left(T_{chill,in} - T_{chill,o} \right) \right\} dt}{t_{cycle}} \qquad (2.16)$$

$$COP = \frac{\int_0^{t_{cycle}} (\dot{m}c_P)_{chill} \left(T_{chill,in} - T_{chill,o} \right) dt}{\int_0^{t_{cycle}} (\dot{m}c_P)_{des} \left(T_{hw,in} - T_{hw,o} \right) dt} \qquad (2.17)$$

Here, t_{cycle} denotes total cycle time.

2.4 Thermally Powered Advanced Adsorption Cooling Cycles

Advanced adsorption cooling cycles are designed either to enhance the performance of basic systems or to use low-temperature-grade heat sources as low as 40°C in combination with a coolant at 30°C. Regenerative systems aim to achieve the former target, and multi-stage systems are designed to achieve the latter. However, as this field has become very large and there are a huge number of studies that cannot be concluded entirely in one chapter, selected examples of adsorption cooling systems are presented here. Saha et al. [5, 13, 51–53] proposed two- and three-stage adsorption cycles to use, respectively, waste heat of temperatures between 50°C and 70°C and between 40°C and 60°C with a coolant at 30°C. Figure 2.2a and b shows the schematic and the pressure–temperature–concentration diagrams of the two-stage adsorption cooling cycle. In this cycle, the evaporation temperature lift ($T_{cond} - T_{evap}$) is divided into two smaller lifts. The refrigerant pressure is therefore raised in two progressive steps: from the evaporator pressure to an intermediate pressure and from the intermediate pressure to the condenser pressure. This makes it possible to use a low-temperature heat source such as solar energy or waste heat (see Figure 2.2b).

In another endeavour, the same authors [54] investigated a regenerative multi-bed dual-mode adsorption system, as shown in Figure 2.3. These cycles have a couple of objectives: the first one is to decrease the peak

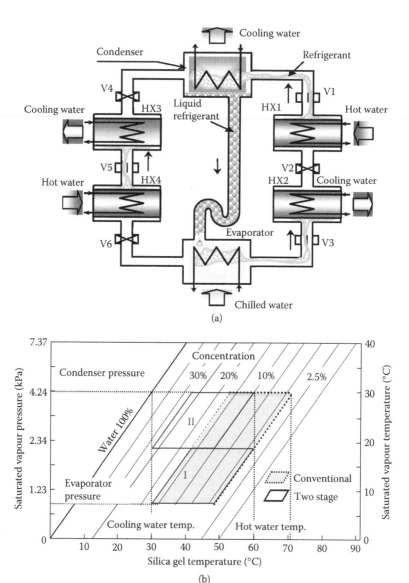

FIGURE 2.2

(a) Schematic diagram of a two-stage silica gel/water adsorption chiller and (b) conceptual pressure-temperature-concentration diagram for single- and two-stage silica gel/water adsorption cycles. (From Saha et al. *Renew. Energ.*, 23(1), 93–101, 2001.)

temperature of both the condenser and the evaporator outlets, and the second objective is to improve the recovery efficiency of waste heat to the cooling load. The authors have reported that the dual-mode cycle is capable of effectively using low-grade waste heat sources of temperatures between 40°C and 95°C as driving heat sources along with a coolant at 30°C. The

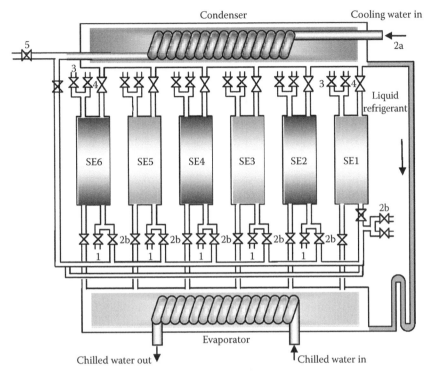

FIGURE 2.3
Schematic diagram of the multi-bed adsorption chiller. (From Saha et al. *Int. J. Refrig.*, 26(7), 749–757, 2003.)

chiller in the three-stage mode is operational with a heat source and heat sink temperature difference as small as 10 K.

To improve the performance of the adsorption system, Miles and Shelton [35] introduced a two-bed thermal wave system using AC/ammonia as an adsorbent/refrigerant pair. In the thermal process, only a single heat transfer fluid loop exists. Figure 2.4 shows a schematic diagram of the proposed cycle. A reversible pump is placed in the closed heat transfer fluid loop to invert the circulation of fluid flow. It is reported that the proposed system achieves a cooling COP as high as 1.9. Similar cycles have been investigated theoretically by Sward et al. [20].

The concept of direct contact condensation and evaporation has been used by Yanagi et al. [55] to develop an innovative silica gel adsorption refrigeration system. As reported by Saha et al. [56], the concept of direct contact during evaporation and condensation (because of the elimination of metallic frames) decreases the temperature difference between hot and cold fluids, which increases heat and mass transfer. In this cycle, the main components are a pair of sorption elements neighbouring on a pair of spray nozzles working as either a condenser or an evaporator housing in the same vacuum chamber. As

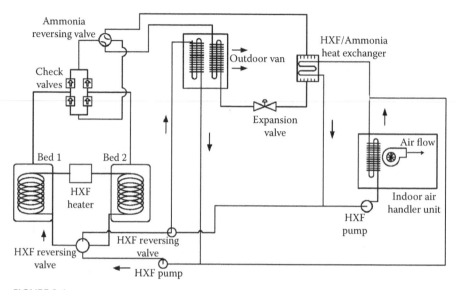

FIGURE 2.4
A thermal wave solid sorption heat pump in cooling mode. (From Miles, D.J., and S.V. Shelton, *Appl. Therm. Eng.*, 16(5), 389–394, 1996.)

can be seen from Figure 2.5, the water vapour evaporating directly from the surface of the sprayed water jet is adsorbed by HX1 in the adsorption mode, whereas the desorbed water vapour from HX2 is condensed on the surface of the sprayed water in the desorption mode. The rated cooling capacity of the Mayekawa pilot plant (Atlanta, USA) is 11.5 kW. The chiller has a COP value of 0.58; for hot, cooling and chilled water, inlet temperatures are 70°C, 29°C and 14°C respectively. The delivered chilled water temperature was reported at 9°C for a 10-minute duration of the adsorption/desorption cycle.

The mass recovery scheme is used to increase the cooling capacity of adsorption cooling systems. The ideology of this technique is discussed in detail by Qu et al. [57] and can be explained as follows: (1) pressure in the desorber reactor increases during the preheating process and nearly reaches the value of condenser pressure at the end of the desorption process; (2) at the same time, adsorber pressure decreases during the pre-cooling process and is nearly equal to the evaporator pressure at the end of the adsorption process; (3) both of the adsorption reactors can be simply interconnected, and the refrigerant vapour flows from the desorber (high-pressure bed) to the adsorber (low-pressure bed); (4) the process continues until the pressures in both reactors are nearly equal; and, (5) finally, both of the adsorption reactors are disconnected and continued in heating and cooling processes as the basic adsorption cooling cycle. Figure 2.6 shows the conceptual pressure-temperature-concentration (Dürhring) diagram of the basic mass recovery cycle [58]. As can be seen from Figure 2.6a, the mass of refrigerant circulated in the system using the mass recovery technique is greater than that

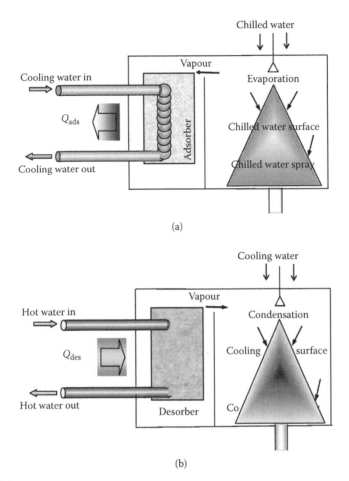

Chilled water

Vapour

Cooling water in

Q_{ads}

Adsorber

Evaporation

Chilled water surface

Chilled water spray

Cooling water out

(a)

Cooling water

Vapour

Hot water in

Q_{des}

Condensation

Cooling surface

Desorber

Co

Hot water out

(b)

FIGURE 2.5
Schematic of the direct contact adsorption cycle (a) adsorption cycle and (b) desorption cycle. (From Yanagi et al. *Proceedings of the International Sorption Heat Pump Conference*, 671–676, 1999.)

of conventional adsorption cooling systems. This leads to a higher cooling output. Figure 2.6b shows a schematic of the basic mass recovery adsorption refrigeration cycle.

A considerable number of researchers have investigated the performance of adsorption cooling systems using the mass recovery technique. The following are some representative examples: Akahira et al. [59] investigated the performance of a silica gel/water adsorption cooling system with a mass recovery process using an experimental prototype machine. They studied the effect of operating conditions of the proposed system on specific cooling effects and COP. Results of this study show that the SCP of the mass recovery cycle is superior to that of the conventional cycle. Moreover, the mass recovery cycle is effective with a low-temperature heat source. Khan et al. [60] theoretically

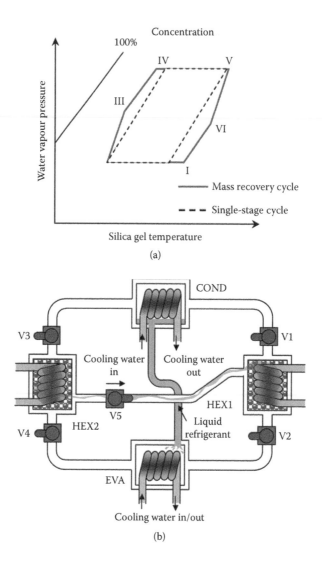

FIGURE 2.6
(a) Conceptual Dühring diagram of the mass recovery adsorption refrigeration cycle and (b) schematic of the basic mass recovery adsorption refrigeration cycle. (From Alam et al. *Renew. Energ.*, 29, 1461–1475, 2004.)

studied the performance of a solar or waste heat–driven three-bed adsorption cooling cycle using a mass recovery scheme. The proposed chiller is driven by exploiting solar/waste heat of temperatures between 60°C and 90°C with a cooling source at 30°C for air-conditioning purposes. They reported that the cooling effect, as well as the solar/waste heat recovery efficiency, of the chiller with the mass recovery scheme is superior to that of a three-bed chiller without mass recovery for heat source temperatures between 60°C and 90°C. But

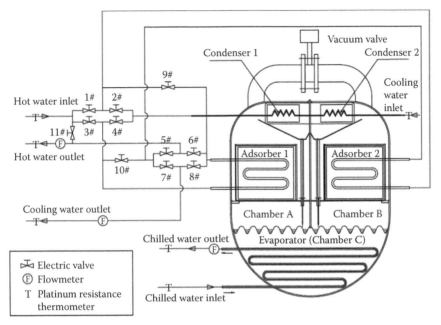

FIGURE 2.7
Schematic diagram of the silica gel/water adsorption chiller. (From Wang et al. *Int. J. Refrig.*, 28, 1073–1083, 2005.)

the COP of the proposed chiller is higher than that of the three-bed chiller without mass recovery when the heat source temperature is below 65°C.

Wang et al. [61] proposed a new design for the silica gel/water chiller. The chiller consists of three chambers: A, B and C, as shown in Figure 2.7. Two of them (A and B) are adsorption/desorption chambers, and the third one (C) is a heat pipe chamber that functions as the evaporator. Both chambers A and B contain one adsorber, one condenser and one evaporator (half the water-evaporating surface of the third chamber). In this system, water is the refrigerant and the heat media of the heat pipe is methanol. To fulfil the mass recovery scheme, a vacuum valve is installed between the two main chambers. The system design and working principles are described in detail in the literature [61]. Wang et al. [62] built a prototype of this chiller and tested its performance. The authors reported that the system COP is 0.38 and it can produce a cooling capacity of 7.15 kW for hot and cold water temperatures of 84.8°C and 30.6°C respectively. The outlet chilled water temperature is 11.7°C.

A recent trend in adsorption cooling is to use newly synthesised refrigerants to replace R134a in mobile air-conditioning (MAC) systems. Hydrofluoroolefins (HFOs) such as HFO-1234ze(E) and HFO-1234yf are newly synthesised refrigerants that have zero ozone depletion potential (ODP) and a low global warming potential (GWP) in the range of less than 10, and it is expected that these two refrigerants will be used as next-generation coolants for mobile air conditioning systems [63]. Jribi et al. [63] theoretically investigated the dynamic behaviour of a four-bed adsorption chiller using highly porous AC of type Maxsorb III as the adsorbent and R1234ze(E), whose GWP is as low as 9, as the refrigerant. They reported that with 80 kg of Maxsorb III the system is able to produce 2 kW of cooling power at a driving heat source temperature of 85°C. There is an ongoing project at Warwick, Coventry, United Kingdom, that is aimed at investigating various conceptual designs of novel hybrid refrigeration and heat pump systems. The hybrid systems use an environmentally friendly refrigerant, R723 (azeotropic mixture: 40% dimethylether and 60% ammonia), that is compatible with copper nickel alloy ($CuNi_{10}$): it is a combined sorption and vapour compression refrigeration machine driven by a dual source (heat and/or electricity) [64]. The use of the dual source makes the proposed systems highly flexible and energy efficient when operating. To maximise the energy efficiency, Saha et al. [65] recently patented an AACD cycle, which is introduced in Section 2.5.1.

2.5 Advanced Adsorption Cooling cum Desalination Cycle

2.5.1 Working Principle of the AACD Cycle

The adsorption cycle uses the physisorption process between an adsorbent and an adsorbate to produce cooling power from the evaporation of saline water in the evaporator and potable water in the condenser. These useful effects are achieved by the amalgamation of 'adsorption-triggered-evaporation' and 'desorption-resulted-condensation' processes driven by low-temperature hot water that can be extracted from industrial waste heat or solar energy [65–67].

The schematic of an innovative two-evaporator and three-reactor AACD cycle is illustrated in Figure 2.8 [65]. The AACD cycle comprises three adsorbent beds, that is, one condenser and two evaporators, in which the evaporators are assembled to operate as a single adsorption device where two temperature levels of cooling (both sensible and latent) are generated while, concomitantly, fresh or potable water is produced at the condenser. The high- and low-pressure evaporators are in thermal and mass communication with the beds, operating in tandem but in a predetermined manner with an externally supplied heat source.

The inventive step lies in the unique arrangement of the three-bed and two-evaporator design, and when it is operated with an optimal cycle time the production yields of desalted water and the cooling effects are almost tripled when compared with conventional (single-evaporator type) two-bed or four-bed adsorption cycles. It is therefore an object of the present system

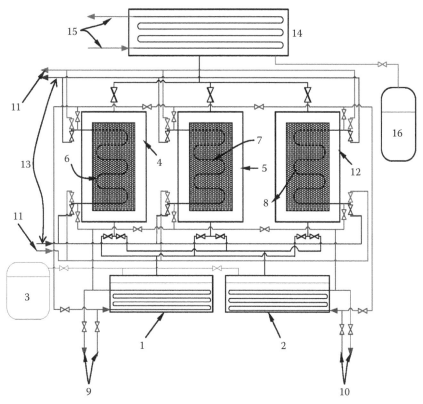

FIGURE 2.8
Schematic diagram of an advanced adsorption cooling cum desalination system. (From Saha et al. Desalination system and method, PCT/WO 2011/010205 A1, 2011.)

to show the three-bed-assisted advanced adsorption cycle that produces optimal cooling and desalination, with only a low-temperature heat input varying from 60°C to 80°C.

How does the proposed cycle work? During vapour communication or uptake between a bed (containing unsaturated adsorbent) and either the low-pressure or the high-pressure evaporator, the high-pressure evaporator produces a cooling stream at temperatures near ambient temperatures, typically from 18°C to 30°C, which can be used directly for sensible or process cooling. The low-pressure evaporator, on the other hand, generates a stream of coolant at 4°C–10°C that is useful for air conditioning or dehumidification applications. Owing to these external thermal loads, two different vapour pressures are maintained in the evaporators. A desorption process refers to the removal of adsorbed water vapour from the adsorbent by the application of heat, which is supplied from a coolant heat source or some other means. The desorbed vapour is condensed in the condenser, and the liquefied condensate, which is of distilled water quality, is collected in the collection tank.

2.5.2 Mathematical Modelling of the AACD Cycle

The mathematical model of a three-bed two-evaporator adsorption cycle for cooling and desalination is developed based on adsorption isotherms, kinetics, mass and energy balances between the sorption elements and the evaporator and the condenser. Type RD silica gel is used as the adsorbent. The Dubinin–Astakhov (D-A) equation is used to calculate the uptake of water vapour by silica gel at a specific temperature and pressure. The transient uptake by silica gel can be obtained using the well-known LDF equation (Equation 2.6).

The overall mass balance of the cycle is given by

$$\frac{dM_{s,evap}}{dt} = \dot{m}_{s,in} - \dot{m}_{d,cond} - \dot{m}_{brine} \tag{2.18}$$

Here, $M_{s,evap}$ is the sea water revenue in the evaporator, $\dot{m}_{s,in}$ is the rate of feed sea water, $\dot{m}_{d,cond}$ is the mass of potable water extracted from the condenser and $\dot{m}_{brine}$ is the mass of concentrated brine rejected from the evaporator. It should be noted here that the concentrated brine is discharged periodically depending on the concentration level in the evaporator. The energy balance of the evaporator in communication with the adsorber (master or slave) is written as follows:

$$\left[c_{p,s}\left(T_{evap}\right) M_{s,evap} + c_{p,HX} M_{HX,evap} \right] \frac{dT_{evap}}{dt}$$
$$= h_f\left(T_{Feed}\right)\dot{m}_{s,in} - h_{fg}\left(T_{evap}\right) M_{sg}\left(\frac{dq_{M_{ads}}}{dt} + \delta \frac{dq_{S_{ads}}}{dt} \right) \tag{2.19}$$
$$+ U_{evap} A_{evap}\left(T_{chilled,in} - T_{chilled,out}\right) - h_f\left(T_{evap}\right)\dot{m}_{brine}$$

The value of δ in this equation is either 1 or 0 depending on the communication of the slave adsorber with the evaporator. During the switching period, the value of δ is 0 since only the master adsorber is connected with the evaporator.

Similarly, the energy balance of the condenser is given by

$$\left[c_p\left(T_{cond}\right) M_{cond} + c_{p,HX} M_{HX,cond} \right] \frac{dT_{cond}}{dt}$$
$$= -h_f\left(T_{cond}\right)\dot{m}_{d,cond} + h_{fg}\left(T_{cond}\right) M_{sg}\left(\frac{dq_{M_{des}}}{dt} + \varepsilon \frac{dq_{S_{des}}}{dt} \right) \tag{2.20}$$
$$+ U_{cond} A_{cond}\left(T_{cond,in} - T_{cond,out}\right)$$

The value of ε is 1 when the slave desorber bed is connected with the condenser; otherwise its value is 0.

The energy balance of the master adsorber/desorber bed is given as follows:

$$\left(M_{sg}c_{p,sg} + M_{HX}c_{p,HX} + M_{abe}c_{p,a}\right)\frac{dT_{M_{ads}/M_{des}}}{dt} = \pm Q_{st}\left(T_{M_{ads}/M_{des}}, P_{evap/cond}\right)$$

$$M_{sg}\frac{dq_{M_{ads}/M_{des}}}{dt} \pm \dot{m}_{cw/hw}c_{p,cw/hw}\left(T_{M_{ads}/M_{des}}\right)\left(T_{cw/hw,in} - T_{cw/hw,out}\right) \tag{2.21}$$

The energy balance of the slave adsorber/desorber bed is written as follows:

$$\left(M_{sg}c_{p,sg} + M_{HX}c_{p,HX} + M_{abe}c_{p,a}\right)\frac{dT_{S_{ads}/S_{des}}}{dt} = \pm Q_{st}\left(T_{M_{ads}/M_{des}}, P_{evap/cond}\right)$$

$$M_{sg}\frac{dq_{S_{ads}/S_{des}}}{dt} \pm \dot{m}_{cw/hw}c_{p,cw/hw}\left(T_{S_{ads}/S_{des}}\right)\left(T''_{cw/hw,in} - T_{cw/hw,out}\right) \tag{2.22}$$

The third term in the left-hand side of Equations 2.21 and 2.22 represents the thermal mass contribution by the adsorbed phase, and $T''_{cw/hw,in}$ is the temperature of the cooling or hot water that comes out from the master adsorber/desorber bed. The isosteric heat of adsorption, Q_{st}, is calculated as follows [68]:

$$Q_{st} = h_{fg} + E\left\{-\ln\left(\frac{x}{x_m}\right)\right\}^{1/n} + Tv_g\left(\frac{\partial P}{\partial T}\right)_g \tag{2.23}$$

where v_g is the volume of the gaseous phase and h_{fg} is the latent heat.

The outlet temperature of the water from each heat exchanger is estimated using the LMTD method, and it is given by

$$T_{out} = T_0 + (T_{in} - T_0)\exp\left(\frac{-UA}{\dot{m}c_p(T_0)}\right) \tag{2.24}$$

Here, T_0 is the temperature of the heat exchanger. The energy required to remove the water vapours from the silica gels, master and slave desorption here, $\left(Q_{M_{des}/S_{des}}\right)$, can be calculated by using the inlet and outlet temperatures of the heat source supplied to the reactors, and this is given by

$$Q_{M_{des}/S_{des}} = \dot{m}_{hw}c_{p,hw}\left(T_{hw,in} - T_{hw,out}\right) \tag{2.25}$$

where $\dot{m}_{hw}$ and $c_{p,hw}$ indicate the mass flow rate and the specific heat capacity, respectively, of the heating fluid.

Here, the heat of evaporation $\left(Q_{\text{evap}}\right)$ and the condensation energy $\left(Q_{\text{cond}}\right)$ are given by

$$Q_{\text{evap}} = \dot{m}_{\text{chilled}} C_{p,\text{chilled}} \left(T_{\text{chilled,in}} - T_{\text{chilled,out}}\right) \tag{2.26}$$

$$Q_{\text{cond}} = \dot{m}_{\text{cond}} C_{p,\text{cw}} \left(T_{\text{cond,out}} - T_{\text{cond,in}}\right) \tag{2.27}$$

It is noted that the roles of the reactors are switched for adsorption or desorption process according to the cycle time allocation. The evaporator and condenser units communicate with at least one adsorber or desorber in every cycle operation, thus providing continuous cooling energy and potable water.

Finally, the performance of the AACD cycle is assessed in terms of SCP and specific daily water production (SDWP). The SDWP and SCP of the cycle are defined as follows:

$$\text{SCP} = \int_0^{t_{\text{cycle}}} \frac{Q_{\text{evap}} \tau}{M_{\text{sg}}} \, dt \tag{2.28}$$

$$\text{SDWP} = \int_0^{t_{\text{cycle}}} \frac{Q_{\text{cond}} \tau}{h_{\text{fg}} \left(T_{\text{cond}}\right) M_{\text{sg}}} \, dt \tag{2.29}$$

The mathematical modelling equations of the AACD cycle are solved using Gear's BDF method from the International Mathematics and Statistics Library (IMSL) linked by the simulation code written in Fortran, and the solver uses double precision with a tolerance value of 1×10^{-8}. With the proposed system, performance of the adsorption cycle is presented in terms of two key parameters: SCP and SDWP.

2.5.3 Performance Results of the AACD Cycle

Figure 2.9 shows the simulated temperature–time histories of the adsorber and desorber beds 4, 5 and 12 of the AACD cycle, which are shown schematically in Figure 2.8. The simulation of the innovative adsorption cycle is done by using the Fortran IMSL function. A set of modelling differential equations are solved using Gear's BDF method. From the simulation, it is found that desorption occurs at temperatures ranging from 60°C to 80°C. The low- and high-pressure bed temperatures vary from 33°C to 40°C.

Figure 2.10 shows the temperature–time histories of the condenser and the low-pressure and high-pressure evaporators (1 and 2) of the embodiment of the AACD cycle. It is observed from the present simulation that the temperature of the low-pressure evaporator, 1, ranges from 5°C to 7°C, which is very prominent for air-conditioning applications. The temperature of the high-pressure evaporator varies from 20°C to 22°C, which is good for sensible cooling. The specialty of the present system is that it decreases the peak evaporation temperature as opposed to conventional one-evaporator-type adsorption chillers.

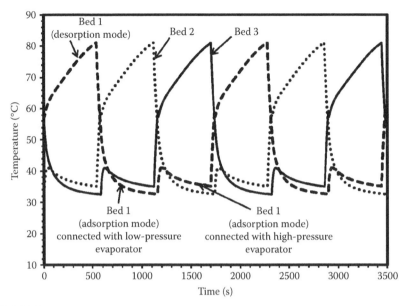

FIGURE 2.9
Temperature profiles of the adsorber and desorber beds of the advanced adsorption cooling cum desalination cycle.

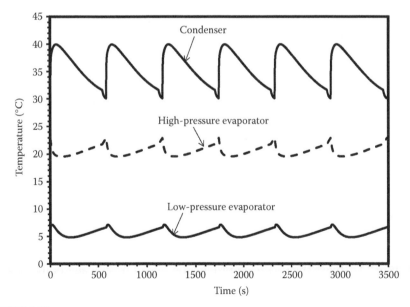

FIGURE 2.10
Temperature profiles of the condenser and the low-pressure and high-pressure evaporators of the advanced adsorption cooling cum desalination cycle.

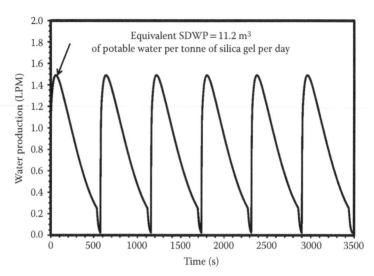

FIGURE 2.11
Freshwater production rate of the present advanced adsorption cooling cum desalination cycle.

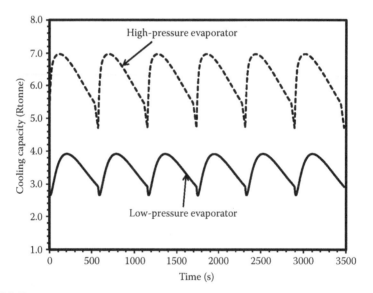

FIGURE 2.12
Effective and sensible cooling capacities of the advanced adsorption cooling cum desalination cycle.

Figure 2.11 shows the predicted production rate of fresh water of the AACD cycle. The freshwater production rate in terms of SDWP is shown in Figure 2.11, and the predicted SDWP is 11.2 m³ of fresh water per tonne of silica gel per day.

Finally, Figure 2.12 shows the effective and sensible cooling capacities as a function of operating time according to one embodiment of the present invention. The cycle-averaged sensible cooling capacity is 6 Rtonne, and the effective cooling capacity is 3.5 Rtonne.

2.6 Conclusions

The salient points of this chapter can be summarised as follows:

The silica gel/water-based two-stage adsorption cycle, which is powered by waste heat of temperatures between 50°C and 70°C with a coolant at 30°C, can produce cooling energy.

The silica gel/water-based multi-bed, dual-mode cycle can effectively use low-grade waste heat of temperatures between 40°C and 95°C along with a coolant at 30°C.

The cooling COP of the AC/ammonia-based thermal wave cycle is as high as 1.9. The silica gel/water-based direct contact condensation and evaporation cycle can significantly improve heat and mass transfer.

The proposed AACD cycle describes the use of two evaporators and three adsorption beds or reactors in a thermally driven adsorption cycle, in which both evaporators are connected with the reactors at two different pressure levels. The AACD cycle is capable of producing (1) chilled water at 4°C–10°C with a varying cooling capacity range of 3–4 Rtonne per tonne of silica gel, and (2) cooling water at 20°C–25°C with a cooling capacity ranging from 5 to 7 Rtonne per tonne of silica gel. The former is suitable for space cooling, and the latter is suitable for process cooling. Simultaneously, the proposed AACD cycle produces an SDWP of 11.3 m^3 per tonne of silica gel per day at rated operating conditions. The AACD cycle seems to be the most energy efficient adsorption cooling cum desalination system. Its widespread dissemination may eventually lead to the achievement of a carbon-neutral sustainable society.

Acknowledgements

The authors express their gratitude towards Prof. K.C. Ng and Dr. K. Thu of the National University of Singapore, Prof. S. Koyama of Kyushu University, Japan, and Dr. A. Chakraborty of Nanyang Technological University, Singapore for their help and enlightening advice.

Nomenclature

A	Surface area (m^2)
A_0	Coefficient in Equation 2.2 (kg/kg)
A_1	Coefficient in Equation 2.2 (kg/[kg K])
A_2	Coefficient in Equation 2.2 (kg/[kg K])
A_3	Coefficient in Equation 2.2 (kg/[kg K])
B_0	Coefficient in Equation 2.3 (–)
B_1	Coefficient in Equation 2.3 (1/K)
B_2	Coefficient in Equation 2.3 (1/K)
B_3	Coefficient in Equation 2.3 (1/K)
COP	Coefficient of performance (–)
D_s	Surface diffusivity (m^2/s)
D_{s0}	Pre-exponential constant (m^2/s)
E_a	Activation energy (J/mol)
F_0	Geometric parameter used in the LDF model (–)
h	Enthalpy (J/kg)
$k_s a_v$	Particle mass transfer coefficient (1/second)
K_0	Pre-exponential constant used in Equation 2.4 (1/Pa)
M	Mass (kg)
m	Mass (kg)
$\dot{m}$	Mass flow rate (kg/s)
P	Equilibrium pressure (kPa)
P_s	Saturation pressure (kPa)
Q_{st}	Isosteric heat of adsorption (kJ/kg)
q^*	Adsorption equilibrium uptake (kg/kg)
q	Instantaneous adsorption uptake (kg/kg)
q_m	Monolayer capacity (kg/kg)
R	Gas constant (kJ/[kg·K])
R_p	Radius of adsorbent particle (m)
SCE	Specific cooling effect (kJ/kg)
T	Temperature (K)
t	Time (second)
t_1	Dimensionless constant used in Equation 2.4 (–)
t_{cycle}	Cycle time (second)
U	Overall heat transfer coefficient (W/[m^2·K])

Superscripts

cond	Condenser
evap	Evaporator

Subscripts

ADS	Adsorbent
ads	Adsorption
c	Critical

chill	Chilled water
eff	Effective
f	Liquid phase
g	Gaseous phase
HX	Heat exchanger
i	Adsorption or desorption bed
in	Inlet
j	Cooling source or heating source
liq	Liquid
o	Outlet
s	Saturation
sg	Silica gel
v	Vapour
w	Water

References

1. B.B. Saha, E. Boelman, T. Kashiwagi, Computer simulation of a silica gel water adsorption refrigeration cycle – the influence of operating conditions on cooling output and COP, *ASHRAE Transactions*, 101(2), 348–357 (1995).
2. E. Boelman, B.B. Saha, T. Kashiwagi, Experimental investigation of a silica gel water adsorption refrigeration cycle – the influence of operating conditions on cooling output and COP, *ASHRAE Transactions*, 101(2), 358–366 (1995).
3. X.Q. Zhai, R.Z. Wang, Experimental investigation and theoretical analysis of the solar adsorption cooling system in a green building, *Applied Thermal Engineering*, 20(1), 17–27 (2009).
4. R.J.H. Grisel, S.F. Smeding, R. de Boer, Waste heat driven silica gel/water adsorption cooling in trigeneration, *Applied Thermal Engineering*, 30(8–9), 1039–1046 (2010).
5. B.B. Saha, T. Kashiwagi, Experimental investigation of an advanced adsorption refrigeration cycle, *ASHRAE Transactions*, 103(2), 50–58 (1997a).
6. B.B. Saha, A. Chakraborty, S. Koyama, K. Srinivasan, K.C. Ng, T. Kashiwagi, P. Dutta, Thermodynamic formalism of minimum heat source temperature for driving advanced adsorption cooling device, *Applied Physics Letters*, 91, 111902 (2007).
7. K.C. Ng, Recent developments in heat-driven silica gel-water adsorption chillers, *Heat Transfer Engineering*, 24(3), 1–3 (2003).
8. K.C.A. Alam, M.Z.I. Khan, A.S. Uyun, Y. Hamamoto, A. Akisawa, T. Kashiwagi, Experimental study of a low temperature heat driven re-heat two-stage adsorption chiller, *Applied Thermal Engineering*, 27(10), 1686–1692 (2007).
9. A. Sakoda, M. Suzuki, Fundamental study on solar powered adsorption cooling system, *Journal of Chemical Engineering of Japan*, 17(1), 52–57 (1984).
10. A. Sakoda, M. Suzuki, Simultaneous transport of heat and adsorbate in closed type adsorption cooling system utilizing solar heat, *Journal of Solar Energy Engineering*, 108, 239–245 (1986).

11. H.T. Chua, K.C. Ng, A. Malek, T. Kashiwagi, A. Akisawa, B.B. Saha, Modeling the performance of two-bed, silica gel-water adsorption chillers, *International Journal of Refrigeration*, 22(3), 194–204 (1999).

12. H.T. Chua, K.C. Ng, W. Wang, C. Yap, X.L. Wang, Transient modeling of a two-bed silica gel–water adsorption chiller, *International Journal of Heat and Mass Transfer*, 47(4), 659–669 (2004).

13. B.B. Saha, A. Akisawa, T. Kashiwagi, Silica gel water advanced adsorption refrigeration cycle, *Energy*, 22(4), 437–447 (1997b).

14. B.B. Saha, S. Koyama, T. Kashiwagi, A. Akisawa, K.C. Ng, H.T. Chua, Waste heat driven dual-mode, multi-stage, multi-bed regenerative adsorption system, *International Journal of Refrigeration*, 26(7), 749–757 (2003).

15. D.I. Tchernev, Solar air conditioning and refrigeration systems utilizing zeolites, *Proceeding of the Meetings of Commissions E1-E2*, Jerusalem, Israel, 209–215 (1979).

16. G. Alefeld, H.C. Bauer, P. Mailer-Laxhuber, M. Rothmeyer, A zeolite heat pump heat transformer and heat accumulator, *Proceedings of International Conference on Energy Storage*, Brighton, UK, 61–72 (1981).

17. M. Tather, A. Erdem-Senatalar, The effects of thermal gradients in a solar adsorption heat pump utilizing the zeolite–water pair, *Applied Thermal Engineering*, 19(11), 1157–1172 (1999).

18. M. Tather, B. Tantekin-Ersolmaz, A. Erdem-Senatalar, A novel approach to enhance heat and mass transfer in adsorption heat pumps using the zeolite–water pair, *Microporous and Mesoporous Materials*, 27(1), 1–10 (1999).

19. L.Z. Zhang, Design and testing of an automobile waste heat adsorption cooling system, *Applied Thermal Engineering*, 20(1), 103–114 (2000).

20. B.K. Sward, M.L. Douglas, F. Meunier, Adsorption heat pump modeling: the thermal wave process with local equilibrium, *Applied Thermal Engineering*, 20(8), 759–780 (2000).

21. H.M. Lai, An enhanced adsorption cycle operated by periodic reversal forced convection, *Applied Thermal Engineering*, 20(7), 595–617 (2000).

22. D.C. Wang, Z.Z. Xia, J.Y. Wu, Design and performance prediction of a novel zeolite–water adsorption air conditioner, *Energy Conversion and Management*, 47(5), 590–610 (2006).

23. J.J. Guilleminot, A. Choisier, J.B. Chalfen, S. Nicolas, J.L. Reymoney, Heat transfer intensification in fixed bed adsorbers, *Heat Recovery Systems and CHP*, 13(4), 297–300 (1993).

24. M. Pons, D. Laurent, F. Meunier, Experimental temperature fronts for adsorptive heat pump applications, *Applied Thermal Engineering*, 16(5), 395–404 (1996).

25. M. Pons, J.J. Guilleminot, Design of an experimental solar powered, solid adsorption ice maker, *Journal of Solar Energy Engineering-Transactions of the ASME*, 103(4), 332–337 (1986).

26. N. Douss, F. Meunier, Effect of operating temperatures on the coefficient of performance of active carbon-methanol systems, *Heat Recovery Systems and CHP*, 8(5), 383–392 (1988).

27. R.E. Critoph, Performance limitations of adsorption cycles for solar cooling, *Solar Energy*, 41(1), 21–31 (1988).

28. N. Douss, F. Meunier, Experimental study of cascading adsorption cycles, *Chemical Engineering Science*, 44(2), 225–235 (1989).

29. E.E. Anyanwu, U.U. Oteh, N.V. Ogueke, Simulation of a solid adsorption solar refrigerator using activated carbon/methanol adsorbent/refrigerant pair, *Energy Conversion and Management*, 42(7), 899–915 (2001).
30. E.E. Anyanwu, C.I. Ezekwe, Design, construction and test run of a solid adsorption solar refrigerator using activated carbon/methanol as adsorbent/adsorbate pair, *Energy Conversion and Management*, 44(18), 2879–2892 (2003).
31. L.W. Wang, J.Y. Wu, R.Z. Wang, Y.X. Xu, S.G. Wang, X.R. Li, Study of the performance of activated carbon–methanol adsorption systems concerning heat and mass transfer, *Applied Thermal Engineering*, 23(13), 1605–1617 (2003).
32. I.I. El-Sharkawy, M. Hassan, B.B. Saha, S. Koyama, M.M. Nasr, Study on adsorption of methanol onto carbon based adsorbents, *International Journal of Refrigeration*, 32(7), 1579–1586 (2009).
33. R.E. Critoph, Forced convection enhancement of adsorption cycles, *Heat Recovery Systems and CHP*, 14(4), 343–350 (1994).
34. R.E. Critoph, Forced convection adsorption cycles, *Applied Thermal Engineering*, 18(9–10), 799–807 (1998).
35. D.J. Miles, S.V. Shelton, Design and testing of a solid-sorption heat-pump system, *Applied Thermal Engineering*, 16(5), 389–394 (1996).
36. Z. Tamainot-Telto, R.E. Critoph, Adsorption refrigerator using monolithic carbon-ammonia pair, *International Journal of Refrigeration*, 20(2), 146–155 (1997).
37. Z. Tamainot-Telto, R.E. Critoph, Monolithic carbon for sorption refrigeration and heat pump applications, *Applied Thermal Engineering*, 21(1), 37–52 (2001).
38. R.E. Critoph, S.J. Metcalf, Specific cooling power intensification limits in ammonia–carbon adsorption refrigeration systems, *Applied Thermal Engineering*, 24(5–6), 661–678 (2005).
39. R.Z. Wang, J.P. Jia, Y.H. Zhu, Y. Teng, J.Y. Wu, J. Cheng, Q.B. Wang, Study on a new solid absorption refrigeration pair: active carbon fiber–methanol pair, *Trans. of ASME*, 119, 214–218 (1997).
40. L.L. Vasiliev, D.A. Mishkinis, A.A. Antukh, L.L. Vasiliev Jr, Solar–gas solid sorption heat pump, *Applied Thermal Engineering*, 21(5), 573–583 (2001).
41. L.L. Vasiliev, Solar sorption refrigerators with dual sources of energy, *Proceedings of the International Sorption Heat Pump Conference*, Shanghai, China, 26–33 (2002).
42. M. Kumita, S. Mori, T. Yokogoshiya, S. Otsubo, Adsorption equilibria for activated carbon fiber/alcohol pairs and their applicability to adsorption refrigerator, *Journal of Chemical Engineering of Japan*, 36(7), 812–818 (2003).
43. I.I. El-Sharkawy, K. Kuwahara, B.B. Saha, S. Koyama, K.C. Ng, Experimental investigation of activated carbon fibers/ethanol pairs for adsorption cooling system application, *Applied Thermal Engineering*, 26(8–9), 859–865 (2006).
44. I.I. El-Sharkawy, B.B. Saha, K.C. Ng, A. Chakraborty, Adsorption equilibrium measurement methods. In: *Advances in Adsorption Technology*, ed. B.B. Saha and K.C. Ng, Chapter 5, Nova Science Publishers, New York, 165–200 (2011).
45. I.I. El-Sharkawy, B.B. Saha, S. Koyama, K.C. Ng, A study on the kinetics of ethanol-activated carbon fiber: theory and experiments, *International Journal of Heat and Mass Transfer*, 49(17–18), 3104–3110 (2006).
46. B.B. Saha, I.I. El-Sharkawy, A. Chakraborty, S. Koyama, S.H. Yoon, K.C. Ng, Adsorption rate of ethanol on activated carbon fiber, *Journal of Chemical & Engineering Data*, 51(5), 1587–1592 (2006).

47. B.B. Saha, I.I. El-Sharkawy, A. Chakraborty, S. Koyama, Study on an activated carbon fiber-ethanol adsorption chiller: Part I—system description and modelling, *International Journal of Refrigeration*, 30(1), 86–95 (2007).
48. B.B. Saha, I.I. El-Sharkawy, A. Chakraborty, S. Koyama, Study on an activated carbon fiber-ethanol adsorption chiller: Part II—Performance evaluation, *International Journal of Refrigeration*, 30(1), 96–102 (2007).
49. J. Tóth, State equations of the solid–gas interface layers, *Acta Chimica Academiae Scientiarum Hungaricae*, 69, 311–338 (1971).
50. B.B. Saha, A. Chakraborty, S. Koyama, Y.I. Aristov, A new generation cooling device employing $CaCl_2$-in-silica gel-water system, *International Journal of Heat and Mass Transfer*, 52, 516–524 (2009).
51. B.B. Saha, K.C.A. Alam, A. Akisawa, T. Kashiwagi, Two-stage non-regenerative silica gel water adsorption refrigeration cycle, *Proceedings of 2000 ASME Advanced Energy System Division*, Orlando, USA, 40, 65–69 (2000).
52. B.B. Saha, A. Akisawa, T. Kashiwagi, Solar/waste heat driven two-stage adsorption chiller: the prototype, *Renewable Energy*, 23(1), 93–101 (2001).
53. B.B. Saha, E.C. Boelman, T. Kashiwagi, Computational analysis of an advanced adsorption-refrigeration cycle, *Energy*, 20(10), 983–994 (1995).
54. B.B. Saha, S. Koyama, T. Kashiwagi, A. Akisawa, K.C. Ng, H.T. Chua, Waste heat driven dual-mode, multi-stage, multi-bed regenerative adsorption system, *International Journal of Refrigeration*, 26(7), 749–757 (2003).
55. H. Yanagi, T. Asano, K. Iwase, F. Komatsu, Development of adsorption refrigerator using a direct contact condensation and evaporation on sprayed water, *Proceedings of the International Sorption Heat Pump Conference*, Munich, Germany, 671–676 (1999).
56. B.B. Saha, K.C.A. Alam, Y. Hamamoto, A. Akisawa, T. Kashiwagi, Sorption refrigeration/heat pump cycles, *Transactions of the JSRAE*, 18(1), 1–14 (2001).
57. T.F. Qu, R.Z. Wang, W. Wang, Study on heat and mass recovery in adsorption refrigeration cycles, *Applied Thermal Engineering*, 21, 439–452 (2001).
58. K.C.A. Alam, A. Akahira, Y. Hamamoto, A. Akisawa, T. Kashiwagi, A four-bed mass recovery adsorption refrigeration cycle driven by low temperature waste/renewable heat source, *Renewable Energy*, 29, 1461–1475 (2004).
59. A. Akahira, K.C.A. Alam, Y. Hamamoto, A. Akisawa, T. Kashiwagi, Experimental investigation of mass recovery adsorption refrigeration cycle, *International Journal of Refrigeration*, 28, 565–572 (2005).
60. M.Z.I. Khan, B.B. Saha, K.C.A. Alam, A. Akisawa, T. Kashiwagi, Study on solar/waste heat driven multi-bed adsorption chiller with mass recovery, *Renewable Energy*, 32, 365–381 (2007).
61. D.C. Wang, Z.Z. Xia, J.Y. Wu, R.Z. Wang, H. Zhai, W.D. Dou, Study of a novel silica gel-water adsorption chiller. Part I. Design and performance prediction, *International Journal of Refrigeration*, 28, 1073–1083 (2005).
62. D.C. Wang, Z.Z. Xia, J.Y. Wu, R.Z. Wang, H. Zhai, W.D. Dou, Study of a novel silica gel-water adsorption chiller. Part II. Experimental study, *International Journal of Refrigeration*, 28, 1084–1091 (2005).
63. S. Jribi, B.B. Saha, S. Koyama, A. Chakraborty, K.C. Ng, Study on activated carbon/HFO-1234ze(E) based adsorption cooling cycle, *Applied Thermal Engineering*, 50, 1570–1575 (2013).

64. Z. Tamainot-Telto, S.J. Metcalf, R.E. Critoph, Investigation of activated carbon-R723 pair for sorption generator, *Proceedings of Heat Powered Cycles Conference*, Berlin, Germany, Paper No. 600, (2009).

65. B.B. Saha, K.C. Ng, A. Chakraborty, K. Thu, Desalination system and method, PCT/WO 2011/010205 A1, (2011).

66. I.I. El-Sharkawy, K. Thu, K.C. Ng, B.B. Saha, A. Chakraborty, S. Koyama, Performance improvement of adsorption desalination plant: experimental investigation, *International Review of Mechanical Engineering (I.RE.M.E.)*, 1(1), 25–31 (2007).

67. K. Thu, K.C. Ng, B.B. Saha, A. Chakraborty, S. Koyama, Operational strategy of adsorption desalination systems, *International Journal of Heat and Mass Transfer*, 52(7–8), 1811–1816 (2009).

68. A. Chakraborty, B.B. Saha, S. Koyama, K.C. Ng, On the thermodynamic modeling of isosteric heat of absorption and comparison with experiments, *Applied Physics Letters*, 89, 171901 (2006).

3

Optimisation of Adsorption Dynamics in Adsorptive Heat Transformers: Experiment and Modelling

Yuri I. Aristov

CONTENTS

3.1   Introduction 64
3.2   Experimental Study of Adsorption Dynamics in AHT 66
      3.2.1   Exponential Kinetics 67
      3.2.2   Shape of the Adsorption Isobar 68
      3.2.3   Adsorption versus Desorption Time 69
      3.2.4   Effect of the Boundary Conditions...70
      3.2.5   Effect of the Adsorbent Grain Size...70
      3.2.6   Effect of the Number of Layers...71
      3.2.7   Effect of the S/m Ratio...73
      3.2.8   Grain-Size-Sensitive and Grain-Size-Insensitive Modes 74
      3.2.9   Driving Force for Ad-/Desorption Process....................................75
3.3   Numerical Modelling of Adsorption Dynamics......................................81
      3.3.1   Mono-layer Configuration...82
              3.3.1.1   Distributions of Temperature and Pressure Inside
                        the Adsorbent Grain 84
              3.3.1.2   Temporal Evolution of the Uptake/Release
                        Curves..86
              3.3.1.3   Driving Force for Heat and Mass Transfer...................88
      3.3.2   Multi-layer Configuration..95
              3.3.2.1   Model Description 95
              3.3.2.2   Simulation of a Multi-layer Configuration....................97
3.4   Dynamic Optimisation of Adsorptive Heat Transformation
      Cycles... 101
3.5   Conclusions.. 103
3.6   Nomenclature... 103
Acknowledgements 105
References.. 105

3.1 Introduction

Adsorption technology has already become a key tool that is used pervasively in industry for well-established applications, such as gas separation and purification. Impressive progress has recently been made in a relatively new, although quite close to practice, application of adsorbents for heat transformation that includes cooling, heat pumping and heat amplification. The current state of the art in this specific application was presented in a book (Saha and Ng 2010) and review articles (Critoph and Zhong 2005, Aristov 2007, Kim and Infante Ferreira 2008, Vasiliev et al., 2008, Wang et al., 2009, Choudhury et al., 2010). Various adsorption heat transformers (AHTs) (chillers, heat pumps and amplifiers) have been developed during the past few years, and several chillers have now passed over from the prototype stage to small serial production (Wang and Oliveira 2006, Jakob and Kohlenbach 2010). Despite significant progress achieved, there is still a lot of room for improving this technology (Ziegler 2009), first of all, for enhancing the specific cooling/heating power (SCHP) and, hence, reducing the AHT size. The main challenge is dynamic optimisation of the integrated unit 'Adsorbent—Heat Exchanger' (Ad-HEx) to enhance heat and mass transfer processes and harmonise adsorbent properties with cycle boundary conditions and HEx design (Aristov 2009).

For analysing the dynamic behaviour of AHTs, it is obligatory to (1) consider the adsorbent and heat exchanger as an integrated unit rather than two separate matters, and (2) study ad-/desorption dynamics under the real operating conditions of AHTs (Aristov 2009). To fit these requirements, a novel experimental method of dynamic study of Ad-HEx units, the so-called large temperature jump method (LTJM), was suggested (Aristov et al., 2008). It closely imitates conditions of the AHT isobaric stages when de-/adsorption is initiated by a fast jump/drop of the temperature of a heat exchanger wall which is in thermal contact with the adsorbent. This method provides a simple and efficient tool to investigate the effect of the adsorbent nature (Glaznev and Aristov 2010a, Glaznev et al., 2010), its grain size (Glaznev and Aristov 2010b, Gordeeva and Aristov 2011), cycle boundary conditions (Glaznev and Aristov 2010b, Gordeeva and Aristov 2011, Veselovskaya and Tokarev 2011), presence of a non-adsorbable gas (Glaznev and Aristov 2008, Glaznev et al., 2010, Okunev et al., 2010), heating rate (Aristov et al., 2008), adsorption uptake (Glaznev and Aristov 2010b), shape of the adsorption isobar (Glaznev et al., 2009), adsorption driving temperature (Glaznev and Aristov 2010b, Gordeeva and Aristov 2011, Veselovskaya and Tokarev 2011) and the number of the adsorbent layers (Glaznev and Aristov 2010a, b, Aristov et al., 2012). In the first part of this chapter, we summarise and illustrate the main experimental findings and general regularities of adsorption dynamics which have been revealed so far, as well as discuss some recommendations which could improve the dynamics of real AHT units. This dynamic study was performed for promising adsorbents of water – Fuji silica RD, silicoaluminophosphate FAM-Z02 and SWS-1L (silica modified by calcium chloride) – as well as selected adsorbents of methanol and ammonia.

Two main concepts of the Ad-HEx configuration were clearly differentiated: (1) n-layers of loose adsorbent grains contacted with HEx fins (Lang et al., 1999), and (2) the fin's surface coated with a consolidated sorbent layer (Bauer et al., 2009). Although the coating approach looks very encouraging, so far there is no experimental data which demonstrate that it is certainly superior to the loose grains concept. Moreover, experimental data obtained by LTJM demonstrated that quite acceptable SCHP values of 1–2 kW/(kg adsorbent) can be, in principle, obtained for a multi-layer configuration of loose grains (Aristov 2009). In our opinion, a choice between the loose grains and coated surface concepts is still open and needs more theoretical and experimental analysis. In this chapter, we focus on the adsorption dynamics for the loose grains configuration (Figure 3.1b), which is simple in realisation and ensures good intergrain vapour transport (Raymond and Garimella 2011). On the other hand, it is assumed to suffer from poor heat transfer both in the adsorbent layer and between the adsorbent grains and the HEx surface. Because of this, the layer should not be too thick and its optimisation is strictly necessary. The loose grains configuration has been successfully realised in several prototypes (Saha et al., 2001, Tamainot-Telto et al., 2009, Freni et al., 2012b) as well as commercial adsorption chillers (Matsushita 1987).

A more sophisticated study of ad-/desorption dynamics was performed at Moscow State University by a numerical simulation of coupled heat and mass transfer in an Ad-HEx unit (Okunev et al., 2008a). For a mono-layer Ad-HEx configuration ($n = 1$), we have reported below an effect of the adsorption isobar shape on the dynamics of water sorption, and investigated the formation of a quasi-stationary regime that is established after a short transient period. The main features of these two modes, such as the energy balance, coupling

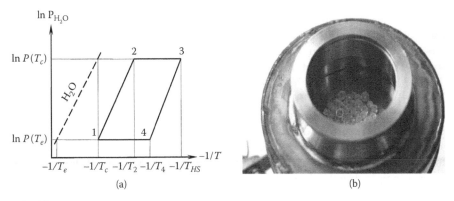

FIGURE 3.1

(a) Basic 3T cycle of AHT in the Clapeyron diagram and (b) the measuring cell with loose silica grains.

of heat and mass transfer and instant cooling power, have been comprehensively studied. For a multi-layer configuration, a novel numerical approach has recently been realised at the Istituto di Tecnologie Avanzate per l'Energia (ITAE-CNR) to describe the adsorption dynamics and extract the heat and mass transfer coefficients (Freni et al., 2012a).

Finally, several recommendations followed from this experimental and numerical study were checked with a prototype of an adsorptive chiller at ITAE-CNR (Italy) (Aristov et al., 2012, Sapienza et al., 2012) and the University of Warwick (UK) (Veselovskaya et al., 2010). These tests clearly demonstrated that the dynamic performance of AHT can be significantly improved by a proper time organisation of the AHT cycle as predicted by the LTJM tests.

The aim of this chapter is to highlight the most important findings and general regularities that have been revealed as a result of the systematic experimental and numerical studies of the adsorption dynamics of water, methanol and ammonia in AHT systems.

3.2 Experimental Study of Adsorption Dynamics in AHT

A basic three-temperature (3T) AHT cycle consists of two isosters and two isobars (Figure 3.1a) and is uniquely characterised by a set of the three boundary temperatures (T_e, T_c, T_{HS}), where T_e is the evaporator temperature, T_c is the condenser temperature and T_{HS} is the temperature of the external heat source. The uptake (release) temporal evolution $q(t)$ uniquely characterises the main dynamic features of isobaric stages '4 → 1' and '2 → 3' of AHT. These stages make the main contribution to the coefficient of performance (COP) and the SCHP, while the isosteric stages '1 → 2' and '3 → 4' are fast and less significant as a rule.

Ad-/desorption during isobaric stages of the AHT cycle is initiated by a fast cooling/heating of metal surface, which transfers heat to or from the adsorbent layer mounted on it. The new LTJM (Aristov et al., 2008) was intently developed to imitate in a laboratory the real AHT conditions and to get dynamic information appropriate to a real AHT cycle. Adsorption was initiated by the temperature drop from T_4 down to T_1 ($T_4 \rightarrow T_1$) at almost constant pressure $P(T_e)$ (Figure 3.1a). Desorption was initiated by the temperature jump from T_2 down to $T_3 = T_{HS}$ ($T_2 \rightarrow T_3$) at almost constant pressure $P(T_c)$ (Figure 3.1a). A detailed experimental procedure of LTJM can be found elsewhere (Aristov et al., 2008, Glaznev and Aristov 2010b) together with principle drawbacks of the common procedure for dynamic analysis of AHT performance (Aristov 2009).

As mentioned earlier, LTJM has been systematically applied for the loose grains configuration to investigate how the dynamics of water, methanol and ammonia adsorption is affected by various factors. The main experimental findings on AHT adsorption dynamics are as follows:

1. The initial stage of ad-/desorption process follows exponential kinetics (Aristov 2009, Glaznev and Aristov 2010b, Gordeeva and Aristov 2011, Veselovskaya and Tokarev 2011).

2. A concave adsorption isobar is profitable for fast desorption, while a convex one for fast adsorption (Okunev et al., 2008a, 2011, Glaznev et al., 2009).

3. The isobaric adsorption stage of AHT cycle is commonly slower than the isobaric desorption stage (Aristov 2009, Glaznev and Aristov 2010a, b).

4. The ad-/desorption dynamics is invariant with respect to the ratio S/m = 'heat transfer surface'/'adsorbent mass' (Aristov 2011).

5. The grain-size-insensitive regime is found, for which adsorbent grains of a different size result in the same dynamic performance (Glaznev and Aristov 2010a, Glaznev et al., 2011).

6. The instant driving force for ad-/desorption is the temperature difference between the plate and the grains, DTD = $T_3 - T_g(t)$ or $T_g(t) - T_1$ (Glaznev and Aristov 2010b, Gordeeva and Aristov 2011, Veselovskaya and Tokarev 2011).

3.2.1 Exponential Kinetics

It is really amazing that the quite complex and coupled process of adsorption, heat and mass transfer, ultimately results in simple exponential dependence of the sorption uptake/release on time (Figure 3.2):

$$q(t) = q(0) \pm \Delta q \, \exp\left(\frac{-t}{\tau}\right)$$
(3.1)

This equation describes the ad-/desorption dynamics by a *single* characteristic time τ and an appropriate rate constant $K = 1/\tau$. This permits stupendous simplification of the analysis and easy quantitative comparison of various adsorbents, boundary conditions, Ad-HEx configurations and so on. Equation 3.1 is quite universal: it is valid for all tested adsorbents (eight materials) and adsorbates up to 80–100%, 65–100% and 50–100% of the final conversion for $n = 1$, 2 and 4, respectively (Glaznev and Aristov 2010a, Glaznev et al., 2011, Aristov et al., 2012). The evolution of the tail can be slower than exponential so that for any adsorbent and boundary conditions, a ratio of the characteristic times $\tau_{0.9}/\tau_{0.8}$ is typically 1.3–1.7 (Glaznev and Aristov 2010b, Gordeeva and Aristov 2011, Veselovskaya and Tokarev 2011).

Recommendation 1. We would suggest to restrict the duration of AHT isobaric stages by time $\tau_{0.8}$ (for $n = 1$) or even less (for $n > 1$). This would allow

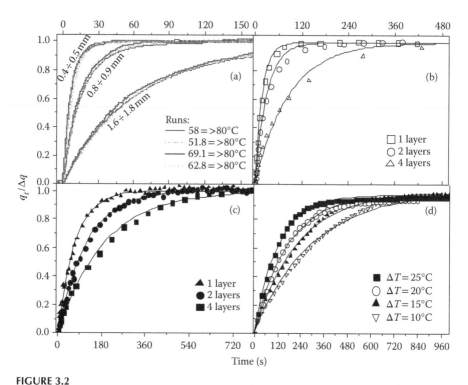

FIGURE 3.2

(a) Dimensionless uptake/release curves of the water desorption from silica Fuji RD at a three-grain size, various boundary temperatures and mono-layer configuration (Glaznev and Aristov 2010b) (b) the water adsorption on silica Fuji (0.4–0.5 mm) at various n (Aristov 2011) (c) the methanol adsorption on LiCl/silica at various n (Gordeeva and Aristov 2011) (d) the ammonia adsorption on $BaCl_2$/vermiculite at $P = 5.8$ bar and various DTD = ΔT (Veselovskaya and Tokarev 2011). Lines indicate exponential fits.

avoiding a dramatic drop of the cycle SCHP at longer times at the expense of just a little reduction of the cycle COP.

3.2.2 Shape of the Adsorption Isobar

The effect of the shape of equilibrium adsorption isobar on the dynamics of isobaric water sorption was experimentally studied by Glaznev et al. (2009). Composite sorbent SWS-1L was used for these tests as its water adsorption isobars have segments with both convex and concave shapes. The conclusion drawn from those experiments was that under the same boundary conditions, desorption is faster than adsorption for a concave isobar segment and vice versa for a convex one. Okunev et al. (2008a) and Glaznev et al. (2009) proposed that the most challenging shape of the adsorption isobar is a step-like isobar, because it ensures a maximal average DTD between the grains and HEx fins during isobaric AHT stages. The effect of the isobar shape on the dynamics of isobaric stages of AHT cycles was comprehensively investigated by its mathematical modelling by Okunev et al. (2011) (see Section 3.3).

3.2.3 Adsorption versus Desorption Time

If one compares the characteristic times of ad-/desorption for all tested cycles (Table 3.1) (Glaznev and Aristov 2010b, Gordeeva and Aristov 2011, Aristov et al., 2012, Sapienza et al., 2012), the desorption runs (2–3, Figure 3.1a) are faster than appropriate adsorption runs (4–1) by a factor of 2.2–3.5. In our

TABLE 3.1

The Characteristic Sorption Times τ and $\tau_{0.8}$, and the Specific Powers W_{max} and $W_{0.8}$ Absorbed/Released in Evaporator/Condenser at Various n (Adsorbent Mass $m =$ const $= 0.314$ g)

Grain Size (mm)	Run (°C)	N	τ (s)	$\tau_{0.8}$ (s)	W_{max} (kW/kg)	$W_{0.8}$ (kW/kg)
0.2–0.25	50.0 → 30	1	13.3	25.6	12.3	3.70
		2	17	30.9	9.6	3.17
		4	26	69.1	6.3	1.41
		8	44	139	3.7	0.65
	58.0 → 80	1	6.2	9.8	26.4	3.70
		2	6.2	10.4	26.4	3.17
		4	12	23.5	13.6	1.41
		8	21	61.7	7.8	0.65
0.4–0.5	50.0 → 30	1	18	31.1	9.1	3.15
		2	30	55.3	5.5	1.70
		4	86	175	1.9	0.52
	58.0 → 80	1	6.7	10.5	24.4	3.15
		2	13.5	21.9	12.1	1.70
		4	42.1	75	3.9	0.52
0.8–0.9	50.0 → 30	1	47.3	86	3.5	1.15
		2	100	180	1.65	0.51
		4	200	391	0.82	0.22
	58.0 → 80	1	16.5	27.2	9.9	1.15
		2	45	78.2	3.65	0.51
		4	115	209	1.4	0.22
1.6–1.8	50.0 → 30	1	182	300	0.90	0.32
		2	316	553	0.52	0.16
		4	588	1053	0.28	0.08
	58.0 → 80	1	64	106	2.55	0.32
		2	169	275	0.97	0.16
		4	310	520	0.52	0.08

Source: From Glaznev, I.S., et al. Water sorption dynamics in adsorption chillers: A few layers of loose Fuji silica grains, Proceedings of the VIII Minsk International Seminar 'Heat Pipes, Heat Pumps, Refrigerators, Power Sources', Minsk, Belarus, Sept. 12–15, 2011, 1: 161–8.

opinion, this is due to higher average temperature and vapour pressure during the desorption stage (Figure 3.1a). A concave shape of the sorption isobar is also profitable for the desorption process (Glaznev et al., 2009). This substantial difference in the durations of ad-/desorption isobaric phases helps to come to the conclusion (Aristov et al., 2012) that commonly settled equal durations ($t_s = t_a$) do not result in optimal AHT performance.

Recommendation 2. The duration of the isobaric adsorption phase should be prolonged at the expense of shortening the isobaric desorption phase. This would allow a more complete adsorption stage, thus increasing the amount of adsorbate exchanged, the cycle COP and SCHP as well. The first confirmation of this idea was obtained by Aristov et al. (2012).

3.2.4 Effect of the Boundary Conditions

In this study, we have fixed the regeneration temperature T_{HS} at 80°C and considered the four boundary sets $(T_e, T_c, T_{HS}) = (5, 30, 80), (10, 30, 80),$ (5, 35, 80) and (10, 35, 80), which are typical for adsorptive air-conditioning driven by low-temperature heat (Pons et al., 1999). We have found that, for a mono-layer configuration, the cycle boundary conditions just weakly affect the dimensionless kinetic curves, which is somewhat unexpected (Figure 3.2a). In this case, the specific cooling power depends mainly on the adsorbent grain size.

3.2.5 Effect of the Adsorbent Grain Size

For a mono-layer configuration, both adsorption and desorption become faster for smaller grains of Fuji silica RD (Figures 3.2 and 3.3), and two regions can be selected:

1. For large grains ($d = 2R_g = 0.8$–0.9 and 1.6–1.8 mm), the characteristic time strongly depends on the grain size for both adsorption and desorption runs (Figure 3.3b): $\tau \sim R_g^2$. For these grains, the τ value does not depend on the rate of heating the metal support. This regime is adsorbent sensitive.

2. For small grains (0.2–0.25 and 0.4–0.5 mm), the sorption process is fast and its time is comparable with the time of metal plate heating. In this limit, the time τ is not adsorbent-specific, but rather reflects the fin heating scenario. As a result, the dependence $\tau(R_g)$ almost vanishes (Figure 3.3b).

A crossover between the two limiting modes nominally corresponds to the silica grains of approximately 0.55 mm, which can be considered as an upper limit of the optimal grain size for a mono-layer configuration of silica Fuji RD loose grains.

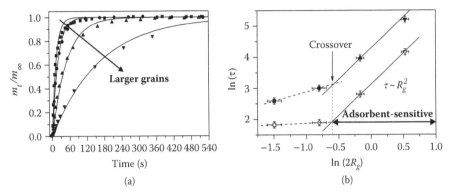

FIGURE 3.3

(a) The uptake curves for a mono-layer of loose Fuji silica grains of different size. Solid lines indicate exponential approximation (b) Exponential time τ versus grain size $2R_g$ in logarithmical scale. Solid lines indicate linear approximation $\ln \tau = \ln \tau_0 + 2p \ln R_g$. Adsorption run $50 \geq 30°C$ (solid symbols) and desorption run $58 \geq 80°C$ (open symbols). Grain size: 0.2–0.25 mm (■), 0.4–0.5 mm (●), 0.8–0.9 mm (▲), 1.6–1.8 mm (▼).

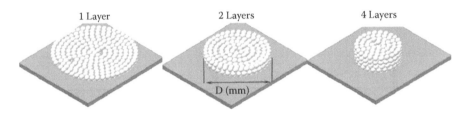

FIGURE 3.4

Schematics of the loose-grain configurations with $n = 1, 2$ and 4.

3.2.6 Effect of the Number of Layers

For a mono-layer configuration, the COP is not optimal because the mass of the adsorbent housed in the mono-layer is smaller than the inert masses of the AHT unit. Because of this, configurations with the number n of loose grains layers more than one are more realistic. On the other hand, the whole layer should not be too thick and optimisation of the n number and the grain size is strictly necessary. We have conducted LTJM analysis of the ad-/desorption dynamics for n-layers of a loose grains of silica Fuji RD with $n = 2$ and 4. This case reflects a realistic situation of a compact heat exchanger of a finned flat-tube type in which 2–4 adsorbent grains are housed in a 1–2 mm gap between the fins (see, e.g. Aristov et al., 2012). We have selected for the measurements only one boundary set typical for cycles driven by low-temperature heat: adsorption run $50 \geq 30°C$ and desorption run $58 \geq 80°C$. The grains of Fuji silica-type RD of four sizes are tested as fractions 0.2–0.25, 0.4–0.5, 0.8–0.9 and 1.6–1.8 mm. The dry sample weight was maintained constant, so that the diameter $D \sim R_g^{-0.5}$ of the circle covered by the grains and the heat transfer surface $S = \pi D^2/4 \sim R_g^{-1}$ are reduced accordingly (Figure 3.4).

For all experimental runs, a gradual rise of the characteristic ad-/desorption time with increasing layer thickness is observed as seen from Figure 3.5 and Table 3.1. The kinetics curves for the smallest grains (0.2–0.25 mm) reveal the least sensitivity to the layer thickness. It is especially representative for the desorption run: the kinetics at $n = 1$ and 2 virtually coincide (Figure 3.5b). This may be because the vapour transfer in such small grains is very fast, and the water adsorption is controlled by the heating

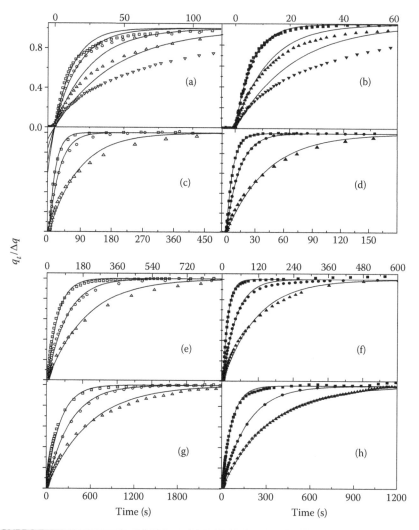

FIGURE 3.5

Dimensionless uptake/release curves of water ad-/desorption for $n = 1$ (∀), 2 (−), 4 (8) and 8 (∇) for adsorption (a, c, e, g) and desorption (b, d, f, h) runs. Grain size: 0.2–0.25 mm (a, b), 0.4–0.5 mm (d, c), 0.8–0.9 mm (e, f) and 1.6–1.8 mm (g, h). Solid lines present exponential approximation.

scenario of the system 'metal plate/adsorbent'. At $n = 1$ and 2, the time of the plate heating/cooling (~10 seconds) is deemed to take the major process time. Further on, we consider only configurations for which the adsorption time is larger than the time of the plate heating. Table 3.1 summarises the characteristic times τ as well as $\tau_{0.8}$ that corresponds to reaching 80% of the equilibrium uptake.

3.2.7 Effect of the *S/m* Ratio

As a first approximation, all the characteristic times linearly grow with the number of layers (not presented), $\tau = An$ (Glaznev and Aristov 2010a) (except for the case of the smallest grains at $n = 1$ and 2, mentioned previously). In these experiments, the adsorbent mass was fixed ($m = 0.314$ g) so that the contact area $S \sim 1/n$ (Figure 3.4). Therefore, the dependence of the rate constant $K = 1/\tau$ on S is expected to be a straight line. In fact, we revealed the linear dependence $\ln K = B \ln S$ with the slopes $B_{0.5} = 0.85$, $B_{0.8} = 1.0$ and $B_{0.9} = 1.05$ rather close to 1.0 (Glaznev and Aristov 2010a). Hence, the rate constant increases proportionally to the area through which the heat is transferred between the plate and the adsorbent layer: $K \sim S$.

On the other hand, at a fixed contact area S the characteristic time grows with the adsorbent mass m as seen from Figure 3.6a, so that the rate constant $K \sim 1/m$. Thus, if one keeps the S/m ratio constant, the kinetics does not change (Figure 3.6b); hence, it is invariant with respect to this ratio: $K \sim S/m$.

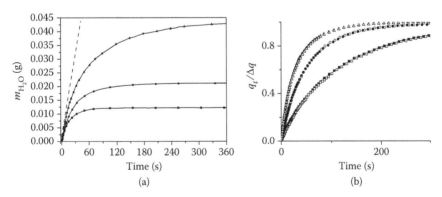

FIGURE 3.6
(a) Uptake curves of water adsorption on Fuji silica grains for $n = 1$ (■), 2 (●) and 4 (▲); run $50 \geq 30°C$, $d = 0.4$–0.5 mm. $S = 13.4$ cm². Dashed line indicates the initial adsorption rate equal for all runs (b) dimensionless uptake/release curves of water adsorption for grain size 0.4–0.5 (△, ▲), 0.8–0.9 mm (△, ▲), and desorption for grains of 0.8–0.9 mm size (○, ●). Empty symbols correspond to the adsorbent mass of 157 mg; filled symbols to 314 mg. For each grain size, the ratio S/m is maintained constant.

3.2.8 Grain-Size-Sensitive and Grain-Size-Insensitive Modes

We have calculated the maximal power generated during ad-/desorption in the evaporator/condenser

$$W_{max} = \frac{\Delta H_e \Delta q}{\tau} \qquad (3.2)$$

and plotted it against the S/m ratio for all the runs (Figure 3.7). The heat of water evaporation is taken as $\Delta H_e = 2480$ J/kg.

For adsorption runs and relatively small grains ($d \leq 0.5$ mm), there is a tendency for linear dependence of the power W_{max} on the invariant S/m: W_{max} (kW/kg) = $(2.5 \pm 0.5)\ S/m$ (m²/kg) (Figure 3.7a). For the largest grains of 1.6–1.8 mm, the power significantly reduces, yet being proportional to the S/m ratio: W_{max} (kW/kg) = $(0.85 \pm 0.1)\ S/m$ (m²/kg) (Figure 3.7a). For the intermediate grain size of 0.8–0.9 mm, W_{max} (kW/kg) = $(1.65 \pm 0.25)\ S/m$ (m²/kg) (Figure 3.7a). Thus, for large grains ($d \geq 0.8$ mm) the slope is proportional to $1/d$.

The same trend is revealed for desorption runs: the majority of experimental points at $d \leq 0.5$ mm lie close to the straight line W_{max} (kW/kg) = (5.7 ± 1.0) S/m (m²/kg), whereas at $d = 1.6$–1.8 mm the power is much smaller: W_{max} (kW/kg) = $(2.2 \pm 0.2)\ S/m$ (m²/kg) (Figure 3.7b). Several experimental points corresponding to the intermediate grain size and $S/m \approx 1$ m²/kg lie between these two curves, thus demonstrating transient behaviour. It is surprising that for small grains the maximal power at the desorption stage almost does not depend on the grain size. Indeed, for one layer of 0.8–0.9 mm grains, two layers of 0.4–0.5 mm grains and four layers of 0.2–0.25 mm grains, the power varies within just 10% (Figure 3.7b), being somewhat larger for smaller grains. Let us note that for all these configurations $S/m = $ const $= 2.1$ m²/kg.

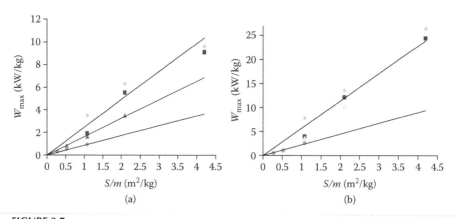

FIGURE 3.7
The maximal power W_{max} against the S/m ratio for (a) adsorption and (b) desorption runs. Silica Fuji RD grains of 0.2–0.25 mm (◆), 0.4–0.5 mm (■), 0.8–0.9 mm (▲) and 1.6–1.8 mm (●). Temperature drop (50°C → 30°C).

Thus, for Fiji silica grains of $d < 0.5$–0.8 mm and an S/m value greater than approximately 1.0–1.5 m^2/kg, the grain-size-insensitive regime is established for which loose silica grains of different size result in the same dynamic behaviour if the ratio S/m is constant. This regime can be also called an adsorbent-insensitive or lumped regime. Probably, in this case, the ad-/desorption dynamics is controlled by the heat transfer between the metal and the adsorbent, while the ad-/desorption itself is fast enough to readily adapt to variations in the heat transfer. The grain-size-sensitive or adsorbent-sensitive regime is realised for larger adsorbent grains, $d \geq 0.8$ mm, for which the adsorption rate is not sufficiently fast to follow the heat flux changes that lead to the reduction of the total rate. Recommendation 3: the size of the loose adsorbent grains has to be selected in such a way that to realise the adsorbent-insensitive mode. Thus, it is not necessary to precisely select the adsorbent grain size within the mentioned range; however, smaller grains seem to be somewhat preferable (Figure 3.7), until they begin creating significant interparticle mass transfer resistance and the pressure gradient along the layer (Ruthven 1989, Raymond and Garimella 2011).

3.2.9 Driving Force for Ad-/Desorption Process

The metal plate is a source (sink) of heat of desorption (adsorption), and the maximal adsorption power is linked with the appropriate heat flux between the plate and the grains as

$$\frac{\Delta H_{ads} m \Delta q}{\tau} = \alpha S \Delta T_{max} \tag{3.3}$$

where α is the heat transfer coefficient, S is the heat transfer area and ΔT_{max} is the maximal DTD. Hence, the characteristic adsorption time

$$\tau = \frac{\Delta H_a m \Delta q}{\alpha S \Delta T_{max}} = \left(\frac{\Delta H_a \Delta q}{\Delta T_{max}} \right) \left(\frac{m}{\alpha S} \right) \tag{3.4}$$

depends on the thermodynamic parameters (ΔH_a, Δq and $\Delta T_{max} = T_3 - T_2$ or $T_4 - T_1$) that can be taken from the cycle diagram (Figure 3.1a) and the parameters of Ad-HEx (α, S/m), which affect the adsorption dynamics. Hence, the value of W_{max} calculated from the initial slope of uptake (release) curves is expected to be proportional to ΔT_{max} as was experimentally confirmed for all studied adsorbents, adsorbates and boundary conditions (more than 15 tests; see Figure 3.8). The heat transfer coefficient $\alpha = (W_{max} \, m)/(S \, \Delta T_{max})$ can be obtained from the slope of the function $W_{max}(\Delta T_{max})$:

1. For water, $\alpha \approx 100$ W/(m$^2 \cdot$ K) for adsorption and $\alpha \approx 250$ W/(m$^2 \cdot$ K) for desorption (Figure 3.8a) (Glaznev and Aristov 2010b).

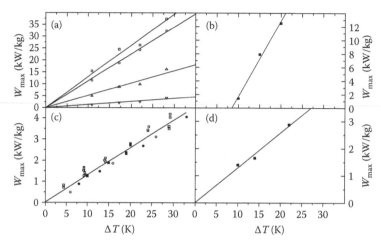

FIGURE 3.8

The specific power W_{max} versus the driving temperature difference. (a) Water desorption from the Fuji silica with the rain size 0.2–0.25 mm ($\square$), 0.4–0.5 mm ($\bigcirc$), 0.8–0.9 mm ($\triangle$), 1.6–1.8 mm ($\triangledown$) (Glaznev and Aristov 2010b) (b) methanol adsorption on LiCl/silica with the grain size 0.4–0.5 mm (Gordeeva and Aristov 2011) (c) ammonia adsorption on BaCl$_2$/vermiculite at different ammonia pressures: $\blacksquare$, 5.8 bar; $\square$, 6.9 bar; $\bullet$, 12.2 bar; $\bigcirc$, 15.8 bar (Veselovskaya and Tokarev 2011) (d) ammonia adsorption on BaCl$_2$/vermiculite in the AC prototype (Veselovskaya et al., 2010).

2. For methanol, α = 75–100 W/(m$^2\cdot$K) for adsorption on grains of 0.4–0.5 mm (Figure 3.8b) (Gordeeva and Aristov 2011).

3. For ammonia, α = 115 W/(m$^2\cdot$K) for the grains of 0.5–1 mm, and 90 W/(m$^2\cdot$K) for 1–2 mm grains (Veselovskaya and Tokarev 2011).

Thus, the α value for adsorption runs is close to (100 ± 20) W/(m$^2\cdot$K) for any adsorbent and adsorbate being approximately twice larger for desorption. These values are much larger than those commonly used in modelling of AHT dynamics (40–60 W/[m$^2\cdot$K]) (Zhang and Wang 1999, Yong and Sumathy 2002). Possible reasons of this enhancement, among which are heat pipe or re-condensation heat transfer effect (Zhang and Wang 1999) and the convective Stefan flux (Okunev et al., 2010) that can occur during the LTJM experiments, need special consideration.

Composites 'salt in porous matrix' sorb water (methanol, ammonia) due to the chemical reaction of adsorbate with the confined salt. In this case, the sorption rate depends on the two driving forces: (1) the driving force for chemical reaction, which is the difference $\Delta(\Delta F)$ between the current sorption potential $\Delta F(P, T) = -RT \ln(P/P_0)$ and the sorption potential at the final state $\Delta F(P_e, T_1)$ or $\Delta F(P_c, T_3)$ (Gordeeva and Aristov 2011), and (2) the driving force for heat transfer DTD. Due to this, the line $W_{max}(\Delta T_{max})$ for the LiCl/silica composite intersects the X-axis at $\Delta T_{max} \approx$ 8°C > 0 (Figure 3.8b), which is a supercooling temperature necessary to drive the gas–solid reaction LiCl + 3H$_2$O = LiCl·3H$_2$O. For the confined salt, the supercooling is expected to be lower than that for the bulk one, although no experimental data in bulk are available for direct comparison. This can also be important for other new composites 'salt in porous matrix' suggested for AHT (Aristov 2013).

We have compared the dynamics of ammonia adsorption on the BaCl$_2$/vermiculite composite of 1–2 mm grains in the laboratory LTJM unit (Veselovskaya and Tokarev 2011) and in a prototype adsorptive chiller with approximately the same ratio *S/m* built at the University of Warwick (Veselovskaya et al., 2010). It has been found that the dependence of initial power on ΔT_{max} for this prototype is described by the same straight line (Figure 3.8d) as for the LTJM tests (Figure 3.8c) with similar heat transfer coefficient $\alpha = (90 \pm 15)$ W/(m$^2 \cdot$ K). Evidently, the LTJM dynamic data can be used for Ad-HEx scaling-up and optimisation of AHT working cycles if one properly takes into account the invariant *S/m*.

A number of experimental data have demonstrated that the initial specific cooling power can be very high, for example 10 kW/kg or more. Of course, this power is observed only during a quite short initial period of the water interacting with the adsorbent. At longer time, the power is gradually reduced according to the decrease in the temperature difference between the plate and the adsorbent layer. This difference approaches zero at the equilibrium, which is reached faster for layer that exchanges smaller mass of water (0.012 g at $n = 1$ as compared with 0.024 and 0.048 g for $n = 2$ and 4, respectively) (Figure 3.6a). A similar conclusion was made by Glaznev et al. (2010; see Figure 4 therein and appropriate text). Although it takes more time to reach adsorption equilibrium for thicker layers, the absolute power that is released during water adsorption at $n = 4$ is larger than that at $n = 1$ and 2 at any time (Figure 3.6a).

From a practical point of view, it is reasonable to restrict the duration of AHT isobaric stages by time $\tau_{0.8}$ or even lesser. Similar estimations have been performed for the specific cooling power that corresponds to 80% of the final conversion and its cycle average value $W_{0.8}$ (Table 3.1). Neglecting duration of the isosteric stages, the latter can be calculated as

$$W_{0.8} = \frac{0.8 \Delta q \Delta H_e}{\tau_{0.8}^{ads} + \tau_{0.8}^{des}} \tag{3.5}$$

At the lumped regime, the cycle average power $W_{0.8}$ is found to be a linear function of the *S/m* ratio as well: $W_{0.8}$ (kW/kg) = (0.75 ± 0.15) *S/m* (m^2/kg), and the cooling power generated in the evaporator is a unique function of the HEx surface area: $W_{0.8}$ (W) = (750 ± 150) *S* (m^2) ≈ 750 *S* (m^2). It is significantly smaller than the maximal cooling power for short periods of time: W_{max} (W) = (2400 ± 500) *S* (m^2) ≈ 2400 *S* (m^2).

As the specific power is proportional to the ratio *S/m* in the lumped mode, it is convenient to use this ratio to assess the degree of dynamic perfection of

TABLE 3.2

Parameters of the Selected Ad-HEx Units Taken from Literature

HEx Type	Finned Flat Tube	Finned Tube, Loose Grains	Plate	Finned Tube, Compact Layer
Dimensions (mm)	$257 \times 170 \times 27$	See P. 79	$150 \times 150 \times 150$	See P. 80
Metal mass, M (kg)	0.636	—	9	6.08
Overall volume, V (dm³)	1.1	33.3[a]	3.37	8.6
Adsorbent mass, m (kg)	0.4	20	0.75	1.75
Mass metal/mass adsorbent, M/m	1.81	—	12	3.5
Heat transfer surface, S (m²)	1.66	35.4	1.35	1.7
Ratio S/V (m²/dm³)	1.51	1.06	0.40	0.20
Ratio S/m (m²/kg)	~4	1.77	1.80	0.97
Specific power, W_p (kW/kg)	0.4	0.2	0.6	
1.3–2.6^b	1.2[c]			
0.3^d				
0.12^e				
LTJ-specific power, $W_{0.8}$ (kW/kg)	2.5	1.0	1.1	0.6
Ratio $W_{exp}/W_{0.8}$	0.15	0.2	0.5	0.2

[a] Volume available for adsorbent.
[b] Maximal power.
[c] Instantaneous power in the condenser.
[d] Instantaneous power in the evaporator.
[e] Average cycle power.

an Ad-HEx unit: the larger is this ratio, the higher power per unit adsorbent mass can be obtained. To characterise a heat exchanger itself, the ratios S/V (m²/dm³) and S/M (m²/kg) can be used, where V and M are the HEx volume and mass, respectively. Evidently, for making a compact Ad-HEx, both ratios have to be maximised. We have assessed the values of S/m and S/V for selected laboratory AC prototypes of four different types for which the necessary data are available in literature (Table 3.2):

1. A compact heat exchanger of a finned flat-tube type tested in ITAE-CNR (Messina, Italy) with a distance between the flat tubes of 10 mm and between the fins of approximately 1 mm (Figure 3.9a) (Aristov et al., 2012). Adsorbent grains are contacted mainly with fins that are secondary heat transfer elements. The heat transfer distance is only 0.5–1 mm, while the vapour has to penetrate in the narrow slits ($1 \times 10 \times 27$ mm³) through the maximal distance

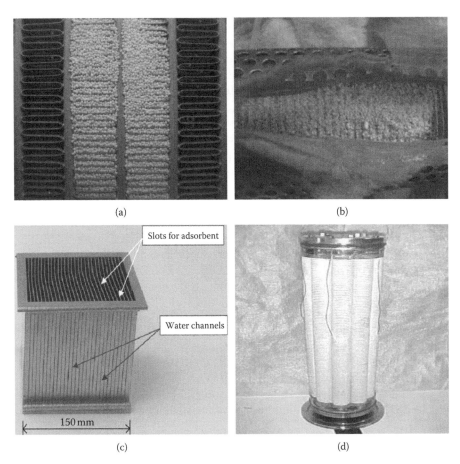

FIGURE 3.9
View of the analysed Ad-HEXs taken from (a) Aristov et al. (2012) (b) Gong et al. (2011) (c) Veselovskaya et al. (2011) and (d) Freni et al. (2007).

of 13.5 mm. This well-designed heat exchanger has $S/m \approx 4$ m^2/kg, $S/M = 2.6$ m^2/kg, $S/V = 1.51$ m^2/dm^3 and a ratio (adsorbent mass/ HEx mass) > 0.5 (Aristov et al., 2012).

2. A finned tube heat exchanger: each tube is 1230 mm long with a space between fins of 2.5 mm (Gong et al., 2011). Each fin is 0.3 mm thick and 23 mm high (Figure 3.9b). The composite adsorbent with grain diameter of 0.5–1 mm is filled between the fins. The adsorbent density is about 600 kg/m^3 and the mass is 22 kg. The heat transfer area in each finned tube is 35.4 m^2.

3. A plate heat exchanger is made of a nickel-brazed stainless steel and designed as 29 layers of active carbon each 4 mm thick (Figure 3.9c). The loose adsorbent grains are directly contacted with the flat-tube

plate, which is a primary heat transfer element, and no secondary elements, such as fins, are installed. The heat conduction path length through the adsorbent is approximately 2 mm, which, together with a larger heat conductivity of a carbonaceous adsorbent, enables rapid temperature cycling (Tamainot-Telto et al., 2009, Veselovskaya et al., 2010). Although the mass transfer occurs through the narrow slits of 4 mm over the maximal distance of 75 mm, it does not limit the process rate due to high $P(NH_3) = (1–10 \text{ bar})$.

4. Figure 3.9d shows the Ad-HEx that consists of a finned tube heat exchanger with the space between fins filled with a compact layer of SWS-1L containing 25 wt.% of bentonite clay as a binder (Freni et al., 2007). Each tube is 560 mm long with the space between fins of 2 mm. Mass of the adsorbent is 1.75 kg. The total heat transfer area is 1.7 m². During the adsorption phase, the instantaneous useful effect was 0.3 kW/kg (adsorbent + binder), while the instantaneous power supplied during desorption was about 1.2 kW/kg. Interestingly, the prototype desorption power was much higher than the adsorption one in accordance with the predictions of Figure 3.7. The cooling power calculated considering the total cycle time (which is similar to $W_{0.8}$) was 0.12 kW/kg.

Unfortunately, we could not make this analysis for commercial AC units having an adsorbent layer consolidated with HEx-fins that have appeared in the market (Bauer et al., 2009, Jakob and Kohlenbach 2010). To the best of our knowledge, the input data necessary for such study are not available in literature.

The LTJ approach naturally takes into account restrictions of the adsorption rate caused by limited heat transfer between the metal support and the adsorbent layer as well as in the layer, and by intraparticle mass transfer resistance. On the other hand, the LTJ tests are performed under somewhat ideal conditions because there is no limitation of interparticle mass transport, as the mass transfer surface is quite large and the layer is thin. Another advantage is a rapid change of the temperature of the metal support that is directly heated/cooled by a heat carrier fluid. Because of this, the process-specific power is rather high: under the lumped mode, the ratio $S/m = 1–4 \text{ m}^2/\text{kg}$ can ensure the $W_{0.8}$ values of 0.7–3.0 kW/kg. However, the average cycle powers obtained for real AC prototypes are 2.5–7 times lower than those measured in the LTJ tests (Table 3.2). This significant difference is due to several imperfections and pitfalls of either hardware or process organisation rather than due to the adsorbent properties that are already accounted for in the LTJ tests. Indeed, these tests have clearly demonstrated that known adsorbents, such as silica Fuji RD, FAM-Z02 and SWS-1L, are able themselves to consume or generate much larger power (Figure 3.7) for a simple configuration of loose adsorbent grains and ΔT_{max} only 20°C. By the way, the consolidated adsorbent layer considered by Freni et al. (2007) provides the same specific power as a layer of loose grains studied by Gong et al. (2011) at similar fin geometry. Probably, in the denser adsorbent layer, the enhancement of heat transfer is nearly compensated by the reduction in the vapour transport.

The Ad-HEx developed by Tamainot-Telto et al. (2009) is closer to the optimal performance: it possesses the experimental power that is approximately half of the predicted one. The peculiarity of its construction is the absence of any fins and a direct contact of the adsorbent with a primary heat transfer element, namely, with the metal plate that is immediately cooled/heated by water. This manner of heat transfer is similar to that in LTJ experiments. High specific power of this unit was reached at the expense of relatively low COP (0.2–0.25) due to a large ratio $M/m = 9$ (Tamainot-Telto et al., 2009). For the other units, the ratio M/m is typically 2–5 (Table 3.2), and COP = 0.4–0.6.

Our dynamic analysis confirmed that there is plenty of room for technological improvement of AHT units. Careful revision of the AHT construction and process parameters has to be made in order to increase the AHT-specific power. At least, but not last, several possible drawbacks are as follows:

- The heat transfer through the secondary fin surface seems to be no efficient enough and should be enhanced. Commonly, Ad-HEx units of AHT are based on commercial HEx units available in the market and specifically developed for air-conditioning or heat dissipation from engines to ambient air. They should be re-examined for maximizing heat transfer to/from adsorbent bed.

- A finned tube configuration with a granulated adsorbent layer may suffer from intergrain mass transfer resistance that significantly increases for smaller grains and longer fins. Additional mass transfer resistance can be due to a metal net that is used to fix adsorbent grains inside the fins.

- Low speed and/or improper distribution of cooling/heating liquid so that the heating/cooling rate restricts the unit power.

- Reduction of the adsorption rate due to the presence of residual air (Glaznev et al., 2010).

- Evaporator/condenser power does not fit the power of the ad-/desorption process.

- Non-optimised durations of the AC cycle or/and its individual phases (Aristov et al., 2012).

3.3 Numerical Modelling of Adsorption Dynamics

Several models of non-isothermal adsorption presented in literature are mainly focused on two extreme cases: *a single adsorbent grain* and *a porous adsorbent bed* composed of many layers of the grains. For both these cases, analytical solutions were obtained if a pressure gradient is a driving force of the adsorption process, that is a fast and small variation of the vapour pressure over the adsorbent. A review of the analytical models can be found in Ruthven (1984), while the numerical ones in Sun and Meunier (1987). Unfortunately, these models are not applicable to describe isobaric cooling/heating processes occurring during the ad-/desorption phases of an AHT cycle initiated by a temperature jump/drop of HEx walls.

Moreover, the porous adsorbent bed models (Ruthven 1984) consider a porous layer as a homogeneous medium for heat and mass transfer. As a consequence, single effective thermal conductivity and mass diffusivity are used in heat and mass balance equations. A similar homogeneous approach was used for describing a non-isothermal adsorption process described by Sun and Meunier (1987). This lumped approach can be realised only for a relatively thick layer which contains at least 10–20 grains in the direction perpendicular to the metal surface. For AHT, more realistic is the case of low number of adsorbent grains ($1 < n \leq 4$–10).

Therefore, mathematical models adapted to the actual AHT conditions, like temperature jump and low n-number, are highly desirable for further study and optimisation of the adsorption dynamics in AHTs. It is performed below in this section by using a Fickian diffusion model of combined heat and mass transfer in (1) a single adsorbent grain, means, for a *mono-layer* configuration, and (2) in a *multi-layer* of loose grains with $1 < n \leq 4$.

3.3.1 Mono-layer Configuration

In the first model, a spherical adsorbent grain is in thermal contact with a metal plate subjected to a fast temperature jump/drop. Combined heat and mass transfer in a single adsorbent grain was described by the following system of differential equations:

1. The energy balance equation for a single adsorbent grain including the heat of adsorption:

$$\rho c_p(T,q)\frac{\partial T}{\partial t} - \frac{\rho \Delta H_a}{\mu_{H_2O}}\frac{\partial q}{\partial t} = \lambda \Delta T \qquad (3.6)$$

at $0 < r < R_g$, with the relevant initial and boundary conditions

$$T(r,0) = T_0$$

$$\frac{\partial T(0,t)}{\partial r} = 0$$

$$\alpha\left[T_f - T(R_g,t)\right] = \lambda\frac{\partial T(R_g,t)}{\partial r}$$

where ρ is the grain density, λ is its thermal conductivity, ΔH_a is the adsorption heat, and μ_{H_2O} is the molecular mass of water.

2. The mass balance equation including the diffusion of water vapour inside the grain and sorption processes:

$$\frac{\partial C}{\partial t} = D\Delta C - \frac{\rho}{\mu_{H_2O}\varepsilon}\frac{\partial q}{\partial t} \qquad (3.7)$$

$$C(r,0) = \frac{P}{RT_0}$$

$$C(R_g,t) = \frac{P}{RT(R_g,t)}$$

$$\frac{\partial C(0,t)}{\partial r} = 0$$

We assumed the local adsorption and thermal equilibrium in each point of the grain. The pressure P inside the grain was calculated as $P(r, t) = RT(r, t) C(r, t)$. The system of differential equations presented in Equations 3.6 and 3.7 was numerically solved by the method of runs and iterations including an implicit finite difference method in order to obtain the radial and temporal distributions of temperature $T(r, t)$, concentration $C(r, t)$ and pressure $P(r, t)$ in the vapour phase, as well as water concentration $q(r, t)$ in the adsorbed phase. Information about these distributions is unique as they can hardly be measured experimentally. The average values of temperature $T_{av}(t)$, pressure $P_{av}(t)$ and uptake/release $q(t)$ were obtained by radial averaging these distributions. More details about the model can be found elsewhere (Okunev et al., 2008a).

This model was first used for describing experimental data obtained by LTJM (Aristov et al., 2008), and extracting parameters that characterise heat and mass transfer in the composite sorbent $CaCl_2$/silica gel. Then, it was applied for the analysis of how a residual gas (air or hydrogen) affects the water sorption dynamics on the same sorbent (Okunev et al., 2010). Finally, a similar model was used to describe the water sorption dynamics initiated by a jump of the vapour pressure over this sorbent (Okunev et al., 2008b).

Here the model of Okunev et al. (2008a) is used to answer the questions 'Which isobar shape is optimal for AHT?' and 'How the driving forces for heat and mass transfer are being developed after the initial jump/drop of metal support temperature?' (Okunev et al., 2011). The analysis has been performed for a relatively small temperature jump/drop (60°C ↔ 70°C) and a set of model isobars (stepwise, linear and exponential; see Figure 3.10). The step position is chosen at T_{st} = 60.5, 62, 65, 38 and 69.5°C (Okunev et al., 2011). Adsorbent grain size $2R_g$ is fixed at 1.5 mm.

The adsorbent specific heat as a function of the uptake and temperature was taken from Aristov (2012). The vapour diffusivity in the grain pores

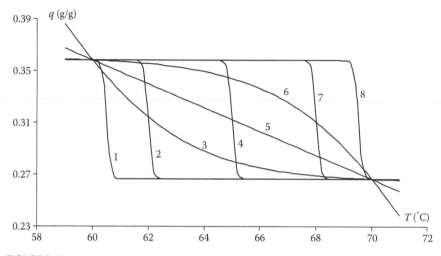

FIGURE 3.10
Model isobars of water adsorption at $P(H_2O) = 56.5$ mbar: (1) step at 60.5°C; (2) step at 62°C (3) exp (concave) (4) step at 65°C (5) linear (6) exp (convex) (7) step at 68°C (8) step at 69.5°C.

$D = 3.0 \times 10^{-6}$ m²/s is fixed the same as in the work of Okunev et al. (2008a). The coefficient α of heat transfer between the grain surface and the vapour is 60 W/(m² · K). These three values are equal to those of the composite sorbent SWS-1L which is considered to be promising for AHT (Aristov et al., 2002, Critoph and Zhong 2005, Wang et al., 2009).

3.3.1.1 Distributions of Temperature and Pressure Inside the Adsorbent Grain

The temperature is found to be constant within the grain (not presented), but is essentially changing with time (Figure 3.11a). The thermal Biot number $Bi = \alpha R_g/\lambda$ represents the ratio of the internal thermal resistance R_g/λ of the grain to its effective external surface resistance $1/\alpha$ (λ is the grain thermal conductivity). It is equal to 0.04 << 1, which means that the heat transfer inside the particle is enough rapid and a uniform temperature profile may be assumed along the grain radius.

Temporal evolution of the average grain temperature T_{av} can be divided into two parts (see Figure 3.11a for the step at 62°C):

1. A very fast drop of the temperature from the initial temperature $T_{in} = 70°C$ down to approximately 62°C for adsorption, and a very fast rise of the temperature from the initial temperature $T_{in} = 60°C$ again to approximately 62°C for the desorption. This fast process takes just 2–3 seconds for desorption and 10–12 seconds for adsorption, and is associated with the heating/cooling of the adsorbent grain without appreciable contribution of the adsorption (or desorption) process itself.

2. A gradual change of the grain temperature that reaches its final temperature T_f after approximately 2000 seconds for adsorption and approximately 700 seconds for desorption. Thus, the second part is much longer that the first one. During both the adsorption and the desorption processes, the grain temperature is essentially in a non-steady state, which indicates the governing role of the heat transfer from the plate to the grain (or back) at all times.

Variation of the grain temperature immediately affects the pressure evolution inside the grain. In contrast to the temperature, the pressure is non-uniform along the grain radius during both adsorption and desorption (Figure 3.12), which is due to the slow intragrain vapour transfer. The average pressure increases in time, then passes a maximum and finally decreases approaching the pressure outside the grain, $P = 56.5$ mbar (Figure 3.11b). For desorption run 60°C → 70°C at $t = 0$, the driving force for heat transfer is maximum and proportional to $\Delta T_{dr} = \Delta T_{max} = T_f - T_{in} = 10°C$, while the driving force for mass transfer equals zero. This initiates the mentioned fast heating of the grain up to approximately 62°C, means, to the step temperature T_{st} that reduces the driving temperature difference down to $\Delta T_{dr} \approx T_f - T_{st} = 8°C$. The mentioned jumping of the grain temperature stimulates desorption of water molecules first to the gas phase inside the pores. As desorption is fast, for short periods of time the diffusional flux of vapour out of the grain is not sufficient to compensate the pressure increase; thus, a gradient of pressure appears inside the grain (Figure 3.12a). This gradient generates the driving force for mass transport, which, as a first approximation, is proportional to the slope of the $P(r)$ curve near the grain external surface (at the dimensionless radius approaching 1). After a short transient period of approximately 10 seconds, this slope gradually decreases with time (Figure 3.12a). Hence, the driving forces for heat and mass transfer interactively reduce while the water

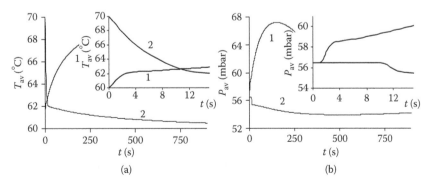

FIGURE 3.11
Temporal evolution of the average grain temperature (a) and the average pressure (b) inside the grain during ad-/desorption processes. Step at 62°C. Inserts demonstrate the evolution of the average temperature and pressure for short process times.

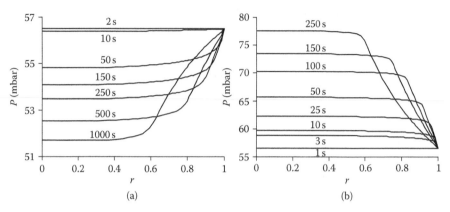

FIGURE 3.12
Radial distribution of the pressure inside the grain during adsorption (a) and desorption (b) processes. $T_{st} = 62°C$. $r = 0$, the grain centre; $r = 1$, the grain external surface.

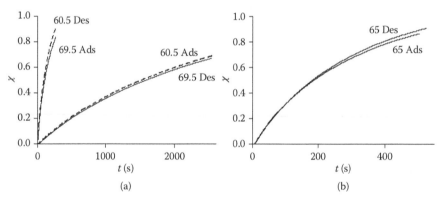

FIGURE 3.13
Dynamics of the water adsorption and desorption for (a) appropriate exponential approximations (dashed lines) and (b) various isobar shapes (solid lines). Legends are on the graphs.

desorption proceeds under the steady-state mode. The same is valid for the adsorption process (Figure 3.12b). Thus, the heat and mass transfer processes in the system involved are inevitably coupled and strongly affect each other.

3.3.1.2 Temporal Evolution of the Uptake/Release Curves

All uptake curves are presented as the dimensionless conversion degree χ versus the process time t (Figure 3.13) which gives information about the time necessary for reaching equilibrium rather than about the absolute sorption rate. The calculated uptake/release curves have been approximated by the function $\chi(t) = 1 - \exp(-t/\tau)$, which is defined by a single characteristic time τ. For the step-like and linear isobars, the exponential approximation is quite precise until $\chi = 0.7$–0.8 for both adsorption and desorption runs (Figure 3.13). Meaningful deviations (up to $\Delta\chi = 0.05$) are observed only for the concave exponential isobar during the adsorption process and for the convex exponential isobar during the desorption process (both not presented). At conversions $\chi > 0.7$–0.8, the kinetics calculated for step-like isobars are faster than the exponential one and the process is completed in a finite time (Figure 3.13) that resembles the shrinkage core kinetics. Thus, the major part (approx. 70–80%) of the uptake/release curves can be satisfactorily described by just one characteristic time.

Although the near-exponential character of the adsorption dynamics is poorly sensitive to the shape of water adsorption isobar, the characteristic time τ does strongly depend on this shape (Table 3.3). The fastest adsorption processes is observed for the isobar step at 69.5°C ($\tau = 150$ seconds). This can be rationally understood if one assumes that the instant process driving force is the temperature difference $\Delta T_{dr}(t) = T_{av}(t) - T_f = T_{av}(t) - 60°C$ between the current temperature of the adsorbent grain T_{av} and the final temperature of the metal support $T_f = 60°C$. In this case, the adsorption rate is expected to be maximal at $T_{st} = 69.5°C$ because the adsorption heat is removed at the maximal temperature difference. Accordingly, the fastest desorption corresponds to the step at $T_{st} = 60.5°C$ ($\tau = 133$ seconds) because, in this case, the driving force for the heat supply $\Delta T_{dr} = T_f - T_{av} = 70°C - T_{av}$ is maximal.

The convex exponential isobar (Figure 3.10) also provides a quite fast adsorption process ($\tau = 172$ seconds), while the concave one ensures a fast desorption ($\tau = 150$ seconds). For the linear isobar, both adsorption and desorption are 1.6–1.7 times slower than the most optimal case (Table 3.3). Tentatively, an optimal shape of a water adsorption isobar could give, in general, acceleration of the isobaric stage by a factor of 1.5–3 with respect to non-optimised isobars.

TABLE 3.3

Characteristic Time (in seconds) of Water Ad-/ Desorption at Various Isobar Shapes

Isobar Shape	Adsorption	Desorption
Step at 60.5°C	2200	133
Step at 62°C	619	160
Step at 65°C	269	258
Step at 68°C	176	625
Step at 69.5°C	150	2280
Exponent (concave)	474	157
Exponent (convex)	172	469
Linear	243	225

Source: From Okunev, A.P., et al. Modeling of water sorption dynamics in heat transformers: An optimal shape of the sorption isobar, Proceedings of the VIII Minsk International Seminar 'Heat Pipes, Heat Pumps, Refrigerators, Power Sources', Minsk, Belarus, Sept. 12–15, 2011, 1: 204–11.

The dimensionless kinetic curves are invariant with respect to the absolute value of the difference $|T_{in} - T_{st}|$ (or $|T_f - T_{st}|$). For instance, the water adsorption at $T_{st} = 69.5°C$, when $|T_{in} - T_{st}| = |60°C - 69.5°C| = 9.5°C$, proceeds almost identically with the water desorption at $T_{st} = 60.5°C$, when $|T_{in} - T_{st}| = |70°C - 60.5°C| = 9.5°C$ (not presented). Rough parity of the adsorption and desorption characteristic times τ at equal $|T_{in} - T_{st}|$ values is clearly seen from Table 3.3.

3.3.1.3 Driving Force for Heat and Mass Transfer

The energy balance equation in the grain

$$\frac{4}{3}\pi R_g^3 \rho c_p \frac{\partial T_{av}}{\partial t} = 4\pi R_g^2 \lambda \mathrm{grad} T\left(t, R_g\right) + \frac{4}{3}\pi R_g^3 \frac{\rho}{\mu_{H_2O}} \frac{\partial\left(\Delta H_a q\right)}{\partial t} \tag{3.8}$$

gives a link between the average grain temperature T_{av} and the sorption uptake/release $q(t)$. Here ρ is the adsorbent density and $q(T_{av}, P_{av})$ is the local equilibrium uptake, g H_2O/g (dry adsorbent).

Taking into account the boundary condition

$$\alpha\left[T_f - T\left(R_g, t\right)\right] = \lambda \frac{\partial T\left(R_g, t\right)}{\partial r}$$

and assuming the adsorption heat ΔH_a to be constant in the considered temperature interval, the balance can be re-written as

$$\rho c_p \frac{\partial T_{av}}{\partial t} = \frac{3}{R_g}\alpha\left[T_f - T_{av}\left(t\right)\right] + \frac{\rho}{\mu_{H_2O}}\frac{\Delta H_a \partial(q)}{\partial t} \tag{3.9}$$

This equation links a change in the grain enthalpy (the left term) with the heat flux through the grain external surface and the heat flux released/absorbed due to the water ad-/desorption.

Separate contributions of these three terms at the very initial part of the desorption process clearly demonstrate that for the first 2 seconds the change in the grain enthalpy is much larger than the heat flux that initiates the water desorption. Indeed, a very fast rise of the temperature from the initial temperature $T_{in} = 60°C$ to ~62°C is observed at $t \leq 2$–3 seconds (Figure 3.14b). This fast process is associated with the heating of the adsorbent grain without appreciable contribution of the desorption process itself. Indeed, during this period, the efficient specific heat $c_p = (dQ/dT)/m$ of the grain is almost constant (Figure 3.15a) and close to the specific heat of the grain with the initial water uptake $q = 0.274$ g/g for adsorption and with $q = 0.358$ g/g for desorption. This is because of the specific shape of the sorption isobar (Figure 3.10) for which desorption starts only when the grain temperature

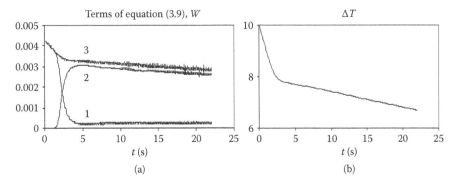

FIGURE 3.14
(a) Energy balance of a single grain calculated by Equation 3.9: (1) change in the grain enthalpy (2) heat power consumed for desorption (in this case $\partial q/\partial t \leq 0$) (3) heat flux through the grain surface (b) temperature difference between the support and the grain as a function of time. Temperature jump 60°C → 70°C. Model isobar 2 on Figure 3.10.

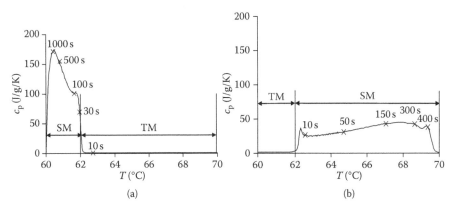

FIGURE 3.15
Efficient specific heat of the grain as a function of the average grain temperature (a) adsorption (b) desorption. $T_{st} = 62°C$. TM and SM conditionally denote the transient and quasi-stationary modes. Numbers indicate the process time.

reaches approximately 62°C. As a result, further heating causes the increase in the vapour pressure inside the grain due to desorption of water molecules first to the gas phase inside the pores. The negative gradient of pressure appears inside the grain, thus creating the driving force for vapour transport towards the external grain surface. While the grain temperature is rapidly increasing, the pressure gradient first decreases for 2–3 seconds, then passes a minimum, and after that gradually increases in time (Figure 3.16a). For the second part of the heating process, the c_p value is much larger than for the first part (Figure 3.15a), because the heat is consumed mostly for desorption of water rather than for heating the grain. The function $c_p(T)$ has a break at $T \approx 62°C$ (Figure 3.15a) for both the adsorption and desorption runs

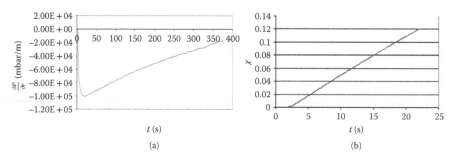

FIGURE 3.16
Temporal evolution of the pressure gradient at the grain external surface (a) and the dimensionless uptake (b). Temperature jump 60°C → 70°C, model isobar 2.

60°C ↔ 70°C which is caused by the step on the isobar of water adsorption at $T_{st} = 62°C$ (Figure 3.10). This break correlates with a break of the functions $T_{av}(t)$ (Figure 3.14b) and $P_{av}(t)$ (Okunev et al., 2011).

After a short transient period of 2–3 seconds, the water desorption starts, and the change in the grain enthalpy becomes much smaller than the heat consumption for the water desorption that gradually reduces in time (Figure 3.14a). Thus, two regimes are observed:

- A transient mode (TM) with a very short desorption time (2–3 seconds): no desorption (Figure 3.16b), fast decrease of the temperature gradient (Figure 3.14b) and origination of the pressure gradient (Figure 3.16a). There is an imbalance between the driving forces for heat and for mass transfer: at $t = 0$ the driving force for heat transfer is maximal, while that for mass transfer equals zero. At this regime, the heat transfer from the plate to the grain plays a governing role and creates the vapour flux inside the grain.

- A quasi-stationary mode (SM) for longer periods of time: quasi-stabilisation of the heat fluxes (Figure 3.14a). The driving forces for both heat and mass transfer interactively reduce under this mode, and are inevitably coupled and strongly affect each other.

The two regimes are also clearly seen on the system phase diagram that links the average grain temperature T_{av} and pressure P_{av} for ad-/desorption runs (Figure 3.17). Initial segments are characterised by almost constant pressure and near-zero ad-/desorption uptake and present the transient mode of the process. The border between these modes can be conditionally fixed at $T \approx 62°C$, which corresponds to the step position on the model isobar. The TM lasts 2–4 seconds for desorption and approximately 30 seconds for adsorption (Figure 3.17). For the latter case, cooling the grain from 70°C down to 62.7°C takes approximately 10 seconds at the initial pressure of 56.5 mbar. After that the water adsorption starts, the pressure goes down and

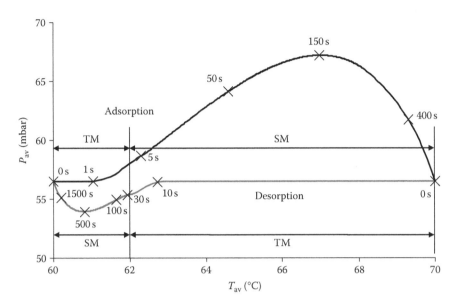

FIGURE 3.17
The average pressure in the grain versus the average grain temperature. Temperature variation 60°C ↔ 70°C, model isobar 2. Numbers indicate the process time.

the temperature reduction becomes much slower due to the heat released by adsorption. The average grain temperature reduces from 62°C to the equilibrium temperature 60°C for approximately 2000 seconds (Figure 3.17). The adsorption heat is removed through the grain external surface with the rate proportional to the instant temperature difference $[T_f - T_{av}(t)]$ as seen from Equation 3.9.

At the SM, the contribution of the grain enthalpy becomes much smaller than the heat consumption for desorption and the heat supply through the grain surface, because $\partial T_{av}/\partial t \approx 0$ (Figure 3.14a). Hence, Equation 3.9 can be simplified as

$$\frac{\rho}{\mu_{H_2O}} \Delta H \frac{\partial q}{\partial t} \approx \frac{3}{R_g} \alpha \left[T_f - T_{av}(t) \right] \qquad (3.10)$$

thus predicting that at the SM, the heat consumption rate is nearly equal to the intensity of heat transfer. Indeed, we have found that the instant desorption rate calculated as

$$\frac{\partial q}{\partial t} \approx \frac{3\mu_{H_2O}}{\Delta H \rho R_g} \alpha \left[T_f - T_{av}(t) \right] \qquad (3.11)$$

is a linear function of $[T_f - T_{av}(t)]$ except for a short transient period t_{tr} right after the temperature jump (Figure 3.18a). The same is valid for the adsorption rate (Figure 3.18b).

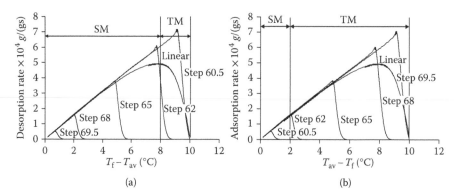

FIGURE 3.18
(a) Desorption and (b) adsorption rates as a function of the instant driving force for the isobar shapes indicated on the graph. The stationary (SM) and transient (TM) desorption modes are shown for $T_{st} = 62°C$.

After establishing the quasi-stationary regime, the adsorption and desorption runs for all equilibrium isobars fall on the same straight line, thus indicating a universal character of the function $W = \alpha[T_f - T_{av}(t)]$ with the constant coefficient α as seen from Equation 3.11. Thus, the rate of ad-/desorption on a single grain increases for smaller grain, larger heat transfer coefficient α and higher temperature difference $[T_f - T_{av}(t)]$. The slope is equal for the adsorption and desorption processes and does not depend on the isobar shape.

The time-averaged temperature difference $[T_f - T_{av}(t)]$ can be considered an efficient driving temperature difference which depends on the conversion degree χ. We have averaged the value of $[T_f - T_{av}(t)]$ from $t = 0$ to $t = \tau_{0.7}$, which corresponds to $\chi = 0.7$, and found that, for the step-like isobars of Figure 3.10, it is proportional to the temperature difference $(T_f - T_{st})$ with the coefficient equal to 0.51 for adsorption and 0.56 for desorption (not presented). Thus, in this case, the efficient driving temperature difference is approximately half of the temperature difference $(T_f - T_{st})$. This confirms the conclusion that for fast isobaric ad-/desorption in AHTs the adsorption isobar should be step-like and the step has to be positioned much different from the final temperature of the metal plate T_f as possible (Okunev et al., 2011).

From what is written earlier, it seems like the process dynamics is dictated merely by heat transfer processes. Since the heat transfer and the mass transfer are inevitably coupled, we elucidate below the contribution of the mass transfer to the overall dynamics. First, the evolution of the average grain temperature is faster at smaller vapour diffusivity D inside the grain: it is quicker approaching the final temperature T_f (Figure 3.19a). That results in a smaller time-averaged difference $[T_f - T_{av}(t)]$ and, hence, in slower overall desorption (Figure 3.19b). Interestingly, the increase of diffusivity by three orders of magnitude (from 3×10^{-8} m^2/s to 3×10^{-5} m^2/s) results in the acceleration of desorption by only 10 times as revealed from the initial slope of the uptake curve (Figure 3.19b). Hence, under

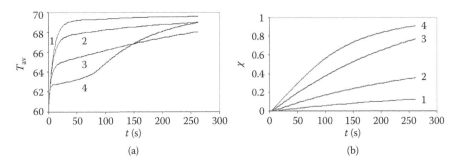

FIGURE 3.19
(a) Average grain temperature and (b) dimensionless desorption uptake as functions of the vapour diffusivity (1) 3×10^{-8} m2/s; (2) 3×10^{-7} m2/s; (3) 3×10^{-6} m2/s; (4) 3×10^{-5} m2/s).

studied conditions, the overall dynamics is more sensitive to the heat transfer parameters rather than to the mass transfer ones. Therefore, the ad-/desorption process is limited mainly by heat transfer between the plate and the grain. The mass transfer becomes a rate-limiting process at an efficient vapour diffusivity D lower than 10^{-9}–10^{-10} m²/s (at $2R_g = 1.5$ mm). In this case, the initial desorption rate W is proportional to the diffusional rate constant $K_{dif} = D/R_g^2$ as seen from Figure 3.20. The initial rate has been determined as a slope of the dimensionless release curve $\chi(t)$ at short process time $t \rightarrow 0$ (Figure 3.19b). In the case of diffusion limitation, the desorption release curves are invariant with respect to the dimensionless time $(K_{dif}\, t) = (Dt/R_g^2)$ (not presented).

The grain considered in this simulation seems to be too large for application in AC ($2R_g = 1.5$ mm). Indeed, it has been recommended (Aristov et al., 2012) that the size of the loose adsorbent grains has to be selected small enough (e.g. $2R_g \leq 0.5$ mm for Fuji silica RD) to realise the so-called adsorbent-insensitive mode. It is evident that for such smaller grains, a crossover to the purely diffusion regime occurs at even lower vapour diffusivity of 10^{-10}–10^{-11} m²/s. Such a slow water motion can relate to activated diffusion in micropores, and is typical for zeolites and other molecular sieves (Ruthven 1984).

Isothermal adsorption dynamics is usually controlled by mass transfer rather than by intrinsic gas adsorption on the surface. For mesoporous solids at low pressure, a Knudsen diffusion commonly is a dominant transport mechanism with the diffusivity $D_{kn} = 4850\, d_p P(T/M)^{1/2}$ m²/s (d_p is the pore size, P is the pressure, T is the temperature and M is the molecular mass) (Ruthven 1984). In our basic model, we have assumed the vapour diffusivity in the grain pores to be $D = 3.0 \times 10^{-6}$ m²/s, as was revealed in our previous simulations. This is a quite typical value that corresponds to the water vapour Knudsen diffusion in cylindrical mesopores of $d = 30$ nm. Here and below, we neglect the pore tortuosity that commonly makes the diffusion process slower by a factor of 2–3 (Ruthven 1984). In the mesopores of 3 nm diameter, the D_{kn} value is equal to 3.0×10^{-7} m²/s, which corresponds to case

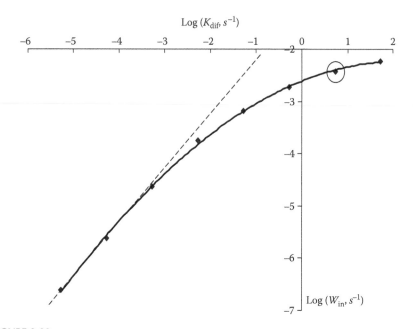

FIGURE 3.20
Initial desorption rate as a function of the efficient diffusion constant as a log–log plot. Dashed line presents the relation $\log(W) = n \log(D)$ with $n = 0.99$. Circle indicates $D = 3.0 \times 10^{-6}$ m^2/s that has been fixed for all other simulations in this chapter.

2 in Figure 3.19a and b. High pore diffusivity of 3.0×10^{-5} m^2/s can be associated either with quite large pores ($d = 300$ nm) or with additional contribution of the water surface diffusion (Medveď and Černý 2011).

Overall, the size of common adsorbent grains used in AHTs is between 0.2 and 1 mm. For a mono-layer configuration of such grains, the heat transfer from the HE wall to the grain is a rate-limiting process at all realistic situations. The pure rate limitation by intraparticle water diffusion is important only at unfeasibly low values of the water vapour diffusivity. This is expected to be even more valid for AHT cycles with methanol and ammonia vapours as working fluids due to more intensive diffusional flux stemmed from a higher gradient of the vapour pressure than for water. Heat transfer problems become more severe for a multi-layer bed of loose grains, which is commonly organised in a way to minimise the intergrain diffusional resistance (Raymond and Garimella 2011).

We have presented earlier the results of numerical modelling for the step-like isobar with $T_{st} = 62°C$ as a representative example. A similar study has been performed for all the isobars displayed in Figure 3.10. Thus, the main recommendation on dynamic optimisation of the adsorbent is to use in an AHT cycle an adsorbent which at the evaporator pressure has a steep isobar with a step positioned at the initial adsorption temperature, while at the condenser pressure has a steep isobar with a step positioned at the initial desorption temperature. Adsorbents with such step-like or sigmoid isobars are known and, in fact, are recognised as promising for adsorptive transformation of low-temperature heat (Aristov 2013).

3.3.2 Multi-layer Configuration

As we have already mentioned, a mono-layer configuration can be practicable only for adsorbent grains that are large enough—tentatively, more than 1 mm in size; otherwise, the ratio (M/m) is too large and unfavourable for getting a high COP. As demonstrated earlier, for very large grains, the grain-size-sensitive regime is revealed, so that the SCHP value reduces as $1/d$. Hence, more encouraging is a multi-layer configuration of relatively small adsorbent grains ($d \leq 0.5$ mm). Here we present a numerical model of non-isothermal water sorption on n layers of loose silica grains with a low n number ($1 \leq n \leq 4$) (Freni et al., 2012a).

Three configurations of the Ad-HEx unit are considered and compared: one, two and four layers of loose grains of Fuji silica-type RD located on a metal support subjected to a fast jump/drop of temperature. The model was used to simulate the ad-/desorption process for a different grain size (0.45, 0.85 and 1.7 mm). Results of these preliminary simulations were validated by comparison with the LTJM experimental data reported by Glaznev and Aristov (2010b).

3.3.2.1 Model Description

The simulated layer configuration consists of n spherical adsorbent grains that are in contact and surrounded by a stagnant water vapour at a constant pressure. Grain #1 is in thermal contact with a metal plate subjected to a fast temperature jump/drop. The reference control volume for the phenomenological description of the studied system is shown in Figure 3.21, which refers to the case of four adsorbent grains.

To develop a consistent model, the following main assumptions are made (Freni et al., 2012a):

1. The adsorbent grain is isotropic and has a uniform porosity.
2. The solid phase is in local thermal equilibrium with the adsorbate in the adsorbed phase.
3. The adsorbent grains are identical in shape, size and properties.
4. The grains have an hexagonal close-packed arrangement.
5. The gas phase is ideal.
6. The water intraparticle diffusivity is temperature independent.
7. Heating/cooling of the metal plate is not simulated and a proper boundary condition is adopted.
8. The vapour phase surrounding the adsorbent grain is stagnant.
9. Radiation heat transfer has been neglected.
10. The simulated control volume has no loss/gain of heat to the ambient.

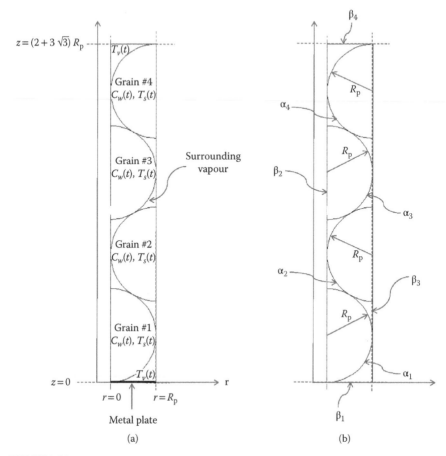

FIGURE 3.21
Schematic diagram of (a) four-adsorbent-grain configuration and (b) pertinent boundaries.

The stagnant vapour assumption can be justified by considering that many previous studies (Di et al., 2001, Yamamoto et al., 2002, Raymond and Garimella 2011) have indicated that the dominant mass transfer resistance in silica gel/water adsorbent beds is an intraparticle diffusion. Therefore, mass transport from the vapour reservoir to the grain surface can be reasonably neglected. This should be especially valid for the specific configuration studied here, namely a thin layer ($n \leq 4$) of relatively large grains (0.45–1.7 mm), as indicated by Ruthven and Lap-Keung Lee (1981).

Energy and mass balance equations, together with appropriate initial and boundary conditions, can be found in Freni et al. (2012). The system of partial differential equations was numerically solved through a finite element method, using the COMSOL Multiphysics® simulation environment, which has already been proved to be a suitable tool for simulation of adsorbent coatings in AHT (Freni et al., 2010, Füldner and Schnabel 2010). The software solves the energy and mass transport equations by the 'Chemical Engineering Module'.

The developed COMSOL tool was validated by considering a simple system with a known dynamic behaviour. Specifically, the model was adapted to the case of isothermal water adsorption on a single spherical grain, which is initiated by a small pressure drop. These specific conditions were experimentally achieved for a Fuji silica RD by the isothermal differential step (IDS) method (Aristov et al., 2006), and the resulting uptake curves were well described by a proper analytical solution (Ruthven 1984) as well as by the suggested COMSOL model (Freni et al., 2012a) (Figure 3.22a). In both cases, the best fitting was achieved with an effective water diffusivity 4.8×10^{-7} m^2/s that is equal to the experimentally measured diffusivity (Aristov et al., 2006).

3.3.2.2 Simulation of a Multi-layer Configuration

Simulations are carried out considering a single grain, two grains and four grains of the adsorbent Fuji silica-type RD (the grain size 0.45 mm), subjected to a temperature drop 50°C → 30°C at $P = 9$ mbar, which is typical for the AHT adsorption stage. The thermophysical properties of Fuji silica-type RD are taken from Chua et al. (2002) and Aristov et al. (2006).

A useful overview of the system thermal evolution after the fast cooling of the metal support is given by the temperature maps calculated for $n = 1, 2$ and 4 (Figures 3.23a–c). Figure 3.23a clearly shows that the cooling process driven by the metal plate at 30°C involves a thermal range from the upper grain to the lower grain, which directly contacts the metal plate. A high maximum thermal range $\Delta T = T_{max} - T_{max} = 5.86°C$ was found for the four-grain configuration, while this thermal range is reduced to $\Delta T = 0.62°C$ for the mono-layer case. In all simulated

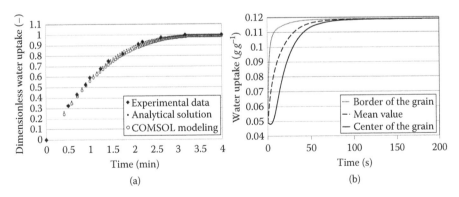

(a)   (b)

FIGURE 3.22

(a) Experimental uptake curve measured by the IDS method [$T = 49°C$, the initial and final pressures are 22.0 and 24.2 mbar, $2R_g = 0.9$ mm (Aristov et al., 2006), compared with the analytical solution (Ruthven 1984) and the COMSOL simulation (Freni et al., 2012a)] (b) time evolution of the local uptake calculated at different positions inside the grain for a one-grain configuration (grain size, 0.225 mm).

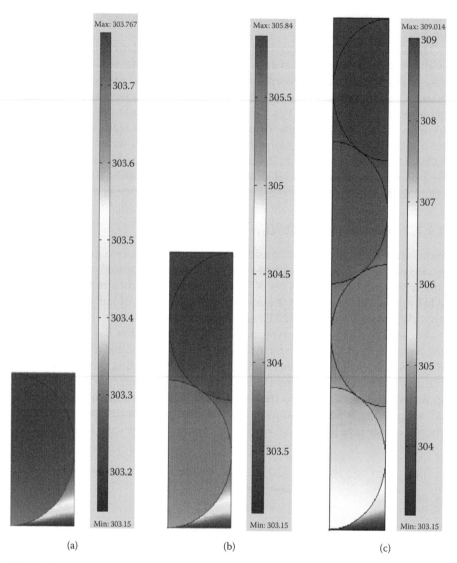

FIGURE 3.23
Grain temperature mapping for 20 seconds of adsorption process calculated for (a) one-grain (b) two-grain, and (c) four-grain configurations (Freni et al., 2012b).

cases, the temperature distribution inside the adsorbent grains is rather uniform, which is confirmed by a small value of the thermal Biot number $Bi = (1/\alpha)R_g/\lambda = 0.02 \ll 1$. Another important dimensionless parameter is the Lewis number $Le = a_s/D$, which is the ratio of the thermal diffusivity $a_s = \lambda/(\rho, c_p) \approx 1.5 \times 10^{-6}$ m^2/s to the vapour diffusivity $D = 4 \times 10^{-7}$ m^2/s taken from Aristov et al. (2006). The large value of $Le \approx 4$ also favours the homogeneous temperature distribution inside the grain. Differently, the water vapour surrounding the grains presented

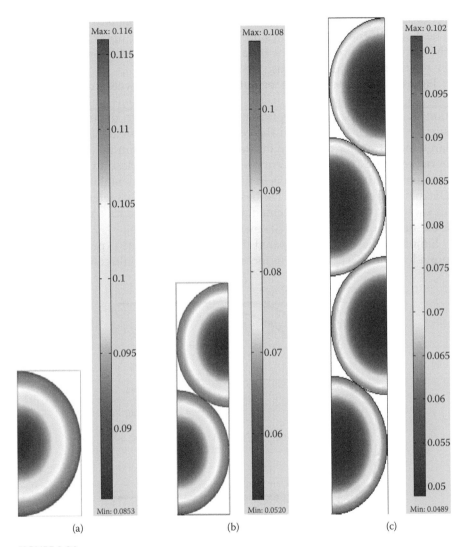

FIGURE 3.24
Water uptake mapping for 20 seconds of adsorption process calculated for (a) one-grain
(b) two-grain, and (c) four-grain configurations (Freni et al., 2012a).

a more pronounced temperature gradient due to the low thermal conductivity of
the vapour phase ($\lambda_v = 0.025$ W/[m·K] at $P = 9$ mbar).

Similar mapping was presented for the water uptake distribution for the
same configurations (Figures 3.24a–c). The obtained uptake maps highlight
that the lower grains, being at a lower temperature with respect to the upper
grains, present a slightly higher mean water loading. Figure 3.24 shows that
the uptake gradient inside the grain (up to 0.06 g/g) is more important than
that along the layer (less than 0.01 g/g). Indeed, the water adsorption front

proceeds radially from the external surface to the internal part of the grains. The centre of the grains always presents a lower water uptake than that of the external surface, due to the intraparticle mass transfer resistance caused by relatively low vapour diffusivity inside the grains.

Figure 3.22b presents the evolution of the local water uptake calculated at different locations inside the grain for the one-grain configuration. As expected, the adsorption rate in the vicinity to the external grain surface is larger than that in the centre of the grain, evidently due to the large mass transfer resistance inside the grain. The mean uptake curve lies between these two limiting curves, and its shape is close to an exponential one (Figure 3.22b), which is in line with the experimental observation of Glaznev and Aristov (2010b). The best fit is found for a characteristic time $\tau = 13.3$ seconds, which is in agreement with the value found for a mono-layer of loose grains of 0.2–0.25 mm (Glaznev and Aristov 2010b).

Simulation results are compared (Figure 3.25a) with the experimental data provided by the LTJM apparatus for one-, two- and four-grain configurations of

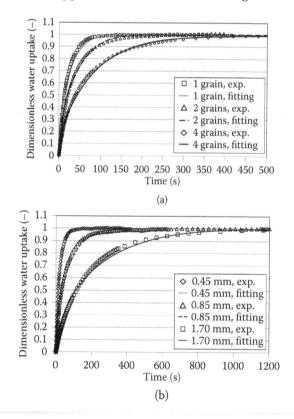

FIGURE 3.25
(a) Experimental and calculated evolution of the average water uptake for *n* layer configurations of Fuji silica RD grains ($d = 0.45$ mm) (Freni et al., 2012a) (b) simulated and experimental dimensionless water uptake curves for different adsorbent grain sizes (adsorption case).

Fuji silica RD (Glaznev and Aristov 2010b). The best numerical fit was achieved by assuming an effective diffusivity of 6×10^{-7} m²/s, which is only slightly larger than the Knudsen diffusivity $D_{kn} = 4.1 \times 10^{-7}$ m²/s in the Fuji silica pores and close to the values $(2.3–6.3) \times 10^{-7}$ m²/s, experimentally determined using the IDS method by Aristov et al. (2006). Our model does not take into account 'heat pipe' or 're-condensation' heat transfer effects (Zhang and Wang 1999), nor the convective Stefan flux generated by the adsorption process (Nusselt 1916).

The ad-/desorption kinetics for different size R_g of the adsorbent grains has been simulated by the developed model and compared (Figure 3.25b) with experimental data obtained using the LTJM apparatus by Glaznev and Aristov (2010b). For all the simulations, the best data fitting is found by adopting $\alpha = 120$ W/(m²·K) and $D = 6 \times 10^{-7}$ m²/s. Maximum deviation between the experimental and simulated curves is about 10% for most of the simulations. The results obtained confirm that the grain size of the adsorbent must be carefully selected when designing AHT adsorbers. Moreover, it was found that the characteristic time for desorption is between about one-third and one-half of the adsorption one; thus, an optimal AHT cycle should be properly designed by taking into account the unequal duration of the adsorption and desorption stages as suggested by Aristov et al. (2012).

In this section, we have reported a mathematical model of coupled heat and mass transfer in the porous layer that consists of a low number of loose adsorbent grains, which is a quite realistic case for AHT practice. The system of partial differential equations has been solved by using the COMSOL Multiphysics simulation environment. Results of this simulation are found to be in reasonable agreement with the experimental data provided by the LTJM for one-, two- and four-grain configurations of Fuji silica RD. All these data confirm that optimisation of the Ad-HEx configuration is of primary importance for maximizing the AHT performance.

3.4 Dynamic Optimisation of Adsorptive Heat Transformation Cycles

Results obtained by LTJM have taught us that the duration of desorption phase of AHT cycle should be shorter than that of the adsorption one (recommendation 2). Hence, the equal duration of adsorption and desorption phases is hardly to be an optimal case, and appropriate reallocation is necessary to give more time for adsorption at the expense of desorption shortening. This has been realised and tested in the single-bed adsorption chiller built in ITAE-CNR (Aristov et al., 2012). All experiments are carried out by fixing the same inlet temperatures of the external heating/cooling fluids $(T_d = 95°C, T_e = 10°C, T_c = 30°C)$.

The prototype tests confirmed that the optimal desorption duration should indeed be two to three times shorter than the adsorption one, and for the optimal ratio t_d/t_a, both the COP (not presented) and SCHP (Figure 3.26a) may increase almost twice as compared with the common case $t_d = t_a$. The sharp dependence COP(t_d) is because, at too short desorption time ($t_d < 80$–90 seconds), the optimal desorption degree cannot be reached, which reduces both the COP and the SCHP. If this time is longer than the optimal one, the duration of the adsorption phase is reduced accordingly, and there is not enough time for the water uptake to recover. For the tested adsorber, the optimal adsorption time can be estimated as 250–280 seconds. For the shorter cycle, both the COP and the SCHP are more sensitive to the choice of durations of the adsorption and desorption stages than for the longer cycle (Figure 3.26a) (Aristov et al., 2012), and a proper choice of desorption time t_d is especially important.

The reallocation of ad-/desorption durations proposed causes a subsequent change in cooling cycle organisation, because each adsorber now is connected with an evaporator longer than the half-cycle time. Let, for instance, the time t_d be twice as short as t_a. In this case, the following cycle rearrangements can be done for a two-bed configuration (Figure 3.26b):

- One bed is connected with the evaporator where cold is produced. At the same time, another bed is under regeneration and is connected with the condenser where heat is rejected.
- Both beds are linked with the evaporator and generate a double-chilling effect.

Thus, cold is continuously generated so that each bed is linked with the evaporator two-thirds of the cycle time t_{cycle} and with the condenser only one-third. In this case, to smooth the cooling effect produced, the chiller could be equipped with an intermediate cold storage unit. To allow the use of a continuous driving heat input, an intermediate heat store or a buffer may

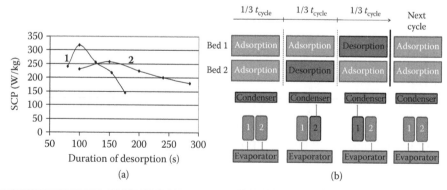

FIGURE 3.26
(a) SCHP as a function of the desorption stage duration. Total cycle time: 385 (1) and 600 seconds (2). (b) Management of a 2-bed cooling unit with reallocated duration of adsorption and desorption stages (Aristov et al., 2012).

also be needed. The three-bed configuration is considered by Aristov et al. (2012). Similar optimisation was performed by Sapienza et al. (2012).

3.5 Conclusions

An Ad-HEx unit is a core of any AHT, and its efficient performance is of prime importance. In this chapter, we present recent advances in understanding and trends in studying the adsorption dynamics. Dynamic experiments have been performed in laboratory Ad-HEx units which closely imitate the conditions of isobaric stages of AHT when desorption/adsorption is initiated by a fast jump/drop of the temperature of a heat exchanger wall that is in thermal contact with the adsorbent layer. All results concern the configuration of Ad-HEx with one, two or four layers of loose adsorbent grains. A mathematical model of coupled heat and mass transfer during the isobaric adsorption process developed at Moscow State University (for a mono-layer configuration) has been used for detailed dynamical analysis and extracting the heat and mass transfer coefficients. Finally, several recommendations followed from the LTJM study have been checked with the absorptive chiller prototypes at ITAE-CNR and the University of Warwick. These tests clearly demonstrated that dynamic performance of AHTs can be significantly improved by a proper management of the AHT cycle (reallocation of relative durations of isobaric adsorption and desorption phases, optimisation of the cycle time and the conversion degree, etc.). The most important findings and general regularities that have been revealed as a result of the LTJM application for systematic studying the adsorption dynamics of water, methanol and ammonia in AHT systems are summarised, illustrated and discussed in this chapter.

3.6 Nomenclature

A	constant, s^{-1}
B	constant, dimensionless
Bi	thermal Biot number, dimensionless
C	specific heat capacity, $J/(kg \cdot K)$; vapour concentration, $kg \cdot m^3$
D	diameter, m; diffusivity, m^2/s
ΔF	adsorption potential, J/mol
H	heat of adsorption/evaporation, J/mol
K	rate constant, s^{-1}
m	adsorbent mass, kg
M	heat exchanger mass, kg

(Continued)

n	number of the adsorbent layers, dimensionless
P	pressure, mbar
q	adsorption uptake (release), g/g
r	radius, m
R	radius, m; universal gas constant, J/(mol · K)
S	surface area, m^2
T	temperature, K, °C
ΔT	temperature change, K, °C
t	time, second
V	volume, m^3
W	power, W; specific power, W/kg
$\overline{W}$	average specific power, W/kg

Greek Symbols

α	heat transfer coefficient, W/(m^2 · K)
χ	dimensionless conversion degree
λ	heat conductivity, W/(m · K)
μ	molecular weight, kg/mol
ρ	density, kg/m^3
τ	characteristic time, second

Subscripts

0	initial
ads	adsorption
av	average
c	condenser
des	desorption
dr	driving
e	evaporator/evaporation
exp	experimental
f	final
g	grain
H$_2$O	water
HS	external heat source
in	initial
max	maximal
p	constant pressure
s	solid
st	step position

Superscripts

| ads | adsorption |
| des | desorption |

Acknowledgements

The author thanks his colleagues Dr. I.S. Glaznev, Dr. L.G. Gordeeva, Dr. M.M. Tokarev, Dr. J.V. Veselovskaya and Dr. I.S. Girnik for their experimental contribution; Dr. B.N. Okunev, Dr. A.P. Gromov and Prof. L.I. Heifetz for mathematical modelling; and the Russian Foundation for Basic Researches (Projects 08-08-00808, 09-08-92604 and 10-08-91156) and the Siberian Branch of the Russian Academy of Sciences (Integration Project N 120A/2012) for partial financial support of this activity.

References

Aristov, Yu.I. 2007. Novel materials for adsorptive heat pumping and storage: Screening and nanotailoring of sorption properties (review). *J. Chem. Eng. Jpn.* 40:1241–51.

Aristov, Yu.I. 2009. Optimal adsorbent for adsorptive heat transformers: Dynamic considerations. *Int. J. Refrig.* 32:675–86.

Aristov, Yu.I. 2011. Optimization of adsorption dynamics in adsorptive heat transformers: Experiment and modeling. Proceeding of the VIII Minsk International Seminar 'Heat Pipes, Heat Pumps, Refrigerators, Power Sources'. Minsk. Belarus. Sept. 12–15, 2011. 1:27–39.

Aristov, Yu.I. 2012. Adsorptive transformation of heat: Principles of construction of adsorbents database. *Appl. Therm. Eng.* 42:18–24.

Aristov, Yu.I. 2013. Challenging offers of material science for adsorption heat transformation: A review. *Appl. Therm. Eng.* 50:1610–1618.

Aristov, Yu.I., B. Dawoud, I.S. Glaznev, and A. Elyas. 2008. A new methodology of studying the dynamics of water sorption/desorption under real operating conditions of adsorption heat pumps: Experiment. *Int. J. Heat Mass Transfer.* 51:4966–72.

Aristov, Yu.I., G. Restuccia, G. Cacciola, and V.N. Parmon. 2002. A family of new working materials for solid sorption air conditioning systems. *Appl. Therm. Eng.* 22:191–204.

Aristov, Yu.I., A. Sapienza, A. Freni, D.S. Ovoschnikov, and G. Restuccia. 2012. Reallocation of adsorption and desorption times for optimizing the cooling cycle parameters. *Int. J. Refrig.* 35:525–31.

Aristov, Yu.I., M.M. Tokarev, A. Freni, I.S. Glaznev, and G. Restuccia. 2006. Kinetics of water adsorption on silica Fuji Davison RD. *Microporous Mesoporous Mater.* 96:65–71.

Bauer, J., R. Herrmann, W. Mittelbach, and W. Schwieger. 2009. Zeolite/aluminum composite adsorbents for application in adsorption refrigeration. *Int. J. Energ. Res.* 33:1233–49.

Choudhury, B., P.K. Chatterjee, and J.P. Sarkar. 2010. Review paper on solar-powered air-conditioning through adsorption route. *Renew. Sust. Energ. Rev.* 14:2189–95.

Chua, H.T., K.C. Ng, A. Chakraborty, N. Oo, and M.A. Othman. 2002. Adsorption characteristics of silica gel + water systems. *J. Chem. Eng. Data.* 47:1177–81.

Critoph, R.E., and Y. Zhong. 2005. Review of trends in solid sorption refrigeration and heat pumping technology. *J. Process Mech. Eng. E.* 219:285–300.

Di, J., J.Y. Wu, Z.Z. Xia, and R.Z. Wang. 2007. Theoretical and experimental study on characteristics of a novel silica gel-water chiller under the conditions of variable heat source temperature. *Int. J. Refrig.* 30:515–26.

Freni, A., B. Dawoud, F. Cipitì, S. Chmielewski, G. Maggio, and G. Restuccia. 2010. Finite element-based simulation of heat and mass transfer process through an adsorbent bed in an adsorption bed in an adsorption heat pump/chiller. Proceedings of the ASME-ATI-UIT 2010 Conference on Thermal and Environmental Issues in Energy Systems. I:489–94.

Freni, A., G. Maggio, F. Cipitì, and Yu.I. Aristov. 2012a. Simulation of water sorption dynamics in adsorption chillers: One, two and four layers of loose silica grains. *Appl. Therm. Eng.* 44:69–77.

Freni, A., F. Russo, S. Vasta, M.M. Tokarev, Yu.I. Aristov, and G. Restuccia. 2007. An advanced solid sorption chiller using SWS-1L. *Appl. Therm. Eng.* 27:2200–204.

Freni, A., A. Sapienza, I.S. Glaznev, Yu.I. Aristov, and G. Restuccia. 2012b. Testing of a compact adsorbent bed based on the selective water sorbent 'silica modified by calcium nitrate'. *Int. J. Refrig.* 35:518–24.

Füldner, G., and L. Schnabel. 2010. Non-isothermal kinetics of water adsorption in compact adsorbent layers on a metal support. COMSOL Conference. CD-ROM Conference Proceedings. ISBN 978-0-9766792-8-8. Paper 318:6.

Glaznev, I.S., and Yu.I. Aristov. 2008. Kinetics of water adsorption on loose grains of SWS-1L under isobaric stages of adsorption heat pumps: The effect of residual air. *Int. J. Heat Mass Transfer.* 51:5823–27.

Glaznev, I.S., and Yu.I. Aristov. 2010a. Dynamic aspects of adsorptive heat transformation. In *Advances in Adsorption Technologies*, eds. B. Saha and K.S. Ng, 107–63. Singapore: Nova Science Publishers.

Glaznev, I.S., and Yu.I. Aristov. 2010b. The effect of cycle boundary conditions and adsorbent grain size on dynamics of adsorption chillers. *Int. J. Heat Mass Transfer.* 53:1893–98.

Glaznev, I.S., I.S. Girnik, and Yu.I. Aristov. 2011. Water sorption dynamics in adsorption chillers: A few layers of loose Fuji silica grains. Proceedings of the VIII Minsk International Seminar 'Heat Pipes, Heat Pumps, Refrigerators, Power Sources'. Minsk, Belarus. Sept. 12–15, 2011. 1:161–68.

Glaznev, I.S., D.S. Ovoshchnikov, and Yu.I. Aristov. 2009. Kinetics of water adsorption/desorption under isobaric stages of adsorption heat transformers: The effect of isobar shape. *Int. J. Heat Mass Transfer.* 52:1774–77.

Glaznev, I.S., D.S. Ovoshchnikov, and Yu.I. Aristov. 2010. Effect of residual gas on water adsorption dynamics under typical conditions of an adsorptive chiller. *Heat Transfer Eng. J.* 31:924–30.

Gong, L.X., R.Z. Wang, Z.Z. Xia, and Z.S. Lu. 2011. Study on an adsorption chiller employing lithium chloride in silica gel and methanol: Design and experimental study. Proceedings of the International Sorption Heat Pump Conference. Padua, Italy. April 6–8, 2011. 329–39.

Gordeeva, L.G., and Yu.I. Aristov. 2011. Composite sorbent of methanol 'LiCl in mesoporous silica gel' for adsorption cooling: Dynamic optimization. *Energy.* 36:1273–79.

Jakob, U., and P. Kohlenbach. 2010. Recent developments of sorption chillers in Europe. *IIR Bulletin.* 34–40.

Kim, D.S., and C.A. Infante Ferreira. 2008. Solar refrigeration options—A state-of-the-art review. *Int. J. Refrig.* 31:3–15.

Lang, R., M. Roth, and M. Stricker. 1999. Development of a modular zeolite-water heat pump. Proceedings of the International Sorption Heat Pump Conference, Munich, Germany. 1999. 611–18.

Matsushita, M. 1987. Adsorption chiller using low-temperature heat sources. *Energy Conserv.* 39:96–106.

Medveď, I., and R. Černý. 2011. Surface diffusion in porous media: A critical review. *Micropor. Mesopor. Mat.* 142:405–22.

Nusselt, W. 1916. Surface condensation of water vapor. *Z. Ver. Deut. Ing.* 60:541–46.

Okunev, B.N., A.P. Gromov, L.I. Heifets, and Yu.I. Aristov. 2008a. A new methodology of studying the dynamics of water sorption/desorption under real operating conditions of adsorption heat pumps: Modeling of coupled heat and mass transfer. *Int. J. Heat Mass Transfer.* 51:246–52.

Okunev, B.N., A.P. Gromov, L.I. Heifets, and Yu.I. Aristov. 2008b. Dynamics of water sorption on a single adsorbent grain caused by a large pressure jump: Modeling of coupled heat and mass transfer. *Int. J. Heat Mass Transfer.* 51:5872–76.

Okunev, B.N., A.P. Gromov, I.S. Glaznev, and Yu.I. Aristov. 2011. Modeling of water sorption dynamics in heat transformers: An optimal shape of the sorption isobar. Proceedings of the VIII Minsk International Seminar 'Heat Pipes, Heat Pumps, Refrigerators, Power Sources'. Minsk. Belarus. Sept. 12–15, 2011. 1:204–11.

Okunev, B.N., A.P. Gromov, V.L. Zelenko, I.S. Glaznev, D.S. Ovoshchnikov, L.I. Heifets, and Yu.I. Aristov. 2010. Effect of residual gas on the dynamics of water adsorption under isobaric stages of adsorption heat pumps: Mathematical modelling. *Int. J. Heat Mass Transfer.* 53:1283–89.

Pons, M., F. Meunier, G. Cacciola, R.E. Critoph, M. Groll, L. Puigjaner, B. Spinner, and F. Ziegler. 1999. Thermodynamic based comparison of sorption systems for cooling and heat pumping. *Int. J. Refrig.* 22:5–17.

Raymond, A., and S. Garimella. 2011. Intraparticle mass transfer in adsorption heat pumps: Limitations of the linear driving force approximation. *J. Heat Transfer.* 133:42001–13.

Ruthven, D.M. 1984. *Principles of Adsorption and Adsorption Processes.* New York: John Wiley.

Ruthven, D.M., and Lap-Keung Lee. 1981. Kinetics and non-isothermal sorption: Systems with bed diffusion control. *AIChE J.* 27:654–62.

Saha, B.B., A. Akisawa, and T. Kashiwagi. 2001. Solar/waste heat driven two-stage adsorption chiller: The prototype. *Renew. Energy.* 23:93–101.

Saha, B., and K.S. Ng (ed). 2010. *Advances in Adsorption Technologies.* Singapore: Nova Science Publishers.

Sapienza, A., I.S. Glaznev, S. Santamaria, A. Freni, and Yu.I. Aristov. 2012. Adsorption chiller driven by low temperature heat: New adsorbent and cycle optimization. *Appl. Therm. Eng.* 32:141–46.

Sun, L.M., and F. Meunier. 1987. A detailed model for non-isothermal sorption in porous adsorbents. *Chem. Eng. Sci.* 42:1585–93.

Tamainot-Telto, Z., S.J. Metcalf, and R.E. Critoph. 2009. Novel compact sorption generators for car air conditioning. *Int. J. Refrig.* 32:727–33.

Vasiliev, L.L., Yu.I. Aristov, and V.E. Nakoryakov. 2008. Chemical and sorption heat engines: State of the art and development prospects in the Russian Federation and the Republic of Belarus. *J. Eng. Thermophys.* 80:19–48.

Veselovskaya, J.V., R.E. Critoph, R.N. Thorpe, S. Metcalf, M.M. Tokarev, and Yu.I. Aristov. 2010. Novel ammonia sorbents 'porous matrix modified by active salt' for adsorptive heat transformation: 3. Testing of 'BaCl$_2$/vermiculite' composite in the lab-scale adsorption chiller. *Appl. Therm. Eng.* 30:1188–92.

Veselovskaya, J.V., and M.M. Tokarev. 2011. Novel ammonia sorbents 'porous matrix modified by active salt' for adsorptive heat transformation: 4. Dynamics of quasi-isobaric ammonia sorption and desorption on BaCl$_2$/vermiculite. *Appl. Therm. Eng.* 31:566–72.

Wang, R.Z., and R.G. Oliveira. 2006. Adsorption refrigeration—An efficient way to make good use of waste heat and solar energy. *Prog. Energy Combust. Sci.* 32:424–58.

Wang, L.W., R.Z. Wang, and R.G. Oliveira. 2009. A review on adsorption working pairs for refrigeration. *Renew. Sust. Energy Rev.* 13:518–34.

Yamamoto, E., F. Watanabe, N. Kobayashi, and M. Hasatani. 2002. Intraparticle heat and mass characteristics of water vapor adsorption. *J. Chem. Eng. Jpn.* 35:1–8.

Yong, L., and K. Sumathy. 2002. Review of mathematical investigation on the closed adsorption heat pump and cooling systems. *Renew. Sust. Energy Rev.* 6:305–37.

Zhang, L.Z., and L. Wang. 1999. Momentum and heat transfer in the adsorbent of a waste-heat adsorption cooling system. *Energy.* 24:605–24.

Ziegler, F. 2009. Sorption heat pumping technologies: Comparisons and challenges. *Int. J. Refrig.* 32:566–76.

4

Mechanisms of Intensive Heat Transfer for Different Modes of Boiling

Victor V. Yagov

CONTENTS

4.1 Introduction 109
4.2 Nucleate Boiling 111
    4.2.1 Main Peculiarities of the Process Revealed in Experiments ... 111
    4.2.2 Basic Mechanism of Nucleate Boiling 113
    4.2.3 An Approximate Model of Nucleate Boiling Heat Transfer 115
    4.2.4 Flow Boiling Heat Transfer 121
4.3 Transition Boiling 123
    4.3.1 Brief Analysis of Experimental Results 123
    4.3.2 Model of Heat Transfer at a High Jakob Number 125
4.4 Film Boiling of Subcooled Liquids 132
    4.4.1 General Characteristics of Film Boiling 132
    4.4.2 New Experimental Results on Cooling of High-
        Temperature Bodies 135
4.5 Concluding Remarks 140
References 141

4.1 Introduction

The term 'boiling' embraces a wide spectrum of different phenomena: nucleation, vapour bubble growth and departure, heat transfer at different regimes of pool boiling, flow boiling in channels under different combinations of controlling parameters, boiling crisis and so on. This chapter is concerned with mechanisms of intensive heat transfer under nucleate, transition and film boiling. This is not a usual approach, because three very different modes of boiling are considered in one chapter based on high intensity of heat transfer as their single general feature.

Nucleate boiling is a widely used process as an integral part of many technologies; its capability to remove high heat fluxes at rather low wall superheats is commonly known and applied for cooling different heat-generating and heat-transferring surfaces. However, till now there is no general understanding of the mechanisms of nucleate boiling heat transfer, and many different empirical correlations are used in practical calculations of heat transfer coefficient (HTC). In this chapter, it is proved that the availability of three-phase boundaries (interlines) is a peculiarity of nucleate boiling, which distinguishes it from other modes of convective heat transfer. The extremely high heat flux density at these boundaries explains the high heat transfer intensity under nucleate boiling and presents a generic feature of this boiling mode. This basic mechanism is considered in developing an approximate theory of nucleate boiling heat transfer.

Transition boiling remains the least examined mode of boiling at present, although it is used rather often in different technologies. During quenching of metallic bodies or cooling of castings, transition boiling occurs inevitably. Post-accident cooling of a nuclear reactor, which also involves transition boiling, is widely investigated using experimental models and simulated in numerical studies. The main difficulties of studying transition boiling are connected with providing steady regimes of heat transfer in experiments. The high intensity of heat transfer at transition boiling is not surprising in itself; one can estimate HTC based on a typical boiling curve. The problem is to reveal main mechanisms of heat transfer in this boiling mode. For rather low reduced pressures, which are typical conditions for experimental studies of transition boiling, an approximate model is developed and presented in this chapter.

Film boiling is characterised by a rather low intensity of heat transfer in the case of saturated liquids. At the same time, this boiling mode can be considered as the clearest one on its mechanisms; a continuous interface provides prerequisites for the theoretical modelling of the process. But the extremely high intensity of heat transfer in film boiling of highly subcooled liquids now has no explanation even on qualitative levels. Many researches believe that the cooling process in quenching is controlled by nucleate boiling; this is very doubtful as the temperature of a cooled surface in the process is usually much higher than the critical temperature of a coolant. Actual scientific investigations on film boiling of subcooled water were undertaken by G. Hewitt, D. Kenning and co-workers (Aziz et al., 1986, Corell and Kenning 1988, Zvirin et al., 1990). They established that HTC at the surface of a copper ball in film boiling at water subcoolings higher than 22 K at atmospheric pressure can be as high as 10 kW/m$^2 \cdot$K or even more. These values are 1–2 orders of magnitude higher than those typical of film boiling of saturated liquids. Based on visual observations, the authors called the aforementioned boiling regime 'microbubble boiling'. However, no hypothesis on mechanisms of heat transfer in these regimes was discussed by the authors (Aziz et al., 1986, Corell and Kenning 1988, Zvirin et al., 1990).

Recently, similar studies of film boiling of subcooled liquids were conducted in our laboratory at the Moscow Power Engineering Institute (MPEI), Moscow, Russia. Two main distinctions are featured in this work. The first is the simultaneous measurement of temperature at several points during the cooling of nickel balls of rather large diameters. The second is using different liquids as coolants in the experiments. The experiments confirmed the existence of intensive cooling regimes during film boiling of subcooled water. But it was unexpectedly revealed that such regimes were not observed in isopropanol even at very high subcoolings. A detailed description of the new experimental results and some qualitative ideas on the mechanism of intensive heat transfer in film boiling of subcooled liquids are given in this chapter.

4.2 Nucleate Boiling

4.2.1 Main Peculiarities of the Process Revealed in Experiments

Nucleate boiling remains a field of heat transfer that is intensively developed both experimentally and theoretically. There are two main reasons for this: (1) the process has a wide spectrum of applications, as mentioned in Section 4.1; and (2) at present there is no commonly accepted theory of the process based on first principles that allows the prediction of boiling heat transfer to the accuracy required in engineering. Nevertheless, our present knowledge on the process is enough not only for using different empirical correlations in calculations of HTC but also for reaching some general qualitative conclusions on its main features.

As early as 1959, it was argued by Labuntsov (2000, p. 83) that developed nucleate boiling is governed by its internal mechanisms and even strong external hydrodynamic effects including mass flow rate and gravitational and other body forces do not affect heat transfer. This idea has been realised in predicting equations by Labuntsov (2000, pp. 104 and 138) for nucleate boiling heat transfer, which are based on approximate but clear physical models. Much experimental evidence has been obtained in favour of this approach. In particular, both at high body forces in fast rotating systems and at microgravity conditions (Straub 2001), HTC in nucleate boiling keeps the values typical for common terrestrial conditions if heat flux density and pressure are the same.

Nucleate boiling heat transfer obviously differs from any single-phase convective heat transfer in its main mechanisms. It is commonly known that in single-phase convection heat transfer, intensity increases with an increase in flow velocity. In nucleate boiling, one can define two characteristic velocities. The first one is the average rate of evaporation:

$$W_0 = q/(h_{LG}\rho_G) \tag{4.1}$$

where q is the heat flux, h_{LG} is the latent heat of evaporation and ρ_G is the vapour density.

This is often used in modified Reynolds or Peclet numbers in dimensionless correlations for nucleate boiling heat transfer. The other actual velocity in nucleate boiling is a vapour bubble growth rate. This value is ordinarily used in more sophisticated approaches to the problem. As external hydrodynamics does not affect heat transfer in developed nucleate boiling, there are no grounds to search any other scale for velocity here. Due to a decrease of vapour specific volume with an increase in pressure, both the aforementioned velocities become very small at high reduced pressures. At the same time, HTC in nucleate boiling increases continuously with pressure. This means that an analogy with single-phase convection in principle cannot explain extremely high values of HTC in pool boiling at high reduced pressures.

Table 4.1 presents several examples of the measured values of HTC in nucleate boiling at high reduced pressures. Very different liquids are considered. Besides HTC, average rate of evaporation and vapour bubble growth rate, the values of thickness of a so-called equivalent liquid film ($\Delta_e = \lambda/\alpha$, where λ is thermal conductivity and α is HTC) are given. The bubble growth rate is calculated according to the Labuntsov (2000, p. 23) formula:

$$R = \sqrt{12 \frac{\lambda \Delta T t}{h_{LG} \rho_G}} \tag{4.2}$$

At the growth time $t = 1$ ms, that is, at the moment close to nucleation, when a bubble grows quickly. It is noteworthy that the bubble growth rate in a uniformly superheated liquid calculated according to the Plesset–Zwick equation for Jacob number $Ja < 1$ is essentially lower than that calculated according to Equation 4.2. The first four lines in Table 4.1 correspond to boiling occurring on the outer surface of horizontal cylinders. (The data of Loebl et al. [2005] given in the fourth line, strictly speaking, relates to a moderate rather than a high reduced pressure.) The last column in Table 4.1 gives values of the liquid velocity U_∞, which can provide the same heat transfer intensity in the case of external cross-flow of a cylinder (by means of single-phase convection), the well-known equation by Zhukauskas being used in the calculations. The fifth line of the table presents the typical value of HTC in helium pool boiling at atmospheric pressure on a copper horizontal surface (Verkin et al., 1987). In this case, a comparison with single-phase convection is given for a jet of liquid helium released from a nozzle of 3 mm in diameter. It is seen in all the aforementioned cases that the actual velocities in boiling at high reduced pressures are several orders of magnitude lower than the velocity of a single-phase liquid, which is needed to get the HTC values observed in the experiments with boiling. The equivalent film thickness is also impressed: it is difficult to conceive of such a thin conductive liquid layer covering an entire surface. In some cases, the calculated values of Δ_e are less than a characteristic size of surface roughness.

Thus, at high reduced pressures actual flow velocities in nucleate boiling are extremely low to provide the observed high intensity of heat transfer by means of any possible mechanism of single-phase convection.

TABLE 4.1

Rounded-Off Experimental Values of HTC in Nucleate Pool Boiling and Characteristic Velocities

Liquid	p/p_{cr}	Heater	q (kW/m²)	α (kW/ m²·K)	Δ_e (μm)	W_0 (m/s)	dR/dt (m/s)	U_∞ (m/s)
Water (Borishanskiy et al., 1961)	0.67	Tube, 6 mm	830	315	1.57	0.008	0.0052	90
Ethanol (Borishanskiy et al., 1961)	0.78	Tube, 5 mm	350	140	0.79	0.009	0.0046	88
R-134a (Kotthoff et al., 2006)	0.80	Tube, 25.4 mm	60	160	0.31	0.0013	0.0018	274
Carbon dioxide (Loebl et al., 2005)	0.19	Tube, 16 mm	82	20	7.5	0.007	0.013	10.7
Helium (Verkin et al., 1987)	0.445	Flat disc	4	10	1.96	0.0116	0.0083	12.8

4.2.2 Basic Mechanism of Nucleate Boiling

There exists a peculiarity of nucleate boiling that can explain the extremely high heat transfer intensity in this process and distinguishes it from any other mode of convective heat transfer. This is the availability of interlines, that is, the boundaries of contact of three phases. A stimulus to understanding the role of evaporation along an interline in boiling was given objectively by the work of Moore and Mesler (1961), in spite of the absence of a direct mention of this effect. In this work, it was first shown that in nucleate boiling a local heat flux under a growing vapour bubble can be two orders of magnitude higher than the average one at the heated wall. A hypothesis on liquid microlayer evaporation as a reason for such peak heat fluxes suggested by Moore and Mesler (1961) has been confirmed by direct observation in some special experiments. The experiments also revealed that at the centre of the bubble base a dry spot exists. This seems to be quite natural. As an active nucleation site is an area of direct vapour–solid contact, the dry spot does exist and its boundary is the line of contact of three phases. As at the interline itself a liquid film thickness tends to zero, the local heat flux at the isothermic heated surface formally tends to infinity if the temperature at the liquid/vapour interface is assumed to be equal to T_s. In one study (Yagov 2009), a subsequence of such an unrealistic assumption is discussed.

An approach to modelling the evaporating liquid film at the interline region developed by Wayner and his co-workers (Wayner 1973, 1978, Wayner et al., 1976) seems to be the most sound. According to Wayner, a thin liquid film near the interline comprises the regions of a non-evaporating adsorbed film, an evaporating film with the essential influence of London–van der Waals forces on liquid evaporation and an intrinsic meniscus. In the latter region, the London–van der Waals forces are negligible and the film thickness is much less than that in the outer part of the meniscus; this means that the intrinsic meniscus presents an area of extremely high intensity of evaporation. All the aforementioned regions are extremely small in comparison with any other linear scales commonly used in heat transfer analysis. According to Wayner's investigations, the width of the entire interline region is on the order of 1 μm or less.

Consequently, a nucleation site is the dry spot and along its boundary a very narrow zone of extremely intensive evaporation exists. Bearing in mind in addition that a typical size of the nucleation site is approximately equal to the equilibrium vapour bubble radius R_*, one can consider the site a point heat sink at the heated surface. At the space between the nucleation sites, convective heat removal from the surface occurs. Based on the aforementioned ideas, the author (Yagov 1988) has suggested that the total heat flux from the wall is the following sum:

$$q = q_1 + q_2 \tag{4.3}$$

where q_1 and q_2 are corresponding heat fluxes due to intensive evaporation at the dry spot boundary and due to convection to liquid between the nucleation sites. The very small fraction of dry spot area in relation to the total heated surface area is in favour of a possibility of using a simple additive law in this case (see Figure 4.1). At that time (1998), the main objective of

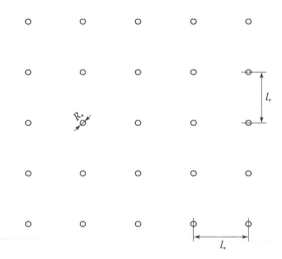

FIGURE 4.1
Schematic of nucleation site distribution.

the author was simply to improve Labuntsov's model of boiling heat transfer (Labuntsov 2000, p. 104). According to this model, nucleation and the very fast initial growth of vapour bubbles can be characterised by a definite pulsating velocity, which is analogous in part to a friction velocity in single-phase turbulent convection. Its averaged value is determined on the basis of equating the total kinetic energy resulting from simultaneously growing bubbles with the kinetic energy resulting from the pulsating motion of liquid with this velocity. The velocity discussed determines a thickness of a liquid layer at the wall, which in turn determines the heat flux density. According to the content of the analysis this heat flux is the convective heat flux q_2, and Labuntsov for the first time obtained an equation for this heat flux directly based on the approximate mechanistic model. This equation is as follows:

$$q_2 = C_1 \frac{\lambda^2 \Delta T^3}{\nu \sigma T_s} \tag{4.4}$$

where C_1 is a numerical constant of the order of magnitude of 10^{-3}, ν is kinematic viscosity, σ is surface tension and $\Delta T = T_w - T_s$ is wall superheat.

The other heat transfer mechanism due to direct liquid evaporation in growing bubbles was not determined in Labuntsov's studies on the basis of any mechanistic model. In his later work of 1972 (Labuntsov 2000, p. 138), this effect was involved by means of the empirical function of vapour and liquid densities ratio. It is clear that the limited possibilities of convection to remove high heat fluxes at high reduced pressures discussed in the Section 4.2.1 are valid also in this case. Although Labuntsov's final equation for nucleate boiling heat transfer is in rather good agreement with numerous experimental data, his model does not give a physically grounded quantitative explanation for the extremely intensive heat transfer in nucleate boiling at high reduced pressures.

Now it is possible to state that the heat flux resulting from intensive liquid evaporation in the vicinity of dry spot boundaries, which was initially considered by the author as the additional one, factually presents the main mode of heat removal and is really a generic feature of nucleate boiling. So if one imagines the circles representing nucleation sites in Figure 4.1 to be points, the two mechanisms of heat removal from the wall can be considered as practically independent from each other, both the heat fluxes in Equation 4.3 being related to the entire area of the heated surface.

4.2.3 An Approximate Model of Nucleate Boiling Heat Transfer

The discussion in Section 4.2.2, undoubtedly, essentially affects the conceptual view of nucleate boiling principal mechanisms. However, the basic steps of analysis and the final equation for HTC remain quite similar to those published in Yagov (1988). The problems of boiling incipience on solid surfaces and of nucleation site density prediction are the most complex in boiling (Yagov 2009). But at our present stage of knowledge, a simple approach by Labuntsov (2000, p. 104) seems to be the most suitable from a practical viewpoint. Assuming that a commercial heated surface presents cavities of any size that correspond to ordinary observed wall superheats and employing dimensional analysis, one obtains the following correlation for nucleation site density:

$$n_F = C_0 R_*^{-2} \tag{4.5}$$

where the equilibrium radius of a vapour bubble is determined by the Laplace equation:

$$R_* = \frac{2\sigma}{\Delta p} \tag{4.6}$$

C_0 is a numerical factor on the order of magnitude 10^{-8} to 10^{-7}.

Obviously, the heat flux due to liquid evaporation at the boundaries of dry spots can be determined as follows:

$$q_1 = Q_{ds} n_F \tag{4.7}$$

where Q_{ds} is the heat removal rate at a dry spot. This quantity was derived on the basis of an approximate model of the interline region. Using the aforementioned results by Wayner, one can present a schematic of the liquid film meniscus in the vicinity of a dry spot as shown in Figure 4.2. A model of the liquid film at the interline region is inevitably strongly idealised. According to the estimations of Yagov (1988), the liquid film thickness at the zone of intense evaporation, δ_m, is on the order of 10^{-8} to 10^{-7} m at ordinary conditions of nucleate boiling. These estimations coincide with

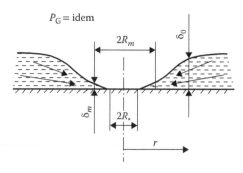

FIGURE 4.2
Schematic of liquid film meniscus at a dry spot boundary.

further numerical simulations. The paper by Stephan and Hammer (1994) was, probably, the first numerical simulation of heat transfer in the so-called 'microregion' that is in the thin liquid film at the boundary of a nucleation site at the heated surface. Later, Stephan and his co-workers further developed this approach (Stephan 2002), in particular, for binary mixture boiling. In the work by Stephan (2002), the maximum local heat flux in propylene/propane mixture boiling is obtained at a distance of about 2×10^{-7} m from the boundary of an adsorbed (non-evaporating) liquid film, where the film thickness is about 1×10^{-8} m. It is clear that for such linear scales any actual heated surface can hardly be considered as the smooth one. The aforementioned scales are less than the height of microroughness in the experiments of Theofanous et al. (2002), where a special nanoscale surface was used. This consideration gives additional arguments in favour of an approximate model of the process.

The approximate model of Yagov (1988) considers the same characteristic regions of the thin liquid film at the vicinity of a dry spot as were later strictly analysed by Stephan and Hammer (1994). However, the aim of the approximate model was not finding the exact solution for some particular conditions but obtaining a general correlation between the controlling parameters and the thermophysical properties of a liquid. According to Wayner's studies, liquid flow in the film is governed mainly by capillary forces, or more exactly by the gradient of the film surface curvature. As vapour pressure is uniform and the meniscus curvature H decreases from its maximal value $1/R_*$ at the interline till $1/\delta_0$ at the outer part of the meniscus, the pressure gradient in the liquid is directed from the dry spot boundary to the thick part of the film. Assuming that the aforementioned curvature variation occurs at the short distance $\Delta r \simeq \delta_m$, one obtains the following equation for the pressure gradient in the liquid film:

$$\frac{dp}{dr} \sim \frac{\sigma}{(R_* \delta_m)}$$

This expression is the only one that is obtained by means of physical estimations. For a known pressure gradient, there exists an analytical solution for radial liquid flow in the film. This solution gives the liquid flow rate per film width unit:

$$G_R = \frac{1}{3} \frac{dp}{dr} \frac{\delta_m^3}{\nu} \sim \frac{\sigma \delta_m^2}{R_* \nu} \tag{4.8}$$

If all the liquid supplied to the interline is evaporated, the heat flux per film width unit corresponds to the aforementioned value G_R:

$$q_R = \lambda \Delta T = G_R h_{LG} \tag{4.9}$$

From Equations 4.8 and 4.9, an important expression for liquid film thickness in the region of strong evaporation follows:

$$\delta_m = C_2 \sqrt{\frac{\lambda \Delta T \nu R_*}{\sigma h_{LG}}} \qquad (4.10)$$

Assuming a linear law of liquid film variation with radial coordinates in the strong evaporation zone, one obtains the heat sink per dry spot:

$$Q_{ds} \sim \lambda \Delta T \delta_m \qquad (4.11)$$

From Equations 4.5, 4.7, 4.10 and 4.11, an equation for the heat flux due to evaporation in the interline region follows:

$$q_1 = C_3 \left(\frac{\lambda \Delta T}{R_*}\right)^{3/2} \left(\frac{\nu}{\sigma h_{LG}}\right)^{1/2} \qquad (4.12)$$

Equation 4.4 was obtained by Labuntsov on the basis of the expressions for isolated bubble growth rate (Equation 4.2) and for nucleation site density (Equation 4.5). It is reasonable to use a more general approach to obtain an equation for the convective constituent of heat flux. According to the schematic in Figure 4.1, the average distance between nucleation sites is obviously

$$l_* = 1/\sqrt{n_F} = C_0^{-1/2} R_*$$

where $C_0^{-1/2} \cong 10^3 - 10^4$.

Assuming that the convective heat flux is determined by the thickness of a conductive liquid layer, Δ_0, which depends on liquid viscosity and characteristic scales of length and velocity, one gets the following expression:

$$\Delta_0 \sim \left(\frac{\nu R_*}{W_0}\right)^{1/2} \qquad (4.13)$$

In Equation 4.13, an analogy with a laminar boundary layer was used, the average distance between nucleation sites and a mean evaporation rate (Equation 4.1) being the length and velocity scales, respectively. This equation is valid for both isolated and coalesced bubble boiling regimes; in the former case, it gives an effective thickness of the conductive liquid layer, and in the latter case, Δ_0 is the liquid macrofilm thickness.

Thus, the convective heat flux in boiling is expressed as follows:

$$q_2 = C_4 \lambda \Delta T \sqrt{\frac{q}{\rho_G \nu R_* h_{LG}}} \qquad (4.14)$$

One can see that this equation transforms into Equation 4.4 if the total heat flux, q, coincides with the convective heat flux and the pressure jump, Δp, at the bubble interface is expressed through ΔT by means of the Clapeyron–Clausius equation.

In the general case, substituting Equations 4.12 and 4.14 into Equation 4.3, one gets a quadratic equation in relation to the total heat flux, q. After solving it, the interpretation of a quantity, R_*, is made. The pressure jump in Equation 4.6, as it was shown in Yagov (1988), in the general case can be presented as follows:

$$\Delta p = \frac{\rho_G h_{LG} \Delta T}{T_s}\left(1 + \frac{h_{LG}\Delta T}{2R_i T_s^2}\right) \tag{4.15}$$

where R_i is a gas constant. It is seen from this equation that at high wall superheats, ΔT, this equation differs essentially from the formula deduced from the Clapeyron–Clausius equation. As convective heat flux q_2 is significant at low reduced pressures, when high wall superheats are observed, in Equation 4.14 R_* is calculated according to Equations 4.6 and 4.15. At high reduced pressures, nucleation site density is very high and a contribution of the first constitutive of heat flux becomes dominant. Under these conditions, the wall superheat is commonly small; therefore, in Equation 4.12 R_* is determined using the Clapeyron–Clausius equation, that is, the expression in brackets in Equation 4.15 is assumed to be equal to unity.

Bearing this consideration in mind, the final equation for nucleate boiling heat transfer can be presented as follows:

$$q = 3.43 \times 10^{-4} \frac{\lambda^2 \Delta T^3}{v\sigma T_s}\left(1 + \frac{h_{LG}\Delta T}{2R_i T_s^2}\right)\left(1 + \sqrt{1 + 800B} + 400B\right) \tag{4.16}$$

where B is a dimensionless parameter expressed as follows:

$$B = \frac{h_{LG}(\rho_G v)^{3/2}}{\sigma(\lambda T_s)^{1/2}} \tag{4.17}$$

Two numerical factors in Equation 4.16 were determined from the best fitting to experimental data.

Figure 4.3 is reproduced from the original paper (Yagov 1988). The calculated curve is built in accordance with Equation 4.16 in the form of dependence of non-dimensional HTC

$$\tilde{\alpha} = \alpha\left(\frac{v\sigma T_s}{q^2\lambda^2}\right)^{1/3}\left(1 + \frac{h_{LG}\Delta T}{2R_i T_s^2}\right)^{-1/3}$$

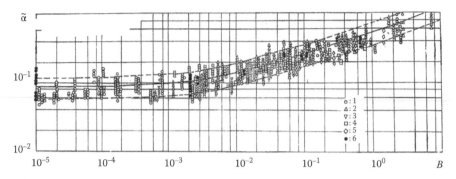

FIGURE 4.3

Comparison of calculations according to Equation 4.16 with experimental data on nucleate boiling heat transfer: (1) water, (2) ethanol, (3) benzene, (4) refrigerants, (5) nitrogen and (6) methane, ethane, ethylene. (Reproduced from Yagov, V.V., *Therm. Eng.*, 35, 65–70, 1988.)

on the non-dimensional parameter B determined according to Equation 4.17. About 3000 experimental points for 12 different liquids (water, ethanol, methane, ethane, ethylene, nitrogen, benzene and different refrigerants) were used for comparison. More than 91% of the points lie within the range ±35% from the calculated line. The parameter B is a strong function of reduced pressure. The experimental data presented in Figure 4.3 relate to reduced pressures at very wide ranges, $p/p_{cr} = 1.8 \times 10^{-4} - 0.94$. As has already been mentioned, when the reduced pressure increases the nucleation site density also increases and the contribution of heat flux q_1 becomes predominant in total heat flux. For a long time after the derivation of Equation 4.16, many new experimental data were compared with prediction calculations in accordance to the equation. In particular, the results for hydrocarbons and fluorocarbons were tested. Certainly in some cases, especially in boiling at heated walls with properties strongly different in comparison with ordinary commercial surfaces, the deviation of experimental and predicted values of HTC essentially occurs. But in general, the comparisons confirmed the rather good predicting capability of the equation.

In the last few years, a substantial upswing in interest was observed in studying the nucleate boiling of carbon dioxide. In one study (Yagov 2009), a comparison of the calculations according to Equation 4.16 with the experimental data of Loebl et al. (2005) on carbon dioxide pool boiling was presented. The experiments were conducted at rather low CO_2 reduced pressures, $p/p_{cr} = 0.1$ and 0.19, if one remembers that the reduced pressure at triple point is 0.07 for this substance. Horizontal tubes of copper, stainless steel and aluminium with an outer diameter of 16 mm and different surface treatments were used as heaters. The experimental points were distributed near the computed curves with scatter that is typical for nucleate boiling, the higher experimental HTCs being observed for surfaces with higher thermal activity and more roughness.

4.2.4 Flow Boiling Heat Transfer

Regularities of nucleate boiling remain the dominant ones at flow boiling under conditions of rather high heat fluxes. As was demonstrated in Yagov (1995), Equation 4.16 describes well the experimental results on heat transfer in the boiling of water and refrigerant R-113 in tubes and rather unique data on the boiling heat transfer of nitrogen in impinging jets. Generally, in flow boiling a total HTC is successfully predicted with the help of a simple interpolation between the HTC values calculated using reliable equations for nucleate boiling (α_b) and for single-phase convection (α_c):

$$\alpha = (\alpha_b^3 + \alpha_c^3)^{1/3} \tag{4.18}$$

HTC at nucleate boiling (α_b) is calculated according to Equation 4.16; at developed turbulent flow, the equation by Petukhov et al. (1986) is recommended for computing HTC at single-phase convection (α_c). When one considers an ordinary boiling curve, a range of heat fluxes, which corresponds to commensurable influences of convective and boiling mechanisms of heat transfer, is very narrow. This is due to the strong dependence of boiling HTC on heat flux. In many cases, in particular, in applications of power engineering, the aforementioned range of regime parameters is not important practically.

According to Labuntsov (2000), interpolating equations similar to Equation 4.18 are valid up to void fraction $\varepsilon \leq 0.7$; at low reduced pressures ($p/p_{cr} < 0.05$), this usually corresponds to small flow qualities ($x < 0.05$). At the same time, in experiments with carbon dioxide nucleate boiling remains a dominant mechanism of heat transfer at $x \leq 0.7 - 0.8$ (Thome and Ribatski 2005, Cheng et al., 2008). In a paper (Yagov and Minko 2011), it was proved that at high reduced pressures, which are a feature of CO_2 flow boiling data, even in annular flow at void fraction $\varepsilon = 0.8 - 0.9$ nucleate boiling occurs inside a thin liquid film. The other peculiarity of the aforementioned experimental data is the rather low level of heat flux densities ($<0.1q_{cr}$ at pool boiling at a given pressure). Under these conditions, the contribution of convective mechanism to total heat transfer is relatively high and can increase with an increase in vapour quality. This means that the convective constituent of HTC has to be modified.

It is well known that in two-phase flow, wall shear stress strongly increases with an increase in flow quality. In homogeneous model approximation, a linear dependence exists between these parameters. It is reasonable to suppose that the intensity of convective heat transfer in two-phase flow also increases with rising vapour quality. At turbulent flow regimes, the interdependence between momentum and energy transfer is determined by the Reynolds analogy. It can be shown that in turbulent flow for liquids with Prandtl number (Pr) ≥ 1, HTC is related to wall shear stress by the following simple correlation:

$$\alpha \sim \tau^{1/2} \tag{4.19}$$

Using the assumption that at high reduced pressures the homogeneous approximation is justified for two-phase flow, one can obtain the following equation for convective heat transfer:

$$\alpha_c = \alpha_0 \left(1 + \frac{x(\rho_L - \rho_G)}{\rho_G} \right)^{1/2} \tag{4.20}$$

where ρ_L is liquid density and α_0 is HTC at single-phase convection at the same mass flux G, which is controlled by $\mathrm{Re}_0 = Gd_h/\mu_L$ and Pr_L. The liquid properties are chosen at the corresponding saturation temperature. The same equation known as the Boyko–Krushilin formula was suggested first for in-tube condensation at turbulent flow regimes.

For $\mathrm{Re}_0 \geq 5000$, α_0 is calculated according to the equation by Petukhov et al.:

$$\mathrm{Nu} = \frac{\mathrm{Re}_0 \, \mathrm{Pr} \, (\xi/8)}{1 + 900/\mathrm{Re}_0 + 12.7\sqrt{\xi/8}(\mathrm{Pr}^{2/3} - 1)} \tag{4.21}$$

where $\mathrm{Nu} = \alpha_0 d_h/\lambda$.

The friction factor is calculated according to the Filonenko formula:

$$\xi = \left(1.82 \lg \mathrm{Re}_0 - 1.64\right)^{-2} \tag{4.22}$$

In the range $2000 \leq \mathrm{Re}_0 \leq 5000$, the modification of Equation 4.21 made by Gnielinski is used.

Thus, convective HTC calculated in accordance with Equations 4.20 through 4.22 is used in Equation 4.18 together with Equation 4.16 to obtain the total HTC in flow boiling at high reduced pressures. Such a calculation method allows the actual increase of HTC with vapour quality that is observed in the experiments at low heat fluxes and high mass flow rates.

In Figure 4.4a, the data of Yun et al. (2005) at low heat fluxes are presented. As shown in Figure 4.4a, essential dependence of HTC on the vapour quality is observed. For this case, $\alpha_b \approx 7.4$ kW/m$^2 \cdot$K. This is close to the experimental values only at low vapour qualities ($x \leq 0.15$); at higher values of x, HTC increases more than two times. Equation 4.18, with Equations 4.20 through 4.22, successfully reflects this effect. Figure 4.4b presents the data of Yun et al. (2005) at the same pressure and channel as Figure 4.4a but at essentially a higher heat flux ($q = 20$ kW/m^2). As a result, the influence of convective contribution in heat transfer is less even at a high mass flow rate, $G = 1500$ kg/m$^2 \cdot$s. A good agreement of the predicted values and the experimental data is observed, with practically all points being in the band $\pm 30\%$ denoted by the dotted lines equidistant from the predicted line.

The vertical dotted lines in Figure 4.4 correspond to a boundary vapour quality, x_b, calculated in accordance with the correlation by Sergeev (2000).

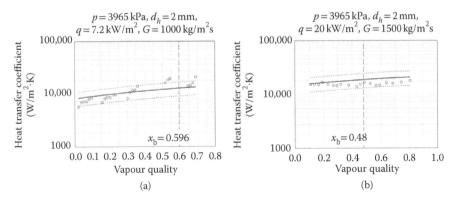

FIGURE 4.4
Heat transfer coefficients at carbon dioxide flow boiling calculated and measured by Yun et al. (2005): (a) low heat flux (b) moderate heat flux.

The correlation was grounded by means of qualitative analysis of main mechanisms in thin liquid films at the annular regime of two-phase flow and justified by wide comparison with, but not only for the water/steam system. The correlation was tested on the data for carbon dioxide, and it was found that it failed to incorporate the actual regularities of the process. This seems to be natural as dry outs in experiments occurred at unusually low heat fluxes, but Sergeev's correlation does not account for the heat flux effect. Qualitative analysis of the data shows that this effect does exist. Besides, some experiments are featured with a gradual rather than an abrupt decrease of HTC with vapour quality, the latter being in contrast to Sergeev's model of dry out.

In the paper by Yagov and Minko (2011), it was shown that Equation 4.18, with Equations 4.16 and 4.20 through 4.22, describes quite satisfactorily not only the experimental data for CO_2 flow boiling in channels 0.6–10 mm in diameter but also the data for water, nitrogen and some refrigerants flow boiling at reduced pressures higher than $p/p_{cr} = 0.2$. It is noteworthy that under the considered conditions, the effect of channel diameter is practically negligible; this is quite natural, as all internal linear scales in two-phase flow at high pressures are much less than the hydraulic diameter of a channel.

4.3 Transition Boiling

4.3.1 Brief Analysis of Experimental Results

Transition (or transient) boiling remains the least examined mode of boiling. This was marked in some very informative review papers (Kalinin et al., 1987, Auracher 1990) more than 20 years ago and may be stated today also. The experiments in transition boiling are difficult enough due to the 'unnatural' dependence of heat flux on wall superheat. Steady regimes of transistion boiling heat transfer of a liquid can be realised if the surface is heated by means of condensation of the vapour of the other substance or by forced convection of liquid metal. Auracher (1990) and his co-workers used the more sophisticated method of transition boiling investigation with an electric current heating the wall under controlled conditions. The wall temperature was controlled by measuring the temperature close to the boiling surface, comparing it with a set point value, feeding the difference signal into the controller and adjusting the power of electric heating according to the controller output (Auracher and Marquardt 2004). Unsteady experiments usually give data on heat transfer that are strongly different in comparison with the results of steady experiments. The large scatter is typical for experimental data in transition boiling.

Nevertheless, now there is enough experimental information to define at least qualitatively the main regularities of transition boiling. In the 1950s, liquid/solid contacts during the process were a subject of much discussion (Westwater 1958). Today, after direct measurements of the liquid–solid contact times there are no doubts not only in relation to their actual existence, but also in relation to their dominant role in heat transfer during transient boiling. Nevertheless, now there is factually no general view on the main regularities of transition boiling heat transfer. It is indicative that in the software for nuclear power station safety for transient boiling heat transfer, a simple interpolation is used between the two points of boiling crises at the $q(\Delta T)$ logarithmic dependence.

In the paper by Auracher and Marquardt (2004), some rather sophisticated experiments with microsensors that lead to experimental $q(\Delta T)$ dependencies along full boiling curves are discussed. These experiments probably offer a chance to develop physically based boiling heat transfer models, which include the phenomena and information explored by the microsensor techniques. In particular, in the study by Auracher and Marquardt (2004), an interfacial area–density model is considered to determine entire boiling curves based on data from optical probes. However, very important information on the internal characteristics of a two-phase system near the heated wall in different modes of boiling (such as void fraction distribution and contact frequency above the surface) cannot in itself lead to the heat transfer model. Besides, transition boiling strongly differs from either nucleate or film boiling regime in practice. Certainly, one can wish that the authors Auracher and Marquardt (2004) are successful in their intention, but even now a rather simple approximate model of transition boiling heat transfer seems to be a more realistic objective.

The experiments on transient boiling of different liquids at sub-atmospheric pressures performed at the end of 1980 in MPEI (National Research University, Moscow, Russia) (Samokhin and Yagov 1988) have stimulated the development of such a model. In the experiments, the boiling surface was heated by means of steam condensation at its finned bottom side; steam was generated in the lower part of the thermosyphon with 2-MPa maximal pressure. This method obtained the boiling curves at steady-state regimes with sections of nucleate and transition boiling. The transition boiling of water, ethanol, isopropanol, *n*-heptane, *n*-hexane and refrigerant R-113 has been studied at pressures of 3–100 kPa. An approximate model of transition boiling heat transfer was presented in Yagov and Samokhin (1990) and later with small corrections in Yagov (1993).

4.3.2 Model of Heat Transfer at a High Jakob Number

In one study (Yagov 1993), a rather exhaustive analysis of the predicting methods for transition boiling heat transfer existing at that time has been given. To the best of the author's knowledge, the state of the art in the field has not essentially changed. This gives the grounding to confine description with the proposed model only and not to dwell on the shortcomings of the previous ones.

The model considers three stages of the process of transient boiling in a chosen area of the heated surface. The first one begins at a moment of liquid/ solid contact; during this stage, a thin liquid layer with a thickness on an order of magnitude of several values of vapour bubble equilibrium radius, R_*, is superheated and plenty of vapour nuclei are activated. Due to high liquid superheat, the born bubbles grow very quickly and coalesce into a rather large vapour volume with the thin liquid film beneath. The duration of this stage of the process is 2–3 orders of magnitude less than the periods of the other two stages. The second stage is a process of liquid macrofilm evaporation; its duration depends on the film thickness and the evaporation rate. This stage determines the liquid/solid contact time, t_L, which is very important. As after full evaporation of the macrofilm pressure in the vapour volume is much greater than that in the surrounding liquid, the large bubble extends (grows) at the surface up to departure; this is the third stage of the process, for which duration is the vapour/solid contact time, t_G. In a study (Pan et al., 1989), the similar three stages of transition boiling are also considered, but their content and contribution in heat transfer is quite different in comparison with our approach (see Yagov and Samokhin [1990] and Yagov [1993] for details). For rather extensive surfaces, it is assumed that a relative fraction of wetted surface (s_L) at a given time moment is equal to a relative liquid/solid time duration, that is,

$$\frac{s_L}{(s_L + s_G)} = \frac{t_L}{(t_L + t_G)}$$

This means that the process is assumed to be an ergodic one.

Now, the aforementioned three stages are considered in detail. At the moment when the liquid touches the superheated wall, all the surface cavities are filled with vapour; so their activation requires only superheating the very thin liquid layer with a thickness of several R_* values, the latter being determined according to Equation 4.6. At high wall superheats, the equilibrium radius of a vapour bubble is very small (usually <1 μm). Estimations show that the time required for the activation of nuclei is on the order of 10^{-6} seconds. One can consider that the activation occurs practically instantly at the moment when the liquid touches the wall. It is easy to understand that the contribution of heat for superheating the thin liquid layer to total heat flux in the consideration is negligible; the opposite conclusion of Pan et al. (1989) is a result of unrealistically using a 'turbulent thermal conductivity' at the wall. The initial growth of a vapour bubble with a size close to R_* is controlled by the inertia effects:

$$R = \sqrt{\frac{2\Delta p}{3\rho_L}} \times t$$

where t is the current growth time and Δp is the difference between the saturated vapour pressure at temperature T_w and the liquid pressure 'in infinity'. At conditions typical for transition boiling, a 10-fold increase in bubble radius from its initial value, R_*, occurs for a time period on the order of 10^{-7} seconds. Assuming that nucleation site density can be estimated in accordance with Equation 4.5 with a slightly higher value of the constant C_0, one can see that plenty of growing bubbles coalesce for a very short time on the order of 10^{-6} seconds. As a result, a large vapour volume (a mushroom-shaped bubble) is formed, and the thin liquid film is restrained beneath this volume (Figure 4.5a). A hemispherical shape is assumed for such a bubble (Figure 4.5b). The average thickness of the restrained liquid film has to be on the order of the size of the vapour bubbles at the moment they coalesce, that is,

$$\delta = C_5 R_* \tag{4.23}$$

The constant C_5 must be on the order of 10–100. Thus, in the proposed model, it is assumed that all the described events that lead to the formation of the mushroom-shaped vapour bubble with the liquid macrofilm underneath it occur immediately at the moment of liquid/solid contact. As mentioned

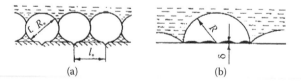

(a)   (b)

FIGURE 4.5
Schematic of (a) vapour bubbles formation at liquid/solid contact and (b) subsequent growth of a hemispherical vapour bubble.

earlier, the duration of the first stage of the considered process is much less than the time length of the other two stages.

During the second stage of the process, after the liquid macrofilm is fully evaporated, much of the heat is removed from the heated wall. Assuming that the hemispherical shape is kept up till full evaporation of the macrofilm is achieved, the bubble radius R_2 at the time length t_L of liquid/solid contact is determined from mass balance:

$$\frac{2}{3}\pi \rho_G R_2^3 = \pi \rho_L R_2^2 \delta$$

Bearing in mind Equation 4.23, this leads to the following equation for R_2:

$$R_2 = C'_5 R_* \frac{\rho_L}{\rho_G} \tag{4.24}$$

The constant C'_5 has the same order of magnitude as C'_5.

At high Jakob numbers $[Ja = \rho_L c_p \, \Delta T/(\rho_G h_{LG})]$, bubble growth rate at the heated surface according to Yagov (2001) is determined as follows:

$$\dot{R} = \frac{3}{8} \frac{R_i^{3/4} T_s^{5/4}}{h_{LG}} \left(\frac{\lambda c_p}{\rho_G} \right)^{1/4} t^{-1/4} \tag{4.25}$$

For obtaining an explicit expression for liquid/solid contact time length, it is necessary to determine the pressure jump in Equation 4.6. In Labuntsov and Yagov (2007), for low reduced pressures and high temperature differences along the saturation line the following equation was proposed:

$$\Delta p = 1.4 \frac{\rho_G h_{LG}^2 (\Delta T)^2}{R_i T_s^3} \tag{4.26}$$

Bearing in mind that at $t = t_L$, $R = R_2$ from Equations 4.6, 4.24 through 4.26 it can be found:

$$t_L = C_6 \left(\frac{\rho_L \sigma}{h_{LG}} \right)^{4/3} \left(\frac{R_i}{\lambda c_p} \right)^{1/3} \left(\frac{T_s}{\rho_G} \right)^{7/3} \frac{1}{(\Delta T)^{8/3}} \tag{4.27}$$

Preliminary estimations give an order of magnitude 10^2–10^3 for the constant C_6. At $C_6 = 500$, Equation 4.27 is compared with the measurement results in Figure 4.6. The experimental data for liquid/solid contact duration at transition boiling of water, ethanol and refrigerant R-113 presented by Kalinin et al. (1987) and the data for water given by Lee et al. (1985)

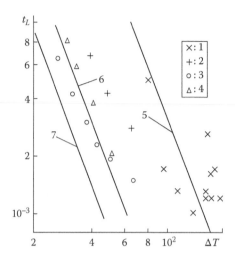

FIGURE 4.6
Time length of liquid/solid contact at transition boiling at atmospheric pressure: (1) water (Lee et al., 1985), (2) water (Kalinin et al., 1987), (3) ethanol (Kalinin et al., 1987), (4) R-113 (Kalinin et al., 1987), (5, 6, 7) calculation in accordance with Equation 4.27 for water, ethanol and R-113, respectively.

are depicted in Figure 4.6. As shown in Figure 4.6, Equation 4.27 correctly reflects the strong dependence of t_L on wall superheat and in general manifests satisfactory agreement with the data. One has to bear in mind the rather wide scatter of experimental data in transition boiling.

The third stage of the process is caused by the elevated vapour pressure in a bubble in comparison with outside liquid pressure after the full evaporation of the liquid film. The high growth rate of the bubble leads to essential excess pressure in the liquid at the bubble interface and consequently inside the bubble. In one study (Yagov 1993), the excess pressure was estimated for water boiling at atmospheric pressure at $\Delta T = 100$ K based on Equation 4.25: at $t = t_L \cdot \Delta p = \Delta p_2 = 1.2 \times 10^4$ Pa. This means that during the third stage adiabatic expansion of vapour bubble has to occur. In some studies (Yagov and Samokhin 1990, Yagov 1993), a solution for underwater explosion was used for this process. In this case, bubble radius growth is determined by the exponent law: $R \sim t^{2/5}$. The excess pressure decreases during bubble expansion and vanishes when the bubble growth rate falls to a terminal rising velocity in the quiescent liquid. This determines the conditions of detachment of a bubble from the wall, ending the third stage of the process. An analysis leads to an equation for vapour/solid contact duration and gives for a ratio of liquid/solid and vapour/solid contact times the following equation:

$$\frac{t_L}{t_G} = \frac{500(\rho_L \sigma T_s)^{25/24} g^{5/8}}{(\lambda c_p h_{LG})^{5/12} \rho_G^{5/3} R_i^{5/24} (\Delta T)^{25/12}} \tag{4.28}$$

Coincidence of constants in Equations 4.27 and 4.28 means that a numerical factor in the expression for t_G was postulated equal to unity. The latter equation allows checking the model by means of comparing the calculations with the data of Dhuga and Winterton (1985, 1986) and Rajabi and Winterton (1988). In these studies, experimental data on a fraction of wetted surface in methanol transition boiling at atmospheric pressure are presented; the results of Dhuga and Winterton (1985, 1986) have been obtained for the cooling process, whereas the data of Rajabi and Winterton (1988) relate to steady conditions. The fraction of wetted surface for an ergodic process obviously can be expressed through the ratio of liquid/solid and vapour/solid contact times as follows:

$$\gamma = \frac{t_L}{(t_L + t_G)} = \frac{(t_L/t_G)}{(1 + t_L/t_G)} \tag{4.29}$$

The comparison is presented in Figure 4.7. One can see reasonable agreement between the data and the calculation results. Thus, the proposed approximate model appears to be apt to describe the internal characteristics of transient boiling of different liquids, the experimental data (Kalinin et al., 1987, Lee et al., 1985, Dhuga and Winterton 1985, 1986, Rajabi and

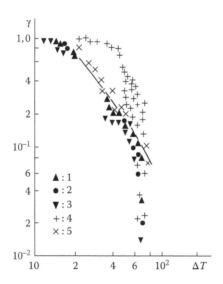

FIGURE 4.7
Variation of a fraction of wetted surface in methanol transition boiling with wall superheat: (1—3) three different runs at steady conditions (Rajabi and Winterton 1988); (4 and 5) cooling processes in the study by Dhuga and Winterton (1985) and Dhuga and Winterton (1986), respectively; (line) calculation in accordance with Equations 4.28 and 4.29.

Winterton 1988) being obtained by different scientific teams. Factually only one empirical numerical factor is used in Equations 4.27 through 4.29. This gives some arguments in favour of the proposed heat transfer model.

By neglecting the heat removal during vapour/solid contact, it is easy to calculate the average heat flux in transition boiling using the aforementioned equations for characteristic time lengths of the process. Obviously, this heat flux is determined by the energy consumed for liquid film evaporation, the thickness of the film being given by Equation 4.23. So one gets the following average heat flux:

$$q = \frac{C_5 \rho_L h_{LG} R_*}{(t_L + t_G)}$$

Using Equation 4.6 and Equations 4.26 through 4.29 in this expression, one finally gets

$$q = 32 \frac{(\rho_L \sigma)^{17/24} R_i^{11/24} T_s^{41/24} g^{5/8}}{\left(\lambda c_p h_{LG}\right)^{1/12} \rho_G^{1/3} (\Delta T)^{17/12} (1 + (t_L/t_G))} \tag{4.30}$$

Equation 4.30 is written with the definite numerical constant being equal to 32, which was fitted to available experimental data. Its magnitude order was correctly predicted during the analysis; this was detailed in Yagov and Samokhin (1990). The other empirical constant is incorporated in the ratio t_L/t_G in the last multiplier in the denominator of Equation 4.30. But below it will be shown that in many cases $(1 + t_L/t_G)$ can be equated to unity. Figure 4.8 presents a comparison of calculation according to Equation 4.30 with the experimental data. The presentation form is adopted from Auracher (1990); from this study, the experimental data of Weber and Johansen (1990) on water transient boiling at forced flow with a mass flow rate of 0.136 kg/m²·s are also used.

As shown in Figure 4.8, there is good agreement between the data and the calculations according to Equation 4.30. Earlier (Yagov and Samokhin 1990), the experimental data of Samokhin and Yagov (1988) on transition boiling at sub-atmospheric pressures were used to determine a numerical factor in Equation 4.30. During computing, it was assumed that the relative time length of liquid/solid contact was $t_L/t_G \ll 1$, so that $(1 + t_L/t_G) \cong 1$. Strictly speaking, this means that Equation 4.30 is valid mainly in the region of transition boiling that is close to the point of minimum heat flux on the boiling curve. The model used an assumption of high Jakob numbers, which allows the application of bubble growth law, Equation 4.25. Practically, Equation 4.25 can be used at $Ja \geq 200$.

The results presented in Figure 4.9 additionally shed light on the problem. The curves 4 for methanol and 5 for water are built according to the

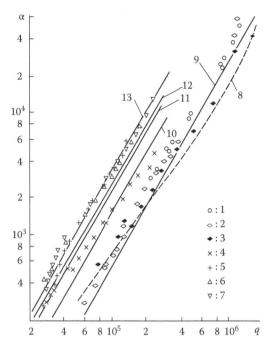

FIGURE 4.8
Heat transfer coefficient at transition boiling: (1—3) water at $p = 0.1$ MPa (Westwater 1958, Abbasi et al., 1989, Auracher 1990); (4) methanol at $p = 0.1$ MPa (Westwater 1958); (5, 6) R-113 at $p = 0.05$, 0.1 MPa; (7) R-114 at $p = 0.3$ MPa, correspondingly (Hess 1973); (8) calculation for water at $p = 0.1$ MPa on the empirical formula of Abbasi et al. (1989); (9—13) calculation on Equation 4.30 for water, methanol and R-113 at $p = 0.05$ MPa, 0.1 MPa, and R-114 at $p = 0.3$ MPa, correspondingly.

calculations of Equation 4.30 at the condition $t_L/t_G = 0$, whereas the curves 6 and 7 correspond to calculations including the actual values of the relative time length of liquid/solid contact, t_L/t_G, in accordance with Equation 4.28. At high wall superheats both pairs of curves merge, but at low ΔT the difference between the pairs becomes essential. As it is seen from Figure 4.8, calculation using the actual value of t_L/t_G leads to better agreement with experimental data, at least, for water transition boiling. The data for methanol manifest the essential difference between steady-state and unsteady transition boiling. In this particular case, the unsteady data are in better agreement with the prediction.

In general, it is possible to state that the proposed model leads to a reasonably accurate predicting equation for heat transfer at transition boiling at low reduced pressures. The applicability of Equations 4.28 and 4.30 is determined by the condition $Ja \geq 200$, but actually the agreement between calculations and the data is good even at $Ja \sim 100$.

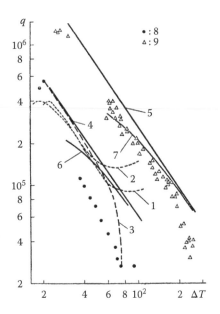

FIGURE 4.9
Heat transfer in transition boiling of methanol and water at atmospheric pressure: (1, 2) smoothed curves of two runs of unsteady cooling of the surface in methanol (Abbasi et al., 1989); (3) calculation using experimental data on a fraction of wetted surface γ for steady transition boiling of methanol (Rajabi and Winterton 1988); (4, 5) calculation for methanol and water on Equation 4.30 at $t_L/t_G = 0$; (6, 7) the same calculations with t_L/t_G computed in accordance with Equation 4.28; (8, 9) experimental data for methanol (Rajabi and Winterton 1988) and for water (Abbasi et al., 1989) at steady boiling.

4.4 Film Boiling of Subcooled Liquids

4.4.1 General Characteristics of Film Boiling

In the past, film boiling of subcooled liquids did not much attract the attention of investigators. For example, Ametistov et al. (1995) do not discuss influence of liquid subcooling on film boiling. The studies (Aziz et al., 1986, Corell and Kenning 1988, Zvirin et al., 1990) mentioned in the beginning of this chapter are cited rather seldom. In a recent study (Bolukbasi and Ciloglu 2007), the liquid subcooling effect on film boiling heat transfer was analysed only for low subcooling values; the authors believe that at $\Delta T_{sub} \geq 40$ K, film boiling does not exist. This conclusion is based on the observation of a very fast cooling process at high subcoolings. However, an intensive cooling process in film boiling of highly subcooled liquids presents the most interesting and practically important phenomenon.

In our research team film boiling of subcooled liquids have been studied since 2007. The schematic of the experimental facility is depicted in

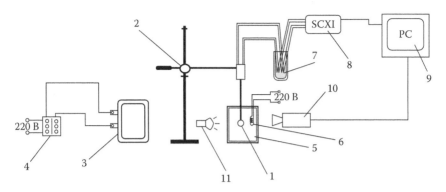

FIGURE 4.10
Schematic of the experimental set-up: (1) test section, (2) rotary-and-fixing device, (3) electric furnace, (4) laboratory autotransformer, (5) thermostat tank, (6) electric heater, (7) Dewar vessel, (8) measurement module, (9) personal computer, (10) high-speed digital camera, (11) light system.

Figure 4.10. Nickel balls of 30- and 45-mm diameter were used as high-temperature bodies. The cooling liquids used were pure water, isopropanol (propanol-2) and their binary mixtures. Each ball was outfitted with several sheathed chromel–alumel microthermocouples: one of them was embedded in the centre of the sphere and three or four thermocouples were mounted on the surface. During an experimental run, a ball was heated in the electric furnace to a temperature much higher than the limiting temperature of liquid state (for water, the initial temperature of the ball was 650°C–720°C). Then the ball was removed by means of the rotary-and-fixing device into the thermostat filled with the cooling liquid at the pre-established temperature. The readings of the thermocouples were registered by a module, NI SCXI-1102, with a frequency of about 100 Hz and were transferred to a personal computer (PC) with a USB interface. The cooling process in the thermostat was filmed using a high-speed digital camera, VSFAST. A detailed description of the experimental procedure and the data treatment is given by Lexin et al. (2009, 2011).

The fragment of video shown in Figure 4.11 gives important information on the cooling process in highly subcooled water ($\Delta T_{sub} \approx 87$ K). The 12 consecutive frames shown in Figure 4.11 cover approximately 3 seconds. The first three frames (the upper row) refer to the regime of stable film boiling. The fourth frame (first frame in the second row) shows the beginning of vapour film surface perturbation. This perturbation grows rapidly: as far as one can estimate using the images, the vapour film disappears for 1 second (four frames). During this time, the surface temperature of the ball decreases from 600°C to 200°C. A rough estimation on the basis of lumped heat capacity gives a heat flux as high as 8–9 MW/m² at the ball surface. It is clear that most of this period corresponds to film boiling of the subcooled liquid as the ball surface temperature exceeds the critical one of water (~374°C).

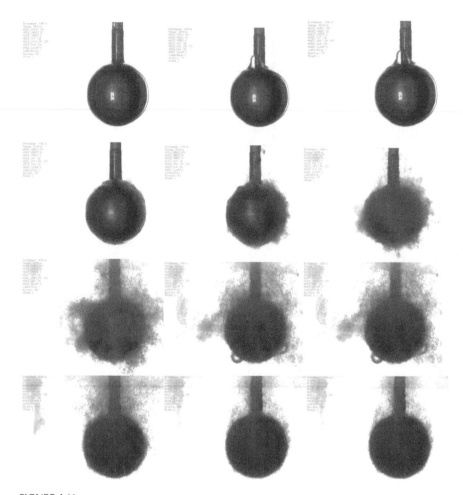

FIGURE 4.11

Visualisation of pool boiling process on the surface of a nickel ball 30 mm in diameter in 12.5°C water (time intervals between the frames are 250 ms).

The remaining frames in Figure 4.11 refer to transient and nucleate boiling. In experiments with significant water subcoolings ($\Delta T_{sub} \geq 50$ K), stable film boiling during the initial period of cooling occurs without the departure of vapour bubbles from the surface of the film. With more subcooling, a higher temperature of intensive cooling process incipience was observed.

Quite a different picture is observed during cooling in saturated or slightly subcooled ($\Delta T_{sub} \leq 5$ K) water. Three typical frames of a video of this process are presented in Figure 4.12. The main part of cooling in this case is film boiling; usually it lasts more than 2 minutes. Figure 4.12a corresponds to film boiling; a stable and rather smooth vapour film on the ball surface can be observed. In the upper part of the ball, near the holder, steam bubbles were formed. The cooling rate in this regime is rather low, approximately

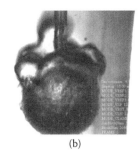
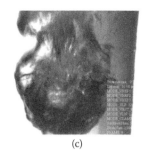

(a)   (b)   (c)

FIGURE 4.12
Typical pictures of boiling at the surface of a nickel ball 30 mm in diameter during cooling in saturated water (a) film boiling (b) transition boiling (c) nucleate boiling.

3.5–4 K/s; the surface heat flux is less than 100 kW/m². When the wall temperature falls below 220°C–250°C the vapour film is strongly disturbed, as shown in Figure 4.12b; this obviously corresponds to the beginning of transition boiling. The cooling rate at this moment increases 2–3 times. Figure 4.12c is a typical view of the cooling process in nucleate boiling. The time length of cooling in transition and nucleate boiling is commonly about 5 seconds.

Cooling in saturated liquids does not give any new information on film boiling; HTCs in this case appear to be in good agreement with calculations according to well-known equations discussed particularly by Ametistov et al. (1995). The main attention in our studies was on the intensive cooling process in highly subcooled liquids.

4.4.2 New Experimental Results on Cooling of High-Temperature Bodies

Basic quantitative results have been obtained with a nickel ball 45 mm in diameter. The junctions of the thermocouples were silver brazed in the centre and at the four points of the ball surface. Each thermocouple cable was mounted in a groove milled on the wall surface; the grooves then were welded with nickel, and the wall surface was carefully treated to restore its sphericity. The results of the primary measurements are saved in PC memory as numerical values of five thermocouple readings at consecutive time moments through an interval of 0.01 second. Figure 4.13 presents a typical picture of one experimental run of the ball cooling in water at 50°C. Every run was repeated under identical conditions; in general, a good reproducibility of the thermograms was observed. Some occasional oscillations of the curves, which are visible in Figure 4.13, were also fixed in the measurements. As thermocouple junctions undergo cyclic heating to very high temperatures and fast cooling, they can be destroyed. This limited the number of repeated experimental runs. Nevertheless, after 5–6 reproductions of the same regime, smoothed, averaged thermograms can be obtained for each thermocouple. These averaged thermograms are, of course, quite similar to the curves shown in Figure 4.13.

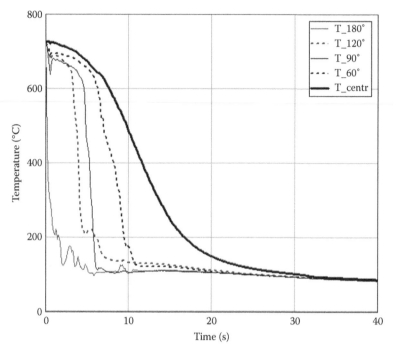

FIGURE 4.13

Typical experimental thermograms of cooling a 45 mm nickel ball in subcooled water ($\Delta T_{sub} = 50$ K; the polar angle for the location of thermocouples is accounted from the top point of the ball).

The discussed regimes are characterised by an extremely high rate of surface cooling. From the analysis of individual thermograms, it was established that in distinct experimental runs the rate of decrease in temperature can be higher than 1000 K/s. As shown in Figure 4.13, for the 45-mm ball, significant non-uniformity of cooling conditions at the surface was revealed. The lower part of the ball surface is cooled earlier and more intensively. A front of fast cooling of the surface then moves upwards, being replaced to an upper thermocouple location for approximately 6 seconds. Naturally, an essential delay time characterises the cooling rate at the ball's centre. As is clear from Figure 4.13, the surface temperature during most of the intensive cooling process remains markedly higher than the superheat-limit temperature of liquids so that the process is controlled by film boiling.

For understanding the main mechanisms of film boiling of subcooled liquids, it seems important to study the regularities of cooling high-temperature bodies in non-aqueous liquids. At atmospheric pressure, a choice of possible coolants is rather scarce due to limitations on chemical stability. In our studies (Lexin et al., 2009, 2011), experimental results with propanol-2 are presented. Its normal boiling point is 82.2°C and its critical

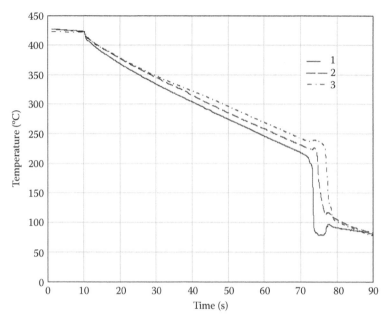

FIGURE 4.14

Thermograms of a 45 mm nickel ball cooling in pure isopropanol at −14°C (ΔT_{sub} = 96 K): (1, 2) surface temperatures at polar angles θ = 120°, 90°, respectively; (3) temperature of the ball centre.

point is 253.6°C. So an initial safe temperature of the ball of 420°C–430°C provides a sufficient range of wall superheats corresponding to film boiling. Unexpectedly, the ball cooling in isopropanol did not include a mode with high intensity of heat transfer in film boiling even at very high subcoolings (up to 96 K). Typical thermograms of cooling of the 45-mm ball at an isopropanol temperature of −14°C are shown in Figure 4.14. They are quite similar to the cooling process in saturated water. A stable regime of film boiling takes longer than 60 seconds; intensive cooling begins only at a surface temperature below 230°C, that is, in transition boiling.

This result stimulated several experiments with water–isopropanol mixtures of different compositions. These experiments revealed that a 50% (weight per cent) mixture approximately presents a boundary between different types of cooling regimes. At isopropanol fractions higher than this boundary, intensive heat transfer in film boiling of subcooled liquids was not observed. (All experiments with mixtures were conducted at practically the same subcooling, ΔT_{sub} = 50 K). In mixtures with isopropanol fractions less than 50%, the fast cooling regime began at a surface temperature, T_w, higher than 350°C, sometimes at the very moment of the ball merging with the liquid.

For quantitative analysis of experimental results, it is necessary to know the heat flux at the ball surface. To determine the heat flux based on the experimental thermograms, one has to solve an inverse heat conduction problem (IHCP). It is well known that the solution of ill-posed (inverse) problems cannot be rigorous. Factually, the estimation of unknown boundary conditions is realised. We used a method based on the numerical solution of the direct unsteady heat conduction problem in two different variants. In the case of uniform heat transfer at the ball surface during the cooling process, a mathematical description of the problem is a spherically symmetric non-steady heat conduction equation. This is the case of cooling in water with subcooling, $\Delta T_{sub} < 20$ K, or in pure isopropanol; the difference in surface temperatures measured by different thermocouples at the same moment did not exceed 10 K in such regimes. The one-dimensional non-steady equation was solved with the so called third kind boundary conditions, that is, at a predetermined HTC. In the first step, a uniform temperature field is assumed in the ball. Then the temperature field inside the ball for some period of time is calculated. The calculation results are checked against the corresponding thermocouple readings; by means of a proper HTC, a choice in interactive regime coincidence of measured and calculated values of temperature is provided. In further calculation steps, the temperature field found in the previous time step is used as the initial one for the following time step. In this way, HTC is determined for the entire cooling process. This method gives good results while spherical symmetry of the temperature field is obeyed.

At strong non-uniformity of heat transfer at the ball surface, such as the one demonstrated in Figure 4.13, the two-dimensional unsteady heat conduction equation was solved. An axial symmetry of the cooling process was assumed in this case, that is, the surface temperature is supposed to be uniform along any circumference at the same polar angle at any time moment. As shown in Figure 4.13, very high temperature gradients on a polar coordinate are observed: whereas the bottom point of the ball is already cooled to about 200°C, the upper part of its surface maintains a temperature of about 650°C. As for the solution itself, the direct axisymmetric unsteady heat conduction problem is solved numerically with the first kind boundary conditions. Based on the averaged experimental thermograms for surface thermocouples, an analytic function was chosen that describes the surface temperature distribution during the cooling process. The hyperbolic tangent on the Fourier number combined with the trigonometric function of a polar angle describes satisfactorily the experimental thermograms. An independent comparison of the measured and calculated thermograms for the ball centre was used as evidence of solution reliability. The aforementioned method of IHCP solution was successfully applied for cooling regimes of the ball in water with subcoolings 50 and 70 K. Results of the calculation of heat flux in different points of the ball surface depending on time are presented in Figure 4.15. Maximal heat fluxes of 2.2×10^6, 3.9×10^6, 4.7×10^6 and 7.4×10^6 W/m^2 were obtained for polar angle values of 60°, 90°, 120° and 180°, respectively.

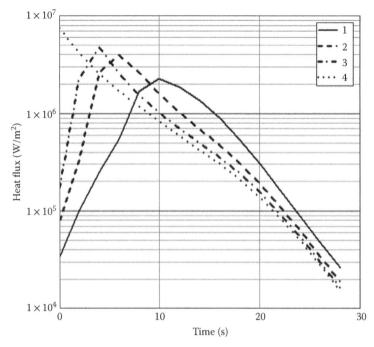

FIGURE 4.15

Heat flux on the surface of a 45 mm nickel ball during cooling in water at 50°C: 1, 2, 3 and 4 denote heat fluxes at $\theta = 60°$, $90°$, $120°$ and $180°$, respectively.

Comparing the results presented in Figures 4.13 and 4.15, it is possible to estimate HTCs in film boiling of subcooled water. At the initial period of cooling (less than 6 seconds), this gives 10–20 kW/(m²·K). This is almost two orders of magnitude higher than that in film boiling of saturated liquids. As the wall temperature significantly exceeds the superheat-limit temperature of liquid, such heat transfer is provided with steam conductivity. The average thickness of a vapour film must be less than 5 μ. A probable mechanism of such intensive heat transfer can be connected with the cavitational collapse of tiny steam bubbles in subcooled liquids near the interface. The high pressure developed in the surrounding liquid close to a bubble surface during cavitation can strongly disturb the film surface and create plenty of areas with very thin vapour film. Small steam bubbles are probably formed at the film surface in areas where the film is extremely thin and heat flux from the wall is very high. Steam microbubbles near the interface were observed in film boiling of subcooled water by Aziz et al. (1986); they are also visible in the frames of video shown in Figure 4.11. Estimation for water has shown that this explanation of the process can be quite realistic. However, unexpected experimental results of the ball cooling in highly subcooled isopropanol do not allow one to build up any quantitative theory of the process.

4.5 Concluding Remarks

It is commonly accepted that high level of heat transfer intensity is determined by means of direct liquid/solid contact. However, now it is clear and is convincingly proved in this chapter that in nucleate boiling this condition is not sufficient. The extremely high values of HTC in nucleate boiling at high reduced pressures cannot be explained by any mechanism of single-phase convection between a heated wall and a liquid. The only mechanism that can explain the extremely high heat transfer intensity in the process discussed is evaporation in the vicinity of interlines, that is, the boundaries of contact of three phases. This peculiarity of nucleate boiling distinguishes it from all other modes of convective heat transfer and can be considered as its generic feature. Based on this idea, an approximate theory of nucleate boiling heat transfer has been developed by the author. The predicting equation following from the theory includes two numerical constants fitted to the experimental data. This equation manifests a good agreement of calculations with the experimental data on nucleate boiling heat transfer of different liquids at a wide range of reduced pressures. On the basis of this equation, a method of calculation of heat transfer in flow boiling at high and moderate reduced pressures has been built up. It was also successfully used in calculations of heat transfer in nucleate boiling of binary mixtures.

The aforementioned analysis of nucleate boiling showed that an equilibrium radius of vapour bubble (R_*) gives the linear scale crucial for the process. This scale obviously determines nucleation site density; it presents a minimal physical size in boiling, in particular, a minimal curvature radius of liquid film surface in the vicinity of a nucleation site. The proposed model of nucleate boiling involved this scale in determining both the constituents of heat flux (Equations 4.12 and 4.14). It is important that R_* is an internal parameter of boiling, which does not depend on any external hydrodynamics.

In transition boiling, the aforementioned linear scale is used as the determining one. A liquid film that remains according to the developed model under a vapour bubble growing after the liquid/solid initial touch at the heated wall is assumed to have a thickness of the order of magnitude of several R_* values. This assumption leads to a reasonable estimation of the time length of liquid/solid contact and finally obtains the predicting equation for transition boiling heat transfer with accuracy to two numerical constants.

As for film boiling of highly subcooled liquids, this process now presents one of the challenges in boiling theory. The experimental data presented in this chapter confirm that high intensity of heat transfer does exist under conditions when direct liquid/solid contact is impossible. For rather large bodies, even those with spherical symmetry (balls), the cooling process is found to be non-symmetric. To explain the extremely high HTCs in film boiling of subcooled liquids, one has to suppose that some effects force a vapour film to become very thin, at least periodically. The cavitational collapse of tiny vapour bubbles near the interface can be considered as one of the possible effects.

References

Abbasi, A., Rajabi, A.A.A., Winterton, R.H.S. 1989. Effect of confined geometry on pool boiling at high temperature. *Exp. Therm. Fluid Sci.* 2: 127–133.

Ametistov, E.V., Klimenko, V.V., Pavlov, Y.M. 1995. *Boiling of cryogenic liquids.* Energoatomizdat, Moscow. p. 400 (in Russian).

Auracher, H. 1990. Transition boiling. Proc. 9th Int. Heat Transfer Conf. Jerusalem, vol. 1, pp. 69–90.

Auracher, H., Marquardt, W. 2004. Heat transfer characteristics and mechanisms along entire boiling curves under steady-state and transient conditions. *Int. J. Heat Fluid Fl.* 25: 223–242.

Aziz, S., Hewitt, G.F., Kenning, D.B.R. 1986. Heat transfer regimes in forced-convection film boiling on spheres. Proc. 8th Int. Heat Transfer Conf. San Francisco, vol. 5, pp. 2149–2154.

Bolukbasi, A., Ciloglu, D. 2007. Investigation of heat transfer by means of pool film boiling on vertical cylinders in gravity. *J. Heat Mass Tran.* 44: 141–148.

Borishanskiy, V.M., Bobrovich, G.I., Minchenko, F.P. 1961. Heat transfer in pool boiling of water and ethanol on the outside surface of tubes, in: *Issues of Heat Transfer and Hydraulics of Two-Phase Media*, edited by S.S. Kutateladze, Gosenergoizdat, Leningrad, pp. 75–93 (in Russian).

Cheng, L., Ribatski, G., Thome, J.R. 2008. New prediction methods for CO_2 evaporation inside tubes: Part II – An updated general flow boiling heat transfer model based on flow patterns. *Int. J. Heat Mass Tran.* 51: 125–135.

Corell, S.J., Kenning, D.B.R. 1988. Film boiling on a molten brass sphere in flowing water. Proc. 2nd UK Nat. Heat Transfer Conf. Glasgow, pp. 1557–1564.

Dhuga, D.S., Winterton, R.H.S. 1985. Measurement of surface contact in transition boiling. *Int. J. Heat Mass Tran.* 28: 1869–1880.

Dhuga, D.S., Winterton, R.H.S. 1986. Measurement of liquid-solid contact in boiling. *J. Phys. E: Sci. Instrum.* 19: 69–75.

Hess, G. 1973. Heat transfer at nucleate boiling, maximum heat flux and transition boiling. *Int. J. Heat Mass Tran.* 16: 1611–1627.

Kalinin, E.K., Berlin, I.I., Kostyuk, V.V. 1987. Transition boiling heat transfer. *Adv. Heat Tran.* 18: 241–323.

Kotthoff, S., Gorenflo, D., Danger, E., Luke, A. 2006. Heat transfer and bubble formation in pool boiling. Effect of basic surface modifications for heat transfer enhancement. *Int. J. Therm. Sci.* 45: 217–236.

Labuntsov, D.A. 2000. *Physical fundamentals of power engineering. Selected works on heat transfer, hydrodynamics and thermodynamics.* MPEI Publisher, Moscow (in Russian).

Labuntsov, D.A., Yagov, V.V. 2007. *Mechanics of two-phase media.* MPEI Publisher, Moscow (in Russian).

Lee, L.Y.M., Chen, J.C., Nelson, R.A. 1985. Liquid-solid contact measurements using a surface thermocouple temperature probe in atmospheric pool boiling water. *Int. J. Heat Mass Tran.* 28: 1415–1423.

Lexin, M.A., Yagov, V.V., Varava, A.N. 2009. Experimental study of heat transfer under conditions of cooling of metal ball. *J. MPEI.* 2: 28–34 (in Russian).

Lexin, M.A., Yagov, V.V., Pavlov, P.A., Zabirov, A.R. 2011. Experimental study of heat transfer at cooling of high-temperature bodies in subcooled liquids. Proc. of 6th Int. Conf. on Transport Phenomena in Multiphase Systems, Ryn, Poland, pp. 301–306.

Loebl, S., Kraus, W.E., Quack, H. 2005. Pool boiling heat transfer of carbon dioxide on a horizontal tube. *Int. J. Refrig.* 28: 1196–1204.

Moore, F.D., Mesler, R.B. 1961. The measurement of rapid surface fluctuations during nucleate boiling of water. *AIChE J.* 7: 620–624.

Pan, C., Hwang, J.Y., Lin, T.L. 1989. The mechanism of heat transfer in transition boiling. *Int. J. Heat Mass Tran.* 32: 1337–1349.

Petukhov, B.S., Genin, L.G., Kovalev, S.A. 1986. *Heat transfer in nuclear power plants.* Energoatomizdat, Moscow (in Russian).

Rajabi, A.A.A., Winterton, R.H.S. 1988. Liquid-solid contact in steady-state transition boiling. *Int. J. Heat Fluid Fl.* 9: 215–219.

Samokhin, G.I., Yagov, V.V. 1988. Heat transfer and critical heat fluxes in boiling of liquids under low reduced pressures. *Therm. Eng.* 35. (*Teploenergetika*, 1988, 2: 72–74, in Russian).

Sergeev, V.V. 2000. Generalization of the experimental data on boiling crisis at upward water flow in channels. *Therm. Eng.* 47: 205.

Stephan, P. 2002. Microscale evaporative heat transfer: modelling and experimental validation, in: *Proc. 12th IHTC*, Grenoble, France.

Stephan, P., Hammer, J. 1994. A new model for nucleate boiling heat transfer. Heat and Mass Transfer 30: 119–125.

Straub, J. 2001. Boiling heat transfer and bubble dynamics in microgravity. *Adv. Heat Tran.* 35: 57–172.

Theofanous, T.G., Dinh, T.N., Tu, J.P., Dinh, A.T. 2002. The boiling crisis phenomenon. Part I: Nucleation and nucleate boiling heat transfer. *Exp. Therm. Fluid Sci.* 26: 775–792.

Thome, J.R., Ribatski, G. 2005. State-of-the-art of two-phase flow and flow boiling heat transfer and pressure drop of CO_2 in macro- and micro-channels. *Int. J. Refrigeration.* 28: 1149–1168.

Verkin, B.I., Kirichenko, Y.A., Rusanov, K.V. 1987. *Heat transfer in cryogenic liquids boiling.* Naukova Dumka, Kiev (in Russian).

Wayner, P.C. Jr. 1973. Fluid flow in the interline region of the evaporating non-zero contact angle meniscus. *Int. J. Heat Mass Tran.* 16: 1777–1783.

Wayner, P.C. Jr. 1978. A constant heat flux model of the evaporating interline region. *Int. J. Heat Mass Tran.* 21: 362–364.

Wayner, P.C. Jr., Kao, J.K., LaCroix, L.V. 1976. The interline heat transfer coefficient on an evaporating wetting film. *Int. J. Heat Mass Tran.* 19: 487–492.

Weber, P., Johansen, K. 1990. Convective transition boiling of water at medium pressure. Proc. of the 9th Int Heat Transfer Conf. Jerusalem, vol. 6, pp. 35–40.

Westwater, J.W. 1958. Boiling of liquids. *Adv. Chem. Eng.* 2: 1–31.

Yagov, V.V. 1988. Heat transfer with developed nucleate boiling of liquids. *Therm. Eng.* 35: 65–70.

Yagov, V.V. 1993. Mechanism of transition boiling of liquids. *J. Eng Phys Thermophys.* 64: 740–751 (in Russian).

Yagov, V.V. 1995. The principal mechanisms for boiling contribution in flow boiling heat transfer, in: J.C. Chen (Ed.), *Convective Flow Boiling*, Taylor & Francis, Washington, DC, pp. 175–180.

Yagov, V.V. 2001. Bubble growth rate at pool boiling in wide range of reduced pressures. *Int. J. Heat Technol.* 2: 17–24.

Yagov, V.V. 2009. Generic features and puzzles of nucleate boiling. *Int. J. Heat Mass Tran.* 52: 5241–5249.

Yagov, V.V., Samokhin, G.I. 1990. Heat transfer mechanism in transition boiling of liquids. *Therm. Eng.* 37. (*Teploenergetika*, 1990, 10: 16–20, in Russian).

Yagov, V.V., Minko, M.V. 2011. Heat transfer in two-phase flow at high reduced pressures. *Therm. Eng.* 58: 283–294.

Yun, R., Kim, Y., Kim, M.S. 2005. Flow boiling heat transfer of carbon dioxide in horizontal mini tubes. *Int. J. Heat Fluid Fl.* 26: 801–809.

Zvirin, Y., Hewitt, G.F., Kenning, D.B.R. 1990. Boiling on free-falling spheres: drag and heat transfer coefficients. *Exp. Heat Tran.* 3: 185–214.

5

A Review of Practical Applications of Heat Pipes and Innovative Application of Opportunities for Global Warming

M. Mochizuki, A. Akbarzadeh and T. Nguyen

CONTENTS

5.1 Introduction 146
5.2 Heat Pipe Revolution 148
5.3 Heat Pipe Applications in Computers and Electronics 149
    5.3.1 Formulation 149
    5.3.2 Examples of Notebook Thermal Solutions 151
        5.3.2.1 Hybrid System 151
        5.3.2.2 Remote Heat Exchanger 152
        5.3.2.3 Vapour Chamber 152
    5.3.3 Examples of Desktop Thermal Solutions 155
    5.3.4 Thermal Solution for Graphics Processors 156
    5.3.5 Thermal Solution Using a Piezo Fan 158
5.4 Thermal Solution Improvement 161
    5.4.1 Improve Heat Spreading from the Processor Die 161
    5.4.2 Remove Integrated Heat Spreader 164
    5.4.3 Heat Pipe Improvement 164
    5.4.4 Vapour Chamber Improvement 165
    5.4.5 Micro-Channel Two-Phase Pump Loop 166
5.5 Clean Energies 167
    5.5.1 Case Study 1: A Heat Pipe Cold Storage System for
        Cooling Data Centres 167
        5.5.1.1 Ice Storage System to Support Main Cooling
            System Failure 169
        5.5.1.2 Pre-Cooler for Chiller Inlet Water 171
        5.5.1.3 Experimental Study: Proof of Concept 173
    5.5.2 Case Study 2: Cold Energy Storage for Agricultural
        Products 175
    5.5.3 Case Study 3: Heat Pipe and Phase Change Material for
        Cooling Concentrated Photovoltaic Technology 177

     5.5.4   Case Study 4: Large-Scale Loop Heat Pipe for Geothermal
           Heat Extraction... 179
     5.5.5   Case Study 5: Bakery Waste Heat Recovery 180
     5.5.6   Case Study 6: Heat Pipe Turbine 182
     5.5.7   Case Study 7: Thermosyphon-Based Thermoelectric
           Generation System for a Solar Pond.. 183
     5.5.8   Case Study 8: Concentrated Solar Thermoelectric
           Generator with PCM Thermal Storage 189
     5.5.9   Case Study 9: Ultra-Large Heat Pipes for Cooling the Earth .. 191
     5.5.10 Case Study 10: Heat Pipe Application for a Nuclear Reactor .. 196
           5.5.10.1 Decay Heat.. 197
           5.5.10.2 Heat Pipe ECCS 198
           5.5.10.3 Gravity-Assisted Water Charging 198
           5.5.10.4 Loop-Type Heat Pipe System....................................... 199
           5.5.10.5 Heat Pipe Thermal Analysis...200
           5.5.10.6 Reactor Thermal Analysis 203
5.6   Concluding Remarks...207
References... 210

5.1 Introduction

For current desktop and server processors, heat dissipation, in general, exceeds 100 W and heat flux could be more than 100 W/cm^2. Passive cooling is no longer appropriate to meet cooling requirements. Other technologies such as liquid cooling, thermoelectric cooling and refrigeration can deliver the required thermal performance, and have been put into practical use for cooling of high-performance computers. However, these cooling options have not yet been widely used because of the integration complexity of the system, limited reliability life data, limited availability of high-volume manufacturing capability and especially higher cost. The cooling technology that is still most widely used in cooling for computers is air cooling, because this is a mature technology with the least operation and lowest costs.

The processor's die surface where the heat is generated is usually small (~1 cm^2). For effective cooling, least temperature gradient would be required between the heat source and the radiating components. The best-known devices for effective heat transfer with the lowest thermal resistance are heat pipes and vapour chambers. Basically, heat pipes and vapour chambers are two-phase heat transfer devices and comprise an evacuated and sealed container that contains a small quantity of working fluid. One end of the container is heated, causing the liquid to vaporise and the vapour to move to the cold end and condense. As the latent heat of evaporation is large, considerable quantities of heat can be transported with a very small temperature difference from end to end. Thus, it is a device of very high thermal conductance. Its equivalent thermal conductivity can be several hundred times than that of a solid copper (Cu) device of the same dimensions.

For data centres, electric power consumption is a major operational cost. As the power supplied to the data centre processing units is ultimately dissipated as heat, a significant fraction of data centre power is used to cool these units (Schmidt et al., 2005; Brill, 2010). It is estimated that for every watt of power consumed by the computing infrastructure, another one-third to half watt is needed to operate the cooling infrastructure. For example, an average data centre with a thermal load output of 8800 kW consumes more than $4 million a year just for cooling purposes. Furthermore, data centre electricity consumption nearly doubles every 5 years. In most countries worldwide, a major portion of electric power is generated from non-renewable energy sources, including coal, gas and nuclear power, which degrade the earth's atmosphere by greenhouse gas emissions. In this regard, energy conservation based on cooling systems for data centres can provide two advantages: first, by reducing electricity consumption and thus operational costs for thermal management and, second, by minimising the carbon emissions in the environment (Patel et al., 2002; Moore et al., 2004; Schmidth et al., 2009). In this chapter, design, thermal analysis and economics of the proposed heat pipe–based cold energy storage system is discussed in detail and compared with the existing chiller using a refrigeration system.

This chapter includes designs, data and discussions of various fan sink air cooling designs showing how the design changed to push the limit of air cooling capability. For example, maximum heat transfer surface area of a heat sink, improved fin shape design to maximise the fin efficiency, optimization of fan air flow pattern, and the use of heat pipes or vapour chambers to enhance heat spreading and heat transfer. The use of a three-dimensional thermal management device not only reduced the physical form factor of the solution but also enabled the processors to operate at their full performance. The utilisation of the two-phase fluid phenomena to spread the heat was a key factor in extending the air cooling limit capability for high-performance computers.

This chapter also presents numerous concepts of using heat pipes to collect the available natural energy and use it for applications without the need for electricity. Some of the concepts have been tried in the past, but have not been widely used in practical applications; however, with current increased concerns regarding global warming environmental issues, these concepts can become a feasible solution.

In response to a recent natural disaster caused by earthquakes and a tsunami in Japan, which threatened the country with a serious nuclear reactor meltdown in Fukushima plant due to the failure of an emergency core cooling system (ECCS) powered by diesel generators, a detailed proposal of a heat pipe–based ECCS design for nuclear power plants will also be presented.

5.2 Heat Pipe Revolution

Heat pipes have been used in various applications for decades. New applications have emerged with time (as shown in the following table). The authors believe that heat pipe could be widely used in preventing global warming in the future.

1970	Aerospace and astronautics
1980	Energy conservation
1990	Industrial use and natural energy utilisation
2000	Computers and electronics cooling
2010	Global warming and the environment

Currently, heat pipes are widely used in the cooling of computers and electronic products. It is believed that approximately 15 million pieces of heat pipes are produced per month worldwide for these applications. There are mainly three basic types of heat pipes (as shown in Figure 5.1): grooved, sintered and composited wick heat pipes.

Figure 5.2 shows an overview of the application of heat pipes. The basic principle is to use the heat pipe to effectively transfer the heat from point A to B, and spread the heat to a larger surface for cooling.

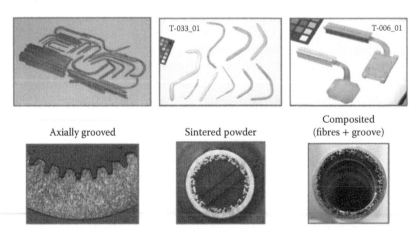

Axially grooved   Sintered powder   Composited
(fibres + groove)

FIGURE 5.1
Types of heat pipes.

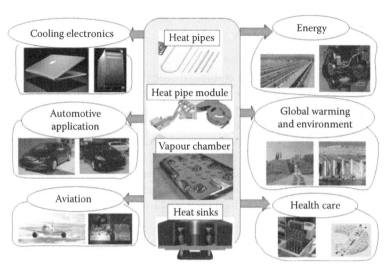

FIGURE 5.2
Heat pipe application.

5.3 Heat Pipe Applications in Computers and Electronics

Heat pipes and vapour chambers have emerged as the most significant technology and cost-effective thermal solution owing to their excellent heat transfer capabilities, high efficiency and structural simplicity. The working fluid is selected based on the operating temperature of the application. In computer applications, the operating temperatures are normally between 50°C and 100°C. For this temperature range, water is the best working fluid. Using heat pipes and vapour chambers to spread and transfer the heat is a key factor for extending the air cooling limit capability for high-performance computers.

5.3.1 Formulation

A basic formulation to calculate the thermal resistance required for the cooling solution is as follows:

$$R_{ja} = R_{jc} + R_{ca} = \frac{T_j - T_a - T_{sys}}{Q} \tag{5.1}$$

where

Q = heat dissipation (W)

R_{ja} = thermal resistance from the CPU die to ambient (°C/W)

R_{jc} = thermal resistance from the CPU die to CPU case surface (°C/W)

R_{ca} = thermal resistance from the CPU case surface to ambient (°C/W)

T_a = ambient temperature (°C)

T_j = junction temperature inside die (°C)

T_{sys} = temperature rise of the ambient inside the system due to other heat-generating components (e.g. hard disk drive and graphics cards) (°C)

In the case of a CPU packaged with an integrated heat spreader (IHS), the R_{jc} varies depending on the type of CPU and the manufacturers. In most cases, the R_{jc} is approximately 0.33°C/W.

The thermal solution provider could control only the R_{ca}. This thermal resistance consisted of the thermal interface material and the cooling solution. Figure 5.3a shows the thermal resistance network, and Figure 5.3b

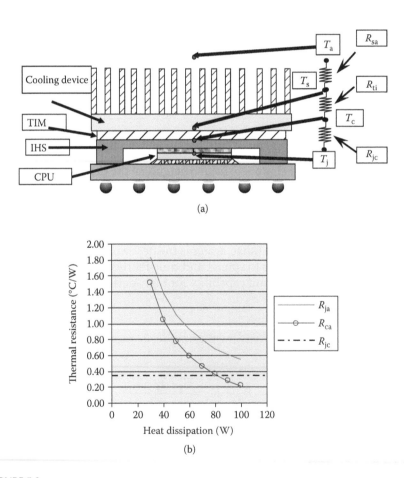

(a)

(b)

FIGURE 5.3

(a) Schematic of thermal resistance (b) R_{ja}, R_{jc} and R_{ca} versus heat dissipation.

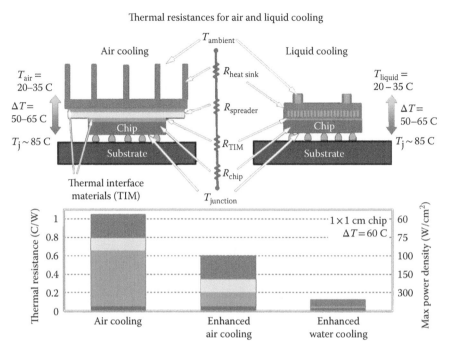

FIGURE 5.4
Thermal performance comparison between air cooling and liquid cooling.

shows the required thermal resistance R_{ja}, R_{jc} and R_{ca} versus heat dissipation. It is assumed that for the majority of CPUs, the junction temperature is 100°C, the outside ambient temperature is 35°C and system temperature rise is 10°C. When the power reaches 100 W, R_{ja} (0.33°C/W) cannot be neglected because the external thermal margin (R_{ca}) almost reaches the limit of air cooling.

Figure 5.4 shows the comparison of general magnitude cooling capability between air and liquid cooling. It shows that the heat transfer capability of liquid cooling can be six times more than that of air cooling.

5.3.2 Examples of Notebook Thermal Solutions

5.3.2.1 Hybrid System

Figure 5.5 shows the hybrid system consisted of heat pipes, die-cast plate, fins and fans (not shown). The two heat pipes were used to spread heat on the aluminium die-cast plate. Aluminium fins attached to the heat pipe and radial fans were used to blow air directly through the fins. The R_{ca} of this system was about 1.2°C/W and this system was capable of dissipating 35 W.

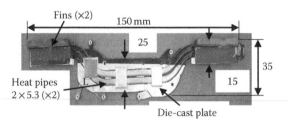

FIGURE 5.5
Hybrid cooling system.

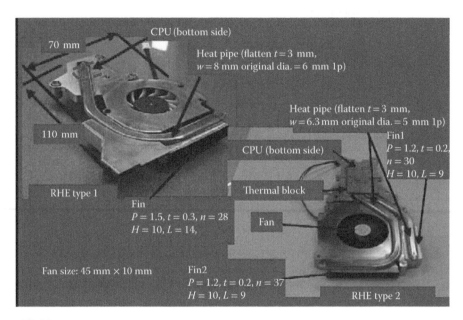

FIGURE 5.6
Remote heat exchanger.

5.3.2.2 Remote Heat Exchanger

Basically, the remote heat exchanger consisted of a heat pipe, a fin and a fan (as shown in Figure 5.6). The fan size was approximately 45 mm × 45 mm and 10 mm thick. The fan airflow was estimated to be 0.15 m³/min at 0 Pa and 90 Pa at 0 m³/min when it was tested in an open environment. The R_{ca} for this design was approximately 1°C/W, with a cooling capacity of about 42 W. Figure 5.7 shows examples of various remote heat exchanger designs.

5.3.2.3 Vapour Chamber

The principle of operation of a vapour chamber is similar to a heat pipe, which is a two-phase heat transfer device. A heat pipe is made of a round

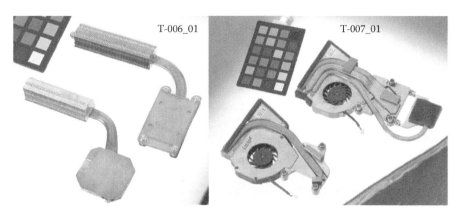

FIGURE 5.7
Examples of various remote heat exchanger designs.

FIGURE 5.8
Examples of various shapes and sizes of vapour chambers.

pipe, and after sealing the ends it can be bent and flattened to the required shape. The vapour chamber container can be made by stamping, cold forging, or machining processes so that the shape is fixed and the container cannot be bent or flattened. Figure 5.8 shows the shapes and sizes of various vapour chambers.

An example of a vapour chamber solution is shown in Figure 5.9. The R_{ca} of this system is about 0.3°C/W. When the boundary conditions are $T_j = 100$°C, $T_a = 35$°C, $T_{sys} = 10$°C and $R_{jc} = 0.33$°C/W, this system can dissipate heat of 87 W.

The performance of a vapour chamber is better than that of a heat pipe because of the two-dimensional heat transfer element. The advantages of the vapour chamber solution compared with the heat pipe solution are given as follows:

- Heat flow is two-dimensional in a vapour chamber, whereas it is one-dimensional in a heat pipe. This gives a vapour chamber higher heat transfer and lower thermal resistance.

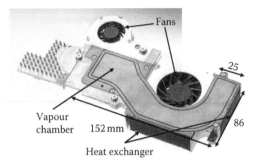

FIGURE 5.9
Vapour chamber solution.

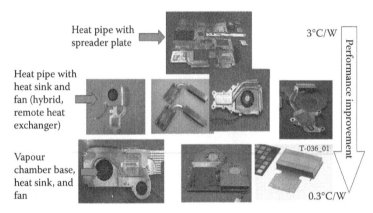

FIGURE 5.10
Thermal solution trends for laptops.

- A vapour chamber has a higher heat flux capability of over 50 W/cm^2.
- A vapour chamber has uniform temperature distribution and large body surface area, so it is excellent for heat dissipation.
- A vapour chamber can be placed directly in contact with the CPU (via thermal interface material), eliminating conducting and contacting resistance of the heat block to which the heat pipe is attached.
- Fins can be attached directly to the vapour chamber, having higher surface of contact; thus, it can reduce contact resistance and increase fin efficiency.

Figure 5.10 shows the thermal solutions trends in laptop PCs. It is evident that to maximise the performance and to push the limit of air cooling capability, the design needs multiple heat pipes or vapour chambers with high-density packed fins and multiple fans.

5.3.3 Examples of Desktop Thermal Solutions

Figure 5.11 shows a summary of design changes to maximise the air cooling capability of desktop PCs. For equal comparison, all the designs are within the boundary of a cooling volume of approximately 90 mm × 90 mm × 65 mm height. The acoustic level at the maximum specification in general is 45 dB at 1 m from the source. The descriptions of the design are given in Figure 5.11:

- Type 1: Normal extrusion heat sink with approximately 7 fins per inch, 1.2 mm fin thickness, 3.5 mm pitch and 30 mm height. The R_{ca} range is 0.4–0.5°C/W.

- Type 2: High-aspect extrusion heat sink. In this design extrusion has been pushed to the limit capability. Fins count approximately 10 fins per inch, 1 mm fin thickness, 2.3 mm pitch and 30 mm height. The aluminium base has a Cu block soldered to minimise the heat spreading resistance. The R_{ca} range is 0.3–0.4°C/W.

- Type 3: The design changes from parallel fin extrusion to radial-type extrusion. The heat sink core has an integrated Cu core to improve heat conduction from base to fins. In general, radial fins could capture the air from the fan better than a parallel plate fin, thus

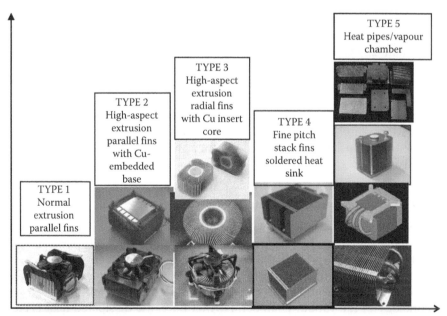

$R_{ca} \sim 0.5°C/W$

Low performance

$R_{ca} \sim 0.2°C/W$

High performance

FIGURE 5.11
Summary of thermal design trends for cooling desktop PCs.

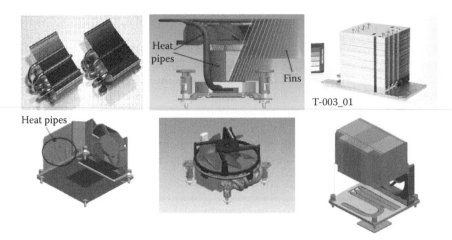

FIGURE 5.12
Examples of heat pipe heat sink thermal designs for cooling desktop PCs.

providing higher fin–air heat transfer coefficient and more efficient cooling. The R_{ca} is approximately 0.3–0.35°C/W.

- Type 4: High-density stack fins are soldered to a metal base. The fin thickness can be as thin as 0.2 mm, and fin gap less than 1 mm. The R_{ca} is approximately 0.25–0.3°C/W.

- Type 5: For the same space constraint, in order to reduce R_{ca} less than 0.25°C/W, it is expected that the base should be a heat pipe or vapour chamber to maximise the heat spreading and heat transfer from source to fins for improvement in cooling.

Figure 5.12 shows an example of how heat pipes are used to improve thermal performance. Heat pipes are used to transfer heat effectively from the heat source to fins, improve the fin efficiency and reduce spreading resistance at the base.

5.3.4 Thermal Solution for Graphics Processors

Figure 5.13 shows an advanced high-performance three-dimensional thermal management device to provide ultra-cooling for high-performance graphics cards. This three-dimensional thermal management device is structured based on spreading the heat dissipation from the graphics processor unit (GPU) in the horizontal plane through the use of the advanced technology of the vapour chamber. Furthermore, the dissipated heat is elevated to the vertical dimension through the use of two auxiliary heat pipes. This advanced thermal solution ensures maximum cooling possibilities for graphics cards in the smallest form factor to deliver the best performance.

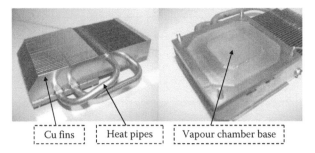

Cu fins | Heat pipes | Vapour chamber base

FIGURE 5.13
Vapour chamber and heat pipe thermal solution for cooling high-performance graphics processors.

The vapour chamber is a two-phase water-working fluid device. It is a totally sealed device, is passive and has no moving parts. The vapour chamber uses the latent heat of vaporisation to transfer the heat, so it has extremely high thermal conductance compared to any known solid metal. Using the three-dimensional management device not only reduces the physical form factor of the solution but also enables the GPU to operate at its full performance. The utilisation of the two-phase fluid phenomena to spread the heat is a key factor for extending the air cooling limit.

The key advantages of this thermal device are as follows:

- Advanced and optimised wick structures inside the vapour chamber to enhance the heat transfer and maximise the heat transfer rate. Heat flux capability is above 75 W/cm^2.
- Thermal spreading resistance in the base which is attached to the die of the graphics card can be reduced by approximately half when compared to the solid Cu base of the same size.
- Combination of vapour chamber with heat pipes to maximise the heat transfer and fin efficiency.
- Overall thermal performance is 0.05°C/W, better than the design with solid Cu base and heat pipe.
- Isothermal surface characteristics of the vapour chamber introduce a uniform heat distribution on the GPU, which maximises the performance of the GPU.

Thermal performance of the combined vapour chamber and heat pipe for cooling a graphics card is shown in Figure 5.14. It is anticipated that the thermal performance can be improved approximately 0.05°C/W. This means that for 100 W input, the processor could operate 5°C lower. Operating the processor at lower temperature would enhance its performance and reliability.

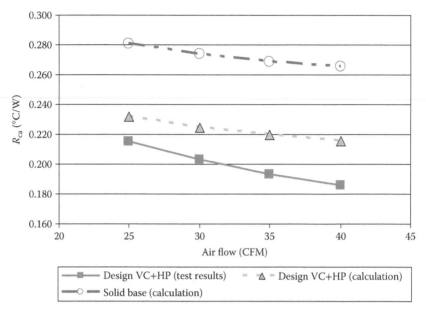

FIGURE 5.14
Thermal performance of vapour chamber and heat pipe thermal solution for cooling high-performance graphics processors.

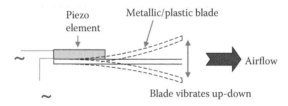

FIGURE 5.15
Basic principle of a piezo fan.

5.3.5 Thermal Solution Using a Piezo Fan

Another air cooling technology is a piezo fan for thermal management of low form factor, low power consumption and low noise devices. Examples of such devices are in the mobile market such as PDAs, phones, music players and portable gaming. The traditional rotary axial fan or blower may not be suitable due to space, power consumption and noise constraints. The basic principle of a piezo fan is shown in Figure 5.15. A long metal or plastic blade is bonded to a piezoelectric material in a cantilever fashion. When current is applied to a piezo layer, the randomly oriented ions in the piezo materials go into alignment, which causes deformation of the piezo layer. Under an alternating current, the blade vibrates back and forth with the same frequency as the alternating current. The vibration generates air flow which can be used for air cooling. The larger the vibration, the larger the air flow and it in turn provides better cooling. The piezo fan can be placed in front of the heat sink to provide cooling air through the heat sink.

Figure 5.16 shows the concept of using a piezo fan with a heat spreader plate and a heat pipe. The piezo fan, axial DC fan and DC blower were tested on the same thermal module for thermal comparison. Thermal test results shown in Figure 5.17 indicate that the piezo fan performed similar to traditional rotational fans. The main advantages of a piezo fan are low power consumption, noise and cost.

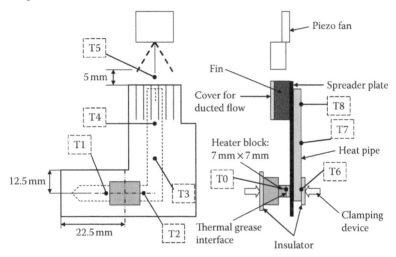

FIGURE 5.16
Thermal solution consisting of heat pipes with spreader plate and piezo fan.

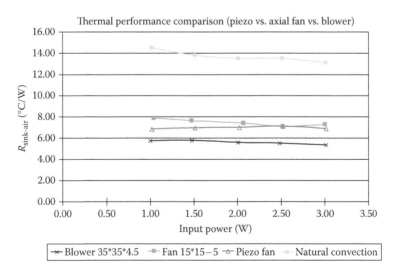

FIGURE 5.17
Thermal performance comparison between piezo fan, axial fan and air blower.

Figure 5.18 shows a novel multiple piezo fan design called 'raked' piezo. In this type of design, in addition to the air flow generation, the blades reduce or destruct the thermal boundary layer proximity to the heat sink fin surfaces and blades. This would improve the heat transfer coefficient and therefore would provide better cooling. The thermal performance of this concept could significantly improved by 40% (as shown in Figure 5.19). The inclination angle of a piezo fan with respect to the heat sink can affect the performance; in this particular case, the inclination angle is 75°. Further study is needed for different geometries of heat sink and piezo fan in order to make a more comprehensive conclusion.

Figure 5.20 shows a different type of piezo fan in the form of round or square. The design is simple and consists of two piezo discs attached to two metal blades. When electrical alternating current is applied, the piezo disks contract and expand, and thus vibrate the blades and induce air flow. Table 5.1 shows the test results comparison between disc piezo fans with a DC blower. Results showed that the air flow of the piezo disc fan is comparable with that of the DC blower, but static pressure is much lower. The power consumption of the piezo disc fan is expected to be lower than that

Raked piezo

Heat sink

FIGURE 5.18
Raked piezo fan.

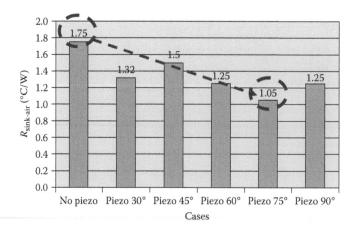

FIGURE 5.19
Thermal performance of raked piezo fan.

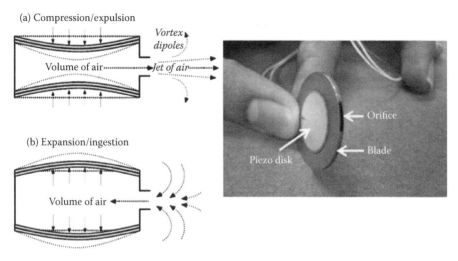

FIGURE 5.20

Piezo fan. (a) Piezo fan in compression mode expelling air (b) piezo fan in expansion mode inhaling air.

TABLE 5.1

Performance Comparison Between Piezo Fan and DC Blower

	Product	Cooling (W)	Size (mm)	Power (mW)	Noise (dBA)	Q (CFM)	P (Pa)
	DCJ	12	$40 \times 40 \times 1.5$	300	28	0.56	10.5
	FAN 1	12	$30 \times 30 \times 3$	200	28	0.55	80

of the DC blower, but the results showed otherwise. Design improvement is required to lower the power consumption. The advantage of the piezo disc fan is the compact slim form factor, which is suitable to apply in electronic devices where space is limited.

5.4 Thermal Solution Improvement

5.4.1 Improve Heat Spreading from the Processor Die

Figure 5.3a is a schematic presentation of the thermal resistance network. In the current technology, the IHS is made up of solid Cu plate, which is interfaced with the CPU's die via thermal grease. The average thermal spreading

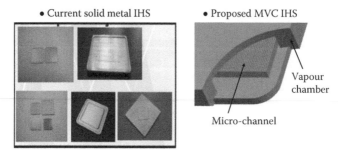

FIGURE 5.21
Micro-channel vapour chamber.

resistance between the die and the IHS (R_{jc} in Figure 5.3b) is a significant ratio to overall thermal resistance.

There are some possibilities to improve the thermal spreading resistance between the CPU's die and the IHS as described below:

- Instead of solid Cu metal IHS, use a two-phase micro-channel vapour chamber (MVC) (as shown in Figure 5.21). Basically, the MVC is an evacuated, sealed container that contains a small quantity of working fluid. Inside the container are skive micro-fin structures that increase the boiling heat transfer area. When one side of the container is heated, it causes the liquid to vaporise and the vapour to move to the cold side and condense. As the latent heat of evaporation is large, considerable quantities of heat can be transported with a very small temperature difference from end to end. Therefore, the MVC-IHS has low spreading thermal resistance, isothermal temperature distribution and no local hot spot compared with solid Cu IHS, where heat spreads solely by thermal conduction.

- Instead of using thermal grease between the CPU's die and IHS, the use of brazing or soldering to minimise the contact resistance should be considered. There would be technical challenges such as thermal expansion coefficient mismatch between mating surfaces and reliability issues such as corrosion and chemical reaction. It is possible that the surface could be plated with materials having a similar thermal expansion coefficient and corrosion resistance.

Figure 5.22 shows an estimate calculation of thermal spreading resistance and comparison between solid Cu-IHS and MVC-IHS for various IHS sizes. The use of MVC-IHS is an advantage when the IHS size is about 10 times the die size based on the assumption that the die size is 10 mm × 10 mm.

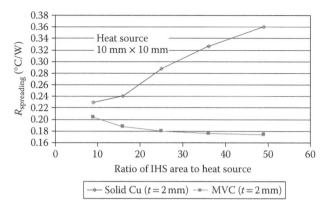

FIGURE 5.22
Thermal spreading resistance comparison between Cu-IHS and MVC-IHS.

FIGURE 5.23
Skive micro-fin structure.

Figure 5.23 shows a picture of the skive micro-fin structure. The manufacturing technique is slicing a block of metal to make the fins. The attributes of the fins are as follows:

- Minimum fin thickness 50 μm
- Minimum fin pitch 100 μm
- Maximum fin height 5000 μm
- Minimum base thickness 300 μm

The advantage of a skive micro-fin is a significant increase in the fin surface area available for heat transfer due to the fine fin thickness and pitch.

The fins and base are made from a solid piece of metal; therefore, there is no fin-to-base thermal contact resistance.

5.4.2 Remove Integrated Heat Spreader

As mentioned earlier, the average thermal spreading resistance between the die and the IHS (R_{jc} in Figure 5.3b) is a significant ratio to overall thermal resistance; therefore, removing the IHS could improve overall thermal performance. The concept of the non-IHS cooling module is shown in Figure 5.24. At present, most of the notebooks use bare die (no IHS), but in the desktop and server processors, IHSs are still used. The reason is likely the reliability issue, because usually the cooling modules for desktop and servers are large and heavy compared to notebook cooling modules. Large and heavy cooling modules would exert higher stress on the processor's die and may cause crack and fatigue failure.

5.4.3 Heat Pipe Improvement

The processor's die surface, where the heat is generated, is usually small, approximately 1 cm^2; therefore, effective cooling requires least temperature gradient between the heat source and radiating components. The best known devices for effective heat transfer with the lowest thermal resistance are heat pipes and vapour chambers. As the power and heat flux of the processor continues to rise, it is most vital to improve the heat pipe maximum heat transfer (Q_{max}) and reduce heat pipe thermal resistance (R_{hp}). There are some possibilities to improve heat pipe characteristics as described below:

- New sintered powder wick heat pipe: The size of powder particles and porosity are the main factors in optimising maximum capillary forces while there is a need to maintain high permeability. Besides, wick thickness and uniformity are key factors for better and stable heat pipe performance.

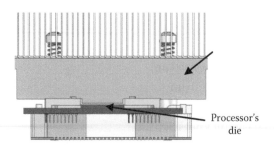

FIGURE 5.24
Thermal solution without IHS.

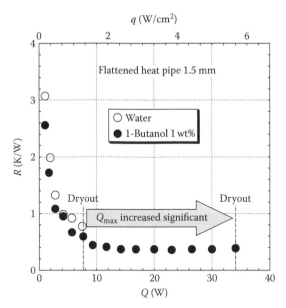

FIGURE 5.25
Heat pipe performance comparison of different fluids.

- New working fluid to improve fluid-wick wettability and boiling heat transfer. Preliminary studies showed that dilute aqueous solutions of high-carbon alcohols present a particular surface tension behaviour that has a strong liquid flow in the nucleation sites during the boiling process. Experimental studies showed that in pool boiling heat transfer, significantly smaller bubble size is formed compared with water. This indicates that the heat transfer performance would be improved. Figure 5.25 shows a significant improvement in maximum heat transfer for 1.5 mm flattened heat pipe. Results showed that heat transfer increased from 8 to 34 W.

5.4.4 Vapour Chamber Improvement

Figure 5.26 shows various ideas to improve the heat transfer performance of vapour chamber. Grooves are added to increase permeability for the liquid to flow easily. Additional screen mesh will increase the capillary force.

Figure 5.27 shows the results of the same vapour chamber configuration; the heat transfer increased from 72 to about 87 W.

Another possibility to improve vapour chamber performance is to consider a higher thermal conductivity material than Cu for the vapour chamber base. For example, the vapour chamber base could be plated with a higher thermal conductivity material such as graphite or diamond. Higher thermal conductivity of the base will help to increase spreading heat transfer and thus will increase the evaporation surface area.

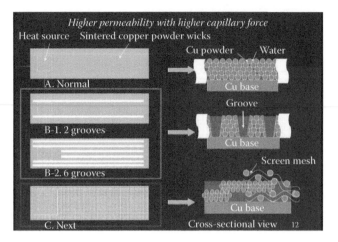

FIGURE 5.26
Vapour chamber improvement ideas.

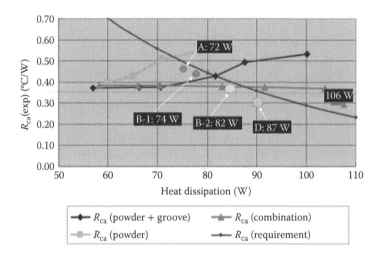

FIGURE 5.27
Vapour chamber performance improvement.

5.4.5 Micro-Channel Two-Phase Pump Loop

Figure 5.28 shows a schematic of the micro-channel two-phase pump loop. Detailed information of the system will not be given here, but just the concept. Typical results indicated that this system could achieve the evaporator thermal impedance of less than 0.1°C·cm²/W at saturated water flow rate less than 50 mL/min. The system is capable of heat flux greater than 50 W/cm².

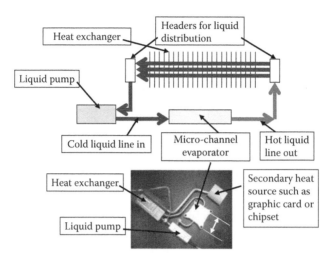

FIGURE 5.28
Schematic micro-channel two-phase pump loop.

5.5 Clean Energies

5.5.1 Case Study 1: A Heat Pipe Cold Storage System for Cooling Data Centres

As described in Section 5.1, a major operational cost of data centres is electric power consumption, and as the power supplied to the data centre processing units is ultimately dissipated as heat, a significant fraction of data centre power is used to cool these units (Schmidt et al., 2005; Brill, 2010). As shown in Figure 5.29, the power consumption is estimated to double every 5 years. The power consumption of data centres in 2001 was about 30×10^9 kWh, and is expected to increase to 120×10^9 kWh in 2011. This is equivalent to the electric power consumption of about 13.2 million houses.

Figure 5.30 shows a typical example of the current practice of hot–cold air flow heat exchange for thermal management of data centres. The authors propose a system which utilises the thermosiphon to capture and store cold energy from the ambient to the storage media. The cold energy can be used for the cooling of the data centre. The concept schematic is shown in Figure 5.31. This system can reduce electrical power consumption related to cooling data centres. The entire system needs to be installed in cold regions at an underground level.

Figure 5.32 shows the working principle of the thermosiphon which can extract heat from high-temperature storage media to low-temperature ambient by means of evaporation and condensation processes. The thermosiphon can transfer heat only when operating in the bottom heat mode in which the

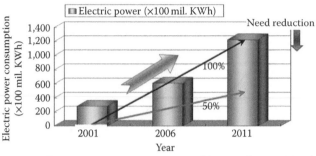

Data centre power consumption will be equivalent to 13.2 mil. houses electric power in 2011.

$$\frac{1200 \times 100 \times 10^6 \text{ KWh}}{2.5 \text{ KWh} \times 10 \text{ h/day} \times 365 \text{ days}} = 13.2 \text{ mil. houses}$$

FIGURE 5.29
Power consumption in data centres.

Hot aisle–cold aisle approach for thermal management of data centre

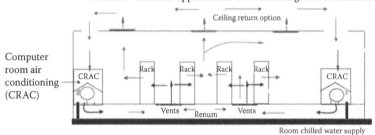

FIGURE 5.30
Typical example of current thermal management system in data centres.

evaporator is below the condenser. When the ambient temperature is higher than the storage media temperature, as, for example, in summer time, the thermosiphon does not function and does not transfer heat. This is because, in this case, the operating condition is in top heat mode in which the evaporator is above the condenser. The thermosiphon has a wick installed in the evaporator to promote thin film evaporation, which enhances the heat transfer coefficient.

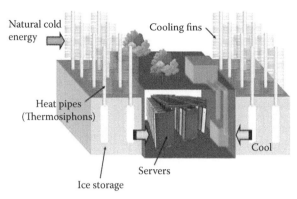

FIGURE 5.31
Concept of cold energy storage for cooling data centres.

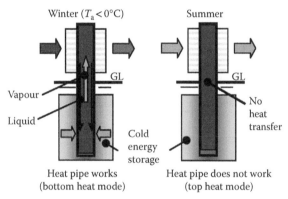

FIGURE 5.32
Thermosiphon diode characteristics.

5.5.1.1 Ice Storage System to Support Main Cooling System Failure

Figure 5.33 shows another concept that used ice storage system to support failure or shut down of the main cooling system. Based on the American Society of Heating, Refrigerating, Air-Conditioning Engineers Standard, the requirement for failure support time is 6 h; therefore, for an 8800 kW data centre, the heat output required is approximately 53 MW·h. The ice storage system needs to support cooling only during times of failure.

Figure 5.34 shows a structural comparison of the current cold water system and the proposed ice storage system. The current system uses a cold water storage tank in which the water temperature is maintained by a chiller. The proposed ice storage system uses thermosiphons to collect and store ambient cold energy for cooling the data centre when the main cooling system shuts down. The advantages of the ice storage system compared to the current cold water system are shown in Table 5.2. Results showed that the ice storage

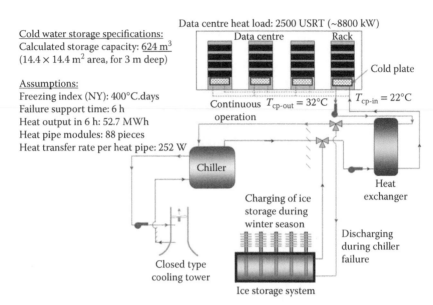

Cold water storage specifications:
Calculated storage capacity: 624 m³
(14.4 × 14.4 m² area, for 3 m deep)

Assumptions:
Freezing index (NY): 400°C.days
Failure support time: 6 h
Heat output in 6 h: 52.7 MWh
Heat pipe modules: 88 pieces
Heat transfer rate per heat pipe: 252 W

Data centre heat load: 2500 USRT (~8800 kW)

Data centre   Rack

Cold plate

Continuous $T_{cp-out} = 32°C$ $T_{cp-in} = 22°C$
operation

Heat
exchanger

Chiller

Charging of ice
storage during
winter season

Discharging
during chiller
failure

Closed type
cooling tower

Ice storage system

FIGURE 5.33
Ice storage system to support data centre failure time.

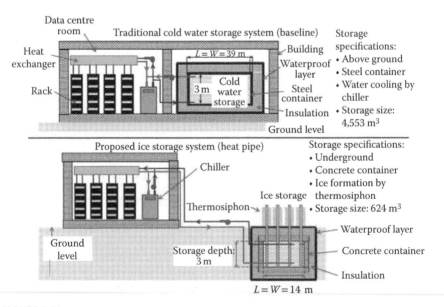

Data centre
room   Traditional cold water storage system (baseline)   Storage
specifications:
Heat $L = W = 39$ m   Building   • Above ground
exchanger   Waterproof   • Steel container
layer   • Water cooling by
Rack   Cold   Steel   chiller
3 m water   container   • Storage size:
storage   4,553 m³
Insulation
Ground level

Proposed ice storage system (heat pipe)   Storage specifications:
• Underground
Chiller   • Concrete container
• Ice formation by
Ice storage   thermosiphon
Thermosiphon   • Storage size: 624 m³

Waterproof layer

Ground   Storage depth: Concrete container
level   3 m
Insulation

$L = W = 14$ m

FIGURE 5.34
Structure comparison between current system (*top*) and the proposed ice storage system
(*bottom*).

TABLE 5.2

Advantages Between the Current System and Proposed Ice Storage System

	Current System: Cold Water Storage	Proposed System: Ice Storage
Size	~4600 m³ (39 m × 39 m × 3 m deep)	~650 m³ (14%) (14.4 m × 14.4 m × 3 m deep)
Energy storage	Same	~63 MW·h (6 hours' cooling)
Construction	Steel (above ground)	Concrete (below ground)
Heat pipes	None	88 heat pipe modules
Power consumption	Running chiller	None
CO₂ emission	13 t (6 hours operating)	None
Cost	~$5 Mil	~$0.2 Mil

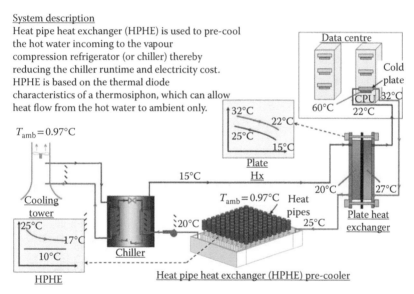

System description
Heat pipe heat exchanger (HPHE) is used to pre-cool the hot water incoming to the vapour compression refrigerator (or chiller) thereby reducing the chiller runtime and electricity cost. HPHE is based on the thermal diode characteristics of a thermosiphon, which can allow heat flow from the hot water to ambient only.

$T_{amb} = 0.97°C$

Cooling tower

Chiller

HPHE

Heat pipe heat exchanger (HPHE) pre-cooler

Data centre

Cold plate

CPU 32°C

Plate Hx

$T_{amb} = 0.97°C$ Heat pipes

Plate heat exchanger

FIGURE 5.35
Pre-cooler for chiller inlet water.

occupied 86% less space, consumed no electric power, caused no CO₂ emission and required minimal cost compared to the current cold water system.

5.5.1.2 Pre-Cooler for Chiller Inlet Water

Another concept for energy conservation is to use a thermosiphon heat exchanger to pre-cool the hot water inlet of the chiller (vapour compression refrigerator), thereby reducing chiller runtime and electricity costs. The system is shown in Figure 5.35. The pre-cooler was designed to reduce the water inlet temperature of the chiller from 25°C to 20°C. The design was based on the weather conditions shown in Figure 5.36 for Poughkeepsie, New York.

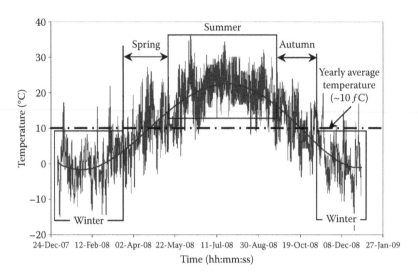

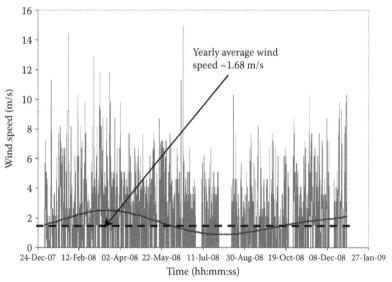

FIGURE 5.36
Hourly temperature and wind speed for Poughkeepsie, New York, 2008.

Figure 5.37 shows a design of the pre-cooler for an 8800 kW data centre. It consisted of 2160 thermosiphon modules placed in a water chamber approximately 60 m in length, 40 m in width and 3.5 m in height. Each thermosiphon has dimensions of 50.8 mm outer diameter and 7 m total length. The 3 m length of the evaporator is submersed in the water cooler chamber, and the 3 m length of the condenser is exposed to ambient for cooling. The condenser has 300 aluminium fins of 300 mm length, 300 mm width and 1 mm thickness.

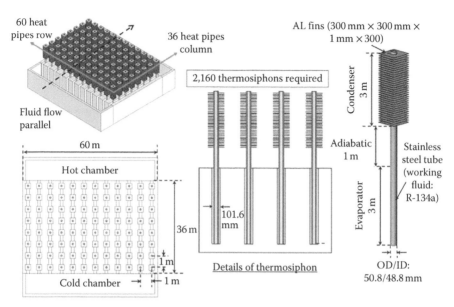

FIGURE 5.37
Pre-cooler design for 8800 kW data centre.

TABLE 5.3

Pre-Cooler Highlights

	2500 USRT Data Centre (~8800 kW)	300 USRT Data Centre (~1060 kW) (12%)
Size of heat exchanger	L 60 m × W 40 m × H 7 m	L 20 m × W 15 m × H 7 m
Cooling capability	Winter: 4400 kW Autumn/spring: 2700 kW Summer: 1100 kW	Winter: 530 kW Autumn/spring: 320 kW Summer: 130 kW
Heat pipes modules	2160	260
Cost savings/year	~$0.6 Mil	~$72 K
Initial cost	~$1.14 Mil	~$130 K
Payback	<2 years	<2 years

Table 5.3 shows some key highlights of the pre-cooler for two cases, a 8800 kW and a 1060 kW data centre. The highest cooling capacity is in winter, where 50% of the data centre cooling is from the pre-cooler, 30% in autumn and spring, and the lowest cooling capability of 12.5% is in summer. The payback time for the pre-cooler is less than 2 years.

5.5.1.3 Experimental Study: Proof of Concept

Figure 5.38 shows a test setup and the experimental facility where the module was tested. A thermosiphon was made of stainless steel, and R-134a was used as

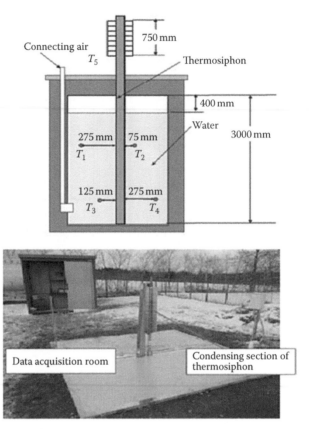

FIGURE 5.38
Pre-cooler design for 8800 kW data centre.

the working fluid. In order to minimise heat leaks from the ambient, the evaporator section of the thermosiphon and the storage tank were buried underground, and the outer surface of the tank was properly insulated. This would help to avoid heat gain from the ground and ambient during the summer season.

Figure 5.39 shows a preliminary test results. The experiment was initially carried out for 25 days; within this period, approximately 113 kg of ice was formed, which was estimated visually by opening the tank and confirming by the readings of the thermocouples installed inside the tank (T1–T4). Experimental results were compared with the predicted ice formation tendency of the heat pipe module using an analytical model which showed good agreement.

Heat transfer from the thermosiphon to the water storage decreased due to the formation of ice on the evaporator surface of the thermosiphon. The thicker the ice, the lower the heat transfer from the thermosiphon to water due to poor thermal conductivity of the ice. To minimise this problem, it is possible to apply high-frequency alternate current to the surface of the evaporator for a short period of time to break up the ice formation. This concept is shown in Figure 5.40.

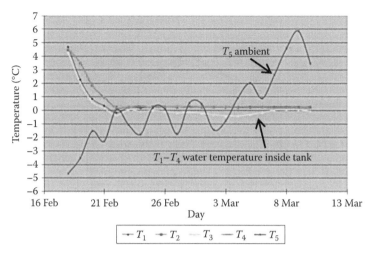

FIGURE 5.39
Experimental test results of ice formation.

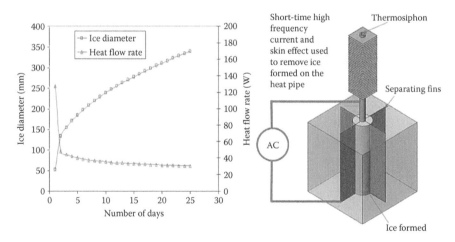

FIGURE 5.40
Concept of ice removal on the surface of the evaporator.

5.5.2 Case Study 2: Cold Energy Storage for Agricultural Products

Figure 5.41 shows the concept of using heat pipe to collect cold energy in the winter season and store it underground to create a permafrost system for storage of agricultural products throughout the year. The prototype shown was built in Hokkaido of Japan, where the freezing index is above 400°C per day. The freezing index is defined as the product of number of days in a year and the average ambient temperature below 0°C. Figure 5.42 shows the freezing indexes for some countries. There are 216 stainless steel heat pipes of 46 mm diameter and 12 m length installed around the store room at 0.5 m intervals. The

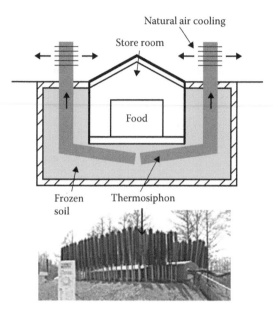

FIGURE 5.41
Permafrost storage system.

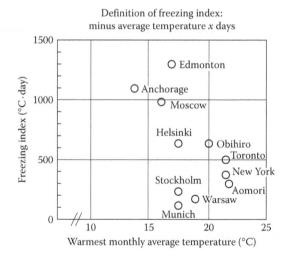

FIGURE 5.42
Freezing index.

store room was 6 m long × 3.6 m wide × 3.6 m high. The heat pipe evaporator is buried in the soil, and the condenser is exposed and cooled by natural ambient. In the winter season, the cold energy system can make a 2 m thick layer of frozen soil and keep the store room below 4°C. The humidity was above 85% throughout the summer. The whole cold energy system is passive; that is, there are no moving parts, there is no electrical consumption and it is reliable and maintenance free. The authors believe that such a system is viable to minimise global warming. The system is applicable in cold regions where the freezing index is at least 400°C per day in order to make it economically viable.

5.5.3 Case Study 3: Heat Pipe and Phase Change Material for Cooling Concentrated Photovoltaic Technology

Concentrated photovoltaic (CPV) technology offers incentives in terms of high conversion efficiency, semiconductor material savings and low unit electricity cost ($/W) as compared to flat-plate photovoltaic (PV) modules. Thermal control of PV cells is necessary to reduce temperature-related performance or efficiency drop and avoid material degradation. A 1°C cooling of PV can help increase energy conversion efficiency in general by 0.5%. This section outlines the concepts of cooling solar CPV modules by means such as heat transport from cells to remote heat sinks by passive thermosiphons or active mechanical pumps, and diurnal storage of waste heat output by cells during the daytime in low-temperature phase change materials (PCMs) and night-time heat dissipation to ambient by means of natural convection and night sky radiation.

Figure 5.43 shows a concept of cooling the solar CPV module. The PV cells are attached to the bottom side of the evaporator of the thermosiphon which is located at the focus line above the parabolic trough. The PV cells are used

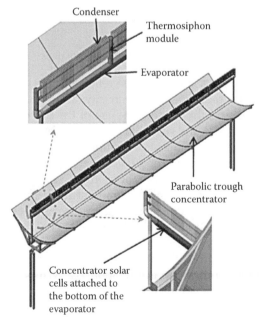

FIGURE 5.43
CPV cooling by thermosiphons with fin heat exchangers.

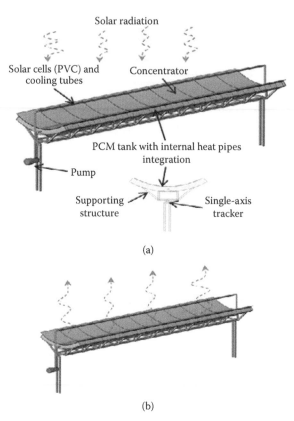

FIGURE 5.44
CPV cooling by use of PCM during (a) day time and (b) night time.

to collect the concentrated solar radiation reflected from the parabolic trough and convert the solar radiation into electricity. The PV cells must be cooled for them to operate, and this is done by use of the thermosiphon to extract the heat from the PV cells and cooling in ambient using fin heat exchangers. Figure 5.44 shows another concept of cooling solar CPV modules by use of PCMs. The heat produced from the PV cells during the daytime is stored in the PCM, which is then cooled at night by night sky radiation and natural convection cooling. The PCM is a low-temperature organic PCM polyethylene glycol (PEG), which has a melting temperature of less than 40°C. In this concept, a pump is required to pump the water to cool the PV cells, and the heated water is cooled by the PCM. In the PCM module, there are heat pipes installed laterally to improve the heat transfer within the PCM, because the thermal conductivity of the PCM is low, about 0.2 W/m·K.

Figure 5.45 shows the relationship between the clear sky temperature and the ambient temperature. This strongly shows the prospective low sky temperature available for radiation cooling. For example, at 15°C ambient temperature and 50% humidity, the sky temperature is 0°C.

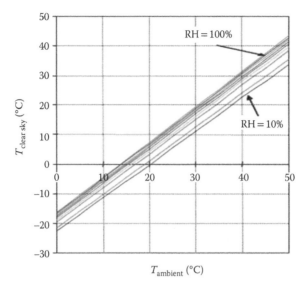

FIGURE 5.45
Sky temperature versus ambient temperature.

5.5.4 Case Study 4: Large-Scale Loop Heat Pipe for Geothermal Heat Extraction

Figure 5.46 shows the concept of using large heat pipes for extraction of geothermal heat. Conventional heat pipes will not work by simply enlarging the heat pipe diameter and length, because the heat load would cause entrainment and flooding phenomena within the heat pipe. With a long heat pipe, heat transfer performance will deteriorate as well due to the difficulty of maintaining uniform liquid film throughout the length of the heat pipe. To address these issues, a new concept in heat pipes was developed in which the flow of vapour and liquid is separated. As shown in Figure 5.46, the liquid flow is separated from the vapour flow. The liquid is fed through a feed tube having a number of nozzles to spray the liquid onto the inner surface of the evaporator of the heat pipe. The spraying of liquid on the evaporator enhances the evaporation heat transfer of the heat pipe. A prototype of this 150 mm diameter × 150 m long loop-type heat pipe with water as the working fluid was developed and installed in Kyushu Island, Japan.

Experimental results are shown in Figure 5.47. The results showed that 65 kW of heat can be extracted at a flow rate of 1.2 L/m and at a working temperature of approximately 70°C. It is estimated that over 100 kW can be obtained at a working temperature of 100°C. Refer to the work of Mashiko et al. (1994a) for a detailed analysis of the prototype structure, calculation and test results.

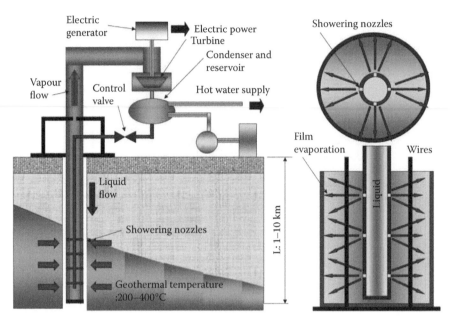

FIGURE 5.46
Large-scale heat pipe for geothermal extraction heat.

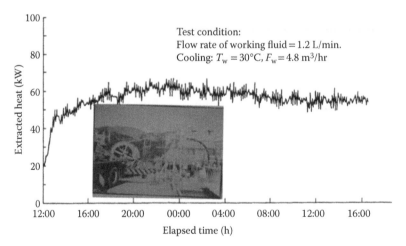

FIGURE 5.47
Test results for geothermal heat extraction by a large-scale heat pipe (150 mm diameter and 150 m long).

5.5.5 Case Study 5: Bakery Waste Heat Recovery

A waste recovery system based on the heat pipe heat exchanger (HPHE) was designed and implemented by RMIT University at Buttercup Bakery, Australia (Dube et al., 2004). In this system, a HPHE was used to recover

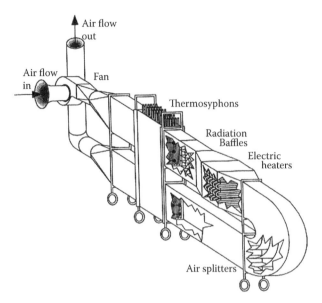

FIGURE 5.48
Bench-type prototype of heat pipe heat exchanged (HPHE)-based waste heat recovery system for bakery.

waste heat from the high-temperature baking oven for heating the low-temperature proofing oven. In Figure 5.48, details of the waste heat recovery system are presented. The temperature of the baking oven's flue gases ranged from 300°C to 350°C and available waste heat was 70–80 kW. For the proofing oven, the energy requirement was between 20 and 45 kW. With an air velocity of 1.5 m/s and heat exchanger effectiveness of approximately 65%, the HPHE-based waste heat recovery system was able to supply all the heat needed by the proofing oven, thus eliminating the need for natural gas heating.

The Buttercup heat exchanger was manufactured locally from steel pipes charged with distilled water. Figure 5.49 shows a schematic of the HPHE. When hot exhaust gases from the baking oven come in contact with the HPHE evaporator containing liquid, the liquid is boiled and the vapour is transferred to the condenser portion by natural convection where it condensed, thereby transferring the energy as latent heat. The heat extracted is fed directly into the proofer oven, conserving energy while providing uncontaminated air. The predicted annual waste heat recovery from an 8 hour shift working 6 days a week was estimated to be approximately 500 GJ.

The total cost of manufacturing and constructing the 50 kW HPHE prototype, excluding data-monitoring operations, was approximately $10,000. Based on annual saving of 500 GJ per year for one shift, cost of natural gas at $4 per GJ and boiler efficiency of 70%, the payback period is less than 3.5 years. For a three-shift operation, the payback period is reduced to 1.5 years.

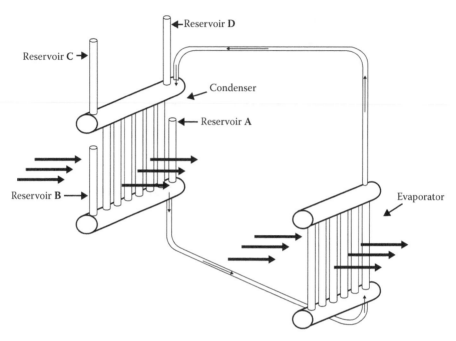

FIGURE 5.49
Schematic of heat pipe heat exchanger.

Additional benefits include more efficient operation of the boilers in terms of lower steam requirements, less gas burnt and less maintenance on the boilers.

5.5.6 Case Study 6: Heat Pipe Turbine

A new concept for power generation from solar, geothermal or other available low-grade heat sources using a heat pipe turbine or a thermosyphon Rankine engine (TSR) was developed and tested (Nguyen et al., 1999). The basis of the engine is the modified thermosyphon cycle, with its excellent heat and mass transfer characteristics, which incorporates a turbine in the adiabatic region. Figure 5.50 shows the concept of the heat pipe turbine that consisted of a closed vertical cylinder with an evaporator, an insulated (adiabatic) section and a condenser. The turbine is placed in the upper end between the evaporator and condenser sections, and a plate is installed to separate the high-pressure region from the low-pressure region in the condenser. Conversion of the enthalpy difference to kinetic energy is achieved through the nozzle. The mechanical energy developed by the turbine can be converted to electrical energy by direct coupling to an electrical generator.

The heat pipe turbine was developed by fabricating and testing a series of prototypes in the RMIT Laboratories, Australia. The aim of the work was to optimise the performance of each successive prototype in order to simplify

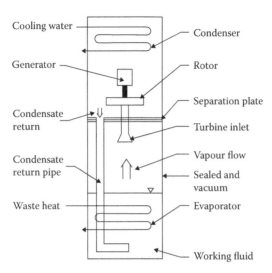

FIGURE 5.50
Heat pipe turbine.

manufacture and to increase the power output. The first prototype turbine rotated at 600–800 rpm, but only very low power output was measured. The second prototype was equipped with an impulse turbine and the heat input was specified as 2 kW. The rotational speed of the turbine was up to 3400 rpm, but the power output was still low. In the third prototype, the reaction turbine was introduced. The turbine was produced as a hollow disc, including two convergent–divergent nozzles at its periphery. The diameter of the heat pipe was 0.16 m and its height 3.15 m. An electrical power output of 5.5 W at 4788 rpm was obtained from a heat input of 4.4 kW. The fourth prototype consisted of a cylinder of height 2.8 m and a diameter of 0.5 m. The turbine configuration was the same as in the third prototype. The heat input was 10 kW and electrical power output of 100 W was obtained at 6000 rpm. Although the efficiency of the proposed heat pipe turbine is low, it has the capability of utilising very low-grade heat and converts it into useful electricity.

5.5.7 Case Study 7: Thermosyphon-Based Thermoelectric Generation System for a Solar Pond

Salinity gradient solar ponds are large bodies of water that act as solar collectors and heat storages. A solar pond is a simple and low-cost solar energy system that collects solar radiation and stores it as thermal energy for a relative longer period. When solar radiation penetrates through the solar pond surface, the infrared radiation component is first absorbed in the surface mixed layer or upper convective zone. However, this heat is lost to the atmosphere through convection and radiation. The remaining radiation will subsequently be absorbed in the non-convective zone and lower convective before the last of the radiation reaches the bottom of the pond. In these ponds, the solution is heavier in the lower region because of a higher salt concentration. As a result, the natural convection that takes place in normal ponds is suppressed. Solar radiation penetrating to the bottom region is thus absorbed there, and temperature of this region rises substantially since there is no heat loss due to convection. The temperature difference created between the top and the bottom of the solar ponds can be as high as 50–60°C. The collected and stored heat can be extracted and used for industrial processes, space heating and even power generation (Akbarzadeh et al., 2005).

A typical solar pond consists of three regions: the upper convective zone (UCZ), the non-convective zone (NCZ) and the lower convective zone (LCZ) (as shown in Figure 5.51). The upper convective zone is the topmost layer of the solar pond. It is a relatively thin layer comprising almost fresh water. The non-convective zone is just below the upper convective zone and has an increasing concentration gradient with respect to the pond depth and relative to the upper convective zone. It also acts as thermal insulation for the bottom layer. The LCZ has the highest salt concentration without any salinity gradient. In solar ponds, if the concentration gradient of the NCZ is great enough, no convective current will occur in this region, and the energy absorbed in the bottom of the pond will be stored in the LCZ. While construction and maintenance of a solar pond of this size is not a problem, the conversion of thermal energy to power is a difficult challenge. Conventional heat engines have too many moving parts and are complex. They are very expensive ($15,000–$20,000/kW) in these small sizes (<0.5 kW for the above example) and difficult to maintain.

In the present research, it is shown that by combining thermosyphon and thermoelectric cells (TECs), it would be possible to utilise the temperature difference existing between the top and the bottom of a solar pond and produce electric power in a fully passive way, that is, no moving parts. In such a scheme, the heat is transferred by the thermosyphon from the lower region

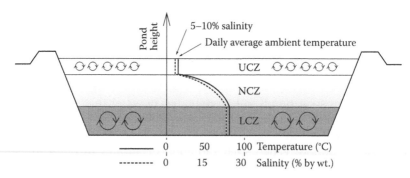

FIGURE 5.51
Salinity gradient solar pond.

of the pond to the hot side of TECs, which maintain good thermal contact with the top of the thermosyphon tube. The cold sides of the cells are in contact with the cold environment of the top layer of the solar pond.

The above proposal utilises thermosyphons, which are highly effective devices for heat transfer, and TECs, which can effectively convert a temperature difference to electric potential and generate power. Both of these devices do not have any macro-scale moving parts and are thus fully passive. Although at present, the efficiency of conversion of heat to electricity by TECs is low (2% for a 50°C temperature difference) and at its best is 10–20% of the Carnot efficiency for the same temperature difference, the availability of the cells and their simplicity suggest that these devices may be very suitable candidates as small and simple energy converters in applications such as small solar ponds. Of course, the developments in recent years in semiconductor materials for TECs and the improvements in their efficiency are promising, indicating that these cells have good potential to be economically viable candidates for conversion of low-grade heat (produced in most solar collectors) into electricity. It should be noted here that the technology for the manufacture of thermosyphons for applications such as above is fully developed and thermosyphons of different sizes can be easily manufactured requiring only a moderate amount of skill.

In remote areas, where the electric grid is not widely available and the sun shines all year round, combined power generation modules based on the small-scale solar pond, thermosyphon and thermoelectric generator (TEG) is one of the viable candidates for providing daily electricity. A TEG has the advantage that it can operate from a low-grade heat source such as waste heat energy. It is also attractive as a means of converting solar energy into electricity. The schematic diagram of the TEG is shown in Figure 5.52. It consists of two dissimilar materials, n-type and p-type semiconductors, connected electrically in series and thermally in parallel. Heat is supplied at the hot side,

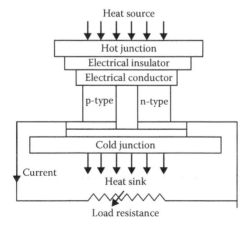

FIGURE 5.52
Schematic diagram of the thermoelectric cell.

while the other end is maintained at a lower temperature by a heat sink. As a result of the temperature difference, the current flows through an external load resistance. The power output depends on the temperature difference, the properties of the semiconductor materials and the external load resistance.

Figure 5.53 shows the proposed power generation unit. The thermosyphon, which is basically an evacuated Cu tube filled with water as the working fluid, is held vertically and is long enough to connect the bottom convective zone of the pond to the top convective zone. A typical temperature profile for a solar pond is also given. It is seen from this profile that the temperature was low and uniform in the top zone. This uniformity was caused by the wind induced wave action near the top. In this zone, the temperature followed closely the daily average ambient temperature. In the middle layer of the pond, where the convective currents were suppressed because of the existence of a strong salinity gradient, the temperature raised continuously until it reached a maximum near the interface with the lower layer. In the bottom layer, the temperature is uniform as in the top layer. The mixing in the bottom zone is primarily due to the absorption of solar radiation by the dark bottom of the pond, and the convection currents were induced. Therefore, the non-convection middle layer, which is sometimes called the gradient layer, separates the upper zone (cold) and the bottom zone (hot). Typical thicknesses for the three layers are 0.2–0.6 m for the top layer, 1–2 m for the middle layer and 0.5–5 m for the bottom layer. The middle section of

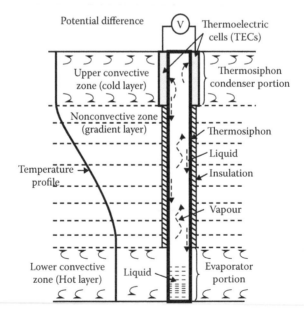

FIGURE 5.53
Schematic of combined thermosyphon and thermoelectric cells as an electricity generation module.

the thermosyphon (adiabatic section) is insulated to prevent heat losses from the thermosyphon to the surrounding water in the gradient layer, which is at a lower temperature than the bottom layer. The TECs are attached to the top part of the thermosyphon (condenser section) with a good thermal bond. The other side of the cells is cooled by the cold and convective currents available in the top convective zone. Therefore, the required temperature difference for power generation is created on the two sides of the TECs.

The working fluid inside the thermosyphon is continuously evaporated in the evaporator, and the resulting vapour travels upward because of the lower pressure in the condenser section caused by a lower temperature. The vapour is then condensed releasing its latent heat, which is transferred to the sides of the TECs attached to the thermosyphon. The resulting condensate travels downward because of gravity. As a result, the two sides of the cells are maintained at different temperatures, and hence an electric potential difference is generated across the cell. On applying an external electric load, an electric current is produced and electric power generated. The produced electric power can be directly used for applications requiring direct current, or converted into alternating current if needed. The electric energy thus produced can be also continuously stored in batteries to provide power to intermittent loads, which may have a higher demand than the capacity of the TECs for a period.

Figure 5.54 shows a schematic of the prototype fabricated for lab testing and system characterisation. Indoor laboratory tests were carried out on a

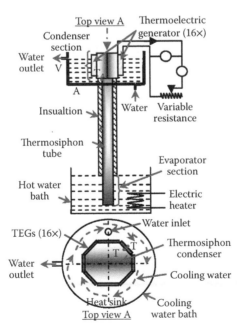

FIGURE 5.54
Prototype to simulate thermosyphon and thermoelectric cells under solar pond conditions.

combined system of thermosyphon and TECs to simulate the operation of the system when installed in the solar pond. For these purposes, a thermosyphon was made from a 100 mm diameter Cu tube with a total length of 2 m. The thermosyphon was charged with water as the working fluid. The TECs were attached to the top part of the thermosyphon. This is the part that will be in the top convective zone of the solar pond, which is the heat sink. TECs were made from bismuth telluride (Bi_2Te_3) with 40 mm × 40 mm area and 3.9 mm thickness. There were 16 cells individually tested by subjecting them to temperature differences by heating one side and cooling the opposite side. As shown in Figure 5.53, the lower section was heated by a hot water reservoir. This resulted in the heating of the inner side of the TECs through the thermosyphon. The outer sides of the cells were cooled by flowing water that simulates the cooling effect of the top layer of the solar pond.

Figure 5.55 shows the power output test results of thermosyphon thermoelectric module (TTM) for various temperature differences across TEG when the water bath temperature was changed from 50°C to 90°C (with the top curve corresponding to 90°C). It can be seen from the graph as expected that the power generated by the TTM module increased with the increase in the temperature across TEG. Here, the temperature of the hot water bath simulated the temperature of the bottom dense layer of the solar pond. By using a number of TECs in series, one would be able to design a TTM suitable for battery charging (12 V) using the heat from the bottom of the solar pond. The output power of the unit can be between 10 and 20 W under normal conditions of the pond. A group of TTMs can produce the power supply for a small house. For example, for a small energy-efficient house, the needed electrical energy can be as low as 2 kW·h/day. Since the TTM works day and night, a total number of 10 units would be required to power the remotely located house without access to grid power.

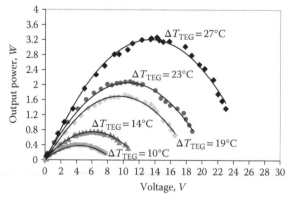

FIGURE 5.55
Power output for various cases of temperature difference across TEG.

5.5.8 Case Study 8: Concentrated Solar Thermoelectric Generator with PCM Thermal Storage

A concentrated thermoelectric generator (CTEG) uses concentrated solar energy as a sustainable heat source and obtains high solar heat flux for increasing the TEC hot-side temperature. The waste heat at the cold side of the TECs must be effectively dissipated to achieve greater temperature difference for higher power generation. Active cooling methods such as the use of an electric fan and water pump to draw power for operating reduced the overall power generated. Hence, passive cooling approaches would seem to be more reliable for a sustainable power generator despite the lower cooling rate. However, passive cooling performance of thermal devices such as heat sinks and heat pipes rely greatly on the ambient conditions and variation. Weather fluctuations such as wind speed and surrounding temperature can significantly affect the natural convection heat transfer performance and could pose a limitation on the passive cooling of concentrated solar system. In order to overcome this drawback, a PCM thermal storage concept has been proposed in this investigation. For transporting large amount of heat from the cold side of TECs to the PCM thermal storage tank, a two-phase closed thermosyphon was used in the design. The combination of thermosyphons and PCM will provide a reliable thermal control system for concentrated thermal electric system by maintaining maximum cell operation temperature below 250°C limit. The objective of the present study is to assess the thermal performance of the CTEG system using thermosyphon as heat-transporting device.

In this proposed design (as shown in Figure 5.56), there are three thermosyphons or gravity-assisted wickless heat pipes implemented into the CTEG-PCM system. Out of the three thermosyphons, two (primary) were used for transferring excess heat from the cold side of the TEG module to the PCM storage tank for storing heat during the day, whereas the third thermosyphon module (secondary) was embedded in the PCM storage tank for transporting heat from the melted PCM to the night time cold ambient, in preparation for the next day

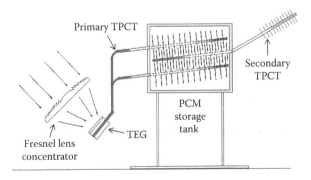

FIGURE 5.56
Schematic of concentrator TEG system with thermosyphon and PCM thermal storage.

cycle. Passive cooling in this system was achieved through repetitive heat charging and discharging from PCM thermal storage by thermosyphon modules.

The solar energy was concentrated on the hot side of the TECs by Fresnel lens concentrator. In this case, two Cu blocks were used to sandwich the TEC to provide necessary temperature difference across TECs. The evaporator sections of two thermosyphons were inserted into cold-side Cu block to acquire excess heat and transfer it to the PCM storage, in which the finned condenser sections of thermosyphons are installed. As mentioned earlier, one thermosyphon has its evaporator section embedded in the PCM tank for removing stored heat from the melted PCM to the cold night-time surroundings. It should be noted that the thermosyphon acts as a thermal diode and allows heat transfer only in one direction, that is, in bottom heat mode (evaporator below condenser configuration).

Figure 5.57 presents the experimental prototype of a thermoelectric generator with thermosyphons and a PCM storage tank. In the lab prototype, only two thermosyphons (one primary and one secondary) were used to validate the passive cooling concept. For the primary thermosyphon, only the condenser section is finned, whereas for the secondary thermosyphon, both the condenser and evaporator are finned. The TECs, as shown in Figure 5.57, were sandwiched between two Cu blocks, one provided with holes for installing heating rods and the other provided with holes to install thermosyphon evaporators. For lab testing, electric cartridge heaters were used as heat simulators. Figure 5.58 shows the Fresnel lens concentrator that will be used to test the proposed system.

The PCM used in the thermal storage tank is paraffin wax. It is non-corrosive, is readily available and has relatively high latent heat storage capacity (140 kJ/kg), which make it attractive to be used in thermal storage. Table 5.4 shows the thermophysical properties of paraffin wax. It is noted that paraffin wax has very low thermal conductivity (0.2 W/m·K), which can affect the heat transfer performance of the thermosyphons. In order to overcome this issue, thermosyphon sections installed inside the PCM tank were adequately finned. These fins help to improve the effective thermal conductivity of the PCM. All gravity-assisted thermosyphons were tilted at 5° angles to facilitate the return of condensate to the evaporator. In addition to

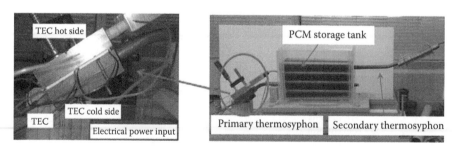

FIGURE 5.57
TEC generator module with thermosyphon and PCM assembly.

FIGURE 5.58
Fresnel lens concentrator with the tracking system and support structure.

TABLE 5.4

Thermophysical Properties of Proposed Paraffin Wax

Description	Value
Melting temperature (°C)	47
Solid density (kg/m³)	880
Liquid density (kg/m³)	760
Latent heat capacity (kJ/kg·K)	140
Specific heat capacity (solid/liquid) (kJ/kg·K)	2.4/1.8
Thermal conductivity (solid) (W/m·K)	0.2

this, PCM storage tank should be elevated above the TEC modules to ensure that thermosyphon condenser section is above the evaporator section.

5.5.9 Case Study 9: Ultra-Large Heat Pipes for Cooling the Earth

Figure 5.59 shows the overall thermal performance of CTEG-PCM system under concentration of 75 suns. The maximum TEC hot side temperature in the TEG module reached 240°C. Hence, for the maximum solar concentration ratio (CR) of 75, the aperture diameter of the Fresnel lens should be sized to 560 mm. The TEC cold side temperature can be maintained at 88°C,

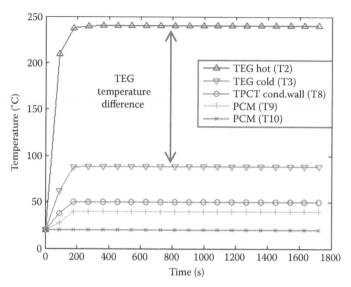

FIGURE 5.59
Thermal performance of CTEG-PCM system (CR = 75).

which corresponds to a constant temperature difference of 152°C across TEC modules. As evident from graph in Figure 5.59, the dissipated heat is absorbed by the PCM as sensible heat for first 200 seconds and thereafter latent heat absorption upon reaching the melting point (47°C). The PCM temperatures (T9 and T10) located at 10mm and 50mm above the primary thermosyphon condenser are designated to capture melting performance during heat absorption. T9 first rises gradually and then become constant during melting process. T10 remained unchanged during the test period of 30 minutes due to the low PCM thermal conductivity. Figure 5.60 presents the effect of concentration ratio on the temperature difference across two TECs and output power. CR of 75 gives maximum output power of 9.5 W at temperature difference of 150 across TEC. The proposed system can utilize photonic energy directly from sun, if the TECs are replaced by PV cells.

Global warming and protection of the earth's environment have been hot topics of concern for all countries. The rise in CO_2 production and thus the rise in the earth's temperature may have been the cause of increased natural disasters in recent years. One idea to prevent the rise in the earth's temperature is to use an ultra-large-scale heat pipe to dissipate the heat generated from people to outer space. Table 5.5 shows heat-transfer capability versus heat pipe size. For example, a 10 km long heat pipe of diameter 1 km is capable of transferring 1×10^{11} W. Figure 5.61 shows an example of energy consumption in Japan. Let us assume that the world's energy consumption is 20 times that of Japan. Calculation shows that a heat pipe of 1 km diameter and 10 km length is required to transfer the heat generated by the world to the outer space.

TABLE 5.5

Heat Transfer Capability versus Heat Pipe Sizes

		Length (Left) (m)	Q: Heat Transfer Rate (K·W)	D: Size of Heat Pipe (m)	A: Cross-Sectional Area (m²)	Q_{ax}: Axial Heat Flux (K·W/m²)	$P = Q \times L$ (K·W·m)	$Z = P/A$ (K·W·m/m²)
Electronics cooling	Personal computer	0.1	0.03	0.006	0.00002826	1062	0.003	106.1571
	Power electronics	0.3	0.1	0.0125	0.00012266	815	0.03	244.586
Heat exchanger	Heat exchanger: Gas air heat exchanger	8.25	58	0.0254	0.00050645	114523	478.5	944810.8
Long heat pipe	Snow melting	5	0.3	0.0254	0.00050645	592	1.5	2961.789
Long-distance heat transfer	Geothermal extract	100	60	0.15	0.0176625	3397	6000	339702.8
	Space heat pipe	10,000	1E+11	1000	785000	127389	1E+15	1.27E+09

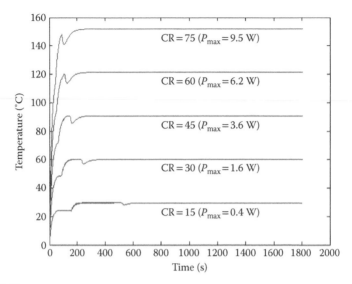

FIGURE 5.60
Temperature difference and output power from two TECs under different ratios.

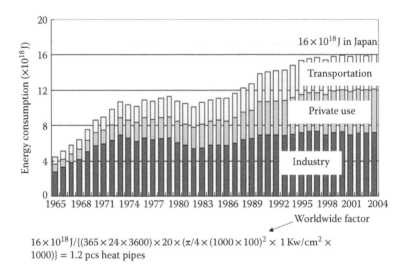

16×10^{18} J/{$(365 \times 24 \times 3600) \times 20 \times (\pi/4 \times (1000 \times 100)^2 \times 1\,\text{Kw/cm}^2 \times 1000$)} = 1.2 pcs heat pipes

FIGURE 5.61
Example of energy consumption in Japan.

Figures 5.62 and 5.63 show the concept and basic design of the ultra-large-scale heat pipe to transfer heat to the outer space. The primary heat pipe is 1 km in diameter and 10 km in height and contains a magnetic fluid, and there are plasma rings on the external body. The primary heat pipe extracts the heat from the ground and transfers the heat upward. There

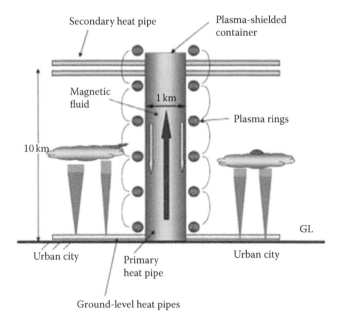

FIGURE 5.62
Concept of ultra-large-scale heat pipe for cooling the earth.

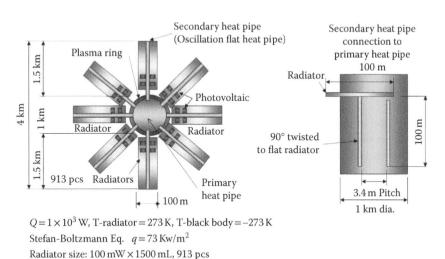

$Q = 1 \times 10^3$ W, T-radiator $= 273$ K, T-black body $= -273$ K
Stefan-Boltzmann Eq. $q = 73$ Kw/m^2
Radiator size: 100 mW $\times$ 1500 mL, 913 pcs

FIGURE 5.63
Design of ultra-large-scale heat pipe for cooling the earth.

are multiple secondary heat pipes attached at the upper end of the primary heat pipe that spread the heat to the radiators. The ultra-large-scale heat pipe for cooling the earth may seem like an impractical dream, but the authors believe that it is feasible.

5.5.10 Case Study 10: Heat Pipe Application for a Nuclear Reactor

Nuclear power has the potential to support the electric energy needs of the growing population. The nuclear energy share in global electricity production is growing fast due to its high energy density, advanced reactor technology, low greenhouse gas emissions, ease of installation and plant expansion. In nuclear power plants, kinetic energy produced by the nuclear fission of the radioactive material (usually uranium-235 or plutonium-239) is converted into heat and thereby to useful electrical power. Nuclear fission provides very high-density energy; for example, 1 kg of U-235 can produce 3 million times the energy generated by the equivalent mass of coal. Japan has 54 nuclear power reactors with a total electric power capacity of 49 GW (30% of the country's demand), and there was a proposal for 19 new reactors with a total capacity of 13 GW to be built in the near future.

The two most commonly used reactors in nuclear power plants are pressurised water reactor (PWR) and boiling water reactor (BWR). PWR pumps high-pressure coolant (water) to the reactor core to extract energy from the nuclear fission reaction. The hot water is then passed through the steam generator, where it heats up the secondary coolant and produces steam, which is passed through turbine to generate electricity. Unlike PWR, in BWR steam (~282°C) is generated in the nuclear core that is directly used to drive the turbine. In case of an accident, BWR is more susceptible to radiation leak than PWR due to direct utilisation of the contaminated steam in the turbine located outside the primary containment. The BWR containment comprises drywell that houses the reactor with a related cooling system and wet well or suppression pool. The suppression pool contains water charge for core cooling during emergency reactor shutdown and for the dumping excess heat (nuclear reaction control) during reactor operation. Figure 5.64 shows a schematic of the BWR-based nuclear power plant that contains a reactor vessel with fuel and control rod assemblies, turbine and generator arrangement, seawater-cooled condenser, suppression pool and, most importantly, electrically driven ECCS with pumps. Nuclear power plants normally use seawater for cooling. In case of any malfunctioning, the nuclear reactors are automatically shut down by using the control rod mechanism. After shutdown, the ECCS is required to transfer and dissipate the residual heat from the core and maintain the reactor temperature within the safer limits (<100°C). The ECCS, which is activated after reactor shutdown, typically uses diesel generators to power a number of pumps for spraying high-pressure water on the hot core. If the active water-cooling system is stopped due to loss of electrical power, then the reactor internal temperature and pressure will build up due to steam formation from the accumulated residual heat causing fuel meltdown (~1800°C) and reactor vessel damage.

One of the worst nuclear accidents at the Fukushima nuclear power plant in Japan was caused by the failure of electric generators for ECCS, and the failure was caused by M9 earthquake and tsunami triggered by it. The

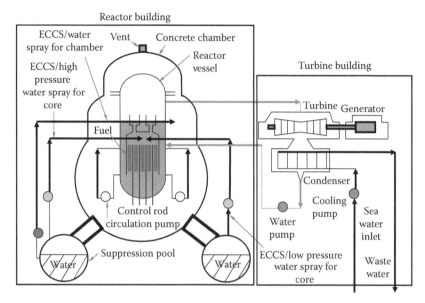

FIGURE 5.64
Boiling water reactor with ECCS system.

system described in Figure 5.64 is similar to the BWR used in the Fukushima nuclear power plant. The active ECCS is not reliable due to its dependence on the electric power and therefore is prone to failure during adverse natural calamities. Passive cooling system is more potential to enhance plant safety through system reliability and eliminate cost associated with the installation, maintenance and operation of the active cooling system that constituted numerous electric pumps with independent and dedicated power supplies. Most of the existing safety (or ECCS) systems are of category D type (need initiate signal, energy from stored sources and limited active components to control) and utilise gravity-assisted fluid (single phase or two phase) circulation through the core to remove the decay heat.

In the present investigation, the heat-pipe-based decay heat removal system (Kaminaga et al., 1988) from the reactor core has been proposed and analysed as a safety system for nuclear reactors. The proposed heat pipe ECCS will operate passively with high runtime reliability.

5.5.10.1 Decay Heat

The decay heat output by the reactor after shutdown process, $P(t)$, can be expressed as follows:

$$\frac{P(t)}{P_o} = 0.066\left[t^{-0.2} - \left(t_s + t\right)^{-0.2}\right] \tag{5.2}$$

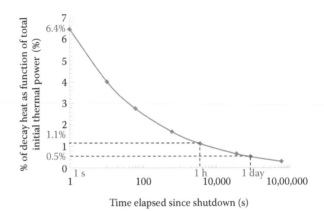

FIGURE 5.65
Decay heat by nuclear reactor, with 1350 MW thermal after shutdown.

It depends on the nominal thermal power before shutdown (P_o), time for which the reactor was in operation before shutdown (t_s) and time since reactor shutdown (t). The decay heat variation after shutdown for a nuclear reactor, similar to Fukushima no. 1 reactor, with 1350 MW thermal and 460 MW electrical output and operating for the last 5 years, is calculated and presented in Figure 5.65. Immediately after reactor shutdown, the decay heat is around 6.4% of total thermal power at normal operating conditions (~87 MW), and it is reduced to 0.5% of total thermal power (~7 MW) after a day and continues to reduce exponentially with time.

5.5.10.2 Heat Pipe ECCS

Figure 5.66 shows the proposed loop-type heat-pipe-based ECCS for BWR. There are two components of the overall ECCS system: (1) gravity-assisted feed water tank and (2) loop-type heat pipe assembly. The concept is to use the elevated water tank for initial flooding of the core with water, by gravity feed, for a specific time (~10 minutes) and passive core cooling using loop-type heat pipe.

5.5.10.3 Gravity-Assisted Water Charging

The feed water flooding of the core using an elevated water tank assists the following:

- Replenishes the coolant inside the core during loss of coolant accident (LOCA)
- Provides initial accelerated cooling of the nuclear core
- Complements additional cooling backup system
- Avoids the Leidenfrost phenomenon during loop-type heat pipe start-up

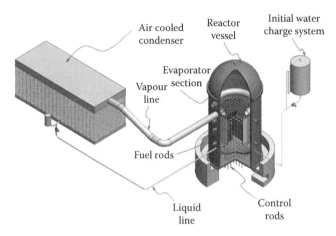

FIGURE 5.66
Proposed heat pipe ECCS with gravity-assisted feed water.

One of the main purposes of the initial water charge is to avoid the Leidenfrost effect, which is a phenomenon where the liquid coming in contact with a surface significantly hotter than the liquid's boiling point produces an insulating vapour layer that keeps the liquid from boiling rapidly, thereby limiting heat transfer from the hot surface to be cooled. Here, the gravity-assisted initial flooding of the core with water will provide cooling for the heat pipe evaporator, thereby limiting the occurrence of the Leidenfrost phenomenon. The feed water tank of 18 m^3 capacity is located 5.1 m above the reactor water level and drains through a 8 cm tube at 10 m/s flow velocity. This will account to 32 t/s of water flow rate for an initial 10 minutes interval. Such a system design will reduce the decay heat to less than 27 MW at the end of the water-charging phase.

5.5.10.4 Loop-Type Heat Pipe System

The loop-type heat pipe system with an evaporator mounted inside the reactor vessel and a condenser installed outside the primary containment building is shown in Figure 5.66. The heat pipe is designed to continually dissipate the decay heat output by the nuclear fuel. The evaporator consists of 62 vertical tubes (0.15 m outer diameter and 6 m long) placed circumferentially around the fuel core and connected via a top and bottom ring-shaped header. The natural-convection-cooled condenser (21 m × 10 m × 5 m) consists of 840 tubes (each 0.15 m outer diameter and 5 m long) with aluminium fins (300 mm diameter and 3 mm thickness at 20 mm pitch) arranged in triangular flow arrangement. Each of these tubes is connected to a top and bottom rectangular header. The heat pipe is made up of stainless steel SUS-316L with internal titanium lining to make the system compatible with water as the working fluid.

5.5.10.5 Heat Pipe Thermal Analysis

In this section, thermal performance of the loop-type heat pipe has been estimated on the basis of the reactor thermal data and heat pipe system geometry. From the reactor's hot water with an initial temperature of 282°C to
ambient air at 40°C, there are a number of conduction- and convection-based
thermal resistances to heat transfer via heat pipe (as shown in a thermal network in Figure 5.67). The conduction heat resistance through a cylindrical
tube is given by Equation 5.3.

$$R_{cond} = \frac{\ln(D_o/D_i)}{2\pi kL} \qquad (5.3)$$

where D_o, D_i and L are the outer diameter, inner diameter and length of the
tube, and k is the thermal conductivity for the tube material (16 W/m·K for
SUS). Convection heat resistance can be expressed as follows:

$$R_{conv} = \frac{1}{hA} \qquad (5.4)$$

where A is the heat transfer surface area and h is the evaporative or condensation (h_e or h_c) heat transfer coefficient or natural convection heat transfer
coefficient (water to evaporator external surface [h_{w-e}] and condenser external
surface to air [h_{c-a}]). These coefficients can be calculated using the following

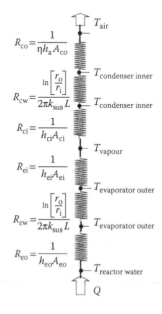

FIGURE 5.67
Heat pipe thermal network.

correlations. The boiling heat transfer coefficient (h_e) in the evaporator can be calculated from Imura's correlation.

$$h_e = 0.32\chi \left(\frac{P_{in}}{P_o} \right)^{0.3}$$ (5.5)

where P_{in} and P_o are the heat pipe inside pressure and ambient pressure, respectively, and

$$\chi = \frac{\rho_l^{0.65} \lambda_l^{0.3} c_{pl}^{0.7} g^{0.2} q_e^{0.4}}{\rho_v^{0.25} L^{0.4} \mu_l^{0.1}}$$ (5.6)

where ρ_l, ρ_v, μ_l, λ_l, c_{pl}, q_e, L and g are liquid/vapour density, liquid dynamic viscosity, liquid thermal conductivity, liquid specific heat, evaporator heat flux and acceleration due to gravity. The condensation heat transfer coefficient (h_c) in the condenser can be calculated from Nusselt's correlation.

$$h_c = 0.925 \left(\frac{\lambda_l^3 \rho_l^2 gL}{\mu_l q_c l_c} \right)^{1/3}$$ (5.7)

where ρ_l, μ_l, λ_l, q_c, L and g are liquid density, liquid dynamic viscosity, liquid thermal conductivity, condenser heat flux, latent heat and acceleration due to gravity.

The water side natural heat transfer coefficient from reactor water to evaporator (h_{w-e}) external surface is estimated from Equations 5.8 through 5.12.

$$h_{w-e} = \frac{Nu_x \lambda_l}{l}$$ (5.8)

where

$$Nu_x = \left[C_1 \left(\frac{v_\infty}{v_w} \right)^{0.21} Ra_x^{1/4} \right]; \ 10^4 \leq Ra_x \leq 4 \times 10^9 - 3 \times 10^{10}$$ (5.9)

$$C_1 = \frac{3}{4} \left(\frac{Pr}{2.4 + 4.9\sqrt{Pr} + 5Pr} \right)^{1/4}$$ (5.10)

$$Ra_x = Gr_x Pr$$ (5.11)

$$Gr_x = \frac{g\beta (T_w - T_\infty) x^3}{v_\infty^2}$$ (5.12)

Subscripts ∞ and w represents the properties of liquid (water) at reactor bulk (T_∞) and evaporator wall (T_w) temperatures. $x = l/2$, where l is the length of the evaporator tube. T, ν, Pr, g and β are temperature, liquid dynamic viscosity, Prandl Number, acceleration due to gravity and thermal expansion coefficient of water (at mean temperature, i.e. $T_m = [T_\infty + T_w]/2$), respectively.

The air side heat transfer coefficient from the condenser external surface to ambient air $(h_{c\text{-}a})$ can be calculated from Equation 5.12.

$$h_{c\text{-}a} = 0.45 \left(\frac{\lambda_a}{D_o} \right) \left(\frac{D_o G_{max}}{\mu_a} \right)^{0.625} \left(\frac{A_f}{A_o^*} \right)^{-0.375} \left(\frac{C_{pa} \mu_a}{\lambda_a} \right)^{-0.375} \tag{5.13}$$

where λ_a, C_{pa}, μ_a, D_o, G_{max}, A_f and A_o^* are air thermal conductivity, air specific heat capacity, air dynamic viscosity, condenser tube outer diameter, air mass flow rate (in kg·m²·h), fin area per metre and non-fin area per meter of condenser tube, respectively. The $h_{c\text{-}a}$ value from Equation 5.13 will be in K·cal/h·°C·m², which can be converted to W/m²·K by multiplying with 1.163. Equation 5.13 is valid for

$$5 < \frac{A_f}{A_o^*} < 12 \tag{5.14}$$

For a finned heat exchanger, the effective finned area for heat transfer is determined by fin efficiency, which depends on the fin properties and fin-to-air heat transfer coefficient $(h_{c\text{-}a})$. For the present heat exchanger design, the annular fin with a rectangular profile was considered for which overall fin efficiency (η_o) can be calculated as follows:

The overall thermal resistance (R_o) of the loop-type heat pipe can be expressed as

$$R_o = \frac{1}{h_{w\text{-}e} A_{eo}} + \frac{\ln(D_{eo}/D_{ei})}{2\pi k_e L_e} + \frac{1}{h_e A_{ei}} + \frac{1}{h_c A_{ci}} + \frac{\ln(D_{co}/D_{ci})}{2\pi k_c L_c} + \frac{1}{h_{c\text{-}a} A_{ct} \eta_o} \tag{5.15}$$

where subscripts eo, ei, co and ci denote the evaporator outer, evaporator internal, condenser outer and condenser internal, respectively. A, D, L, k_e and k_c are area, diameter, length and thermal conductivity of evaporator and condenser tubes (stainless steel), respectively.

Based on the above approach, the results are as follows:

- Heat pipe evaporator boiling coefficient $(h_e) = 14{,}000$ W/m²·K
- Condensation heat transfer coefficient $(h_c) = 5200$ W/m²·K
- Water to evaporator external surface natural heat transfer coefficient $(h_{w\text{-}e}) = 1180$ W/m²·K

- Condenser external surface to ambient air heat transfer coefficient $(h_{c\text{-}a}) = 7.9$ W/m²·K (for air velocity, $V_a = 1$ m/s)
- The overall calculated thermal resistance of the heat pipe (R_o) was 1.44×10^{-5} °C/W

5.5.10.6 Reactor Thermal Analysis

In this section, the reactor vessel is analysed for different core cooling conditions.

5.5.10.6.1 ECCS Failure: No Cooling

If the cooling function of the ECCS has failed, there will be a continuous increase in the temperature of the reactor vessel under the impact of the decay heat released by the nuclear fuel inside the core. Under the no cooling condition, the thermodynamic state of the reactor can be expressed by Equation 5.16.

$$MC_{pw}\left(T_{rf} - T_{ri}\right) = P\Delta t \tag{5.16}$$

where M is the water mass inside the reactor (~200 t), C_{pw} is water specific heat capacity, T_{ri} is the reactor initial temperature (~282°C), T_{rf} is the reactor final temperature after Δt time and P is the rate of decay heat output by reactor, with 1350 MW heat output, as per Equation 5.2. Figure 5.68 shows the transient thermal response of water inside the reactor which will reach nuclear meltdown situation (>1800°C) within a 2 day period due to accumulation of decay heat inside the vessel.

5.5.10.6.2 Heat Pipe ECCS

Heat pipe will be able to provide reliable and passive two-phase heat transfer system for the removal of the decay heat without any failure consequences.

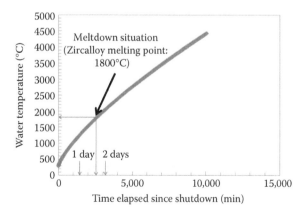

FIGURE 5.68
Transient response of reactor water temperature under ECCS failure condition.

With the proposed loop-type heat pipe system, the overall system energy balance can be expressed by Equation 5.17.

$$MC_{pw}\left(T_{rf} - T_{ri}\right) = \left(P - \dot{Q}_{hp}\right)\Delta t \tag{5.17}$$

where $\dot{Q}_{hp}$ is the heat pipe heat removal rate from the reactor given by Equation 5.18.

$$\dot{Q}_{hp} = \frac{T_{ri} - T_a}{R_o} \tag{5.18}$$

where T_a is the ambient temperature ($\sim40°C$) and R_o is thermal resistance from the reactor coolant to ambient air via the heat pipe system ($1.44 \times 10^{-5}°C/W$). Figure 5.69 shows the transient response of the reactor vessel (or water) temperature, T_{rf}, with the heat pipe heat removal system, as calculated from Equation 5.17. It is observed from the graph that, after the reactor shutdown, the water temperature continued to rise for the first 60 minutes and reached the maximum of 305°C from the initial 282°C, followed by a gradual drop in temperature with time. The initial rise in the reactor temperature was due to larger quantity of decay heat from the reactor core after the shutdown process which exceeded the heat pipe heat dissipation capacity. As the decay heat reduced exponentially with time and dropped below the heat pipe capacity, there was a gentle drop in the reactor vessel temperature. If the overall thermal resistance of the heat pipe unit is doubled ($\sim2.88 \times 10^{-5}°C/W$) by reducing the heat pipe system size, then the extent of temperature rises and time taken to reduce the reactor temperature below a certain value will increase accordingly. In this case, the reactor will achieve the highest temperature of 404°C in 7 hours before commencement of the cooling down operation.

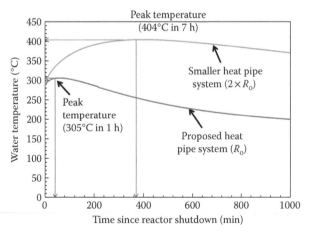

FIGURE 5.69
Transient response of reactor water temperature with loop heat pipe cooling.

Figure 5.70 shows the extended cooling operation for two different heat pipe system sizes. Here, it is estimated that to achieve a temperature below 250°C, the proposed heat pipe system will take less than 7 hours, whereas with a reduced system size, the cooling time will increase to 4.5 days. As safety is one of the prime requirements in the nuclear plants, the heat pipe unit's shorter cooling time will provide a more reliable and secure system.

5.5.10.6.3 Heat Pipe ECCS with Initial Water Charge

The proposed heat-pipe-based ECCS system with initial water charge represents a more advanced and safer ECCS for a nuclear reactor by providing a reliable and passive heat pipe core cooling system and by injecting gravity feed additional water quantity inside the core for 10 minutes, which can replenish the lost coolant and address high-temperature heat transfer limiting issues (Leidenfrost effect). Figure 5.71 shows the thermal model of the proposed system, with its energy balance represented by Equation 5.19.

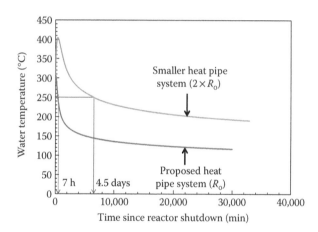

FIGURE 5.70
Extended transient response of reactor water temperature with loop heat pipe cooling.

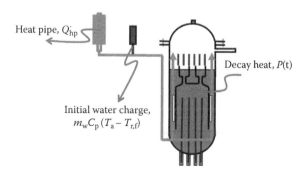

FIGURE 5.71
Thermal model of reactor with heat pipe cooling and initial water charge.

$$MC_{pw}\left(T_{rf}-T_{ri}\right)=\left[\left(P-Q_{hp}^{\cdot}\right)+m_{w}^{\cdot}C_{pw}\left(T_{a}-T_{rf}\right)\right]\Delta t \qquad (5.19)$$

where $m_{w}^{\cdot}$ is the gravity feed water mass flow rate.

Figure 5.72 shows the reactor water temperature response with and without an initial 10 minute water feeding (32 t/s) and with heat pipe cooling. The gravity feed water provided additional cooling capacity for the core, which assists to reduce the highest reactor temperature achieved, provided initial temperature drop after shutdown procedure and reduced reactor cooling time. Figure 5.73 shows the system operation for a larger span of time and that showed with water injection, the reactor can achieve a temperature below 250°C within 4.3 hours, which is 1.6 times less than without water charging. It should be noted that other than reducing cooling time, water feeding provided coolant replenishment and addressed heat transfer limiting issues during heat pipe start-up.

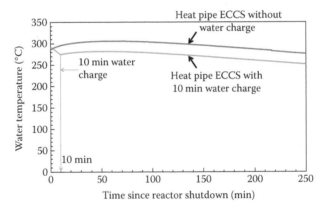

FIGURE 5.72
Reactor water temperature variation with and without initial water charge and heat pipe ECCS.

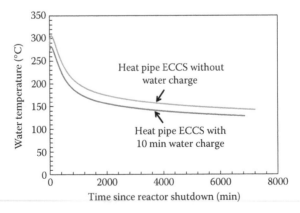

FIGURE 5.73
Extended transient response of water reactor with and without initial water charge and heat pipe ECCS.

In order to validate the concept of cooling the nuclear reactor core using a loop-type heat pipe with initial gravity water charging, a lab prototype with 1:10,000 scale will be fabricated and tested. The proof-of-concept prototype will be expected to transfer heat load up to 3 kW and will be downscaled on the basis of (1) evaporator length to diameter ratio, (2) evaporator active heat flux and (3) evaporator axial heat flux criteria, to simulate experimental conditions close to a real-scale model. The heat pipe container material and working fluid will be similar to the proposed configuration.

5.6 Concluding Remarks

- The computer processor's die surface, where heat generation is usually small and heat dissipation is large, requires a large space for cooling. The most effective way to transfer heat from the heat source to the dissipation area is to use heat pipes or vapour chambers. Using heat pipes in the metal base or vapour chambers as the base helps to reduce thermal spreading at the base. Therefore, the cooling capacity can be increased.

- As the processor power increases and thus the required lower thermal resistance, the tendency of the cooling solution technology to maximise and extend air cooling has moved towards more use of heat pipes and vapour chambers. Processor heat dissipation will continue to increase owing to the demand for higher and faster processors, and therefore to provide adequate cooling, it is essential to continue to research and develop superior heat pipes and vapour chambers having high heat flux capability and minimal possible thermal resistance between heating and cooling ends.

- The thermal resistance from the CPU's die to the IHS comprised a significant ratio of overall thermal resistance. Therefore, there is a need to consider replacing the metal Cu IHS with some other high heat transfer device such as MVC. In addition, direct brazing or soldering the IHS to the CPU's die needs to be considered.

- Heat-pipe-based passive systems can provide reliable and effective thermal control for energy conservation, energy recovery and renewable energy applications.

- HPHE pre-cooler with 118 heat pipes designed for 30°C ambient temperature can effectively dissipate a 30 kW data centre's heat for most of the year. The size of the pre-cooler has a strong dependence on the ambient temperature of the location.

- Heat pipe ice storage system for a 1 MW data centre emergency cooling backup will require only ~68 heat pipes. About 952 kg of ice per heat pipe per winter season can be generated. The cost of the heat-pipe-based ice storage system (0.3–0.68 $/W) is comparable to existing backup technologies ($0.5–$1/W).

- Heat-pipe-based permafrost storage for agricultural products provides a sustainable and environmental friendly approach for cold storages by saving electricity.

- Proposed heat-pipe-based cooling systems utilise natural ambient cold energy, and therefore there is no running cost and no greenhouse gas emissions.

- HPHE with 65% effectiveness provided a waste heat recovery system that was able to supply all the heat needed by the proofing oven, thus eliminating the need for natural gas heating. On the basis of this experimental study, it was observed that the most suitable location for a gas reservoir to collect non-condensable gas in a loop-type HPHE is in the bottom header of the condenser where the condensate exits to the evaporator by gravity effect through an adiabatic pipe.

- Heat pipe turbine to convert available low-grade heat into electricity was developed and tested. The prototype with a cylinder of 2.8 m height and 0.5 m diameter was able to produce 100 W electrical power at 6000 rpm for 10 kW input heat load. Although the efficiency of the proposed heat pipe turbine is low, it has the capability of utilising very low-grade heat and converts it into useful electricity.

- A demonstration system for geothermal heat extraction using large-size heat pipe, 150 mm OD and 150 m long, with showering nozzles for proper liquid distribution in the evaporator was constructed and tested. The heat pipe was able to extract 90 kW heat continuously with 3000 W/m^2 heat flux at the evaporator.

- Combination of thermosyphons and TECs provided a fully passive and simple power supply system for remote area applications using the temperature differences that exist in a typical solar pond. The designed system was able to provide a maximum power point of 3.2 W, which was obtained at 13.4 V and 0.24 A when the temperature difference of 27°C was maintained across 16 TEG modules. In this case, the open-circuit voltage and the short-circuit current values of 26 V and 0.4 A, respectively, were obtained.

- Thermosyphon modules provided an effective heat-transporting system for thermal management of the CTEG module with paraffin wax PCM thermal storage. Based on the simulation model, it was estimated that the designed system is able to maintain a temperature difference of 152°C across the TECs and produced 9.5 W of thermoelectric power at a concentration ratio of 75.

- A novel concept for thermal management of data centres on the basis of heat-pipe-based cold water storage can help to minimise the thermal load on chiller units and thus save electricity and associated costs. It is a compact cold energy storage system for small- and large-size data centres and is a more likely option for locations with yearly ambient temperature below the cold storage water temperature. With the massive electricity usage necessary for today's data centres, the cold-energy-storage-based thermal management system can help to address the energy crisis and global warming faced by the world at present.

- In cold regions, the cold energy available in the cold seasons can be collected and used for some applications without the need for electricity, for example, the cooling of data centres, or cooling and storage of agricultural products, or it can be used in any application that requires some form of cooling.

- Hot energy such as solar heat, geothermal heat, or industrial waste heat can be used for practical applications. For example, solar heat can be used for electricity generation or simply for snow melting on road surfaces.

- On the basis of present research, it is concluded that loop-type heat pipe can be successfully utilised for dissipating decay heat from the nuclear reactor core. The proposed heat pipe ECCS with cylindrical evaporator composed of 62 vertical tubes, each 150 mm diameter and 6 m length, mounted around the periphery of the nuclear fuel assembly, and 21 m × 10 m × 5 m finned condenser installed outside the primary containment was designed to dissipate decay heat from the reactor vessel with 1380 MW full load thermal capacity to ambient air at 40°C. Overall thermal resistance of the heat pipe unit was 1.44×10^{-5} °C/W. Within 6–7 hours, the proposed heat pipe system can reduce the reactor temperature below 250°C from initial working temperature of 282°C. It is recommended to use initial water charge of 32 t/s for 10 minutes to accelerate the cooling time of the core and provide safer ECCS cooling system. With heat pipe ECCS and 10 minutes water charge, the core cooling time was reduced. The proposed heat pipe system can be operated completely passively, which will provide safer operational environment to nuclear power plants.

- Ideas for using heat pipes to make these abundant natural energy sources useful are unlimited. The global warming crisis is real and threatens the earth's environment and inhabitants, so it is a responsibility for all countries and individuals to try to reduce and prevent further damages.

- The authors believe that more development is essential for extremely small- and large-sized heat pipes, that is, micro-scale heat pipes for computers and electronics applications, and large-size heat pipes in kilometres scale for terrestrial and space applications.

References

Akbarzadeh, A., Andrews, J., and Golding, P., 2005, Solar Pond Technologies: A Review and Future Directions, in Y. Goswami (ed.), *Advances in Solar Energy*, Earthscan, London, UK, 233–294.

Brill, K. G., 2010, *Heat Density Trends in Data Processing, Computer Systems, and Telecommunications Equipment: Perspectives, Implications, and the Current Reality in Many Data Centers*, The Uptime Institute, USA.

Dube, V., Akbarzadeh, A., and Andrews, J., 2004, The Effects of Non-Condensable Gases on the Performance of Loop Thermosyphon Heat Exchangers, *Applied Thermal Engineering*, **24**, 2439–2451.

Kaminaga, F., Okamoto, Y., Keibu, M., Ito, H., Mochizuki, M., and Sugihara, S., 1998, Application of Heat Pipe to JSPR Safety, Proceedings of Design Feasibility for JSPR, UTNL-R0229, 31–39.

Nguyen, T., Mochizuki, M., Mashiko, K., Sauciuc, I., Akbarzadeh, A., Johnson, P., Kusaba, S., and Suzuki, H., 1999, Heat Pipe Turbine for Production of Electrical Power from Renewable Sources, Proceedings of the 5th ASME/JSME Joint Thermal Engineering Conference, ASME Paper No: AJTE99-6339, 15–19 March 1999, San Diego, California.

Patel, C. D., Sharma, R., Bash, C. E., and Beitelmal, A., 2002, Thermal Considerations in Cooling Large-Scale High Compute Density Data Centers, Proceeding of the 2002 Inter Society Conference on Thermal Phenomena, IEEE.

Schmidt, R., Iyengar, M., Steffes, J., and Lund, V., 2009, Co-Generation, Grid Independent Power and Cooling for a Data Center, Proceeding of ASME 2009 InterPACK Conference, IPACK, 19–23 July 2009, San Francisco, California.

Schmidt, R. R., Crus, E. E., and Lyengar, M. K., 4/5 July/September, 2005, Challenges of Data Center Thermal Management, *IBM Journal of Research and Development*, **49**, 709.

Bibliography

Akbarzadeh, A., and Wadowski, T., 1996, Heat Pipe Based Cooling Systems for Photovoltaic Cells under Concentrated Solar Radiation, *Applied Thermal Engineering*, **16** (1), 81–87.

Dunn, P. D., and Reay, D. A., 1994, *Heat Pipes*, 4th Ed., Pergamon.

Glasstone, S., and Sesonske, A., 1967, *Nuclear Reactor Engineering*, 2nd Ed., Van Nostrand Reinhold.

Incropera, F. P., and DeWitt D. P., 2002, *Fundamentals of Heat and Mass Transfer*, 5th Ed., Wiley.

International Atomic Energy Agency (IAEA), 2009, *Passive Safety Systems and Natural Circulation in Water Cooled Power Plants*, IAEA-TECDOC-1624, Austria.

Japan Society of Mechanical Engineers, 1986, *JSME Databook: Heat Transfer*, 4th Ed., JSME; Tokyo, Japan.

Mashiko, K., Mochizuki, M., Watanabe, Y., Kanai, Y., Eguchi, K., and Shiraishi, M., 1994, Development of a Large Scale Loop Type Gravity Assisted Heat Pipe Having Showering Nozzles, Proceedings of 4th International Heat Pipe Symposium, 16–18 May 1994, Tsukuba, Japan, 264–274.

Mashiko, K., Ryokai, K., Tsuchiya, F., and Fukuda, M., 1989, Development of an Artificial Permafrost Storage, Proceedings of the ASME/AlChE National Heat Transfer Conference, 5–8 August 1989, Philadelphia, Pennsylvania.

Mochizuki, M., Mashiko, K., Shihyakugari, S., Watanabe, Y., Eguchi, K., Namiki, T., and Shiraishi, M., 1995, A Demonstration Test of a Large Scale Loop Type Heat Pipe With Showering Nozzles, Proceedings of the 9th International Heat Pipe Conference, 1–5 May 1995, Albuquerque, New Mexico.

Mochizuki, M., Nguyen, T., Agata, H., and Kiyooka, F., Advanced Thermal Solution Using Vapor Chamber Technology for Cooling High Performance Desktop CPU in Notebook Computer, The 1st International Symposium on Micro & Nano Technology, 4–7 March 2004, Honolulu, Haiwai.

Mochizuki, M., Nguyen, T., Saito, Y., Kiyooka, K., Wu, X., Nguyen, T., and Wuttijumnong, V., 2007, Advance Cooling Chip by Heat Pipes and Vapor Chamber, 14th International Heat Pipe Conference (14th IHPC), 22–27 April 2007, Florianopolis, Brazil.

Mochizuki, M., Nguyen, T., Saito, Y., Wuttijumnong, V., Wu, X., and Nguyen T., 2005, Revolution in Fan Heat Sink Cooling Technology to Extend and Maximize Air Cooling for High Performance Processor in Laptop/Desktop/Server Application, ASME Summer Heat Transfer Conference & InterPACK '05, 17–22 July 2005, San Francisco, California.

Moore, J., Sharma, R., Shih, R., Chase, J., Patel, C., and Ranganathan, P., 2004, Going Beyond CPUs: The Potential for Temperature-Aware Data Centers, Proceedings of the First Workshop on Temperature-Aware Computer Systems, Munich, Germany.

Nguyen, T., Mochizuki, M., Mashiko, K., and Saito, Y., 2000, Advanced Heat Sink Combined With Heat Pipe for Cooling PC, Proceedings of the 4th JSME-KSME Thermal Engineering Conference, 1–6 October 2000, Kobe, Japan.

Rowe, D. M., 1995, *CRC Handbook of Thermoelectrics*, CRC Press.

Sauciuc, I., Moon, S., Chiu, C., Chrysler, G., Lee, S., Paydar, R., Walker, M., Luke, M., Mochizuki, M., Nguyen, T., and Takenaka, E., 2005, Key Thermal Challenges for Low Form Factor Thermal Solution, Semitherm, March 2005, Dallas, TX.

Singh, R., Akbarzadeh, A., Dixon, C., Riehl, R. R., and Mochizuki, M., 2004, Design, Analysis and Investigation of a Medium Scale Capillary Pumped Loop, Proceedings of the 13th International Heat Pipe Conference, 21–25 September 2004, Shanghai, China, 253–261.

Singh, R., Akbarzadeh, A., and Mochizuki, M., 2004a, Capillary Pumped Loops: An Overview, Proceedings of the 1st International Seminar on Heat Pipes and Heat Recovery Systems, 8–9 December 2004, Kuala Lumpur, Malaysia.

Singh, R., Akbarzadeh, A., and Mochizuki, M., 2004b, Effective Thermal Management Using Capillary Pumped Loop, Proceedings of 1st International Forum on Heat Transfer, Paper No: GS2-23, 24–26 November 2004, Kyoto, Japan, 115–116.

Singh, R., Akbarzadeh, A., and Mochizuki, M., 2008, Thermal Performance of a Capillary Pumped Loop for Automotive Cooling, *Experimental Heat Transfer*, **21** (4), 296–313.

Singh, R., Akbarzadeh, A., Mochizuki, M., Nguyen, T., and Wuttijumnong, V., 2005, Experimental Investigation of the Miniature Loop Heat Pipe with Flat Evaporator, ASME Summer Heat Transfer Conference & InterPACK '05, 17–22 July 2005, San Francisco, CA.

Singh, R., Mochizuki, M., Mashiko, K., and Nguyen, T., 2011a, Heat Pipe Based Pre Cooler and Emergency Cooling Systems for Data Center Energy Conservation, Paper no. S051024, JSME 2011, Proceedings of the Japan Society of Mechanical Engineers Annual Conference, 11–14 September 2011.

Singh, R., Mochizuki, M., Mashiko, K., and Nguyen, T., 2011b, Heat Pipe Based Cold Energy Storage Systems for Data Center Energy Conservation, *Energy*, **36**, 2802–2811.

Singh, R., Mochizuki, M., Mashiko, K., Nguyen, T., Wu, X. P., and Wuttijumnong, V., 2011c, Datacenter Backup Cooling System Using Heat Pipe Based Ice Storage, 48th NHTS, Proceedings of the 48th National Heat Transfer Symposium, 1–3 June 2011, Okayama, Japan.

Singh, R., Tundee, S., and Akbarzadeh, A., 2011d, Electric Power Generation from Solar Pond Using Combined Thermosyphon and Thermoelectric Modules, *Solar Energy*, **85** (2), 371–378.

Tan, L. P., Date, A., Singh, R., and Akbarzedah, A., 2010, Passive Cooling of Concentrated Photovoltaic Cells Using Phase Change Material, Solar2010, Proceedings of the 48th AuSES Annual Conference, 1–3 December 2010, Canberra, Australia.

Tan, L. P., Singh, R., Date, A., and Abkarzadeh, A., 2011, Thermal Performance of Two Phase Closed Thermosyphon in Application of Concentrated Thermoelectric Power Generator Using Phase Change Material Thermal Storage, 10th IHPS, Proceedings of the 10th International Heat Pipe Symposium, 6–9 November, Taipei, Taiwan.

Tsuchiya, F., Ryokai, K., and Mochizuch, M., 1991, A Cold Storage of Agricultural Products Surrounded by Artificial Permafrost Using Heat Pipes, Proceedings of the 6th International Symposium on Ground Freezing, 10–12 September 1991, Beijing, China, 537–542.

Tundee, S., Singh, R., Terdtoon, P., and Akbarzadeh, A., 2008, Power Generation from Solar Pond Using Combined Thermosyphon and Thermo Electric Modules, Proceedings of the SET2008 – 7th International Conference on Sustainable Energy Technologies, 24–27 August 2008, Seoul, Korea.

Wu, X. P., Mochizuki, M., Mashiko, K., Nguyen, T., Nguyen, T., Wuttijumnong, V., Cabsao, G., Singh, R., and Akbarzadeh, A., 2010a, Data Center Energy Conservation by Heat Pipe Cold Energy Storage System, IHTC-14, Proceedings of the International Heat Transfer Conference, Paper No: IHTC14-23128, 8–13 August 2010, Washington, DC.

Wu, X. P., Mochizuki, M., Mashiko, K., Nguyen, T., Wuttijumnong, V., Cabsao, G., Singh, R., and Akbarzadeh, A., 2010b, Energy Conservation Approach for Data Center Cooling Using Heat Pipe Based Cold Energy Storage System, 26th IEEE SEMI-THERM, Proceedings of the Semiconductor Thermal Measurement and Management Symposium, 21–25 February 2010, Santa Clara, CA, 115–122.

Wu, X. P., Mochizuki, M., Mashiko, K., Nguyen, T., Nguyen, T., Wuttijumnong, V., Cabusao, G., Singh, R., and Akbarzadeh, A., 2011, Cold Energy Storage Systems Using Heat Pipe Technology for Cooling Data Centers, *Frontiers in Heat Pipes*, **2**, 013005.

6

Heat Pipes and Thermosyphons for Thermal Management of Solid Sorption Machines and Fuel Cells

L. L. Vasiliev and L. L. Vasiliev Jr.

CONTENTS

6.1   Introduction 214
6.2   Heat Pipes for Fuel Cell Thermal Management 218
    6.2.1   Micro/Miniature Heat Pipes for Fuel Cell Thermal
        Management 220
    6.2.2   Micro Heat Pipe Effect in Porous Media 221
    6.2.3   Mini Loop Heat Pipes 222
    6.2.4   Loop Heat Pipes 223
        6.2.4.1   Mathematical Model 225
        6.2.4.2   Results of Modelling 227
6.3   Vapour-Dynamic Thermosyphons 229
    6.3.1   Heating Devices Based on Vapour-Dynamic
        Thermosyphons 233
    6.3.2   Vapour-Dynamic Thermosyphon as a Thermal Control
        System for Adsorbers of Solid Sorption Heat Pumps 234
    6.3.3   Vapour-Dynamic Thermosyphon as a Thermal Control
        System for Solar Solid Sorption Coolers 234
    6.3.4   Vapour-Dynamic Thermosyphon as a Thermal Control
        System for Snow Melting 237
6.4   New Generation of Heat Pipes and Thermosyphons with
    Nanofluids and Nanocoatings Applications 238
    6.4.1   Nanoporous Coatings of Heat Pipe Evaporators 242
6.5   Sorption Heat Pipes 248
    6.5.1   Air-Conditioning Systems Based on Sorption Heat Pipe
        Components 253
6.6   Conclusions 254
6.7   Acknowledgements 254
References 255

6.1 Introduction

Fuel cells are considered a modern source of heat and power generation and have various advantages in residential applications, such as low emission, low noise level and high potential efficiency (Rogg et al., 2003; Izenson and Hill, 2004; Faghri and Guo, 2005; Garrity et al., 2007). Thermally activated adsorption heat pumping technology is an energy-efficient method for converting low-temperature waste heat into useful energy for use in heating, storage, cooling, freezing and desalination based on the evaporation/condensation process (Saha et al., 2001; Tamainot-Telto et al., 2009). The combination of fuel cell and solid sorption transformer significantly increases fuel cell efficiency (Vasiliev, 2007). The increasing efficiency of new power sources (co-generation, tri-generation systems, fuel cells, photovoltaic systems, etc.) is considered to be achieved with the help of solid sorption heat pumps, refrigerators, accumulators of heat and cold, heat transformers, fuel gas (natural gas and hydrogen) storage systems and efficient heat exchangers (Saha and Ng, 2010).

This nature-friendly technology has been widely discussed since the last decade, and some adsorption machines are already in the market (Wang and Oliveira, 2006; Jakob and Kohlenbach, 2010). Fuel cells and solid sorption transformers with complex compound sorbent materials (active carbon fibre and micro-/nanochemical particles on the surface of the fibre) are now considered together with heat pipe thermal control systems. Recently, fuel cell development, modelling and performance analysis obtained much attention due to their potential for distributed power, which is a critical issue for energy security and environmental protection (Krallik, 2002; Se Min and Vasiliev, 2003; Saraff and Schwendemann, 2004; Faghri, 2005). A generalised diagram of a tri-generation system, based on the application of fuel cells as a source of power in combination with low-grade heat and solid sorption heat pump and heat pipe heat exchangers is shown in Figure 6.1. Low-temperature power systems are generally significantly less expensive to build than high-temperature ones. The overall first law of efficiency for a tri-generation system can be expressed as follows:

$$\eta_{TR} = \frac{(W_e + Q_{heat} + Q_{cold})}{F} \tag{6.1}$$

where W_e is the power generated by the fuel cell; Q_{heat} is the heat produced ($Q_{heat} = Q_{HE} + Q_{AS}$), where Q_{HE} is the heat exhaust of the engine and Q_{AS} is the heat from alternative sources; Q_{cold} is the production of cold; and F is the total fuel consumption.

The fuel cell efficiency is as follows:

$$\eta_{FC} = \Delta G/\Delta H = 1 - T\Delta S/\Delta H \tag{6.2}$$

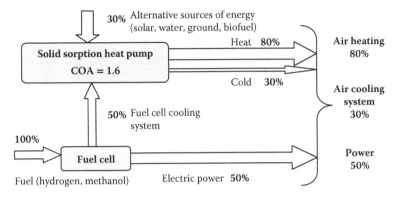

FIGURE 6.1
System of tri-generation based on fuel cell and solid sorption heat pump application, where 100% is the heat equivalent of the fuel and COA is the coefficient of amplification of heat.

where $\Delta G = \Delta H - T\Delta S = -nF\Delta E$.

If we consider a heat engine (gas turbine/electrogenerator) instead of the fuel cell, the efficiency is as follows:

$$\eta = \frac{W}{Q} = \frac{(Q_{SR} - Q_{SK})}{Q_{SR}} = 1 - \left(\frac{Q_{SK}}{Q_{SR}}\right) \qquad (6.3)$$

where Q_{SR} is heat available from the source of the heat engine and Q_{SK} is the heat delivered to the sink of the heat engine.

The most suitable combination is to combine a high-temperature fuel cell (solid oxide fuel cell [SOFC]) with multi-cascading heat pumps and coolers (Vasiliev, 2007). The heat of post-combustion exhaust gases at high temperatures (near 1000°C) from a SOFC post-combustion chamber is recovered by a heat pipe heat exchanger and preheated using the high-temperature reactor of a solid sorption heat pump.

The automotive fuel cells with hybrid power systems functioning in cars, planes and tracks are also very promising; the fuel cells convert fuels directly into electricity without combustion or using mechanical energy. There are zero carbon dioxide emissions from such systems, the wastewater being used to cool the fuel cell stack. In the residential sector with internal combustion engines, polymer electrolyte fuel cells (PEFCs) and SOFCs are welcomed as micro combined heat and power (micro-CHP).

The schematic of a two-adsorber heat pump with heat pipe thermal control is shown in Figure 6.2. This thermal system enables heat pipe heat recovery between adsorbers by the liquid circulating loop and mechanical pump.

The second stream in solid sorption transformer development is related to resorption heat pump and refrigerator technology. Resorption heat pumps and coolers based on reversible solid/gas sorption cycles can have interesting applications for space cooling, when high-temperature waste heat is

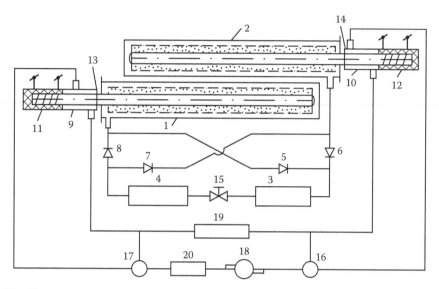

FIGURE 6.2
Two-adsorber heat pump with heat pipe thermal control: (1,2) adsorbers; (3) condenser; (4) evaporator; (5–8) valves; (9,10) liquid heat exchangers; (11,12) heat pipe evaporators; (13,14) copper–water heat pipes; (15) expansion valve; (16,17) reversing valves; (18) liquid pump; (19) liquid flow meter; (20) thermostat.

available and/or the exigencies of a harsh external environment necessitates the thermal control of an object (Lepinasse et al., 2001; Vasiliev et al., 2002).

The first advantage of resorption technology is that it uses nature-friendly refrigerants such as water, ammonia and CO_2 (no CFC, HCFC, HFC), and the second is that resorption transformers are thermally driven and can be coupled with waste heat, solar heat, burning fossil fuels, or biomass. The third advantage of resorption systems (Figure 6.3) is its ability to use a significant number of solid/gas combinations.

The main feature of the sorption heat pump (Figure 6.3) is the application of heat pipes to cool and heat adsorbers. The complex compound inside the adsorbers is made from activated carbon fibre (Busofit) and nano-/microparticle coatings of metal chlorides on the surface of Busofit fibres (Vasiliev et al., 2002).

Resorption systems have a number of advantages. For example, in comparison with existing liquid/gas absorption transformers solid/gas sorption systems have a wider range of working temperatures and have fewer problems of corrosion and crystallisation. Moreover, energy storage capacity is much higher than for liquid absorption heat pumps due to the larger reaction heat in solid/gas sorption systems. The heat/cold can be stored for long periods with low losses (Meunier, 1998).

In conventional resorption systems, major entropy production occurs due to the superheating of vapour during the cold production phase and

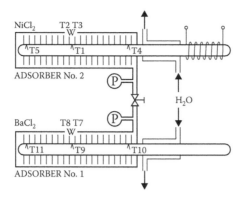

FIGURE 6.3
Sorption heat pump with adsorber 1 (Busofit/BaCl$_2$) and (Busofit/NiCl$_2$), and heat pipes used for the thermal management.

de-superheating of the vapour during the regeneration phase. Thermochemical resorption machines were demonstrated by different authors (Lepinasse et al., 1994; Choi et al., 1996; Goetz et al., 1997) using pairs such as MgCl$_2$/NH$_3$–NiCl$_2$/NH$_3$ and BaCl$_2$/NH$_3$–NiCl$_2$.

Heat pipes are very flexible systems with respect to the effective thermal control of different heat-loaded devices. Their heat transfer coefficient in the evaporator and condenser zones is 10^3–10^5 W/m^2 · K and thermal resistance is 0.01–0.03 K/W; therefore, they lead to smaller areas and masses of heat exchangers. Heat pipes and thermosyphons are easily integrated into fuel cells and simple and multi-stage adsorption systems and ensure the heat and mass recovery from one stage to another. In this chapter, different types of heat pipes are considered, including miniature heat pipes and 'micro heat pipes' (MHPs) for fuel cells, electronic component cooling and space two-phase thermal control systems, loop heat pipes (LHPs), pulsating heat pipes and sorption heat pipes (SHPs). Vapour-dynamic thermosyphons (VDTs) are suggested as the novelty for modern heat exchangers and air-conditioning systems. The new stream in heat pipe technology also includes nanofluid (NF) application and nanocoating (NC) of heat pipe evaporators (Das et al., 2003; You et al., 2003). The generation of engineered nanomaterials represents a major breakthrough in material science and nanotechnology-based materials. Recent advances in nanotechnology have allowed the development of new NFs to describe liquid suspensions containing nanoparticles (NPs) with thermal conductivities that are orders of magnitude higher than those of base liquids and sizes significantly smaller than 100 nm (Vassallo et al., 2004; Bang and Chang, 2005). Considering the rapid worldwide increase in energy demand, intensifying heat transfer processes and reducing energy losses due to ineffective use have become an increasingly important task (Wen and Ding, 2005). NFs and NPs can be applied as the means to increase heat transfer intensity in heat pipes and thermosyphons

(Vasiliev et al., 2004; Mitrovic, 2006). NFs are very stable due to the small size and volume fraction of NPs needed for the enhancement of heat transfer. When NPs are properly dispersed, NFs can offer numerous benefits besides their anomalously high effective thermal conductivity, such as improved heat transfer and stability, microchannel cooling without clogging, the possibility of miniaturising systems scaling and reduction in pumping power, among others (Vasiliev et al., 2007). Thus, NFs have a wide range of industrial, engineering and medical applications in fields such as transportation, micromechanics, heating, ventilating and air-conditioning systems, biomolecule trapping, and enhanced drug delivery. Nanostructures and nanomaterials are becoming more and more commonly used in various industrial sectors like cosmetics, aerospace, communication and computer electronics.

NCs have a grand potential to increase the wettability and heat transfer intensity in small-size heat transfer devices such as miniature heat pipes and miniature heat exchangers. In addition to the technological problems associated with NFs and NCs, there are plenty of unresolved scientific issues that need to be properly addressed. Nucleate boiling of NFs is the subject of numerous studies, which cover questions related to the formation of bubble nuclei in metastable liquids and on micro-/nanoporous coatings. Further detailed investigations are necessary to understand the phenomena of boiling of NFs. In particular, experiments are lacking the effects of NP materials and heat-loaded surfaces on boiling heat transfer from horizontal smooth tubes. The general conclusion is that microparticles and NPs on heat-loaded surfaces initiate bubble generation and enhance two-phase heat transfer (Das et al., 2003; You et al., 2003). Some experimental investigations have revealed that NFs have remarkably higher thermal conductivity and greater heat transfer characteristics than conventional pure fluids (Vassallo et al., 2004; Bang and Chang, 2005). Till now, all the aforementioned experiments with NFs are performed in conventional heat pipes (Wen and Ding, 2005).

6.2 Heat Pipes for Fuel Cell Thermal Management

Fuel cell thermal management can be efficiently performed using heat pipes of different types. For micro/mini fuel cell thermal control, it is efficient to consider the so-called MHP phenomenon (Vasiliev et al., 2007). The MHP phenomenon is often available in nature. For example, there is an analogy between MHP operation and the functioning of a sweat gland (Dunn and Reay, 1976). Open-type MHPs are considered in the literature (Reutskii and Vasiliev, 1981; Vasiliev et al., 1981; Vasiliev, 1993) as a system of thermal control of biological objects and as a drying technology. The MHP concept is interesting to realise for micro/mini fuel cells, which include a fuel cell stack and ancillary systems. Heat pipe fuel cell thermal management can be performed as follows:

1. Micro/miniature heat pipes for micro/mini fuel cells (<10 W)
2. Heat pipes for medium fuel cells (10–100 W)
3. Heat pipes for portable fuel cells (>100 W)
4. Heat pipe systems for stationary fuel cells (stationary electricity generation)

Besides this category of heat pipe application, there are possibilities of applying heat pipes in fuel cartridge thermal control. Heat pipe thermal management is useful for fuel cell heat recovery (co-generation and tri-generation) with a heat pump. The thermal link between a fuel cell stack and a fuel cartridge, or between a fuel cell stack and an energy recovery system (heat pump), can be easily established by heat pipe heat exchangers as they are well suited to this requirement. A fuel-based auxiliary power unit is one possible method of supplying electrical power. But even fuel cells need to improve their efficiency due to the heat dissipated inside them on the level of 50%. It can be realised with the help of solid and liquid sorption transformers. Actual sorption technologies (liquid and solid sorption cycles) have different advantages and drawbacks with respect to their compactness, complexity, cost and the range of working temperatures. A new design of heat pipe–based active thermal control is related with the so-called sorption heat pipes (Vasiliev and Vasiliev, 2005). An SHP includes the advantages of conventional heat pipes and of sorption machines in one unit. Its major advantage is its ability to ensure convective two-phase heat transfer through a capillary porous wick under the pressure drop due to sorbent wick action inside the heat pipe. In an SHP, the same working fluid is used as the sorbate and as a heat transfer medium.

Such a heat pipe includes some basic phenomena interacting with each other: (1) in the sorbent bed, there is a vapour flow (two-phase flow) with kinetic reaction rate and pressure, vapour pressure, geometry, and conductive and convective heat transport with radial heat transfer; and (2) in the condenser and evaporator, there is a vapour flow, liquid flow, interface position, radial heat transfer with kinetic reaction pressure, liquid pressure, vapour pressure, condensation and evaporation, shear stress, geometry, adhesion pressure, convective heat transport and radial heat transfer under the influence of the gravity field. It is convenient to apply a sorption LHP for the whole fuel cell assembly, thermal and water control at higher currents, and excessive water condensation. Heat pipes promote water transport; mitigate flooding and the associated blockage of reactant transport, performance losses and degradation mechanisms; and ensure high Reynolds numbers of two-phase flows.

6.2.1 Micro/Miniature Heat Pipes for Fuel Cell Thermal Management

A thermal fluids management system uses passive approaches for fuel storage and delivery, air breathing, water management, CO_2 release and thermal management (Panchamgam et al., 2008). A fuel cell stack is a complex device, but it consists of some units which are the same. Each elementary fuel cell has several layers (bipolar plate, gas diffusion layers, catalyst layers, membrane) working under the same conditions. Therefore, the thermal management of the device can be performed for the fuel cell stack in general. Different modes of heat pipe–based fuel cell thermal management are shown in Figure 6.4. First of all, the study of heat pipe thermal control is related to two-phase heat transfer and hydrodynamics in the heat pipe evaporator. In particular, for miniature heat pipes a high heat flow rate is present due to the enormous heat flux that occurs in the contact line region of the evaporating curved film (Vasiliev et al., 2007; Panchamgam et al., 2008). Fluid evaporation in the meniscus strongly depends on the heat conductivity of the solid wall and the slip velocity at the liquid–solid interface.

Many micro fuel cells may be placed on 1 cm² of silicon substrate cooled by micro heat pipe array. The cylindrical configuration of micro/mini fuel cells suggested by Yazici (2007) can be conveniently applied with an LHP and annular heat pipe spreaders having minichannels/microchannels as the fuel cell thermal control system. Microchannels in an MHP are used as fluid flow channels with small hydraulic diameters. The hydraulic diameter of MHPs is equal to 10–500 µm. Application of small channels is desirable for two reasons: (1) higher heat transfer coefficient and (2) higher heat transfer surface area per unit flow volume. Micro/miniature heat pipes are to be integrated in the fuel cell configuration, in which grooves for hydrogen and oxygen (tens of micrometres) are disposed in the same substrate (heat pipe envelope) but on opposite planes. A heat pipe–based spreader (plane or cylindrical) has a

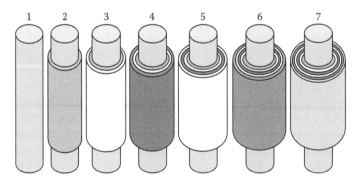

FIGURE 6.4
Heat pipe–based thermal management of a cylindrical fuel cell: (1) loop heat pipe evaporator for heat dissipation in space; (2) anode porous coating; (3) anode porous electrode; (4) membrane; (5) cathode porous electrode; (6) cathode porous coating; (7) outer annular heat pipe (heat pipe spreader).

thickness of 1–3 mm. There are 0.3–0.5 mm vapour channels (grooves) inside with capillary porous deposit on its wall with thickness 30–50 μm consisting from nanoparticles. The spreader (annular heat pipe) ensures good thermal uniformity within the stack. Such a spreader ought to be installed between adjacent fuel cells (Litster and Djilaly, 2008). The LHP withdraws the waste heat from the stack (Figure 6.4). The passages (2) are disposed on the heat pipe outer surface covered by the anode porous electrode (3). The electrode (3) is surrounded by the membrane (4). The membrane is contacting the cathode porous electrode (5). The cathode porous electrode has micro/mini passages (6), which contact the annular heat pipe (7). Visual observation of the heat and mass process in the evaporators of such spreaders testifies that the movement of vapour bubbles in annular minichannels has a complicated character.

6.2.2 Micro Heat Pipe Effect in Porous Media

The MHP concept is typical for capillary open porous systems saturated with liquid and thermally contacted with a heat-loaded wall (Vasiliev, 1993, 1994). Inside the capillary biporous structure (micro- and macropores), two-phase heat and mass transfer occurs with multiple mini menisci of evaporation (interface) on the contact region of the micro- and macropores (Figure 6.5). Macropores are working as vapour channels to transfer the vapour from the meniscus of evaporation to the cold region of the porous body with further vapour condensation. A sintered powder wick can be considered a system with open micro- and macropores. Micropores are used as capillary channels for liquid transport to zones of vaporisation (meniscus). The vapour is generated on the surfaces of menisci in orifices of micropores.

The thickness of a liquid film in zone I of Figure 6.5 is close to the size of the molecular sorption film, and there are no favourable conditions for vaporisation. High intensity of vapour generation (evaporation) occurs in

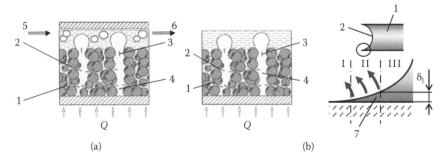

FIGURE 6.5
Schematic of the sintered powder coating in (a) minichannel and (b) liquid pool: (1) micropore; (2) meniscus; (3) macropore; (4) vapour bubble; (5) liquid flow in the minichannel; (6) two-phase flow in the minichannel; (7) zone of evaporation; (Q) heat flow.

zone II of the meniscus. In zone III the liquid film is thick, so its thermal resistance is high. There are a great number of such menisci with the liquid film thickness less than 100 μm over a volume of porous media, so the total area of evaporation is very large. It means that a rise in the heat flux on a heat-loaded wall (within a certain limit) would be possible without increasing the wall temperature. Inside the unit of porous wick volume, there is a constant ratio between micro- and macropores (for one macropore there are some micropores, depending on the wick thickness). The number of active menisci of evaporation depends proportionally on the heat flux number. The heat flow, Q, is equal to the number of menisci, $N_{meniscus}$, multiplied by the latent heat of evaporation, h_{lv}:

$$Q = N_{meniscus} h_{lv} \alpha \ S \Delta T_{evap} \tag{6.4}$$

where α is the local heat transfer coefficient and S is the surface of evaporation, for $\Delta T_{evap} = $ constant.

6.2.3 Mini Loop Heat Pipes

Some types of biporous wicks (metal sintered powder or expanded graphite) are also applicable for mini LHPs (Figure 6.6). Actually, it is most interesting to implement MHPs directly in silicon, or Al_2O_3–Al substrate. Compared to other materials, silicon provides several advantages. It has good heat conductivity (150 W/m · K) and allows one to obtain much smaller devices than those with other metals because of etching process accuracy. Moreover, as an MHP can be machined in the core of the substrate, its thermomechanical constraints are lower compared to other materials. Its thermal expansion coefficient is seven times lower than that of copper and 10 times lower than that of aluminium. The evaporator, a key element of an LHP, determines the serviceability and efficiency of a heat pipe. It consists of the envelope, wick, vapour and liquid lines. The evaporator is joined to a compensation chamber or a

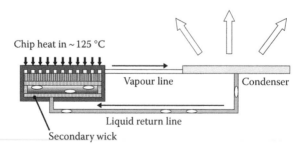

FIGURE 6.6
A mini loop heat pipe with evaporator made from coherent silicon substrate. (From Weislogel et al., 2002).

reservoir into which the liquid from the condenser enters. The heat supplied to the outer surface of the wall penetrates through it and is then transferred through the wick to evaporating menisci. The generated vapour is collected in vapour grooves and is then moved to a condenser by the vapour line.

Liquid inflow into the evaporating zone is realised from the reservoir through the wick. This new technology is a type of microelectromechanical systems (MEMS) process that allows one to 'drill' a pattern of micron-sized holes into a silicon wafer (Weislogel et al., 2002). In fuel cells and heat pipes, the flow characteristics in porous media (gas diffusion layers or capillary wicks) are useful in modelling performance (Cytrynowicz et al., 2002).

Permeability is a parameter that describes the relationship between pressure drop and mass transport through porous media. In heat pipes, effective pore radius is a parameter used to describe the available pressure rise for liquid pumping. In LHPs, one can use an evaporator above the condenser. The vapour flows through the vapour channels towards the condenser, and the liquid goes back to the evaporator due to the capillary pressure head of a porous wick. The assembly of an LHP with some evaporators and condensers is shown in Figure 6.7.

6.2.4 Loop Heat Pipes

LHPs are an attractive alternative for heat regulation (Maydanik et al., 2010). Developed in the Luikov Institute, Minsk, Belarus, the flat aluminium LHP can be conveniently used for semiconductor cooling (Figure 6.8). Such an LHP is more flexible from the point of view of its orientation in space and is more compact.

LHPs possess a unique set of characteristics. First of all, these characteristics include the ability of LHPs to transfer heat efficiently for large distances irrespective of their orientation in the gravity field. Different working fluids can be used, including those with low surface tension and latent heat of vaporisation. LHPs possess high mechanical flexibility owing to the small diameter of vapour and liquid pipes that do not contain a capillary structure. These pipes can be bent in any way. The LHP evaporator and condenser may have dimensions and shapes that adapt easily to the geometrical constraints

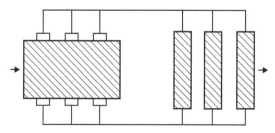

FIGURE 6.7
Loop heat pipe assembly with some evaporators embedded in the fuel cell stack and condensers.

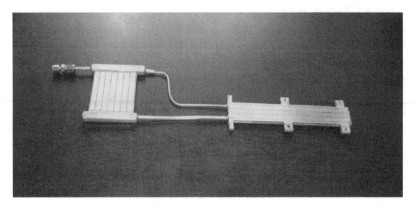

FIGURE 6.8
Aluminium loop heat pipe with propane as the working fluid.

of heat sources and heat sinks. The most convenient wick structure needs to have a large pore diameter layer with high thermal conductivity from the vapour side and a small pore diameter layer with low thermal conductivity from the liquid side. Actually, LHPs are used as thermal control devices in scientific and telecommunication satellites (efficient and flexible thermal link between dissipative elements and radiators). From the thermal modelling point of view, on the one hand, simple spreadsheet analysis is only convenient at the LHP, or thermal subsystem level. On the other hand, a complete thermohydraulic analysis tool could be heavy and time consuming for system-level modelling (ESATAN-FHTS, SINDA-FLUINT).

LHPs with flexible transport lines are needed in some cryogenic applications as thermal links between cryocoolers and the cooled components for a variety of reasons such as vibration isolation, increased thermal transport distance, thermal diode function, and multiple components. One additional technical requirement for cryogenic LHPs is that they should be capable of cooling down from room temperature that is above the critical temperature of the working fluid to the operating cryogenic temperature with the condenser end being cooled by a cryocooler.

Now we consider new LHP aluminium flat panels with minichannels, which are welcomed as innovative heat transfer devices for electronic cooling. The flat minichannel condenser is interesting for the following reasons:

1. In some cases, the main spacecraft radiator is considered as the heat spreader–radiator for an LHP condenser.

2. It is strongly desired that LHP condensers are effective and are compact with a low inner volume and a flat thermal interface.

The analysis of condensation phenomena in aluminium panels with minichannels made from 6063 aluminium alloy (Figure 6.9) is the subject of practical interest when propane is used as the working fluid; recently, there

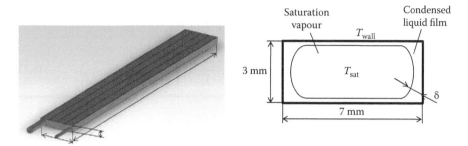

Saturation vapour

Condensed liquid film

T_{wall}

3 mm

T_{sat}

δ

7 mm

FIGURE 6.9
External view and cross section of an aluminium flat condenser.

has been a great deal of interest in the study of Freon fluid flow and heat transfer in minichannels (Wang et al., 2004; Nebuloni and Thome, 2010).

In the design of LHPs, the condenser is made as the aluminium panel with rectangular minichannels (Figure 6.9). One minichannel is connected to the LHP vapour line and another to the LHP liquid line. In the beginning of the LHP operation, the minichannels inside the condenser are filled with the liquid. The vapour generated in the LHP evaporator pushes the liquid from the condenser towards the LHP compensation chamber of the evaporator, simultaneously condensing on the walls of the minichannel. Two scenarios of external heat transfer are interesting: (1) isothermal wall (intense water cooling) and (2) fixed heat transfer coefficient on the cooled condenser side.

6.2.4.1 Mathematical Model

A simplified mathematical model of heat and mass transfer in an LHP condenser obeys the Navier–Stokes equation, the heat energy conservation equation and the continuity equation. To determine the liquid phase in two-phase flow, that is, amount of liquid fraction, the volume of fluid method is used.

6.2.4.1.1 Navier–Stokes and Continuity Equations

The velocity field and pressure in each phase are determined by Navier–Stokes and continuity equations. Considering the presence of a phase transition, these equations are written as follows:

$$\rho\frac{\partial \vec{u}}{\partial \tau}+\rho(\vec{u}\cdot\nabla)\vec{u}=\nabla\cdot\left[-p\vec{I}+\eta\left(\nabla\vec{u}+\nabla\vec{u}^{T}\right)\right]+\vec{F}_{g}+\vec{F}_{st}+F \qquad (6.5)$$

$$\nabla\vec{u}=\frac{S}{\rho} \qquad (6.6)$$

When condensation takes place, that is, $T_v < T_{sat}$, parameter S has the following form (Yang et al., 2008; De Schepper et al., 2009):

$$S = -r_v \alpha_v \rho_v \left| \frac{T_v - T_{sat}}{T_{sat}} \right| \tag{6.7}$$

6.2.4.1.2 Energy Conservation Equation

The temperature field in a condenser is described by the following conservation equation of heat energy:

$$\rho C_p \left(\frac{\partial T}{\partial \tau} + \vec{u} \cdot \nabla T \right) = \nabla \cdot (k \nabla T) + Q \tag{6.8}$$

The heat energy released during condensation is determined by the following expression:

$$Q = r_v \alpha_v \rho_v \left| \frac{T_v - T_{sat}}{T_{sat}} \right| \Delta H \tag{6.9}$$

6.2.4.1.3 Volume of Fluid Method

The tracking of the interface position between vapour and liquid flow was accomplished by solving continuity equations for the volume fractions of different phases. For the liquid and vapour phases, the respective equations are (Yang et al., 2008) as follows:

$$\frac{\partial \alpha_1}{\partial \tau} + \nabla \cdot \left(\vec{v} \alpha_1 \right) = -\frac{S}{\rho_1} \tag{6.10}$$

$$\frac{\partial \alpha_v}{\partial \tau} + \nabla \cdot \left(\vec{v} \alpha_v \right) = \frac{S}{\rho_v}$$

where the effective parameters are determined as follows:

$$\rho = \alpha_1 \rho_1 + \alpha_v \rho_v$$

$$k = \alpha_1 k_1 + \alpha_v k_v \tag{6.11}$$

$$\eta = \alpha_1 \eta_1 + \alpha_v \eta_v$$

6.2.4.1.4 Initial and Boundary Conditions

The initial condition for the Navier–Stokes equation is as follows:

$$v_1|_{t=0} = 0$$

For the heat energy equation,

$$T|_{t=0} = T_{\sin k}$$

$$T_v|_{t=0} = T_{sat}$$

and for the volume fraction equation,

$$\alpha_1|_{t=0} = 1$$

$$\alpha_v|_{t=0} = 0$$

The boundary conditions for the Navier–Stokes equation are

$$v_{wall} = 0\,m/s,\ p_{in} = p_{sat}$$

and the condition for the heat energy equation is

$$T_{wall} = T_{\sin k}$$

6.2.4.2 Results of Modelling

To solve numerically Equations 6.5 through 6.11 with the boundary conditions mentioned in Section 6.2.4.1.4, the finite elements method was used. A grid size of 5000 grid points along a minichannel has been applied. The computational code was written using MATLAB®.

As a result of numerical simulation, the main system characteristics were obtained. The three-dimensional images of liquid- and vapour-phase distributions in the condenser channels for both cases are shown in Figures 6.10 and 6.11. Blue and red colours represent vapour and liquid flows, respectively, whereas the green isosurface illustrates interface position. At the beginning of the experiment condenser channels were completely filled with the liquid. Then the vapour generated in the LHP evaporator enters the condenser, partially blowing away the liquid, and condenses gradually on the cold wall of the minichannel.

As shown in Figure 6.10, vapour condenses considerably faster for the case of isothermal surface of condensation than for the case in which heat transfer is equal to 400 W/m² · K.

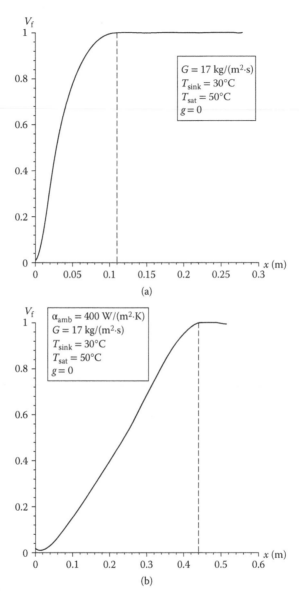

FIGURE 6.10
Three-dimensional image of vapour and liquid distribution inside a loop heat pipe condenser channel: (a) with isothermal surface; (b) with the fixed outside heat transfer coefficient equal to 400 W/m² · K.

The interface positions of single- and two-phase zones can be determined from these dependencies. The region where the heat transfer coefficient is constant corresponds to the single-phase zone, and the region where the heat exchange coefficient is sharply decreasing corresponds to the two-phase

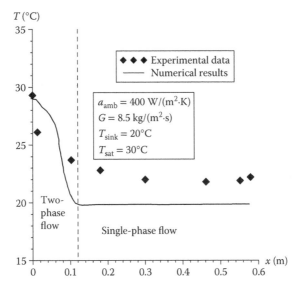

FIGURE 6.11
Experimental and numerical data of temperature distribution in condenser channels.

zone. As the heat transfer coefficient is inversely proportional to the thickness of the liquid layer, we can assume that the thinner liquid layer corresponds to the higher heat transfer coefficient.

As shown in Figure 6.11, there is good agreement between calculated and experimental data for the mass flow $G = 8.5$ kg/m² · s and the vapour temperature $T_{sat} = 30°C$. The difference of a few degrees Celsius is due to the high thermal conductance of the aluminium wall, which is not taken into account in the presented model.

6.3 Vapour-Dynamic Thermosyphons

VDTs are attractive two-phase thermal control types for solid sorption heat transformers. Two-phase conventional thermosyphons and loop thermosyphons have been developed for many years. The most famous publications (Imura et al., 1994; Pioro and Pioro, 1997) were mentioned in some papers. Two-phase loop thermosyphons were used in several applications, such as electronics cooling (Garner and Patel, 2001; Khrustalev, 2002; Khodabandeh, 2005), nuclear power plants (Sviridenko and Shevelov, 2011), waste heat recovery (Dube et al., 2004), asphalt tank heating (Milanez and Mantelli, 2010), and miniature compressors cooling (Possamai et al., 2009; Aliakhnovich et al., 2011), among many others.

VDTs (Vasiliev et al., 1985) were patented in 1985 and are principally distinct from conventional thermosyphons and heat pipes of the same diameter.

A VDT can transfer heat in the horizontal position over long distances. Its condenser is near isothermal with a length of tens of meters (Vasiliev et al., 1986; Vasiliev, 1994). The picture of a VDT is shown in Figure 6.12.

VDT evaporators may have different forms and lengths. The intensification of heat transfer in VDT evaporators is related to the porous coating of the heat-loaded surface (Vasiliev et al., 2012a). A VDT with porous coating of the evaporator ensures a heat transfer enhancement up to five times compared with a plain tube thermosyphon and works without temperature overshoots, which are typical of conventional thermosyphons.

Recent innovations in VDT design related to nanotechnologies anticipate significant impacts on designs of heat pump/cooler and heat/cold accumulators and their thermal solutions. The first main distinction of VDT is the 'long annular gap' between the vapour pipeline and the tubular condenser (tube-in-tube). The vapour generated in the VDT evaporator passes through the long pipeline and enters the annular gap of the condensation zone where it condenses. The vapour pipe is made with a metal of high thermal conductivity. There is intense radial heat transfer with phase change between the vapour pipe and the condenser envelope. The data obtained from a flooded and a partially flooded vapour pipe in confined space (annular channel is 0.2–2 mm thick) testify to the phenomenon of MHP inside a porous structure and that of miniature heat pipe in the annular gap. Visual analysis (glass condenser) and experimental results show that such a combination is favourable for the enhancement of heat transfer in the VDT condenser. The hydraulic diameter of the condenser annular channel has two options: (1) it is more than, or equal to, the capillary constant of the working fluid; and (2) it is less than the capillary constant.

The second VDT fundamental difference consists in the fact that the motion of vapour and liquid flow proceeds being separated (tube in tube).

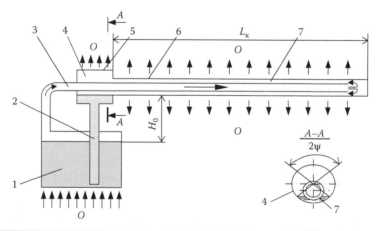

FIGURE 6.12
Vapour-dynamic thermosyphon: (1) evaporator; (2) liquid pipe; (3) vapour pipe; (4) compensation chamber; (5) heat sink; (6) condenser; (7) annular channel; (H_0) hydrostatic pressure drop. (Vasiliev et al., 1985).

This makes it possible to avoid a negative hydrodynamic interaction between the opposite flows of vapour and liquid typical for conventional thermosyphons. When a heat load is supplied to the evaporation zone, the vapour, speeds into the condensation zone, where it condenses and gives up heat to an external heat sink (Figure 6.12).

The condensate returns through a short liquid transport pipe (2) into the evaporator, under the action of hydrostatic pressure, ΔP_g, the difference between the liquid levels in the evaporation and condensation zones:

$$\Delta P_g = g \, (\rho_l - \rho_v)(h_2 - h_1) \tag{6.12}$$

where ρ_l and ρ_v are the liquid and vapour densities, respectively; h_1 and h_2 are the heights of the liquid levels; and g is the gravity constant. The hydraulic head, Δh, is the difference between the liquid levels in the evaporator and the condensate return line. The maximum allowable difference is Δh_{max}, that is, the vertical distance between the evaporator liquid surface and the condenser bottom. The pressure drop in the liquid pipe due to the flow of the working fluid must be compensated by a hydraulic head of condensate above the evaporator. The total pressure drop, ΔP_t, is the summation of the pressure drops due to fluid flow at the evaporator, ΔP_{evap}; the vapour line, ΔP_v; the condenser annular channel, ΔP_{cond}; and the liquid transport pipe, ΔP_l, that is

$$\Delta P_t = \Delta P_{evap} + \Delta P_v + \Delta P_{cond} + \Delta P_l \tag{6.13}$$

The procedure of heat and mass transfer in such VDT annular minichannels is interesting to visualise to be sure in what way the vapour bubbles are interacting with the liquid flow.

There are two options for a VDT condenser design:

1. When the hydraulic diameter is less than the capillary constant, there is no fluid stratification in the channel (Figure 6.13). The transport of the fluid to the annular channel occurs from the vapour transport pipe covered by the microporous coating outside it. The porous coating provides a large surface area and enhances heat transfer inside the annular channel due to mini bubble generation in the pores. The dynamic vapour bubbles pressure on the two-phase flow

FIGURE 6.13
Fluid flow (propane) visualisation in an annular minichannel.

is an additional force that induces liquid influx. Do they serve as mini compressors to push the two-phase flow along the condenser?

2. The hydraulic diameter of the annular channel is more that the capillary constant. The stratification of vapour and liquid flows occurs in the channel (Figure 6.14).

There is intense heat transfer between the liquid flow, the porous coating of the vapour transport pipe (Figure 6.15) saturated with liquid due to capillary forces, and the wall of the VDT condenser (miniature heat pipe effect). The

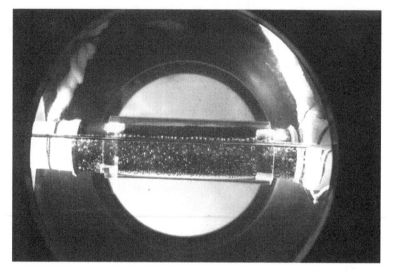

FIGURE 6.14
Fluid flow (propane) visualisation in an annular minichannel: hydraulic diameter is more than capillary constant (two-phase-flow stratification).

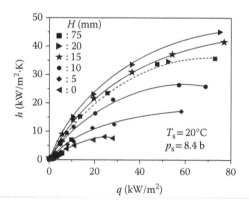

FIGURE 6.15
Heat transfer in the vapour-dynamic thermosyphon annular channel as a function of heat flow in the evaporator.

difference in height between the liquid interface and the top of the vapour transport pipe inside the condenser is denoted as H. At the optimal height, $H = 20$ mm, for propane, heat transfer between the saturated vapour and the condenser wall is near 45 kW/m$^2 \cdot$ K at a heat load equal to 80 kW/m^2.

The main VDT thermal parameters are as follows: the maximum heat load, Q_{max}; the heat flux in the evaporation zone, q; and the thermal resistance, R. To determine the thermal resistance of VDT, the following formula is used:

$$R = (T_e - T_c)/Q \qquad (6.14)$$

where T_e and T_c are average wall temperatures of the evaporator and the condenser, respectively, and Q is the heat load. The VDT wall temperatures of the evaporator and the condenser and the temperatures of the vapour and condensate pipes are necessary to know. Standard copper–constantan thermocouples with data acquisition units were used for the measurements. When the heat flow in a thermosyphon exceeds the heat transfer limit Q_{max} due to pressure drop, the condensate blocks a part of the condenser, increasing the overall thermal resistance of the system.

6.3.1 Heating Devices Based on Vapour-Dynamic Thermosyphons

Long VDTs are used for floor heating in houses, timber drying, roof snow melting and so on. A VDT with a long horizontal condenser (10 m) is shown in Figure 6.16.

VDT thermal resistance R as a function of the heat load is shown in Figure 6.17. The working fluids are water and propane. The VDT envelope is made from stainless steel.

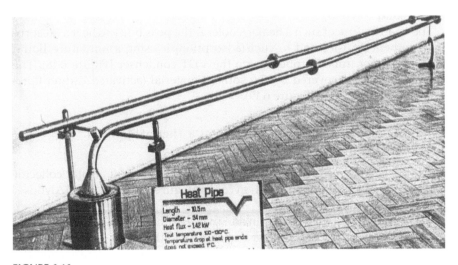

FIGURE 6.16
Vapour-dynamic thermosyphon: the condenser length is 10 m.

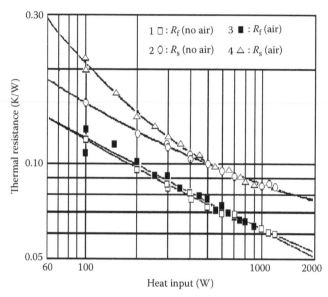

FIGURE 6.17

Thermal resistance, R, of vapour-dynamic thermosyphon as a function of heat load, W: (1) water; (2) propane; (3) water with air; (4) propane with air in the gas trap.

6.3.2 Vapour-Dynamic Thermosyphon as a Thermal Control System for Adsorbers of Solid Sorption Heat Pumps

A solid sorption heater/cooler with VDT thermal control has much potential to be used in air-conditioning and transport cooling applications (Vasiliev et al., 2001a).

The main feature of such a heater/cooler is the possibility to heat a (desorption) sorbent canister and to cool (adsorption) it using a miniature liquid heat exchanger (tube-in-tube) inside the VDT condenser (Figure 6.18). The VDT condenser is covered by the sorbent material (activated carbon fibre) disposed between fins (Figure 6.19).

6.3.3 Vapour-Dynamic Thermosyphon as a Thermal Control System for Solar Solid Sorption Coolers

A VDT was successfully used for the thermal coupling of a solar collector and two adsorbers of a solid sorption cooler (Figures 6.20 and 6.21) (Alyousef et al., 2012).

VDT condensers are disposed inside two adsorbers of a cooler. The working fluid is water. The solar cooler design of a VDT system of thermal management (Figure 6.20) consists of some evaporators and condensers and the vapour and liquid transport mini pipes.

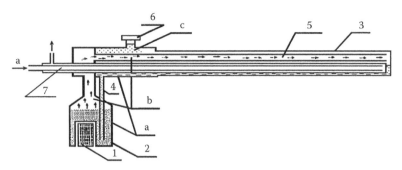

FIGURE 6.18
Vapour-dynamic thermosyphon with a liquid miniature heat exchanger (tube-in-tube) inside: (1) cartridge heater; (2) evaporator; (3) condenser; (4) liquid pipe; (5) vapour pipe; (6) non-condensible gas trap C; (7) liquid heat exchanger; (a) liquid; (b) vapour.

FIGURE 6.19
Vapour-dynamic thermosyphon (stainless steel/water) with a mini-gas burner: sorbent material is disposed between fins.

VDT condensers are switched on and off alternatively by valves (denoted by 2 in Figure 6.20). Such VDTs have low thermal resistance ($R = 0.01 - 0.05$ K/W), and the condenser length is 1 m. The constant heat flow (solar energy) in thermosyphon evaporators is transformed into an intermittent heat flow in condensers (Figures 6.20 and 6.21).

The cooling of the sorbent material inside the adsorbers of the solar cooler was performed by a miniature liquid heat exchanger (Figures 6.18 and 6.21), which was made as a 2 mm SS tube placed inside the annular channel of the VDT condenser.

When the VDT is closed (valve closed), the liquid heat exchanger actively cools the sorbent bed, due to two-phase heat transfer (vapour condensation) inside the annular gap. The time of the adsorption/desorption cycle was 12 minutes (Figure 6.22). During the tests, the temperature of the evaporator

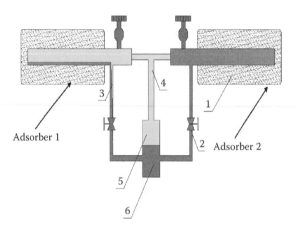

FIGURE 6.20
Two condensers of a vapour-dynamic thermosyphon used to heat two adsorbers of a solar cooler: (1) condensers; (2) valves; (3) liquid line; (4) vapour line; (5) evaporator; (6) liquid pool of the evaporator.

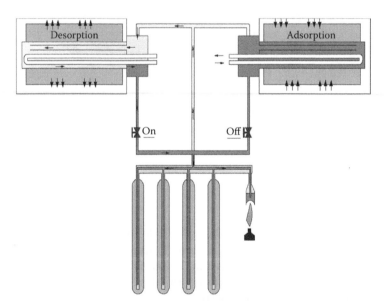

FIGURE 6.21
Solar heater with two adsorbers, solar collectors (with flame as the backup), five evaporators of a vapour-dynamic thermosyphon and two valves.

envelope (curve 1) was constantly near 110°C, whereas the temperature of the adsorber envelope (curves 2 and 3) was periodically changing from an ambient temperature of 20°C (adsorption) to 90°C (desorption). The heat transfer intensity between the thermosyphon condenser and the adsorber is limited by the thermal conductivity of the sorbent bed.

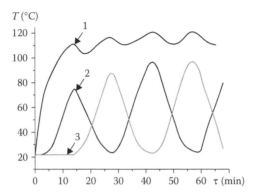

FIGURE 6.22
Temperature evolution of vapour-dynamic thermosyphon: (1) evaporator (2) envelope of adsorber and (3) adsorber.

FIGURE 6.23
Vapour-dynamic thermosyphon with a flexible condenser of small diameter.

Among the additional advantages of VDT is the possibility of using small-diameter flexible condensers (Figure 6.23).

6.3.4 Vapour-Dynamic Thermosyphon as a Thermal Control System for Snow Melting

Snow melting VDT designs are used to heat railway switches, greenhouses, floors of buildings and pavements during winter and so on (Figure 6.24).

The VDT condenser diameter is 16 mm. At an ambient temperature of −7°C to −10°C, a snowfall intensity of 100–150 mm and a wind velocity of 5–10 m/s, a vapour-dynamic heater ensures complete thawing of snow. After the VDT heater is switched on, the snow between the stock and point rail disappears in 1 hour.

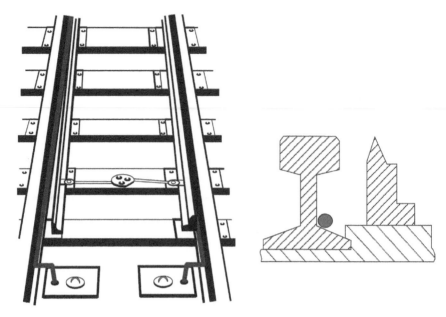

FIGURE 6.24
Vapour-dynamic thermosyphon for snow melting: L – 4.5 m; Q – 2 kW.

6.4 New Generation of Heat Pipes and Thermosyphons with Nanofluids and Nanocoatings Applications

Recent advances in nanotechnology have allowed the development of new NFs to describe liquid suspensions containing NPs with thermal conductivities orders of magnitude higher than those of base liquids and sizes significantly smaller than 100 nm. Considering the rapid increase in energy demand worldwide, intensifying heat transfer processes and reducing energy losses due to ineffective use have become increasingly important tasks. There is some interest in applying NFs and NCs as a means to increase the heat transfer intensity in heat pipes and thermosyphons. NFs are very stable due to the small size and volume fraction of the NPs needed for heat transfer enhancement. When the NPs are properly dispersed, NFs can offer numerous benefits besides the anomalously high effective thermal conductivity such as improved heat transfer and stability, microchannel cooling without clogging, the possibility of miniaturising systems scaling and reduction in pumping power, among others. Thus, NFs have a wide range of industrial, engineering and medical applications in fields such as transportation, micromechanics, heating, ventilating and air-conditioning systems, biomolecule trapping and enhanced drug delivery. Nanostructures and nanomaterials are becoming more and more commonly used in various industrial sectors like cosmetics, aerospace, communication and computer electronics. NCs have a grand potential to increase the wettability and heat transfer intensity in small-size heat transfer devices such as miniature heat pipes and miniature heat exchangers. In addition to the associated technological problems, there are plenty of unresolved scientific issues that need to be properly addressed.

Nucleate boiling of nanoliquids is the subject of numerous studies, which cover questions related to the formation of bubble nuclei in metastable liquids and on micro-/nanoporous coatings. Further detailed investigations are necessary to understand the phenomena of boiling of NFs. In particular, experiments are lacking the effects of NP material and heating surface material on boiling heat transfer from horizontal smooth tubes. The general conclusion is that microparticles and NPs on heat-loaded surfaces initiate bubble generation and enhance two-phase heat transfer. Some experimental investigations have revealed that NFs have remarkably higher thermal conductivity and greater heat transfer characteristics than conventional pure fluids (Das et al., 2003; You et al., 2003; Vassallo et al., 2004). All the aforementioned experiments with NFs in heat pipes were performed with conventional envelopes made from metals (Bang and Chang, 2005; Wen and Ding, 2005).

In some cases, it is interesting to fabricate heat pipe envelopes from transparent material (glass, plastic) and to ensure the heat input to the heat pipe evaporator or thermosyphon by radiation (Figure 6.25).

The bubble generation phenomena in mini volumes and the impulse generated by them are the reason for the circulation of NFs in miniature heat pipe loops (Figure 6.26). This impulse is working as a two-phase mini pump. The bubbles are considered also as a motive force to organise the fluid

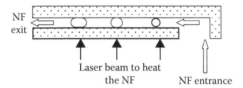

FIGURE 6.25
Schematic of a mini evaporator with transparent walls heated by an impulse laser beam.

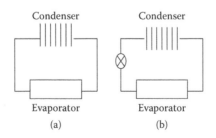

FIGURE 6.26
Schematic of two-phase loop with nanocomposite as the working fluid (a) without mechanical pump, (b) with mechanical pump.

circulation in pulsating heat pipes and loop thermosyphons. Such types of two-phase cooling systems, for example, have a good perspective for space applications in systems of satellite thermal control. Transparent evaporators made from glass or plastic have a real practical interest for mini/micro fuel cell thermal control and photoelectronic components cooling.

Heat pipe transparent condensers with NFs are efficiently cooled by radiation in space. It is interesting to use high-temperature heat transfer devices in power stations as transparent (glass) pulsating heat pipes in air preheaters for furnaces and boilers. One of the major interesting topics is the investigation of the influence of metal oxide NPs (Al_2O_3 NPs) immersed in the fluid on bubble generation and two-phase heat transfer intensification to compare with pure water. As previously shown by Lapotko (Lapotko and Lukianova, 2005; Lapotko et al., 2006), the heat flow generated by light-absorbing NPs (gold NPs) initiates more intense bubble generation when short laser pulses are used as primary sources of thermal energy (Figure 6.27). The temporal scale of photothermal conversion of the energy is limited by the duration of a laser pulse and provides good thermal confinement of the heat released in NPs. The main role of such energy-absorbing NPs is to generate the heat in NFs volume. The limitation of this method for vapour generation is in the delivery of energy to the point of interest: it should be optically transparent to allow optical radiation to reach the NPs. NPs are considered as additional centres of nucleation due to the increased surface of liquid/solid interaction. In such cases, do NPs stimulate the appearance of an earlier threshold of vapour generation? The second aim is to validate this hypothesis and evaluate the influence of 'passive' NPs (Al_2O_3 particles as non-absorbing energy media) as the element of heterogeneity in the fluid on the decrease of the

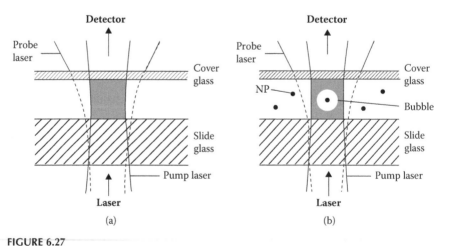

FIGURE 6.27
Visualisation of bubble generation by a short laser pulse (532 nm, 10 ns) in the mini-channel of a flat evaporator with nanofluid (water + Al_2O_3 nanoparticles): (a) control pure water; (b) nanofluid; DAl_2O_3 particle < 220 nm.

energy threshold of bubble generation. Such NPs are considered now as additional centres of nucleation due to the increased surface of liquid/solid interaction (Figure 6.28).

For all studied cases, bubble-specific photothermal (PT) signals (Figures 6.28 and 6.29)—PT responses and PT images—were detected and evaluated. A bubble-specific PT response has a negative symmetrical profile; its front describes bubble expansion and tail describes bubble collapse. The length of bubble-related signal response indicates the bubble lifetime (Figure 6.29, left). In homogeneous media the diameter of the bubbles is much smaller than that for heated volumes (cell or laser beam; Figure 6.29, right). Regardless of the medium, the bubble generation process had statistical nature with the

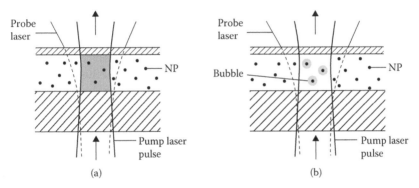

(a) (b)

FIGURE 6.28
Experimental model: (a) laser-induced heating of the volume of liquids with non-absorbing nanoparticles of Al_2O_3; (b) generation of laser-induced bubbles around gold light–absorbing nanoparticles (right); single laser pulse: 532 nm, 10 ns.

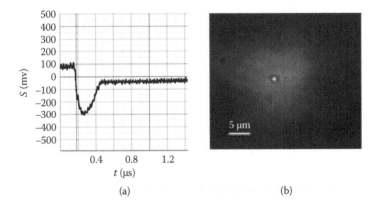

(a) (b)

FIGURE 6.29
Photothermal signals obtained from an individual bubble generated in a transparent microevaporator: (a) Photothermal response with bubble-specific shape and (b) photothermal image. Y axis is for the output of the photodetector (mV).

bubble generation probability PRB from 0 to 1. The energy threshold of bubble generation for light absorbing gold particles is less compared with Al_2O_3 particles and pure water.

6.4.1 Nanoporous Coatings of Heat Pipe Evaporators

Particle deposition on extended surfaces of heat transfer to stimulate the generation of bubbles was studied in the past (Vasiliev et al., 2004, 2007; Mitrovic, 2006). The MHP effect inside the porous structure and the two-phase forced convection in annular minichannels were considered thermodynamically as an efficient means to improve the parameters of mini evaporators. Such an evaporator is used as an effective cooling device for micro- and optoelectronic components (Xie et al., 2003; Vasiliev et al., 2006). The investigation of boiling and evaporation heat transfer in mini grooves on a single horizontal tube (smooth, with porous coating) is a good tool to analyse its cooling efficiency. As a long-term alternative refrigerant, ammonia was used in compact heat exchangers and heat pipes due to its excellent performance, its lack of impact on the environment (ozone depletion potential = 0 and global warming potential < 3) and its physical properties. A number of studies on evaporation phenomena in grooves have been carried out over the past few decades (Holm and Goplen, 1979; Mirzamoghadam and Catton, 1988; Suman et al., 2005). Most investigators have focused their attention on liquid evaporation on menisci formed in smooth grooves with extended thin film, as shown in Figure 6.30a (Xu and Carey, 1990; Stephan and Busse, 1992; Ma and Peterson, 1996).

Intensive evaporation heat transfer occurs on the thin film region that extends from the intrinsic meniscus. However, the extended thin film in the grooves with different sections constitutes only a small portion of the total surface of the grooves. A large fraction of the groove surface is covered by the thick liquid film, or the intrinsic meniscus, where the heat transfer coefficient is particularly low due to the weak thermal conductivity of the liquid or the dry part of the groove and only natural or mixed vapour convection exists between the solid surface and the vapour phase. Compared to heat transfer with liquid/vapour phase changes, the convection heat transfer coefficient in the dry area of the groove surface is insignificant and can be safely neglected. As the heat flux increases, the meniscus in the groove further recedes and the dry area, a region of pronouncedly low heat transfer performance, increases as well.

In 1981, an innovative method was proposed by Vasiliev et al. (Figure 6.30b) to enhance evaporative heat transfer in the grooves of grooved heat pipe (GHP). The surface of trapezoidal grooves (copper GHP) was covered by a thin porous layer of copper sintered powder to ensure an extended surface of evaporation with high heat transfer intensity. Evaporative heat transfer occurs on the meniscus inside the porous coating of the groove (on its bottom and the edge simultaneously). The latter not only improves capillary force action but also considerably extends the surface of evaporation with high heat transfer in comparison with the smooth surface of the groove. It is important to validate the thermal behaviour of grooved heat pipe with porous layer (GHPPL) and smooth GHPs by tests for different porous coatings and materials. To guarantee identical boundary conditions, the GHP samples should be tested simultaneously on the same experimental bench in parallel – evaporator with smooth grooves and evaporator with a nanoporous layer on the grooves at temperatures between −30°C and 70°C. The set of tests was performed with copper sintered powder and Al_2O_3 porous coating on copper and aluminium pipes. After the tests were performed, it was observed that for all temperature ranges the evaporator thermal resistance (R_{ev}) of GHPPLs was low in comparison with the R_{ev} typical of smooth GHPs. The thermal resistance of GHPPLs is 1.3–1.4 times lower (between 0.021 and 0.018 W/K) than that of GHPs (between 0.025 and 0.035 W/K). A detailed analytical model was developed (Wang and Catton, 2011) to predict evaporation heat transfer in a triangular groove. The trapezoid fins disposed between the grooves was covered by a thin porous layer. It was shown that heat transfer in the groove covered by the porous layer is three to six times higher than that in grooves with smooth surfaces. As a result, in the new, advanced design of GHPPL, significant intensification of heat transfer was obtained.

So, the application of NC technology is dependent on improving the cooling capability of GHPPLs. NCs of GHP evaporators formed from microparticles and NPs enhance heat transfer not only in the thin liquid film but also in the liquid pool and flooded surfaces. Thin, porous NCs play the role of additional centres for stable vapour generation, which do not require high superheating of surfaces. The wick (thickness of 25–100 μm) on the surface of mini grooves allows to reduce its thermal resistance and increases the working fluid capillary pressure and permeability at the same time (Vasiliev et al., 2008a,b).

Capillary forces distribute the liquid inside the porous volume of the wick. The surface of the groove edge beyond the zone of the main meniscus turns to be wetted uniformly, and the area of effective evaporation is increased manifold.

Unlike heat transfer with boiling on smooth surfaces, liquid evaporation/boiling on porous deposits (like aluminium oxide NCs) is characterised by the constant sources of nucleation (Figures 6.30 and 6.31). It is due to the limited

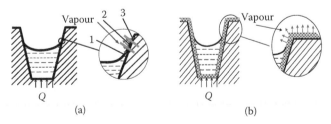

FIGURE 6.30
Evaporation phenomena in trapezoidal grooves: (a) smooth; (b) with porous layer. (1) liquid meniscus; (2) thin film, zone of extensive heat transfer with evaporation; (3) dry zone of the grooves.

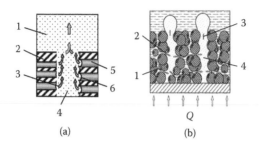

FIGURE 6.31
Model of heat transfer on the bottom and edge of a mini groove with porous coating: (a) upper part of the groove edge: (1) vapour; (2) solid part of the wick; (3,6) vapour stream; (4) macropore free of liquid; (5) micropore with capillary liquid flow (b) bottom part of the groove (liquid pool): (1) micropore; (2) interface meniscus of evaporation; (3) vapour bubble; (4) vapour stream.

number of evaporation menisci inside the porous wick. In porous coatings, the liquid/vapour interface consists of menisci situated inside the macropores and numerous menisci between the macro- and micropores. At the low heat flux the menisci of the evaporation are only in macropores, when the wick is completely saturated with liquid. For such low heat fluxes, the heat transfer is realised by conduction through the wick. Menisci of evaporation in minipores are typical for high heat flows, when the menisci in macropores recede and become open for the vapour flow. For such a case, many nucleation sites (micro menisci develop on the interface between macro- and micropores) are becoming the centres of vapour generation inside the macropores. Following experimental data, the upper part of grooves has more intense heat transfer (Table 6.1) compared to the bottom part (Vasiliev et al., 2010).

The heat flux, q, through the wall of a GHP evaporator can be written as follows:

$$q = \frac{T_w - T_{sat}}{(\delta_{wick}/k_{eff}) + (1/h_e)} \tag{6.14}$$

where $T_w - T_{sat} = \Delta T_t$, which is determined as follows:

$$\Delta T_t = \frac{2\sigma T_{sat}}{h_{lv}\rho_v}\left(\frac{1}{r_v} - \frac{1}{r}\right) \tag{6.15}$$

The schematic of an elementary cell of a capillary porous material and its thermal resistance network is shown in Figure 6.32.

The effective thermal conductivity of porous systems has been a source of interest over the last two centuries. As of now, numerous experimental materials have been devised and a large number of formulae have been put forward to calculate the effective thermal conductivity of porous systems (Luikov et al., 1968). The method of generalised conductivities for the determination of effective thermal conductivity was used in GHP wick analysis. Here, one assumes the complex of Al_2O_3 particles in the elementary cell on

TABLE 6.1

Experimental Values of GHP Evaporator Heat Transfer Coefficients as Functions of Heat Flow

$T_{sat} = -10°C$	Heat flow (W)	40	50	70	80	100	150	170
	h_{smooth} (W/m² · K)	6500	6500	6350	6100	6000	5900	5500
	h_{porous} (W/m² · K)	6000	6250	7000	7300	7600	7000	6400
$T_{sat} = 40°C$	Heat flow (W)	40	50	70	80	100	150	–
	h_{smooth} (W/m² · K)	6600	6700	7100	7300	7700	7100	–
	h_{porous} (W/m² · K)	12,000	12,300	13,000	12,900	12,200	8200	–
$T_{sat} = 70°C$	Heat flow (W)	40	50	70	80	100	120	140
	h_{smooth} (W/m² · K)	–	8500	8500	8400	8200	8000	7800
	h_{porous} (W/m² · K)	–	13,000	13,000	13,100	12,900	11,100	9500

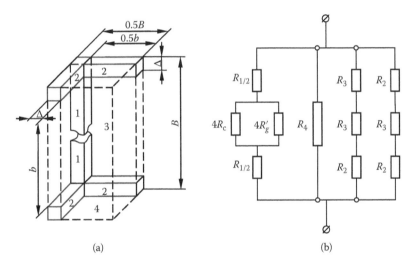

FIGURE 6.32
(a) One quarter of the porous coating elementary cell and (b) diagram of its thermal resistances.

the GHP evaporator to be symmetric and considers only quarters of two particles contacting each other. Let V be the total volume of an elementary cell, V_1 the volume of the solid phase of the elementary cell and V_2 the volume of the vapour phase (macropore). The liquid phase is disposed between two solid particles (in micropore) (Figures 6.30 and 6.31). Following Luikov et al. (1968) one has $\Delta/B = X$ and $H/b = X/(0.5 - X)$. Finally, the wick porosity is considered as $\prod = f(H/b)$.

The effective thermal conductivity, k_{eff}, of the wick is calculated as follows:

$$\frac{k_{eff}}{k_s} = \frac{1}{(1/(H/B)^2) + A} + v_g(1-B)^2 + \frac{2}{(1+s/b)+1/(v_g\,s/B)} \tag{6.16}$$

where

$$A = \frac{1}{\{(k_c/k_s) + [(v'_g/4)(s/B)^2]\} \times 10^3}, \quad B = b + s; s = 2\Delta \quad (6.17)$$

This complex number A characterises the thermal resistance of contact between two Al_2O_3 particles. The term v_g of Equation 6.16 is the ratio of thermal conductivity of the liquid (ammonia) to the solid (in our case, the thermal conductivity of Al_2O_3). It is noted that $v_g = k_1/k_s$; k_c is thermal conductivity of the thermal contact between the particles of Al_2O_3 in vacuum. The thermal conductivity of the porous Al_2O_3 particles, k_s, is equal to 2.1 W/mK. The values of k_{eff} and of thermal resistance calculations were compared with the results of experimentally determined values of the porous coating thermal resistance R_{exp}.

The calculated data of the effective thermal conductivity of the porous deposit saturated with liquid, or vapour, are presented in Table 6.2.

Usually, the main working fluid for aluminium GHPs is ammonia. In Figure 6.33, Table 6.1, heat transfer coefficients for the evaporators of different GHPs are demonstrated. The parameters of GHPs (S1-(S1-1), S2-(S2-1) and S3-(S3-1)) as a function of temperature are obtained for a constant heat load (Vasiliev et al., 2012b). For grooves of different shapes and dimensions, the increase in heat transfer intensity is 1.3–1.6 times. As already mentioned, the heat transfer enhancement depends on the degree of porous surface area on the groove edge. This heat transfer intensity is the highest for GHP S2-1. For GHP S2-1, the largest portion of the total area occupied by the crests of edges is equal to 0.6. For S1-1 surfaces it is 0.43, and for S3-1 surfaces it is 0.33.

Visual analysis and experimental validation of GHP heat transfer intensity using nanoporous technology illustrate the heat transfer enhancement (2.5–3 times) compared to heat transfer on GHPs with smooth grooves and completely modify the hydrodynamics of two-phase flow in grooves. A porous coating with open pores can be considered as a medium in which a significantly large number of MHPs with zones of evaporation and condensation are active (Figures 6.30 and 6.31).

The structure of Al_2O_3 nanoporous coating is shown in Figure 6.34. The rise in heat flux on such coatings may occur with only a slight increase in wall temperature.

TABLE 6.2

Calculated Values of k_{eff} of Al_2O_3 Porous Coating (Thickness of $\delta = 50$ μm) Completely Saturated with Liquid or Vapour Ammonia

Temperature (°C)	−10	30	40
k_{eff}, saturated with liquid (W/mK)	1.14	1.09	1.07
k_{eff}, saturated with vapour (W/mK)	0.646	0.656	0.659

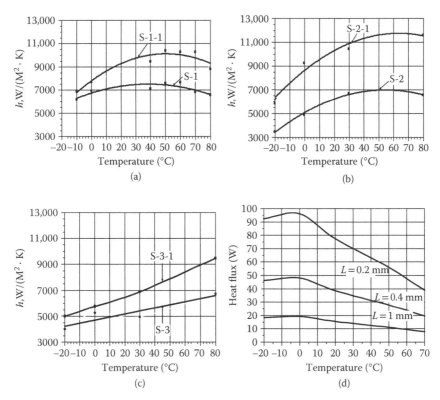

FIGURE 6.33
Heat transfer intensity in the evaporator of GHPs S1 and S1-1, S2 and S2-1, and S3 and S3-1 (ammonia) as a function of saturation vapour temperature: (a) GHP S1 and S1-1; (b) GHP S2 and S2-1; (c) GHP S3 and S3-1; (d) heat flux removed from one edge of the capillary groove with porous coating as a function of temperature. L is the edge width in millimetres; the temperature of ammonia vapour is $-20°C$.

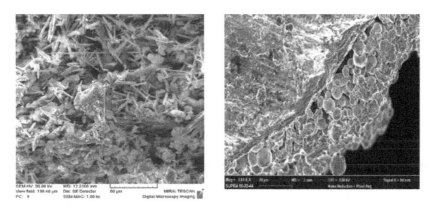

FIGURE 6.34
Aluminium oxide porous coating on the surface of a mini groove.

When determining the required parameters of porous coatings, one should take into account the geometry and dimensions of capillary grooves, vapour temperature and heat load of the evaporator, along with the structural characteristics. A microscale effect is essential inside a porous body, and a miniscale effect is observed in the groove.

6.5 Sorption Heat Pipes

An SHP combines the enhanced heat and mass transfer of conventional heat pipes with the sorption phenomena of a sorbent bed. An SHP can be used as a sorption heat transfer element and be cooled and heated as a heat pipe (Vasiliev and Vasiliev, 2000, 2003; Vasiliev et al., 2001b).

When it is necessary, SHPs are to be used for fuel cell thermal and transport water control in the active mode. 'A 'closed type' SHP' can actively cool the fuel cell stack at heavy power consumption (Figure 6.35).

SHP 'open type' is interesting to apply to cool the total fuel cell assembly, promote water transport and mitigate flooding and the associated blockage of reactant transport, if it is transparent to fluid penetration. Such an SHP is made from permeable materials (a porous fuel cell stack). A water/sorption heat pipe is capable of sucking excess water, stimulating the transport of water in terms of start-up, including freeze-start, transient response and degradation. At high electric currents, excessive water condensation can lead to the 'flooding' of cathode porous layers. SHPs are capable of sucking discrete water droplets through the pores of the gas diffusion layer into the microchannels and transporting them into the heat pipe sorbent bed. The original design of SHP was patented in the Soviet Union (Vasiliev and Bogdanov, 1992).

An SHP has a sorbent bed (adsorber/desorber and evaporator) at one end and a condenser and an evaporator at the other end (Figure 6.35). The basic principle of SHP operation is as follows:

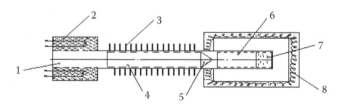

FIGURE 6.35

Sorption heat pipe (Vasiliev and Bogdanov, 1992): (1) vapour channel; (2) sorption structure; (3) finned surface of heat pipe evaporator/condenser; (4) porous wick; (5) porous valve; (6) low-temperature evaporator with a porous wick; (7) working fluid; (8) cold box with thermal insulation.

Phase 1: At the beginning of heat pipe functioning, it is necessary to desorb a sorption structure (2 in Figure 6.35) of SHP due to the absorption of heat of a heat source. During desorption of a sorbent bed, the vapour (1) needs to leave a porous structure (2) and be condensed in the heat pipe evaporator/condenser (3). When the vapour is generated inside the porous structure of a sorbent bed the vapour pressure increases, and the vapour flow enters the condenser to be condensed, releasing heat to the surroundings. A part of the cold working fluid is filtered through the porous valve (5), and it enters the evaporator due to the pressure drop between the hot part of the heat pipe and the evaporator. The other part of the working fluid is returned to the sorbent bed due to capillary forces of the wick (4) and stimulates the procedure of sorbent bed heating (the MHP phenomenon inside the sorbent bed) by the electric heater. When desorption of the sorbent structure is accomplished, the source of energy is switched off, the pressure in the sorbent bed is decreased, and the working fluid is accumulated inside the SHP evaporator.

Phase 2: After phase 1 is completed, the porous valve (5 in Figure 6.35) is opened and the vapour pressure inside the heat pipe is equalised following the procedure of liquid evaporation inside the porous structure of the evaporator (6). During liquid (7) evaporation in the evaporator (6), the air cools inside the cold box (8). When the liquid evaporation is accomplished and the sorbent bed is filled only with vapour, a porous valve is closed and the sorbent bed begins to be cooled by the condenser (3). Thus, phase 2 is completed.

The experimental set-up is shown in Figure 6.36. The SHP is insensitive to some g accelerations, and it is suggested for space and ground applications. This system is composed of an LHP and a solid sorption cooler. The LHP can be transformed into an SHP with the same evaporator, but the SHP has one or more sorption beds in addition.

The SHP extends the limits of two-phase thermal control and ensures successful electronic component cooling even under very harsh environmental conditions (an ambient temperature of 40°C or more) and also ensures deep cooling of space sensors, down to the triple point of hydrogen.

The experimental set-up of an SHP cooler (Figure 6.36) is composed of four main parts: evaporator (I), condenser (II), liquid heat exchanger + compensation chamber (III) and canister with sorbent material (IV). A capillary pumped evaporator is a key element of SHPs. When the SHP evaporator is connected to a canister (IV) (through valve 10) and is disconnected from the condenser (II) (through valve 9), intense fluid superheating occurs inside the wick of the evaporator due to a sharp decrease in pressure (pressure swing in the adsorber/evaporator system). The fast temperature decrease in the evaporator is shown in Figure 6.37. The constant heat input to the evaporator is 300 W. During the first period of time

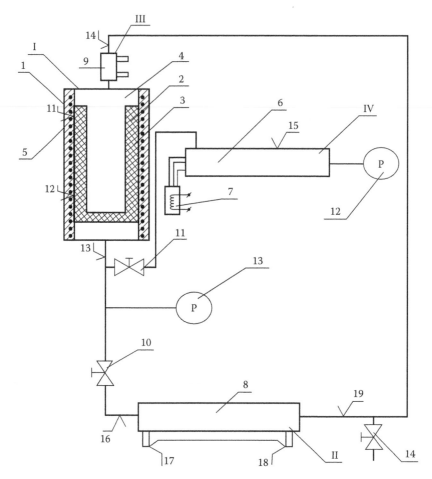

FIGURE 6.36
Schematic of a sorption heat pipe cooler: (I) evaporator; (II) condenser; (III) liquid heat exchanger and compensation chamber; (IV) solid sorption canister (SSC). (1) evaporator envelope; (2) wick; (3) vapour channels; (4) liquid compensation chamber; (5,7) heaters; (6) canister with sorbent bed; (8) condenser; (9) liquid heat exchanger; (10,14) valves; (11) regulated valve; (12,13) pressure sensors; (15–19) thermocouples.

(A in Figure 6.38) (SHP switch on for 4 minutes), saturated liquid ammonia enters the compensation chamber 4.

At the beginning of the experiment, the evaporator 'hot' wall temperature (60°C), vapour outlet temperature (52°C) and liquid entrance tempera-ture (25°C) are constant (Figure 6.37). During the second period of time (B in Figure 6.38) (12 minutes), a sharp heat transfer intensification occurs in the SHP evaporator and the temperature of the evaporator, T_e; the temperature of the liquid in the compensation chamber, T_l; and the vapour outlet temperature, T_v, are becoming near the same. The intense volumetric

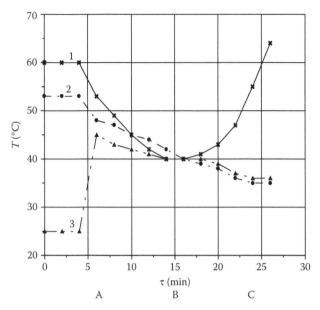

FIGURE 6.37
Temperature field evolution in the loop heat pipe evaporator wall (1) vapour outlet (2) and saturated liquid entrance (3) as a function of time when the solid sorption canister is switched on. (A) constant cooling rate; (B) transient enhanced cooling rate; (C) porous wick partial drying.

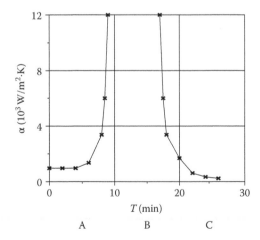

FIGURE 6.38
Heat transfer coefficient α value as a function of time when the loop heat pipe evaporator is cooled by a solid sorption canister: Liquid superheating is reached due to the pressure swing. $Q = 300$ W.

evaporation of the liquid inside the wick porous structure and the vapour condensation inside the compensation chamber (the MHP phenomenon is available inside the wick structure).

During the third period of time (C in Figure 6.38) the liquid inside the compensation chamber is exhausted and the porous structure starts to superheat (heat transfer rate is limited by the thermal conductivity of the wick). The heat transfer evolution during this period of SHP functioning is shown in Figure 6.38. If we compare the heat transfer coefficient for three different cases of the LHP evaporator working with saturated liquid, sub-cooled liquid and superheated fluid (ammonia), we can reach the sharp heat transfer intensification during the liquid subcooling and vapor superheat-ing. For the case of superheating, this heat transfer intensification is more than six times (from 2,000 to 12,000 W/m^2) (Figure 6.38) that of the conven-tional LHP evaporator.

This enhanced mode of LHP evaporator cooling can be near stationary if we use at least two solid sorption canisters that are switched on and off alternatively.

Initially, the heat transfer intensity in the SHP was equal to 3500 $W/m^2 \cdot K$. At least five times more heat transfer intensification was obtained for ammonia boiling in the titanium wick, when a solid sorption cooler was switched on (17,000 $W/m^2 \cdot K$).

6.5.1 Air-Conditioning Systems Based on Sorption Heat Pipe Components

SHPs are the active cooling/heating components for the new type of air-conditioning systems developed in the Luikov Heat and Mass Transfer Institute, Minsk.

Such devices utilise the waste energy of furnaces, driers and boilers to preheat/cool and dry air from the ambient. Solid sorption finned heat pipes are used as the components of the heat exchanger (Figure 6.39). Such a heat exchanger (Figure 6.40) is used to transform the low-grade thermal energy to power the refrigeration system. This air conditioner is devoted to heat

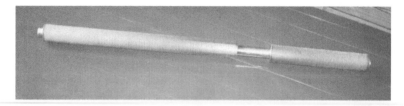

FIGURE 6.39
Sorption heat pipe with adsorber/desorber (active carbon fibre) in one part and evaporator/condenser in another part: the working fluid is ammonia.

FIGURE 6.40
Air-conditioning system with sorption heat pipe heat exchanger to cool and heat the air.

and cool the air. Activated carbon, silica gel, zeolite and calcium chloride are used as adsorbents; methanol, water and ammonia are used as refrigerants.

6.6 Conclusions

Heating and cooling demands by the industrial, commercial and domestic segments of the economy are increasing constantly and presently constitute around 50% of worldwide energy demand (International Energy Agency, 2007). From this point of view, the combination of fuel cells + sorption technologies + heat pipe thermal management is very promising to reduce electricity demand for heating and cooling and reduce the overall energy demand. The heat pipe concept as a thermal control system for solid sorption machines, fuel cells and systems for co-generation and tri-generation is a powerful tool to increase their efficiency. New technologies related to NF and NC applications increasingly improve the heat pipe and solid sorption device parameters. Recent advances in nanotechnology have allowed the development of new NFs and NCs, intensifying heat transfer processes and reducing energy losses. NFs and NCs have been used as the means to increase heat transfer intensity in original designs of heat pipes and thermosyphon evaporators. New, original heat pipes have been invented and suggested for further applications (VDTs, SHPs, mini-grooved heat pipes with nanoporous coatings).

6.7 Acknowledgements

The authors thank their colleagues A.A. Antukh, A.G. Kulakov, D.A. Mishkinis, L.E. Kanonchik, L.P. Grakovich, M.I. Rabetsky and A.S. Zhuravlyov for their experimental contribution in mathematical modelling. The authors thank the Belarus Foundation for Basic Researches (T97-303 dated 1.03.1998. 'Heat Transfer Intensification at Propane Boiling in Porous Coatings of Heat Generation Sources'; T04P-079 dated 03.05.2004. 'Heat and Mass Transfer in Two-Phase Sorption Systems'; T05BR-018 dated 01.04.2005. 'Research of Heat and Mass Transfer Processes in Effective Sorbents and Creation of New Generation of Adsorption Pumps on Their Basis'; T09CO-017 dated 15.04.2009. 'Adsorption Systems of Humidity Maintenance in Museums, Libraries and Archives: New Materials, a Dynamic Principle of Work, Expansion of Parameters of a Microclimate'; Co-operation Agreement INTAS Collaborative Call with CNES and NSAU on Space Technologies Nr 06-1000024-8916 and National Academy of Sciences of Belarus, Russian Academy of Sciences [Integration project]) for partial financial support of this research.

References

Aliakhnovich, V.A., Kireichik, D.G., Vasiliev, L.L., Konev, S.V., Zikman, A.B. 2011. Cooling system for hermetic compressor based on the loop thermosyphon. Proceedings of the 6th International Seminar on Heat Pipes, Heat Pumps, Refrigerators, Power Sources, 12–15 September (2), Minsk, Belarus, pp. 101–110.

Alyousef, Y., Antukh, A., Tsitovich, A., Vasiliev, L. 2012. Three adsorbers solar cooler with sorbent bed and heat pipe thermal control. *Applied Thermal Engineering* 38: 124–130.

Bang, I.C, Chang, S.H. 2005. Boiling heat transfer performance and phenomena of Al_2O_3–water nano-fluids from a plain surface in a pool. *International Journal of Heat and Mass Transfer* 48: 2407–2419.

Choi, H.K., Neveu, P., Spinner, B. 1996. System modeling and parameter effects on design and performance of ammonia based thermochemical transformer. Proceedings of the International Absorption Heat Pump Conference, Montreal, Canada, 17–20 September, 505–512.

Das, S.K, Putra, N., Roetzel, W. 2003. Pool boiling characteristics of nano-fluids. *International Journal of Heat and Mass Transfer* 46: 851–862.

Dube, V., Akbarzadeh, A., Andrews, J. 2004. The effects of non-condensable gases on the performance of loop thermosyphon heat exchangers. *Applied Thermal Engineering* 24: 2439–2451.

Dunn, P.D., Reay, D.A. 1976. *Heat Pipes*. Oxford: Pergamon Press.

Faghri, A. 2005. Micro heat pipe embedded bipolar plate for fuel cell stacks. US Patent 2005/0026015 A1, 3 February 2005, Mansfield, CT.

Faghri, A., Guo, Z. 2005. Challenges and opportunities of thermal management issues related to fuel cell technology and modeling. *International Journal of Heat and Mass Transfer* 48: 3891–3920.

Garner, S., Patel, C. 2001. Loop thermosyphons and their applications to high density electronics cooling. Proceedings of IPACK'01 The Pacific Rim/ASME International Electronic Packaging Technical Conference and Exhibition, 8–13 July, Kauai, Hawaii, p. 15782.

Garrity, P.T., Klausner, J.F., Mei, R. 2007. A flow boiling microchannel evaporator plate for fuel cell thermal management. *Heat Transfer Engineering* 28(10): 877–884.

Goetz, V., Spinner, B., Lepinasse, E. 1997. A solid-gas thermochemical cooling system using $BaCl_2$ and $NiCl_2$. *Energy* 22: 49–58.

Holm, F.W., Goplen, S.P. 1979. Heat transfer in the meniscus thin film transition region. *ASME Journal of Heat Transfer* 101: 543–547.

Imura, H., Takeshita, K., Doi, K., Noda, K. 1994. The effect of the flow and heat transfer characteristics in a two-phase loop thermosyphon. Proceedings of the 4th International Heat Pipe Symposium, Tsukuba, Japan, pp. 95–106.

International Energy Agency. 2007. Renewables for heating and cooling – the untapped potential. *Joint report for the IEA Renewable Energy Technology Deployment Implementing Agreement and the Renewable Energy Working Party*, Paris, France. July, 2007.

Izenson, M.G., Hill, R.W. 2004. Water and thermal balance in PEM fuel cells. *Journal of Fuel Cell Science and Technology* 1: 10–17.

Jakob, U., Kohlenbach, P. 2010. Recent developments of sorption chillers in Europe. IIR Bulletin 34–40.

Khodabandeh, R. 2005. Pressure drop in riser and evaporator in an advanced two-phase thermosyphon loop. *International Journal of Refrigeration* 28(5): 725–734.

Khrustalev, D. 2002. Loop thermosyphons for cooling of electronics. Proceedings of the 18th SEMI-THERM Symposium, San Diego, CA, pp. 145–150.

Krallik, J.H. 2002. Cooling method and apparatus for use a fuel cell stack. US Patent 6,355,368 B1, 12 March 2002.

Lapotko, D., Lukianova, E. 2005. Laser-induced micro-bubbles in cells. *International Journal of Heat and Mass Transfer* 48 (1): 227–234.

Lapotko, D., Lukianova, E., Oraevsky, A. 2006. Selective laser nano-thermolysis of human leukemia cells with microbubbles generated around clusters of gold nanoparticles. *Lasers in Surgery and Medicine* 38 (1): 631–642.

Lepinasse, E., Goetz, V., Crozat, G. 1994. Modeling and experimental investigation of a new type of thermochemical transformer based on the coupling of two solid-gas reactions. *Chemical Engineering and Processing* 33: 125–134.

Lepinasse, E., Marion, M., Goetz, V. 2001. Cooling storage with resorption process. Application to a box temperature control. *Applied Thermal Engineering* 21: 1251–1263.

Litster, S., Djilaly, N. 2008. Performance analysis of microstructured fuel cells for portable applications. In: *Mini-Micro Fuel Cells*, ed. S. Kakac, A. Pramuanjaroenkij and L. Vasiliev, Springer Science + Business Media B.V., pp. 47–74.

Luikov, A.V., Shashkov, A.G., Vasiliev, L.L., Fraiman, Yu, E. 1968. Thermal conductivity of porous systems. *International Journal of Heat and Mass Transfer* 11: 117–140.

Ma, H.B., Peterson, G.P. 1996. Experimental investigation of the maximum heat transport in triangular grooves. *ASME Journal of Heat Transfer* 118: 740–746.

Maydanik, Y., Vershinin, S., Pastukhov, V., Chernyshova, M., Sarno, C., Tantolin, C., Passove, C. 2010. Cooling system for an aircraft electronic box. *Heat Pipe Science and Technology* 1 (3): 251–260.

Meunier, F. 1998. Solid sorption heat powered cycles for cooling and heat pumping applications. *Applied Thermal Engineering* 18: 715–729.

Milanez, F.H., Mantelli, M.B.H. 2010. Heat transfer limit due to pressure drop of a two-phase loop thermosyphon. *Heat Pipe Science and Technology: An International Journal* 1 (3): 237–250.

Mirzamoghadam, A., Catton, I. 1988. A physics model of the evaporating meniscus. *ASME Journal of Heat Transfer* 110: 201–207.

Mitrovic, J. 2006. How to create an efficient surface for nuclear boiling. *International Journal of Thermal Science* 45: 1–15.

Nebuloni, S., Thome, J.R. 2010. Numerical modeling of laminar annular film condensation for different channel shapes. *International Journal of Heat and Mass Transfer* 53: 2615–2627.

Panchamgam, S.S., Chatterjee, A., Plawsky, J.L., Wayner, P.C., Jr. 2008. Comprehensive experimental and theoretical study of fluid flow and heat transfer in a microscopic evaporative meniscus in miniature heat exchanger. *International Journal of Heat and Mass Transfer* 51 (21–22): 5368–5379.

Pioro, L.S., Pioro, I.L. 1997. *Industrial Two-Phase Thermosyphons*. New York: Begell House, inc. Library of Congress Cataloging-in-Publication Data.

Possamai, F.C., Setter, I., Vasiliev, L.L. 2009. Miniature heat pipe as compressor cooling devices. *Applied Thermal Engineering* 29: 3218–3223.

Reutskii, V.G., Vasiliev, L.L. 1981. The basis of plant tissue thermal control. *Reports of the Academy of Sciences of Belarus (in Russian)* 24(11): 1033–1036.

Rogg, S., Höglinger, M., Zwittig, E., Pfender, C., Kaiser, W., Heckenberger, T. 2003. Cooling modules for vehicles with a fuel cell drive. *Fuel Cells* 3: 153–158.

Saha, B., Ng, K.S. (Eds). 2010. *Advances in Adsorption Technologies*. Singapore: Nova Science Publishers.

Saha, B.B., Akisawa, A., Kashiwagi, T. 2001. Solar/waste heat driven two-stage adsorption chiller: the prototype. *Renewable Energy* 23 (1): 93–101.

Saraff, D.B., Schwendemann, J.T. 2004. Flat fuel cell cooler. US Patent 6,817,097 B2, 16 November 2004.

Se Min, O., Vasiliev L. 2003. Heat pipe and method of manufacturing the same. US Patent 20030141045 A1, 31 July 2003.

Stephan, P.C., Busse, C.A. 1992. Analysis of the heat transfer coefficient of grooved heat pipe evaporator wall. *International Journal of Heat and Mass Transfer* 35 (2): 383–391.

Suman, B., Sirshendu, D., DasGupta, S. 2005. Transient modeling of micro-grooved heat pipe. *International Journal of Heat and Mass Transfer* 48: 1633–1646.

Sviridenko, I., Shevelov, D. 2011. Autonomous thermosyphon system for WWER-1000 pressurizer cooldown. Proceedings of the 8th International Seminar on 'Heat Pipes, Heat Pumps, Refrigerators, Power Sources', 12–15 September (2), Minsk, Belarus, pp. 71–78.

Tamainot-Telto, Z., Metcalf, S.J., Critoph, R.E. 2009. Investigation of activated carbon-R723 pair for sorption generator. Proceedings of Heat Powered Cycles Conference, Berlin, Germany, Paper No. 600.

Vasiliev, L.L. 1993. Open–type miniature heat pipes. *Journal of Engineering Physics and Thermophysics* 65(1): 625–631.

Vasiliev, L.L. 1994. Heat pipe technology in CIS countries. Proceedings of the 4th International Heat Pipe Symposium, 16–18 May, Tsukuba, Japan, pp. 12–24.

Vasiliev, L. 2007. Solid sorption heat pumps for tri-generation. *Archives of Thermodynamics* 28 (3): 15–28.

Vasiliev, L.L., Bogdanov, V.M. 1992. USSR Patent 174411 'Heat pipe', B. I. No. 24, 30 June 1992.

Vasiliev, L.L., Vasiliev, L.L., Jr. 2000. Two phase thermal control system with a loop heat pipe and solid sorption cooler. *SAE Technical Paper Series* 01-2492.

Vasiliev, L.L., Vasiliev, L.L., Jr. 2003. Sorption heat pipe – A new thermal control device for space applications, CP654, space technology and applications. Proceedings of the International Forum – STAIF, M.S. El-Genk (Ed.), pp. 71–79.

Vasiliev, L., Vasiliev, L., Jr. 2005. Sorption heat pipe – A new thermal control device for space and ground application. *International Journal of Heat and Mass Transfer* 48: 2464–2472.

Vasiliev, L., Grakovich, L., Khrustalev, D. 1981. Low temperature axially grooved heat pipe. Proceedings of the 4th International Heat Pipe Conference, London, 337–348.

Vasiliev, L.L., Morgun, V.A., Rabetsky, M.I. 1985. US Patent 4554966, 26 November 1985.

Vasiliev L.L., Rabetsky, M.I., Kiselev, V.G. 1986. Vapordynamic thermosyphons. *Journal of Applied Thermophysics* 52 (1): 62–66.

Vasiliev, L.L., Mishkinis, D.A., Antukh, A.A., Vasiliev, L.L., Jr. 2001a. A solar-gas solid sorption refrigerator. *Adsorption* 7: 149–161.

Vasiliev, L.L., Mishkinis, D.A., Antukh, A.A., Vasiliev, L.L., Jr. 2001b. Solar-gas solid sorption heat pump. *Applied Thermal Engineering* 21: 573–583.

Vasiliev L.L., Mishkinis, D.A., Antukh, A.A., Kulakov, A.G., Vasiliev, L.L., Jr. 2002. Multisalt carbon portable chemical heat pump. Proceedings of the International Sorption Heat Pump Conference, 24–27 September, Shanghai, China, pp. 463–468.

Vasiliev, L., Zhuravlyov, A., Shapovalov, A., Litvinenko, V. 2004. Vaporization heat transfer in porous wicks of evaporators. *Archives of Thermodynamics* 25 (3): 47–59.

Vasiliev, L., Zhuravlyov, A., Shapovalov, A.L., Vasiliev, Jr. 2006. Microscale two phase heat transfer enhancement in porous structures. Proceedings of the 13th International Heat Transfer Conference, 13–18 August, Sydney, Australia.

Vasiliev, L., Lapotko, D., Lukianova, E., Zhurablyov, A., Shapovalov, A., Vasiliev, L., Jr. 2007. Two phase heat transfer enhancement in micro channels and heat pipe evaporators with nano porous structures. Proceedings of the 14th International Heat Pipes Conference (14th IHPC), 22–27 April, Florianopolis, Brazil.

Vasiliev, L., Grakovich, L., Rabetsky, M., Romanenkov, V., Vasiliev, L. Jr., Ayel, V., Bertin, Y., Romestant, C., Hugon, J. 2010. Grooved heat pipes with a nanoporous deposit in an evaporator, *Heat Pipe Science and Technology, An International Journal* 1: 219–236.

Vasiliev, L., Zhuravlyov, A., Zhapovalov, A. 2012a. Heat transfer in mini channels with micro/nano particles deposited on a heat loaded wall. *Journal of Enhanced Heat Transfer* 19 (1): 13–24.

Vasiliev, L.L., Grakovich, L.P., Rabetsky, M.I., Vasiliev, L.L., Jr., 2012b. Grooved heat pipes evaporators with porous coating. Proceedings of the 16th International Heat Pipe Conference (16th IHPC), 20–24 May, Lyon, France.

Vassallo, P., Kumar, R., Amico, S. 2004. Pool boiling heat transfer experiments in silica-water nano-fluids. *International Journal of Heat and Mass Transfer* 47: 407–411.

Wang, J., Catton, I. 2011. Enhanced evaporation heat transfer in triangular grooves covered with a thin fine porous layer. *Applied Thermal Engineering* 21 (17): 721–1737.

Wang, R.Z., Oliveira, R.G. 2006. Adsorption refrigeration – An efficient way to make good use of waste heat and solar energy. *Progress in Energy and Combustion Science* 32: 424–458.

Wang, H.S., Rose, J.W., Honda, H. 2004. A theoretical model of film condensation in square section horizontal microchannels. *Journal of Chemical Engineering Research and Design* 82 (A4): 430–434.

Weislogel, M.M., Golliher, E., McQuillen, J., Bacich, M.A., Davidson, J.J., Sala, M.A. 2002. Recent results of the microscale pulse thermal loop. In: *Int. Two-Phase Thermal Control Technology Workshop* 24–26 September. Newton White Mansion, Mitchellville, MD.

Wen, D., Ding, Y. 2005. Experimental investigation into the boiling heat transfer of aqueous based γ alumina nanofluids. *Journal of Nanoparticles Research* 7: 265–274.

Xie, X., Lee, H., Youn, W., Choi, M. 2003. Nano-fluids containing multiwalled carbon nano-tubes and their enhanced thermal conductivities. *Journal of Applied Physics* 94 (8): 4967–4971.

Xu, X., Carey, V.P. 1990. Film evaporation from a micro-grooved surface – An approximate heat transfer model and its comparison with experimental data. *Journal of Thermophysics and Heat transfer* 4: 512–520.

Yang, Z., Peng, X.F., Ye, P. 2008. Numerical and experimental investigation of two-phase flow during boiling in a coiled tube. *International Journal of Heat and Mass Transfer* 51: 1003–1015.

Yazici, M.S. 2007. Passive air management for cylindrical cartridge fuel cells. *Journal of Power Sources* 166: 137–142.

You, M., Kim, J.H, Kim, K.H. 2003. Effect of nanoparticles on critical heat flux of water in pool boiling heat transfer. *Applied Physics Letters* 83: 3374–3376.

7

Modelling of Heat and Mass Transfer in Sorption and Chemisorption Heat Converters and Their Optimisation

L. L. Vasiliev, O. S. Rabinovich, N. V. Pavlyukevich and M. Yu. Liakh

CONTENTS

7.1 Introduction 259
7.2 Numerical Models of Adsorption and Phase Transitions in
    Single-Unit Adsorber of a Heat Converter 260
    7.2.1 Discrete and Continual Approaches.. 260
    7.2.2 Physical and Chemical Sorption Models................................... 261
    7.2.3 Models for Condensation, Evaporation and Multi-Phase
          Transport in an Adsorber 262
    7.2.4 Geometric Scheme of an Adsorber and Its Interaction with
          Other Elements of a Heat Converter 266
    7.2.5 Integrated Models of Adsorption Heat Converters 268
    7.2.6 Validation of the Model.. 269
          7.2.6.1 Comparison with Two-Adsorber Heat Upgrading
                   Experiment... 270
    7.2.7 Example of Program Usage: Two-Channel Chemical-
          Adsorption System with a Thermal Wave 271
7.3 Optimisation of a Sorption Cooler with a Composite Sorbent 272
    7.3.1 Definition of the System.. 272
    7.3.2 Effects of Capillary Condensation on the Characteristics
          of Heat Converters 273
7.4 Concluding Remarks...280
References...281

7.1 Introduction

The history of adsorption heat converters (heat pumps and refrigerators) dates back to a relatively distant past – the middle of the nineteenth century (Carre 1860) – and it is much earlier for chemisorption (Faraday 1823).

At present, such devices have become increasingly important due to a growing demand for energy saving and for the use of renewable or low-grade thermal energy sources. In the development of this important area, diverse novel ideas have emerged related to new sorbents, designs of adsorbers along with the invention of new thermal cycles and methods for their operation control. Experimental investigation of these ideas is combined with computer modelling, which appears to be a powerful and convenient method for the analysis and optimisation of numerous physicochemical processes that occur in adsorption heat convertors. Computer modelling permits fast verification of various convertor designs and combinations of sorbents with different properties. Besides, it enables a designer to compare the efficiency of any of the two options: (1) to use separate evaporators and condensers or (2) to incorporate the condensation and evaporation of sorbate directly into adsorbers. In particular, the latter possibility seems promising for achieving a lowest possible temperature in the refrigeration cycle and maximising the specific cooling power due to the use of additional heat of phase transition (Endo and Komori 2005, Yong and Wang 2007, Alyousef et al., 2012).

This study has two objectives. The first one is to elaborate and verify a convenient computer tool for modelling adsorption heat converters of diverse types and designs on the basis of modern understanding of chemical and physical processes in these devices. The developed heat and mass transfer model includes such specific features as combination of different types of sorbents (exhibiting physical and chemical adsorption) in a single adsorber, heat and mass exchange among adsorbers, condensers and evaporators in multi-unit devices, and the possibility of successive adsorption and capillary condensation in a single adsorber. The second objective is to apply the developed computer code for the analysis and optimisation of several prospective adsorption heat converters.

7.2 Numerical Models of Adsorption and Phase Transitions in Single-Unit Adsorber of a Heat Converter

7.2.1 Discrete and Continual Approaches

An adsorber is a main elementary unit of an adsorption or chemisorption heat converter, and its operation should be considered as a problem related to the multi-phase flows in porous media accompanied by phase and chemical transitions. Generally speaking, there are two basic approaches to the description of such problems, which use continual or discrete models.

In the continual description, the gaseous, adsorbed and liquid phases are treated as interpenetrating continua occupying certain fractions of the pore space (Kaviany 1999). This approach can be successfully used when the characteristic length scale of pores is much smaller than the length scale of phase or chemical transition zones, as well as that of a thermal zone, and any instability of boundaries between zones with no different phase compositions is absent. In such a case of length scale separation, conventional volume averaging can be applied to formulate governing equations for the multi-phase flows and other processes in an adsorber.

The alternative, discrete approach is usually used when the conditions for length scale separation in a particular problem are not satisfied, and it is necessary to consider the transfer processes at the pore scale level. The most developed method for discrete description of processes is based on the pore network models (see the reviews in Pratt [2010, 2011] and references cited therein).

The goal of this work is to develop a comprehensive but, at the same time, efficient and robust numerical tool for modelling different types of adsorption heat converters as a whole device. The tool is destined for fast engineering and design analysis and optimisation. To attain the goal, we concentrate on the continual models that allow easily varying different converter parameters of a system under investigation and permit obtaining a desired result via a shortest way. Of course, we realise that a number of complex and virtually important problems, which can be treated by discrete models, are left beyond our consideration.

A basic model of adsorption heat converter, which is under consideration here, should include models of a few common units comprising the device; the main element is an adsorber unit that will be described in the following sections in more detail.

7.2.2 Physical and Chemical Sorption Models

We consider a general case when a composite sorbent has a complex structure and consists of two or more components that exhibit a physical or chemical type of sorption. At present, models for both of the mentioned sorption processes are available in scientific literature.

At first, let us consider a conventional model of physical sorption. The equilibrium capacity of sorbent (grams of adsorbed matter per 1 g of sorbent) in the case of adsorption of gases can be expressed by the Dubinin–Astahov equation (Dubinin and Astakhov 1971):

$$a_{eq} = \frac{W_0}{b} \exp\left(-B\frac{T^2}{\beta^2} \left[\ln\left(\left(\frac{T}{T_{cr}}\right)^2 \frac{P_{cr}}{P_g} \right) \right]^2 \right) \tag{7.1}$$

where T_{cr} and P_{cr} are critical temperature and pressure for sorbate, W_0 is the limiting volume of adsorption space, B is the structural constant related to the size and distribution of micropores, b is the van der Waals constant, β is the affinity coefficient, T is the current temperature and P_g is the gas (vapour) pressure.

The equations for the kinetics of physical adsorption have the following form (Babenko and Kanonchik 2000):

$$\frac{da}{dt} = W_{ad} = K_{s0} \exp\left(-\frac{E}{R_g T}\right)\left(a_{eq} - a\right) \tag{7.2}$$

where W_{ad} is the adsorption rate, E is the activation energy, R_g is the universal gas constant, K_{s0} is the pre-exponential factor, a is the current value of adsorption capacity filling and t is the time.

The kinetics of chemical sorption processes is more complex. For this case, the following equation has been used for the reactions of synthesis and decomposition respectively (Wang et al., 2010):

$$W_{syn} = a_0 \frac{dx}{d\tau} = A a_0 (1-x)^b \left(1 - \frac{P_s}{P_g}\right); \quad W_{dec} = a_0 \frac{dx}{d\tau} = A a_0 x^b \left(1 - \frac{P_s}{P_g}\right) \tag{7.3}$$

where W_{syn} is the synthesis rate, W_{dec} is the decomposition rate, x is the degree of reaction completion, $x = a/a_0$, a_0 is the maximal value of chemisorption capacity and P_s is the saturation pressure (the pressure of saturated vapour).

The saturation pressure is given by the Clausius–Clapeyron equation:

$$R_g \ln\left(\frac{P_s}{P_0}\right) = -\frac{\Delta H}{T} + \Delta S_0 \tag{7.4}$$

where P_0 is the saturation pressure at reference temperature T_0, ΔH is the evaporation enthalpy (assumed to be constant) and $\Delta S_0 = \Delta H / T_0$ is the entropy of the phase transition.

Generally, chemical adsorption of gas sorbate can proceed via different reaction pathways, thus forming different chemical compounds. The equilibrium state for each type of reaction obeys Equation 7.4 with relevant sets of parameters. The developed programme can be adapted to multi-reaction sorption as well.

7.2.3 Models for Condensation, Evaporation and Multi-Phase Transport in an Adsorber

As mentioned in Section 7.1, modern tendencies in adsorption heat converter design include the use of adsorption processes combined with capillary condensation. When a gas condenses within a porous medium, the problem of heat and mass transport becomes much more complicated because one should consider interrelated flow (filtration) of different phases accompanied by heat transfer and complex interplay between temperature, phase pressures, liquid and gas volume fraction and capillary effects. There are several approaches for treating multi-phase transport in a porous medium, which are discussed in literature in much detail (e.g. Dullien 1992, Kaviany 1999, and more recent books Chen et al., 2006, Pinder and Gray 2008). One of the ways to describe wetting and drying in porous media was developed by A.V. Luikov using the ideas of non-equilibrium thermodynamics (Luikov 1966, 1975). The main driving potentials in Luikov's wetting and drying model are the temperature and specific moisture content in a capillary–porous medium. Luikov's approach features simplicity and a high degree of abstraction. However, it faces difficulties at describing processes in adsorbers because there is a necessity to somehow determine the coefficients of non-equilibrium heat and mass transfer, and, essentially, to incorporate adsorption and desorption processes into the model.

A more constructive approach for multi-phase transport in sorbents is based on the so-called multi-phase filtration models (Kheifets and Neimark 1982, Barenblatt et al., 1990, Dullien 1992, Kaviany 1999, Chen et al., 2006, Pinder and Gray 2008). In this approach, volume averaging is used and the filtration flux of each phase is determined by the pressure gradient and local relative permeability of a porous medium for this phase, which, in turn, depends on local phase composition. Since the multi-phase filtration approach is used as a starting point for analysing the adsorber operation, we describe it in more detail in the following paragraphs.

Within the framework of the multi-phase filtration approach, when the gaseous and liquid phases coexist in a porous medium, the mass transfer should obey two equations describing the mass conservation and chemical/phase conversion:

$$\varepsilon \frac{\partial \theta_g \rho_g}{\partial t} + \nabla \left(\rho_g \vec{v}_g \right) = J_{lg} - (1 - \varepsilon) \rho_s W_{ad} \tag{7.5}$$

$$\varepsilon \frac{\partial \theta_l \rho_l}{\partial t} + \nabla \left(\rho_l \vec{v}_l \right) = -J_{lg} \tag{7.6}$$

where θ_g and θ_l are saturation of a porous medium by gas and liquid respectively ($\theta_g + \theta_l = 1$), ρ_i is the density of ith phase, v is the filtration velocity, J_{gl} is the evaporation rate of the liquid phase, W_{ad} is the adsorption rate (of physical or chemical nature) and ε is porosity.

Besides, it is necessary to supplement the model with an equation for the temperature of a porous medium. Here, we use the so-called one-temperature approximation, that is temperatures of porous skeleton, gas and liquid are assumed to be equal (because the adsorption processes are quite slow):

$$\left(\varepsilon \theta_g \rho_g c_{pg} + \varepsilon \theta_f \rho_f c_{pf} + (1 - \varepsilon) \rho_s c_{ps} + (1 - \varepsilon) \rho_s ac_{pa} \right) \frac{\partial T}{\partial t}$$

$$+ \left(v_g \rho_g c_{pg} + v_f \rho_f c_{pf} \right) \nabla T = \nabla \left(\lambda_{eff} \nabla T \right) - J_{lg} Q_{lg} + (1 - \varepsilon) \rho_s Q_s W_{ad} \tag{7.7}$$

In addition to notations of Equations 7.5 and 7.6, the new symbols in Equation 7.7 have the following meaning: c_{pi} is the heat capacity of the ith component, subscript s denotes the sorbent and Q_s and Q_{gf} are the heats of sorption and gas–fluid phase transition respectively.

The velocities of filtration fluxes for gaseous and liquid components are defined by the law of filtration that, in the simplest case, is expressed as Darcy's law:

$$v_g = -\frac{K_0 K_g}{\mu_g} \nabla P_g$$
$$v_1 = -\frac{K_0 K_1}{\mu_1} \nabla P_1$$

(7.8)

where P_i is the pressure of the ith phase (gas or liquid), μ is the viscosity, K_0 is the permeability of a porous medium and K_g and K_1 are the relative permeability for gas and liquid respectively.

Finally, to complete the description of multi-phase filtration model, the evaporation/condensation rate should be introduced using the Hertz–Knudsen formula:

$$J_{lg} \approx \frac{\left(P_e(T) - P_g\right)}{\sqrt{2\pi\left(R_g T / M\right)}} \varepsilon S_{gl}$$

(7.9)

where P_e is the equilibrium vapour pressure over the liquid meniscus, which is defined by the Kelvin equation:

$$P_e = P_s(T)\exp\left(-\frac{P_\sigma V_1}{R_g T}\right)$$

(7.10)

where $P_s(T)$ is the saturation vapour pressure over a flat liquid surface; V_1 is the molar volume of the liquid; P_σ is the capillary pressure, that is the difference between the pressure in the liquid and the gaseous phase at their interface; and $P_\sigma = P_g - P_1$. In the case of uniform size distribution of the capillaries, the capillary pressure is determined as $P_\sigma = 2\sigma\cos\phi/r$, where r is the mean meniscus radius of curvature and σ is the liquid/vapour surface tension.

Real porous media and, specifically, sorbents used in adsorption techniques are much more complex and have a wide pore size distribution. This feature has an essential effect on the capillary pressure and the nature of adsorption kinetics. First, filling of pores by a liquid, as the result of liquid filtration and vapour condensation, proceeds from small to large pores successively. Hence, such effective characteristics as the relative permeability or capillary pressure depend on the liquid saturation θ_1: $K_g(\theta_1)$, $K_1(\theta_1)$, $P_\sigma(\theta_1)$. In particular, the capillary pressure is expressed via a mean meniscus curvature $\sqrt{\varepsilon/K_0}$ and the Leverett function $J(\theta_1)$ as follows (Kaviany 1999):

$$P_\sigma = 2\sigma \cos\phi \sqrt{\frac{\varepsilon}{K_0}} J(\theta_1) \tag{7.11}$$

Analysis of specific features of sorbents and some quantitative estimations have shown that the complex system of equations for multi-phase filtration can be substantially simplified in many cases of adsorption heat conversion. These estimations are discussed later.

First, it is necessary to point out that pores in a sorbent can be subdivided into three wide ranges (Dubinin 1966, 1970, 1972): (1) micropores with the effective size less than 2 nm, which provide multi-layer and volume adsorption; (2) mesopores with the effective size ranging approximately from 2 to 50 nm, where capillary condensation is possible; and (3) macropores larger than 50 nm, whose surface area is not large, hence adsorption in these pores is insignificant, but they provide main channels for the transport of gaseous phase through the sorbent.

In this chapter, we restrict ourselves to the consideration of the case when the temperature and pressure in an adsorber lie within the gaseous domain of the thermodynamic phase diagram. The latter can be an important requirement for the technical usage of an adsorber because total condensation of a sorbate gas in all pores may lead to serious problems related to drastic slowing down of the mass transfer in the adsorber during the stage of evaporation. In the considered case, condensation may have only the capillary nature and proceed only in mesopores where the capillary pressure reduces the liquid–vapour equilibrium pressure (in accordance with Equation 7.9) sufficiently to initiate condensation. The above is the first simplification in the considered model.

For example, the estimation of the capillary pressure for ammonia at the mean mesopore radius 10^{-8} m gives $P_\sigma \approx 4.7$ MPa, which, in turn, reduces the equilibrium saturation pressure approximately by 10%. However, this capillary pressure is rather high and can be still higher than the driving pressure between adsorbers in a heat converter and is by many orders of magnitude higher than the pressure that is needed for gas filtration in the sorbent. Thus, we conclude that in this case filtration of a liquid in mesopores is impossible and Equation 7.6 can be reduced to the form

$$\varepsilon_{me} \frac{\partial \theta_1 \rho_1}{\partial t} = -J_{lg} \tag{7.6'}$$

It should be noted that porosity ε_{me} in Equation 7.6' relates to mesopores.

The second simplification also relates to the filtration process, namely to the filtration of gas. Simple estimation of the time of pressure relaxation in the sorbent gives the value of less than 1 second, so we reach the conclusion that the gas filtration in an adsorber occurs in a quasi-steady regime, when the time-derivative term in Equation 7.5 is negligible with respect to other terms and can be removed:

$$\nabla\left(\rho_g \vec{v}_g\right) = J_{lg} - \varepsilon_s \rho_s W_{ad} \tag{7.5'}$$

where filtration velocity $\vec{v}_g$ is defined by Darcy's law (Equation 7.8) and ε_s is the volume fraction occupied by the solid sorbent.

Nevertheless, the non-stationary form of Equation 7.5 can be retained when the relaxation method is used for the solution of finite-difference equations:

$$\left(\varepsilon_{me}\frac{\partial \theta_g \rho_g}{\partial t} + \left(1 - \varepsilon_s - \varepsilon_{me}\right)\frac{\partial \rho_g}{\partial t}\right) + \nabla\left(\rho_g \vec{v}_g\right) = J_{lg} - \varepsilon_s \rho_s W_{ad} \tag{7.5''}$$

From the standpoint of three pore-size ranges, the energy equation (Equation 7.7) should be rewritten in the following way:

$$\left[\left(\varepsilon_{me}\theta_g + 1 - \varepsilon_s - \varepsilon_{me}\right)\rho_g c_{pg} + \varepsilon_{me}\theta_l \rho_l c_{pl} + \varepsilon_s \rho_s c_{ps} + \varepsilon_s \rho_s a c_{pa}\right]\frac{\partial T}{\partial t}$$
$$+ \left(v_g \rho_g c_{pg} + v_f \rho_f c_{pf}\right)\nabla T = \nabla\left(\lambda_{eff}\nabla T\right) - J_{lg}Q_{lg} + \varepsilon_s \rho_s Q_s W_{ad} \tag{7.7'}$$

where ε_s, ε_{mi} and ε_{me} are the relative volumes of a porous medium occupied by solid sorbent, micropores and mesopores respectively.

For the case of capillary condensation in mesopores, the following algorithm for the solution of transport equations in an adsorber can be used, which is based on the combined explicit–implicit finite difference scheme:

1. All the source terms, transport coefficients and other liquid saturation dependent functions are calculated by the use of parameters taken from the previous time step.
2. The liquid saturation θ_l at the new time step is found from Equation 7.6' and the gas saturation θ_g is calculated as $1 - \theta_l$.
3. The temperature field described by Equation 7.7' is calculated using the implicit finite-difference method.
4. The gas pressure at the new time step is found from Equation 7.5' using the explicit finite-difference scheme and then the gas velocity is calculated with the use of the first of Equation 7.8.

7.2.4 Geometric Scheme of an Adsorber and Its Interaction with Other Elements of a Heat Converter

The above-presented model of an adsorber describes only one of the elements (units) comprising a heat converter device. For modelling a heat converter as a whole, we consider a simplified scheme of an adsorber unit, as represented as a sketch in Figure 7.1.

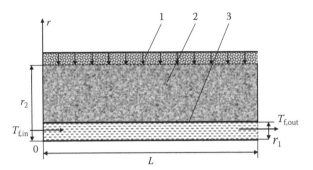

FIGURE 7.1
The sketch of adsorber unit. (1) the channel for sorbate supply; (2) the sorbent; and (3) the heat carrier channel.

The cylindrical reactor has a coaxial design and includes an external jacket and a central tube. The space between the jacket and the tube is filled with a porous sorbent of a certain composition. The latter can be monotype or consists of a mixture of sorbents featuring different sorption mechanism, capacity and kinetic properties. Two basic design variants for the supply of sorbate and heat carrier are used. In the first variant, the sorbate is fed through the external jacket whose internal wall is perforated. Then, the impermeable central tube serves for the forced circulation of the heat carrier having temperature T_f. In the second variant, the heat carrier is supplied through the external jacket, which should be impermeable in this situation, and the gas sorbate is fed through the perforated central tube. In numerical modelling, the first design of the adsorber was considered. The simplification is that we consider variations of all the parameters over a radial coordinate in a cylindrically symmetric adsorber unit neglecting heat/mass transfer in the axial direction.

The inlet temperature of the heat carrier, $T_{f,in}$, is always equal to the temperature of the attached heat reservoir. The outlet temperature of the heat carrier is determined from the heat balance equation:

$$\alpha_{he,1} S \left(T_{r_1} - T_{f,in} \right) = G_f c_{p,f} \left(T_{f,out} - T_{f,in} \right) \tag{7.12}$$

where S is the heat exchange surface between the sorbent and heat carrier and $\alpha_{he,1}$ is the heat exchange coefficient.

The heat exchange coefficient also enters the boundary condition for Equation 7.7′, which is expressed as the following:

$$\alpha_{he,1} \left(T_{r_1} - T_{f,in} \right) = \lambda_{eff} \left. \frac{\partial T}{\partial r} \right|_{r_1} \tag{7.13}$$

Taking into account slow inlet/outlet gas flow at the other boundary of sorbent (i.e. at the outer surface of the adsorber unit), heat exchange between the sorbent and the environment is described using the Newtonian law with coefficient $\alpha_{he,2}$:

$$\alpha_{he,2}\left(T_{r_2} - T_0\right) = \lambda_{eff} \left.\frac{\partial T}{\partial r}\right|_{r_2} \qquad (7.14)$$

where T_0 is the room temperature.

When the heat carrier is fed through the internal tube, coefficient $\alpha_{he,2}$ can be assumed zero.

The adsorber unit described above is connected to other elements of the heat transformer. When the adsorber exchanges gas with an attached unit (e.g. another adsorber, condenser or evaporator), the variations in the pressure in the adsorber caused by changes in temperature and gas content in the latter can be found from the integral mass balance:

$$\frac{dm_g}{d\tau} = w - K_v\left(P_a - P_{connect}\right) \qquad (7.15)$$

where $m_g = \int_V (1 - \varepsilon_s)\rho_g dV$ is the total amount of gaseous sorbate in the adsorber and $w = \int_V \left(-\varepsilon_s \rho_s W_{ad} + J_{lg}\right) dV$ is the integral gas source.

The term $K_v\left(P_a - P_{connect}\right)$ in Equation 7.15 denotes the mass flux of the gas from the considered adsorber with pressure P_a to another unit with pressure $P_{connect}$.

The gas density in pores is determined by the Mendeleev–Clapeyron equation for an ideal gas:

$$\rho_g = \frac{P\mu}{RT} \qquad (7.16)$$

7.2.5 Integrated Models of Adsorption Heat Converters

A complete computer model of physical sorption and/or chemisorption heat converter should include, along with the description of adsorbers, particular models for such elements as condensers, evaporators, heat reservoirs and control valves. An adsorber is connected to other units with the use of control valves. The latter are characterised by hydraulic resistance K_v, so that each condenser or evaporator is maintained at a given temperature $T_{ev/cond}$ and pressure $P_{ev/cond}$, which satisfy the equation of liquid/vapour phase equilibrium similar to Equation 7.4.

Specifications of all possible elements of a heat converter are presented in Table 7.1; they can be considered as unit 'bricks' for designing, modelling, and testing the adsorption-chemical heat converters.

TABLE 7.1

List of 'Bricks' for the Heat Converter Design Set

Elements of Heat Converter	Parameters of an Element
Adsorber S	$r_1, r_2, L, p_s, \lambda_{eff}, \alpha_i, \varepsilon,$ $C_p, T_{in}, Q, \Delta H, \Delta S$
Condenser/ evaporator Cond/Evap	$T_{ev/cond}, P_{ev/cond}$
Heat reservoir HT/LT	T_{hr}, G_f, C_{pf}
Control valve	$K_v,$ $G = K_v \Delta P$

The developed software package is considered as a tool for studying and analysing different design concepts of a heat converter and its operation parameters. This tool offers a choice of a model from the list presented in Table 7.2. These models allow us to investigate a variety of 'classical' designs described in literature. For example, Model 6 relates to the idea of successive amplification of heat-upgrading thermal wave in the channel with high-temperature (HT) adsorbers. This design provides a substantial improvement of the specific heat production (SHP) and the temperature of heat upgrading.

To adapt the software for modelling heat conversion systems with other configurations, only minor modification to the program codes is needed.

7.2.6 Validation of the Model

The developed computer programs were tested in regard to two aspects: (1) sensitivity of the numerical results to the choice of spatial grid and time step used when solving the equations, and (2) agreement with the known experimental data.

TABLE 7.2

Available Heat Converter Schemes

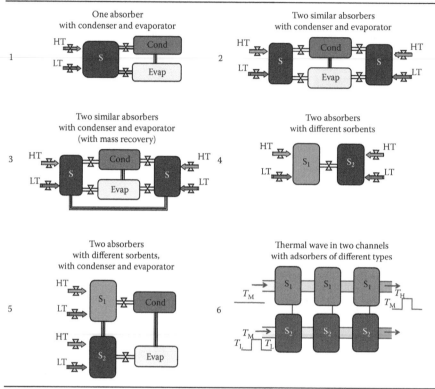

7.2.6.1 Comparison with Two-Adsorber Heat Upgrading Experiment

The results of modelling the heat upgrading process in the system composed of two different adsorbers (using chemisorption salts $MnCl_2$ and $CaCl_2$) were compared with the experimental data presented in Wang et al. (2008) (see also Figure 7.2).

It is seen that the durations of the heat converter operation cycle obtained in modelling and experiment are very close. There is a certain difference between modelling and experiment for the initial stage of the process, which can be explained by a discrepancy in the initial conditions for the compared cases. There is also a discrepancy between calculated and experimental results for a low-pressure stage of the process: the calculated temperature in the adsorber with low-temperature (LT) salt ($CaCl_2$) sharply drops down to the temperature of the heat carrier whereas the corresponding curve for the experimental temperature is much smoother. These disagreements can be related to the limitations of a 1D model of the adsorber, which does

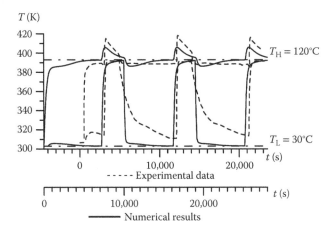

FIGURE 7.2
Comparison of temperature evolution in the modelled two-adsorber system with the experimental results from Wang et al. (2008); upper lines relate to the high-temperature adsorber and lower lines relate to the low-temperature one.

not take into account the longitudinal thermal conductivity in the reactor. Nevertheless, these disagreements are not high, and we can conclude that the developed model has an acceptable level of adequacy.

7.2.7 Example of Program Usage: Two-Channel Chemical-Adsorption System with a Thermal Wave

As an example of numerical modelling with the developed software codes we present the results for the heat upgrading process in the two-channel chemical-adsorption system that corresponds to Model 6 in Table 7.2. The considered system includes the HT adsorbers of ammonia, which are filled with $MnCl_2$, and the LT adsorbers filled with $CaCl_2$. The heat carrier at the inlet of an HT channel is fed from an HT reservoir with a constant temperature T_H. The temperature of the heat carrier in the LT channel is switched from LT T_L at the beginning of the process to T_H, and then again to T_L. The thermal wave in the HT channel is amplified while the wave passes through a series of adsorber pairs (Figure 7.3).

Table 7.3 shows that including a new pair of adsorbers into the system leads to the increase in the main heat conversion characteristics (i.e. the coefficient of performance, COP; SHP; and maximal temperature difference between a heat carrier and sorbent, ΔT), although the rate of this increase gradually drops down.

This example shows that the developed approach to the numerical modelling of adsorption heat converters is rather simple but efficient, thus allowing the user to examine and compare a wide variety of practically important designs.

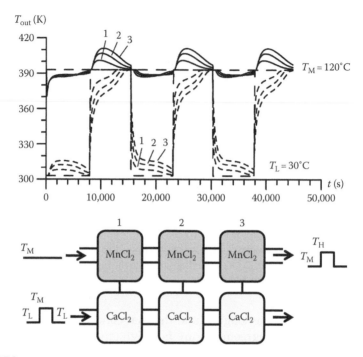

FIGURE 7.3

Amplification of heat upgrading wave in the two-channel systems with three pairs of adsorbers. The numbers identifying the curves correspond to the number of the adsorber pair.

TABLE 7.3

Modelling of Heat Upgrading with the Use of Thermal Wave

Characteristics of Heat Upgrading	One Pair of Adsorbers	Two Pairs of Adsorbers	Three Pairs of Adsorbers
COP ideal	0.54	0.54	0.54
COP	0.35	0.42	0.44
SHP (Wt/kg)	11.9	22.9	32.7
ΔT_{max} (K)	8	13.8	18

7.3 Optimisation of a Sorption Cooler with a Composite Sorbent

7.3.1 Definition of the System

The second example presented in this chapter relates to the modelling and optimisation of solid sorption cooler based on a composite sorption bed (Alyousef et al., 2012). We consider a somewhat simplified design scheme of the device described in the cited work, which consists of two adsorbers with different capacity filled with different composite sorbents. The first sorbent (in the HT adsorber) is a mixture of activated carbon fibre (ACF) 'Busofit' and $MnCl_2$ microcrystals deposited on the filament surface while the second sorbent (in the LT adsorber) is a mixture of ACF and $BaCl_2$ microcrystals. Ammonia is used as a working fluid. Two different salts together with ACF are used as the chemisorption material with shifted chemical equilibrium to increase the sorption capacity of the sorbent bed and raise the heat conversion efficiency. This design scheme corresponds to Case 4 in Table 7.2 because it does not contain a special condenser and evaporator. Nevertheless, as is discussed below, the design of adsorbers permits taking advantage of the phenomenon of capillary condensation of ammonia in the mesopores of ACF in the LT adsorber.

The adsorption and thermophysical parameters of the sorbents are presented in Table 7.4. The mass flow rate of heat carriers in both of the adsorbers is $G_f = 0.003$ kg/s, and the heat capacity of heat carriers is $c_{p,f} = 4200$ J/kg. The geometrical parameters of the adsorbers for the basic variant are chosen as follows:

HT adsorber: $r_1 = 0.008$ m, $r_2 = 0.0245$ m, $L = 2$ m;

LT adsorber: $r_1 = 0.008$ m, $r_2 = 0.0245$ m, $L = 1$ m.

First, let us compare the experimental data from Alyousef et al. (2012) for the cooling stage of the heat conversion cycle with the calculations performed using the developed software package. Figure 7.4 illustrates the results of the comparison. Taking into account serious difficulties in an accurate evaluation of boundary conditions on the external surface of the adsorber as well as the errors induced by the 1D approximation, the comparison shows a rather good coincidence of theoretical predictions with the experiment.

7.3.2 Effects of Capillary Condensation on the Characteristics of Heat Converters

At the end of this chapter, we show how the developed software package for modelling adsorption heat converters can be applied to evaluate the effect of capillary condensation on the efficiency of heat converters with composite sorbents. In this study, it is necessary to answer the following questions:

1. What conditions are needed for the capillary condensation to occur? Which parameters are responsible for the capillary condensation?
2. What are the consequences of capillary condensation? How does it affect the efficiency of heat conversion?

Numerical modelling was based on the geometrical and physical parameters close to the experimental conditions of work (Alyousef et al., 2012), which was discussed in the previous section. The length of the HT

TABLE 7.4

Adsorption and Thermophysical Properties of Sorbents and the Sorbate (NH_3)

Property	HT Adsorber (Busofit + $MnCl_2$)	LT Adsorber (Busofit + $BaCl_2$)
ρ_s (kg/m³)	285.07	362.27
c_s (J/[kg·K])	1154.12	923.6
c_a (J/[kg·K])	1657.12	1469.76
c_g (J/[kg·K])	2289	2289
E	0.75	0.75
ε_{meso}	0.065	0.065
a_{oSalt}	0.27	0
$a_{maxSalt}$	0.81	0.65
$a_{oBusofit}$	0.19	0.19
$a_{maxBusofit}$	0.38	0.38
f_{Salt}	0.48	0.44
$f_{Busofit}$	0.52	0.56
K_o (m²)	1.10^{-8}	1.10^{-8}
	$MnCl_2$	$BaCl_2$
a_{syn}	0.001019	0.0125
a_{dec}	0.0028	0.0195
b_{syn}	1.185	2.104
b_{dec}	1	1.005
Q_{salt} (J/kg)	2,790,000	2,220,000
ΔH (J/mol)	47,416	37,665
ΔS (J/[mol·K])	228.07	227.25
Ammonia properties		
ΔH (J/mol)	23,366	
ΔS (J/[mol·K])	192	
Q_{lg} (J/kg)	1,375,000	
c_l (J/[kg·K])	4826.8	
ρ_l (kg/m³)	595.4	
μ_g (Pa·s)	9.99×10^{-6}	
μ_l (Pa·s)	0.000,126	
σ (N/m)	0.023,321	
	Busofit	
$Q_{Busofit}$ (J/kg)	3,300,000	
K_{so} (s⁻¹)	0.075	
E/R_g	2467	
B/β^2 (K⁻²)	1.72×10^{-6}	
W_o/b	0.38	
Leverett's J-functions		
For condensation	$J(\theta_1) = 8 - 8 \cdot \theta_1^{0.58}$	
For evaporation	$J(\theta_1) = \theta_1^{-0.5} - 1$	

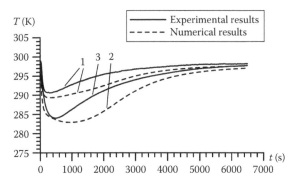

FIGURE 7.4
Comparison of numerical results with experiment (Alyousef 2012) on the cooling stage in the heat converter containing two adsorbers with composite sorbents. (1) Temperature of heat carrier at the outlet of LT adsorber; (2) average calculated temperature within the LT adsorber; (3) experimental temperature at the external surface of LT adsorber.

adsorber, L, the external radius of LT adsorber, r_2 and temperatures of both of the heat reservoirs, T_H and T_L, were chosen as the governing parameters that can influence the occurrence of capillary condensation in the LT adsorber.

For the situation when mesopores in the LT adsorber can be completely filled with liquid ammonia (the basic set of parameters in Table 7.4), the occurrence of capillary condensation is controlled by switching the heat carrier flows through the HT adsorber from a heat reservoir. Here under 'switching' we mean changeover of the heat carrier supply to the HT adsorber: from an HT or LT reservoir. This switching means the interruption of condensation in the LT adsorber and the start of sorption in the HT adsorber. The graphs for temperature and mass of condensed ammonia in the adsorbers for three different conditions of heat carrier switching are presented in Figure 7.5.

The switching of heat carrier flow is performed (*a*) just before the onset of capillary condensation, (*b*) after filling of 50% adsorber volume with liquid ammonia (in this case, mesopores are filled completely in the half of adsorber volume, overall filling of the LT adsorber by the condensate $\Theta_l^* = 0.5$) and (*c*) after complete filling of adsorber with liquid ammonia ($\Theta_l^* = 1$). In Figure 7.5, it is seen that the amount of condensate in the LT adsorber is not high, no more than 25% of full adsorbed and condensed ammonia. The period of condensation (III) and evaporation (IV) is not long as well. Variations in the detailed distribution of ammonia in the adsorbers are shown in Figure 7.6. Thermodynamic parameters for the considered case correspond to the situation when activated carbon does not virtually participate in the heat conversion as a sorbent. Here, it acts only as a material that provides mesopores for capillary condensation.

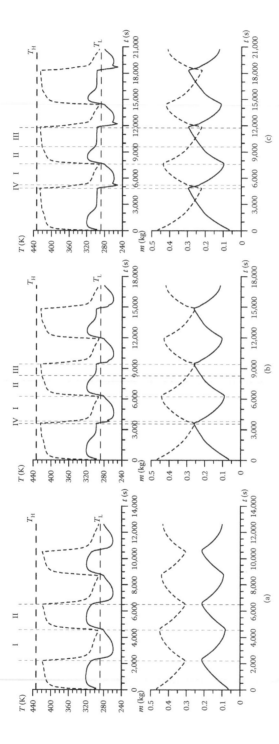

FIGURE 7.5

Variations in time of the volume-averaged temperature and sorbate mass content in the adsorbers. Plots *a, b* and *c* correspond to different conditions of switching the heat carrier flows in the HT adsorber: (a) the switching takes place just before the beginning of capillary condensation in the LT adsorber; (b) switching takes place after 50% volume filling of the LT adsorber; and (c) switching after 100% filling. The basic parameters are $L_1 = 2$ m, $L_2 = 1$ m, $R_1 = 0.0245$ m, $R_2 = 0.0245$ m, $R_0 = 0.008$ m, $T_H = 433$ K and $T_L = 288$ K. Dashed curves relate to the HT adsorber and solid lines to the LT adsorber. Roman numerals: I, desorption in the HT adsorber and sorption in the LT adsorber; II, vice versa; III, capillary condensation in the LT adsorber; IV, capillary evaporation in the LT adsorber.

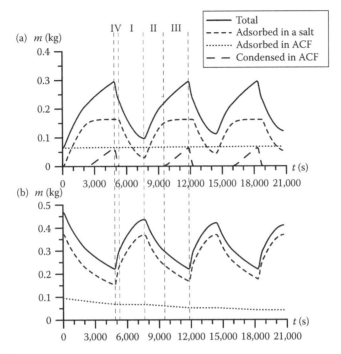

FIGURE 7.6
Variations in detailed distribution of ammonia in (a) the LT adsorber and (b) the HT adsorber.

It is interesting to trace how the changes in operation regimes (switching heat carrier) influence the main characteristics of heat conversion. The results of numerical simulation for different variants of heat carrier flux switching are presented in Figure 7.7.

It is seen that the main result of transition to the regimes with capillary condensation is an improvement in the cooling effect (an increase in the average difference of cooling temperatures, $<\Delta T>$; for LT heat carrier it is about 3 K). This effect is accompanied by a slight increase in COP (no more than 4%) and a simultaneous decrease in the specific cold production (SCP) rate (due to an increase in the thermodynamic cycle duration). Such a small contribution of capillary condensation to the efficiency parameters can be connected with a low volume fraction of mesopores in ACF. The capillary condensation can play a more essential role in heat conversion if a sorbent with a higher volume fraction of mesopores is used.

The hypothetic sorbent with volume fraction of mesopores $\varepsilon_{meso} = 0.13$ (two times larger than in the experimental one) was chosen for further modelling. The results presented below give useful information about the effects of other parameters on capillary condensation in the LT adsorber. Figures 7.8 through 7.11 illustrate changes in the main heat conversion parameters with varying the length of the HT adsorber L_H, external radius of the LT adsorber r_2 (at retention of the LT sorbent volume) and temperatures of HT and LT heat reservoirs.

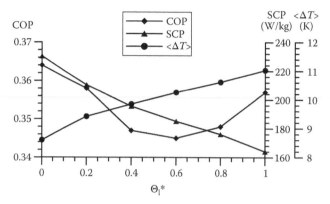

FIGURE 7.7
Performance characteristics of heat conversion (COP, SCP rate and time-averaged cooling of heat carrier in the LT adsorber, <Δ*T*>) for the different filling degrees Θ_i^* of the LT adsorber volume (mesopores) by liquid ammonia.

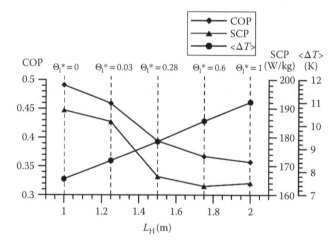

FIGURE 7.8
Performance characteristics of heat conversion (COP, SCP rate and time-averaged cooling of heat carrier in the LT adsorber, <Δ*T*>) for different lengths of the HT adsorber.

It should be noted that switching of heat carrier fluxes in these simulations was done when the amount of ammonia in the HT adsorber (in adsorbed state plus in gaseous state in pores) ceases changing (with the variation of less than 0.001 g per integration time step). From the graphs it can be seen that the adsorber volumes should be matched for achieving a lowest cooling temperature. In other words, as is clear from Figure 7.8, the volume of the HT adsorber must be sufficient to provide capillary condensation in the LT adsorber in order to reach a maximal cooling effect but, in order to prevent a decrease in COP, it should not exceed the optimal value. Further improvement in the cooling effect can be attained by increasing the external radius of the LT adsorber, retaining its volume constant (Figure 7.9). Finally, Figures 7.10 and 7.11 also demonstrate that when varying the temperature of HT or LT heat reservoirs the best results (with regard to attaining both maximal cooling effect and maximal COP) are obtained at temperatures that provide complete filling of the LT adsorber with capillary condensate and lie as close as possible to the domain of incomplete filling. Physical reasons for these effects are rather transparent. If, at some T_H, the condensation of ammonia in the LT adsorber starts, further increase in T_H leads to both additional heat consumption for heating the HT adsorber and lowering the cooling effect in the LT adsorber. This is connected with incomplete heat

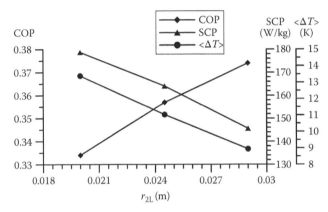

FIGURE 7.9
Performance characteristics of heat conversion (COP, SCP rate and time-averaged cooling of heat carrier in the LT adsorber, <ΔT>) for different external radii of the LT adsorber.

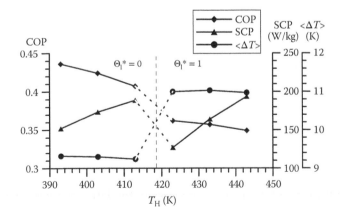

FIGURE 7.10
Performance characteristics of heat conversion (COP, SCP rate and time-averaged cooling of heat carrier in the LT adsorber, <ΔT>) for different temperatures of the HT heat reservoir ($\alpha = 250$ W/m$^2 \cdot$K).

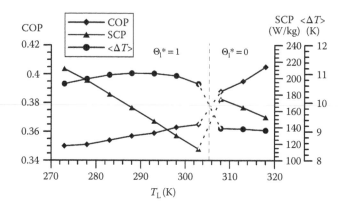

FIGURE 7.11
Performance characteristics of heat conversion (COP, SCP rate and time-averaged cooling of heat carrier in the LT adsorber, $<\Delta T>$) for different temperatures of the LT heat reservoir ($\alpha = 250$ W/m$^2 \cdot$ K).

exchange in the LT adsorber because of too fast a generation of cold in the latter, so that the heat carrier does not have enough time to reduce its temperature due to a limited value of heat exchange coefficient. It is necessary to outline that conditions for the onset of capillary condensation depend on the heat exchange coefficient between the sorbent and heat carrier.

On the other hand, the decrease in T_L below a certain threshold value, which is needed for the onset of condensation, lowers the cooling effect and simultaneously decreases COP. This also can be attributed to incomplete heat exchange between the heat carrier and sorbent in the LT adsorber. On the whole, these effects are the reflection of a general thermodynamic rule: maximal efficiency of heat converter (as well as of any heat machine) can be attained in the regimes where the heat exchange proceeds close to the thermodynamic equilibrium.

7.4 Concluding Remarks

The basic distinctive feature of the developed model for adsorption/chemisorption heat conversion is the consideration of the capillary condensation effect. In many practically important cases, this phenomenon occurs in the mesopores of sorbent whereas the transport of gaseous sorbate occurs via its filtration through macropores. As a rule, the process of gas filtration is not a rate-limiting stage for heat conversion; the latter is determined by the kinetic and heat transfer conditions. When the aforesaid mechanism takes place, the equation for liquid sorbate saturation in a porous sorbent can be substantially simplified and reduced to an ordinary differential equation with a source term depending on local parameters of sorption medium.

A maximal cooling efficiency for the sorption heat convertor is attained under conditions (heat reservoir temperatures, sorption capacity ratio of adsorbers, etc.) that correspond to the boundary of a parametric domain where complete capillary condensation occurs in the whole volume of an LT adsorber. Besides, the heat transfer in the adsorbers should be carried out under quasi-equilibrium conditions to achieve a maximal COP, that is slow; thus a designer should, as usual, seek a compromise between high efficiency and SCP rate.

It is important that the cold production rate in the considered heat converters depends not only on the process parameters (temperatures of heat reservoirs, thermodynamics and kinetics of sorption, mass flow rates of heat carriers) but also on the heat exchange coefficients between the heat carrier and sorbent. It means that optimal efficiency of heat conversion can be achieved by optimisation of the adsorber design.

The performed numerical research has shown that the positive impact of capillary condensation on the efficiency of adsorption cooling can be substantially improved by the use of a sorbent with a high volume fraction of mesopores.

References

Alyousef, Y., Antukh, A.A., Tsitovich, A.P. and Vasiliev, L.L. 2012. Three adsorbers solar cooler with composite sorbent bed and heat pipe thermal control. *Appl. Therm. Engng* 38: 124–130.

Babenko, V.A. and Kanonchik, L.E. 2000. Mathematical modeling of a cylinder with a sorbent and natural gas. *J. Engng. Phys. Thermophys.* 73: 516–529.

Barenblatt, G.I., Entov, V.M. and Ryzshik, V.M. 1990. *Theory of Fluid Flows Through Natural Rocks*. Kluwer Academic Publishers, Dordrecht; Boston.

Carre, F. 1860. *British Patent 2503*; US Patent 30201.

Chen, Z., Huan, G. and Ma, Y. 2006. *Computational Methods for Multiphase Flows in Porous Media*. SIAM, Philadelphia.

Dubinin, M.M. 1966. Porous structure and adsorption properties of active carbons. In *Chemistry and Physics of Carbon* 2:51, ed. P.J Walker Jr. Taylor & Francis: New York.

Dubinin, M.M. 1970. Surface and porosity of adsorbents. In *Fundamental Problems of the Theory of Physical Adsorption*, Eds. M.M. Dubinin and V.V. Serpinskii. Nauka: Moscow (in Russian).

Dubinin, M.M. 1972. *Adsorption and Porosity*. Nauka: Moscow (in Russian).

Dubinin, M.M. and Astakhov, V.A. 1971. Description of adsorption equilibria of vapors on zeolites over wide ranges of temperature and pressure. *Adv. Chem. Ser.* 102: 69–85.

Dullien, F.A.L. 1992. *Porous Media: Fluid Transport and Pore Structure*. Second Edition. Academic Press: San Diego.

Endo, A. and Komori, A. 2005. JP2005127614.

Faraday, M. 1823. On the condensation of several gases into liquids. *Phil. Trans. R. Soc. Lond.* 113: 189–198.

Kaviany, M. 1999. *Principles of Heat Transfer in Porous Media.* Second Edition. Springer-Verlag: New York.

Kheifets, L.I. and Neimark, A.V. 1982. *Multiphase Processes in Porous Media.* Khimia: Moscow (in Russian).

Luikov A.V. 1966. *Heal and Mass Transfer in Capillary-Porous Bodies.* Pergamon Press: Oxford.

Luikov, A.V. 1975. Systems of differential equations of heat and mass transfer in capillary-porous bodies. *Int. J. Heat Mass Transfer* 18: 1–14.

Pinder, G.F. and Gray, W.G. 2008. *Essentials of Multiphase Flow and Transport in Porous Media.* John Wiley & Sons, Inc. Hoboken.

Pratt, M. 2010. Pore network models for the study of transfers in the porous wick of loop heat pipes. *Heat Pipe Sci. Technol.* 1(2): 129–149.

Pratt, M. 2011. Heat and mass transfer in capillary structures. In *Application to Loop Heat Pipes*, ed. L.L. Vasiliev. VIII Minsk International Seminar 'Heat Pipes, Heat Pumps, Refrigerators, Power Sources', Minsk, Belarus, September 12–15.

Wang, C., Zhang, P. and Wang, R.Z. 2008. Investigation of solid–gas reaction heat transformer system with the consideration of multistep reactions. *AIChE J.* 54(9): 2464–2478.

Wang, C., Zhang, P. and Wang, R.Z. 2010. Performance of solid–gas reaction heat transformer system with gas valve control. *Chem. Engng. Sci.* 65(10): 2910–2920.

Yong, L. and Wang, R.Z. 2007. Adsorption refrigeration: A survey of novel technologies. *Recent Pat. Engng.* 1: 1–21.

8

<hr style="width:30%">

Sorption Systems with Heat Pipe Thermal Management for Hydrogenous Gas Storage and Transportation

L. L. Vasiliev and L. E. Kanonchik

CONTENTS

8.1 Introduction 284
8.2 Development and Testing of Sorbents 287
8.3 Design and Numerical Study of a Thermally Regulated Vessel
    with a Carbon Sorbent and a Hydrogenous Gas 292
    8.3.1 Mathematical Model of Adsorbed Vessels with Heat Pipes 293
        8.3.1.1 Calculated Cells and Boundary Conditions for a
               Cylindrical Vessel 296
        8.3.1.2 Calculated Cells and Boundary Conditions for a
               Flat Sectional Vessel 297
        8.3.1.3 Solution Algorithm 299
    8.3.2 Results of Numerical Modelling of Thermally Regulated
        Vessels with Sorbents and Hydrogenous Gases 300
        8.3.2.1 Results of a Numerical Study on Hydrogen Filling
               and Extraction for a Cylindrical Vessel 301
        8.3.2.2 Results of a Numerical Study on Gas Filling of a
               Flat Sectional Vessel 305
        8.3.2.3 Results of a Numerical Study on Gas Extraction
               from a Flat Sectional Vessel 309
    8.3.3 Advantages of a Combined System of Hydrogenous Gas
        Storage 313
8.4 Conclusion 315
Nomenclature 316
References 317

8.1 Introduction

It is known that nearly all of our vehicles currently run on either gasoline or diesel fuel, which is dangerous from the point of view of environmental protection. This situation requires alternative energy carriers to be developed to promote future energy security. Hydrogen and natural gas have the potential to be very attractive alternative energy carriers. These two gases are clean, efficient and derived from diverse domestic resources, such as fossil resources. Natural gas (methane) is one of the primary energy sources, and it will maintain this position for at least the next 20 years. Huge reserves of natural gas are stored as gas hydrates under water and in a permafrost zone. These reserves exceed all other energy resources of Earth, including oil, coal and nuclear fuels. In addition, natural gas is a potential source for large-scale production of pure hydrogen as a result of catalytic conversion of methane. The successful development of new hydrogen technologies, such as fuel cells, transport systems on hydrogen and methane and sorption heat pumps, has shown that the use of hydrogenous gas results in qualitatively new solutions of ecological and power problems. One of the ways of gradual penetration of hydrogen as a fuel for internal combustion engines is its application in dual-fuel vehicles (hydrogen–gasoline, hydrogen–methane etc.) with a sorption system of gas storage. Such an approach yields a minimum increase of weight at the expense of the storage system and does not require application of electromotors, commutators and heavy accumulators. In service of the dual-fuel motor transport, a change is possible from hydrogen fuel to hydrocarbon fuel or to their mixture, depending on their accessibility. The application of combined (hydrogen and methane) gas storage systems is a promising alternative. In fuel cell automobiles, hydrogen can be disposed in a container in the adsorbed state or made immediately from accumulated methane.

The main advantage of hydrogen and natural gas use is associated with ecological cleanliness. They are free of toxic substances that are added, for example, to gasoline to increase the octane number. Among the disadvantages of hydrogen and methane are low densities and small volumetric heats of combustion, which lead to the necessity of storing them in large-sized vessels and cylinders. The realisation of any scheme of hydrogenous gas usage in transport depends on making a cost-effective system of gas storage. The successful development of a better system of hydrogen storage presumes active thermal management and special properties of materials capable of efficiently storing hydrogen. Alongside a large total yield of hydrogen, a high density of energy storage is required for minimisation of volume and/or system weight. Ranges of pressure and temperatures should also correspond with service conditions on transport. The potential alternative to well-known storage technologies like cryogenic liquid and compressed gas is hydrogen storage in a coupled state by solid sorbents [1–3]. Metal hydride–based hydrogen storage systems offer higher volumetric density and better safety compared with conventional methods [4–6]. Unfortunately, the storage of gas as metal hydrides necessitates the application of a quite heavy reservoir that does not fulfil the weight constraints related to mobile applications. Many metals and alloys can also reversibly absorb large amounts of hydrogen, but the absorption energy is in the range of 50–100 kJ/mol. For good hydrogen storage, the enthalpy for hydride reactions should be as low as possible. However, none of them is known for the mobile storage pressure–temperature range with ΔH near 15–24 kJ/mol of hydrogen [7]. Hydrogen physisorption is considered one of the most promising storage technologies to satisfy the U.S. Department of Energy (DOE) goals. In contrast to the chemisorption in metal hydrides, the phenomenon of physical adsorption of undissociated hydrogen molecules on the surface of microporous carbon fibres or particles is a good direction to improve the situation. This possibility of reaching high sorption capacities is due to the ability of carbon to be prepared in a very fine powdered or fibre form with a highly porous structure and due to specific interactions between carbon atoms and gas molecules. The total amount of stored hydrogen strongly depends on the pore geometry and pore size distribution as well as the storage pressure and temperature. It is noted that natural gas is made up of about 95% methane – a gas that cannot be liquefied at ambient temperature. Microporous active carbon adsorbs hydrogen molecules and keeps them by surface intermolecular forces. However, at room temperature, these forces are weak, as they are hindered from thermal agitation of molecules. To store a considerable amount of hydrogen, the carbon samples should be cooled down to cryogenic temperatures. In most cases, the cryogenic hydrogen or methane storage system is economically unprofitable. An essential shortage of this mode is the necessity of using special heat-insulating, hermetic containers to maintain the cryogenic temperature. Hence, it is necessary to select microporous carbon sorbents, capable of retaining hydrogen and methane at low and room temperatures in the middle pressure range. Efficient hydrogen storage technology is an important factor in the application of hydrogen as the fuel for the future. Sorption of hydrogen usually takes place in micropores. Macropores have no practical influence on sorption capacity, as they are only important for gas compression and for sorption/desorption reaction rates. Because of its low density, high surface area and significant micropore volume, low cost [8–10] and accessibility at the industrial level, activated carbons are considered good candidates for hydrogen storage systems.

Recently, studies have focused on searching for high-capacity sorbent materials to be used at room temperature [11]. For conventional activated carbon, the hydrogen uptake is proportional to its surface area and pore volume, whereas, unfortunately, a high hydrogen adsorption capacity (4–6 wt%) can be obtained only at cryogenic temperatures [12]. Many improvements have been obtained using microporous carbonaceous materials with extremely high adsorbing properties for storage of different gases. Zhou et al. [8] studied the hydrogen storage in AX-21 activated carbon and found it to have a

gravimetric storage density of 10.8 wt% at 77 K and 6 MPa. Xu et al. [14] investigated the hydrogen storage capacity of various carbon materials, including
activated carbons, carbon nanohorns, carbon nanotubes and graphitic carbon
nanofibres, at room temperature and liquid nitrogen temperature. At room
temperature, the gravimetric storage density of hydrogen on carbon materials was less than 1 wt% and a super-activated carbon, Maxsorb, had the best
hydrogen storage factor of 0.67 wt%. By lowering the adsorption temperature
to 77 K, hydrogen storage capacity significantly increased and Maxsorb could
store a large amount of hydrogen, 5.7 wt%, at a low pressure, 3 MPa. The
narrow range of micropores in the Maxsorb was favorable to adsorption of
hydrogen molecules that indicate physical nature of the gas adsorption on
the carbon materials. The experiments performed with single-walled and
multi-walled carbon nanotubes to date are not so optimistic [9, 13, 14] and
showed that the amount of storage hydrogen is less than 1 wt% at ambient
temperature and 10 MPa. The cooling of nanotubes down to the liquid nitrogen temperature is a good way to increase its sorption capability to 4–8 wt%.
On the other hand, additional problems that may affect the storage capacity of an adsorption storage system are thermal effects that occur during filling and discharging operations. Released sorption heat causes an increase
in sorbent bed temperature, which results in lower storage capacity. In the
technical gas storage application, only a part of the stored mass of hydrogen
can be extracted from the vessel. Theoretical and experimental investigations
[15–17] have shown that the value of sorption heat has a particularly significant effect during the sorption stage of a sorbent bed. Lamari et al. [18] and
Delahaye et al. [19] have investigated the dynamic thermal behaviour of a storage system on ambient temperature and relatively high operational pressures
(5–15 MPa). Lara et al. [16] analysed the heating dynamics during hydrogenous gas (methane) charging in an activated carbon–packed bed storage tank.
The sorption energy of different carbon materials is only 2–10 kJ/mol. Results
showed that the adsorption process is responsible for about 24% of the temperature increase even for moderately adsorbing activated carbon. Analysis of
the results, reported in the literature, showed that it is difficult to achieve the
target of 6.5 wt% even at 77 K only by the physisorption of hydrogen.

In this chapter, we suggest a new composite sorbent material that combines
the advantages of metal hydrides (high sorption capacity) and the fast dynamics of activated carbon fibre for high-performance hydrogen storage. Another
interrelated chapter objective is thermally controlled hydrogen storage (heat
pipes) system development for dual-fuel (hydrogen and natural gas) automobiles and a fuel cell vehicle. The successful development of a better system of
hydrogen storage presumes active thermal regulation and special properties
of materials capable of storing hydrogen efficiently. Beside activated carbon
fibres (Busofit-M8), the inexpensive, microporous activated carbon obtained by
the thermal treatment of raw materials (wood, sawdust, cellulose, straw, paper
for recycling, peat etc.) after impregnation is especially attractive. In this way,
a number of prospective sorbents were developed in Belarus: the activated

carbon fibre Busofit-M8 (a product of pyrolysis of impregnated cellulose) and the activated carbon made from waste products of wood 'WAC 3-00' [20]. A heat pipe (two-phase loop thermosyphon) is suggested as the heat exchanger for the hydrogen storage vessel. Gas storage was based on combined hydrogen adsorption and compression at moderate pressures (3–6 MPa) and cryogenic temperatures. Modelling of the thermodynamic process of gas charge/discharge was carried out in a flat rectangular vessel with heat pipes thermal management to choose the best performance. The efficiency of the composite sorbent material application was experimentally validated.

8.2 Development and Testing of Sorbents

Impregnated cellulose-based raw material is the attractive host material for the manufacture of special activated carbons for ammonia, methane and hydrogen storage systems with high microporosity, specific surface area and narrow micropore size distribution. In our experiments, some samples of activated carbon Busofit, obtained by a new technology, were investigated. They have been prepared by selective thermal processing at high temperatures near 850°C. Additional activation of the samples was carried out in the presence of carbonic gas. In this way, some of the carbon atoms were removed by gasification, which yields a very porous structure. Numerous pores and cracks were formed in the carbon material, increasing the specific surface area due to the growth of micropore volume. The carbon fibre Busofit can be considered a universal microporous adsorbent with a pore diameter of 1–2 nm and, at the same time, as a material with high gas permeability. Figure 8.1a and b shows the texture of Busofit-M8, performed as a loose fibre bed (or felt) and as monolithic blocks with binder to have a good effective thermal conductivity. The porous texture of different materials was analysed using nitrogen physisorption at 77 K and up to a pressure of 0.1 MPa. Table 8.1 summarises the data, obtained with High Speed Gas Sorption Analyzer NOVA 1200. As an example, Figure 8.2 shows the experimental isotherms of the adsorption and desorption of hydrogen on Busofit-M8 and WAC 3-00 (activated carbon material made from waste wood) at the liquid nitrogen temperature. Absence of an appreciable hysteresis confirms that reversible physisorption exclusively takes place with all investigated materials. Following our analysis, it was found that the activated carbon Busofit-M8 is one of the best sorbent materials used in the liquid nitrogen temperature region. The large number (volume) of micropores (>1 mL/g) and advanced surface area ($S_{BET} = 1939$ m^2/g) are the reasons for its good sorption capacity. Adsorption properties at high pressure were measured using a volumetric weight method. A free gas volume inside the experimental reactor was determined using helium technology. At a pressure 6 MPa, due to the physical sorption of Busofit-M8 the capability of hydrogen storage achieves 0.042 kg/kg.

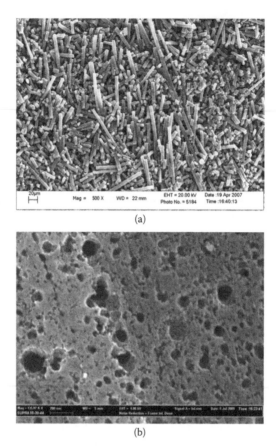

FIGURE 8.1

Photograph of the activated carbon fibre Busofit-M8: (a) micro- and mesoporous surface structure of additionally activated fibre, multiplied by 135,000; (b) briquette Busofit-M8, multiplied by 500.

TABLE 8.1

Textural Characteristics and Hydrogen-Physisorption Capacities at 77 K and 0.1 MPa for Investigated Activated Carbon Fibres

Sorbent	a (kg/kg)	a_v (mL[STP]/g)	S_H (m²/g)	S_{BET} (m²/g)	S_{DR} (m²/g)	V_{DR} (mL/g)	R_{DR} (nm)
Busofit-M2	0.0179	203.9	465	1702	2507	0.89	4.15
Busofit-M4	0.0198	225.1	536	1715	2547	0.9	4.2
Busofit-M8	0.0223	252.9	571	1939	2985	1.04	5.1
WAC 3-00	0.0195	221.1	575	1383	2142	0.74	5.0
207C	0.0184	209.2	502	1300	1944	0.69	4.1
Norit Sorbonorit-3	0.0171	193.8	458	1361	2044	0.73	5.0
Sutcliff	0.0208	236.6	527	1925	2864	1.02	5.36

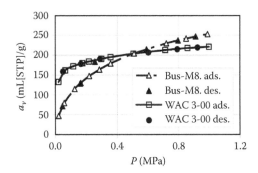

FIGURE 8.2
Isotherms of hydrogen adsorption and desorption on carbon materials at temperature 77 K, measured by the High Speed Gas Sorption Analyzer NOVA 1200.

If we combine the carbon fibre material with metal hydride microparticles, disposed on its surface in the same volume, we can solve the problem of efficient gas storage and transportation. The total volume, V, associated with an activated carbon adsorbent may be split into its components:

$$V = V_c + V_\mu + V_v + V_{void}$$

where V_c is the volume of the carbon atoms of which the adsorbent is composed, V_μ the microporous volume, V_v the meso- and macroporous volume and V_{void} the space inside the vessel free from the adsorbent bed. The latter volume, V_{void}, can be minimised by compression sorbent (making of sorbent blocks).

A composite sorbent bed was performed as a set of micro- and nanoparticles of metal hydrides attached to a filament surface to increase the total sorption capacity of the sorbent bed. Activated carbon fibre with metal hydride particles on its surface can be considered as a promising material for hydrogen storage. We summarise the high heat of chemical reaction and sensible heat of physical adsorption to provide high storage capacity for the sorbent bed. Busofit, as a fast sorbent material, starts to react with hydrogen in the early stage of the heating/cooling time (up to 300 seconds) and accomplishes its action after the chemical reaction of the metal hydride is finished. Because of the high density of metal hydrides, the system is very compact. Metal hydride applications are best suited for systems where power and volume are more critical than weight. For example, a hydrogen compressor or reservoir for hydrogen storage based on reversible solid sorption is attractive if the process ensures high gas density storage. For this, the reactor must contain a large quantity of reactive mixture and the kinetics of the gas–solid reaction must be fast. Heat and mass transfer ought to be high, which implies good porosity and high thermal conductivity of the sorbent bed. For this reason, the metal hydride reaction beds inside the storage vessel are temperature regulated by heat transfer devices – heat pipes. The addition of metal hydride to activated carbon fibre improves the performances of the gas storage vessel. The inclusion of micro- and nanoparticles of a metal hydride in the porous matrix formed by carbon fibres improves the heat transfer and kinetics in the sorbent bed. Successful application in practical systems requires particular properties of the hydride alloys. Good alloy candidates ($LaNi_{4.5}Al_{0.29}Mn_{0.21}$, $Ti_{0.9}Zr_{0.10}V_{0.43}Fe_{0.09}Cr_{0.09}Mn_{1.5}$) must have proper temperature and pressure intervals, high total uptake of hydrogen and high energy density of the material. The pressure plateau must also be in the range of system operational conditions.

The experimental set-up is shown in Figure 8.3. The experimental reactor allows the analysis of hydrogen sorption phenomena at pressure levels up to 6 MPa. After each test, the reactor was placed in a furnace filled with inert gas for sorbent bed regeneration. A cylindrical vessel of 0.023 m inner diameter and 0.3 m length was used to simulate full-scale conditions of the experiment. This experimental rig, made from stainless steel, served as a simulator of the real vessel in the ratio 1:100. This vessel was filled with the sorbent and disposed on an electronic balance, and connected through a valve with the hydrogen storage container by a flexible coil (a thin stainless steel capillary). The vessel has an electric heater and a set of thermocouples for temperature registration of the sample during gas adsorption/desorption. There are two independent modes of sorption capacity registration: (1) by gas volume control and (2) by gas mass control. The hydrogen storage capacity of the sorbent material was controlled as the vessel mass increased/decreased in time. The final storage capacity of the sorbent bed was determined using a controlled volume by gas pressure change inside the vessel at room temperature. The advantage of such an experimental set-up is the possibility of performing the control of gas vessel mass before and after gas adsorption. During the tests,

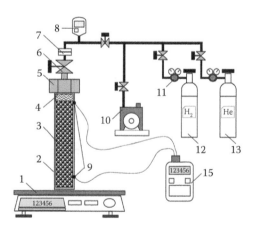

FIGURE 8.3

Experimental set-up: (1) high-precision electronic balances; (2) reactor; (3) sorbent bed; (4) filter (to avoid sorbent particles migration into the system); (5) reactor cap screw; (6) valve; (7) quick-release coupling; (8) pressure sensor; (9) thermocouples; (10) vacuum pump; (11) pressure reducer; (12,13) cylinders with hydrogen and helium; (15) temperature indicator.

the precisely calibrated pressure sensor controlled the pressure inside the vessel (up to 6 MPa) with an accuracy near 1 kPa. Such a combined system of measurements of hydrogen storage capability allows to work with large amounts of sorbent (50–400 g). Before performing the experiments, the samples were 'filled' several times with hydrogen. Then hydrogen was removed from the vessel by heating it up to 500 K. The system was connected to a vacuum pump to reduce the pressure down to 0.7 Pa for some hours. After this procedure was accomplished, the experimental set-up was ready.

The experiments were performed with four different samples: metal hydride $La_{0.5}Ni_5Ce_{0.5}$, activated carbon fibre Busofit-M8 (briquette), and two composite sorbents COM 10–80 (10% Busofit-M8 + 80% $La_{0.5}Ni_5Ce_{0.5}$ + 10% binder) and COM 40–50 (40% Busofit-M8 + 50% $La_{0.5}Ni_5Ce_{0.5}$ + 10% binder). Figure 8.4a and b shows hydrogen isosteres of sorption by two composites (COM 40–50 and COM 10–80) at the temperature interval 273–373 K and the pressure interval 0.1–6 MPa. Figure 8.5a and b demonstrates the evolution of mean temperature and the amount of stored hydrogen in the sorbent bed for the four samples (Busofit-M8, MH, COM 10–80 and COM 40–50) during the adsorption cycle at pressure 3.5 MPa. As follows from the analysis (Figures 8.4 and 8.5), at the beginning of the adsorption/desorption cycle of hydrogen by the sorbent bed the sorption capacity and the mean temperature of Busofit-M8, COM 10–80 and COM 40–50 increase very fast and have the same tendency of temperature evolution. The metal hydride (MH) $La_{0.5}Ni_5Ce_{0.5}$ starts to react, when the sample temperature approaches 350 K. The maximum of the temperature rise (310 K) for Busofit-M8 is reached at 100 seconds, for COM 40–50 (330 K) at 500 seconds, for COM 10–80 (365 K) at 250 seconds and for MH (370 K) at 600 seconds. The adsorption/desorption rate decreases when the time of the adsorption cycle reaches 300–400 seconds for Busofit-M8 and the reaction time 550 seconds for MH. The maximum hydrogen amount was accumulated by the samples COM 10–80 (1.3 wt%)

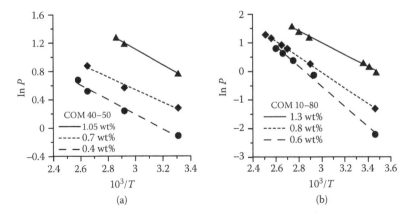

FIGURE 8.4
Hydrogen sorption isosteres typical for composites (a) COM 40–50 and (b) COM 10–80.

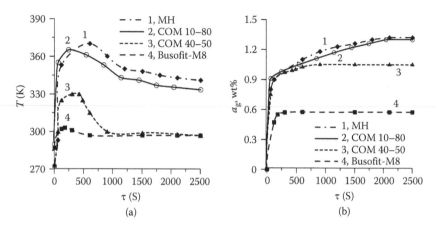

FIGURE 8.5
Evolution of the (a) mean temperature and (b) gravimetric storage density, a_g (wt%), of hydrogen during the adsorption/desorption cycle for the activated carbon fibre Busofit-M8, metal hydride $La_{0.5}Ni_5Ce_{0.5}$ and composites COM 40–50 and COM 10–80 at pressure 3.5 MPa.

and metal hydride $La_{0.5}Ni_5Ce_{0.5}$ (1.32 wt%). These samples were more strongly heated. The maximum temperature rise of the sorbent bed was determined for COM 10–80 and MH. It means the isosteric heat of hydrogen sorption for Busofit-M8 or a combination of Busofit-M8 and $La_{0.5}Ni_5Ce_{0.5}$ is lower compared with that of pure metal hydride. The time to reach equilibrium state varied from 600 seconds (for the carbon sample Busofit-M8) to 2100 seconds (for the metal hydride $La_{0.5}Ni_5Ce_{0.5}$). The slow rate of adsorption/desorption was detected for COM 10–80 and the metal hydride $La_{0.5}Ni_5Ce_{0.5}$. The mean time of full hydrogen adsorption (equilibrium state) equal to 900 seconds was observed for the composite COM 40–50. It means that the permeability and kinetic of adsorption of Busofit-M8 and the composite COM 40–50 is higher compared with that of the metal hydride $La_{0.5}Ni_5Ce_{0.5}$. The amount of accumulated hydrogen for COM 40–50 is equal to 1.05 wt%, which is 15% lesser than that of MH. Therefore, we obtained a fast sorbent bed (compound 40% Busofit-M8 + 50% $La_{0.5}Ni_5Ce_{0.5}$ + 10% binder) with relatively good sorption capacity. The mass of stored hydrogen is nearly the same as that of metal hydride, but the cycle of adsorption/desorption is three times lesser.

8.3 Design and Numerical Study of a Thermally Regulated Vessel with a Carbon Sorbent and a Hydrogenous Gas

Adsorbed hydrogenous gas storage and transportation technology recently became competitive with the compressed gas method due to its high energy density capability. With the reduction of storage pressure through the use of the adsorbed technology, it is possible to use various vessel geometries that endure the high pressures created by the compressed gas technology. This imparts to the design greater versatility.

To enhance the performance and thermodynamic efficiency of the gas storage vessel, a heat pipe thermal control system was suggested. It should be mentioned that heat or cold must be delivered as evenly as possible to all sorbent portions placed in the vessel. The currently employed methods of heating using electrical heaters, water coils, hot exhaust gas tubes and so on do not provide required heat uniformity in the vessel and have very low efficiency. To complicate matters, only direct blowing of the vessel or use of expensive and complex water heat exchangers or special refrigerators can deliver the cold. The heat pipes proposed as a basic element of the thermal management for the storage vessels are known in principle. Heat pipes are highly reliable and efficient heat transfer devices considered for many terrestrial and space applications. This device uses the latent heat of vaporisation (condensation and evaporation) of a working fluid to transfer relatively large amounts of energy over long distances with small temperature drops. There are several Japanese applications (05-092369, 60042867) and patents (JP59003001, JP622204099) and also U.S. (4599867) and Belarus patents (BY, INt.Cl. FO2M 21/4, No. a 19991158) concerning the use of heat pipes for hydrogen storage tanks.

The analytical justification of the vessel operational performances for transportation is a necessary part of investigations in hydrogenous gas sorption storage system design and development. Two variants have been designed and numerically investigated: an ordinary cylindrical vessel and a flat sectional vessel.

8.3.1 Mathematical Model of Adsorbed Vessels with Heat Pipes

A two-dimensional transient model was developed to analyse the influence of thermal control on the operating characteristics of sectional storage vessels with hydrogen and heat pipes. The mathematical model of heat transfer and gas sorption processes in the vessel is based on the following assumptions: (1) the gas in the cylinder is ideal; (2) the porous medium is assumed to be homogenous and isotropic; (3) the temperature of the solid phase is equal to the temperature of the gas phase at each point, because of the high coefficient of the volumetric heat transfer between them; (4) heating and cooling of the sorbent material is carried out by a heat pipe with inner heat transfer coefficient, α_{HP}, equal to 10^3–10^4 W/m^2·K, which is uniform along the surface and high in comparison with the thermal resistance of an interface heat pipe–sorbent bed [21]; and (5) all sorbent elements between the heat pipe fins are under the same conditions, and the resulting dependences for the sorbent bed can be obtained by simply summing over N identical cells. In accordance with these assumptions, the mathematical model of the adsorbed vessel is described by the energy conservation equation, the equations of mass balance and momentum balance, the approximate kinetic equation and the equation of equilibrium state [22–24].

Equation of isothermal adsorption: the adsorption properties of activated carbons are determined with the accuracy sufficient for practical purposes by the potential theory of adsorption developed by Dubinin [25]. Physisorption on microporous carbons can be described using the Dubinin–Radushkevich equation:

$$a_{eq} = \frac{W_0}{v_a} \exp\left\{ -\left[\frac{R_\mu T \ln\left(P_{sat}/P\right)}{E} \right]^2 \right\}$$ (8.1)

The equilibrium state equation (8.1) includes the saturation pressure P_{sat}. Since hydrogen (or methane) sorption isotherms are measured within the temperature and pressure intervals comprising the regions of supercritical states (for hydrogen, $T_{kr} = 33.24$ K and $P_{kr} = 1.298$ MPa), the notion of saturation pressure loses its physical meaning. In the work by Agarwal [26], saturation pressure is determined by the following formula:

$$P_{sat} = P_{cr} \left(\frac{T}{T_{cr}} \right)^2$$ (8.2)

Approximate kinetic equation: many authors [15, 27] have mentioned that any finite rate of sorption is associated with temperature changes in the sorbent volume. Therefore, in mathematical model formulation, it is necessary to take into account the fact that sorption is a non-equilibrium process. High rates of desorption of molecules from microporous adsorbents are the necessary condition for the efficient operation of the hydrogenous gas sorption storage system on vehicles. For this reason, the gas sorption storage system needs to be calculated taking into account its kinetic characteristics. The process of desorption of molecules from bidispersed adsorbents such as activated carbons involves the following stages: escape of adsorbate molecules from active centres on the adsorbent surface, diffusion of molecules in the initial porous structure, diffusion in secondary pores and removal of the desorbate from the gas phase. Estimation of the contribution of rates of individual processes to the total rate of desorption, performed on the basis of analysis of experimental data, has shown [28] that the velocity of molecular escape from the adsorbent surface is of crucial importance. The rate of mass transfer in adsorbent granules increases due to the migration of molecules over the surface of the pores. This transfer is given the name 'surface diffusion'. In practice, the dynamics of sorption is described by the Gluckauf equation [29], in which the motive force of the intra-diffusion adsorption process is determined as the difference of adsorbate concentrations in the solid phase:

$$\frac{da}{d\tau} = \beta_k \left(a_{eq} - a \right)$$ (8.3)

where β_κ is the kinetic coefficient. Sakoda and Suzuki [30] recommend the assumption that the kinetic coefficient, β_κ, in the approximate equation of the kinetics of sorption on activated carbons is equal to the total coefficient of mass transfer, $\beta_\tau = 15D_s/R_p^2$. The coefficient of surface diffusion, D_s, in a microporous sorbent is related to its temperature by the dependence $D_s = D_{s0}$ $\exp[-E/(R_\mu T)]$, where D_{s0} is a phenomenological constant. For the sorption processes of filling the storied system with a gas and its discharge, we use the kinetic equation of sorption in the following form:

$$\frac{\partial a}{\partial \tau} = \left(a_{eq} - a \right) \cdot K_{s0} \exp\left(-\frac{E}{R_\mu T} \right) \tag{8.4}$$

where $K_{s0} = 15D_{s0}/R_p^2$, D_{s0} is a phenomenological constant and R_p is the average particle radius. Thus, to describe the non-equilibrium desorption occurring during the discharge from a cylinder it is necessary to use two equations. One of them (Equation 8.1) describes the dependence of the equilibrium value of adsorption, a_{eq}, on the pressure and temperature, and the other (Equation 8.4) describes the time by which the non-equilibrium adsorption lags behind the equilibrium one, a_{eq}. The relation of magnitude of adsorption, a, to pressure in Equation 8.4 is considered as the dependence of a on $a_{eq}(p, T)$.

The *energy conservation equation* (in the vector form regardless of the selection of a frame of axes) is as follows:

$$\left(\varepsilon \rho_g C_g + \rho C + \rho a C_a \right)\frac{\partial T}{\partial \tau} + \nabla \cdot \left(-\lambda \nabla T + \rho_g C_g T\mathbf{u} \right) = q_{st}\rho \frac{\partial a}{\partial \tau} + \zeta T\varepsilon \frac{\partial P}{\partial \tau} \tag{8.5}$$

The heat of the phase transition of sorption can be calculated using the Clapeyron–Clausius equation with some corrections. In accordance with the study by Young and Growell [31], the isosteric sorption heat of a gas mole is determined by the following formula:

$$q_{st} = \left(1 - \frac{V_a}{V_g} \right) TR_\mu z_g \left[\frac{\partial \ln P}{\partial \ln T} \right]\Bigg|_{a=\text{const}} \tag{8.6}$$

For the ideal gas ($z_g = 1$, $V_a \ll V_g$), the isosteric heat is calculated from the following relation:

$$q_{st} = R_\mu T \left[\frac{\partial \ln P}{\partial \ln T} \right]\Bigg|_{a=\text{const}} \tag{8.7}$$

The limits of applicability of Equation 8.6 depend on the degree to which the gas can be approximated as ideal, that is, on the ratio between the partial molar volumes of the adsorbate and the gas and on the compressibility coefficient. In the case of a non-ideal gas, the estimation obtained with Equation 8.7 should be considered as being approximate.

Mass balance in the vector form is as follows:

$$\frac{\partial(\varepsilon\rho_g)}{\partial\tau} + \nabla(\rho_g\mathbf{u}) + \nabla(-D\nabla a) = -\frac{\partial}{\partial\tau}(\rho a) \tag{8.8}$$

Momentum balance in the vector form is as follows:

$$\rho_g\frac{\partial\mathbf{u}}{\partial\tau} + \frac{\eta}{K}\mathbf{u} = \nabla\left[-PI + \eta\left(\nabla\mathbf{u} + (\nabla\mathbf{u})^T\right)\right] \tag{8.9}$$

Equations 8.1 through 8.9 define four variables, T, P, a and $\mathbf{u}$, that are dependent on space coordinates and instant of time. For them, the initial conditions are as follows:

$$T(\tau_0) = T_0, \ P(\tau_0) = P_0, \ \mathbf{u}(\tau_0) = 0, \ a(\tau_0) = a_{eq}(T_0, P_0) \tag{8.10}$$

8.3.1.1 Calculated Cells and Boundary Conditions for a Cylindrical Vessel

Figure 8.6 shows the calculated element of a cylindrical vessel for the storage of adsorptive hydrogenous gas. The analysed element of the sorbent bed is bounded by the envelope ($r = R$) of the vessel, the wall of the finned heat pipe ($r = R_0$) and the planes of symmetry passing through the centre of the fin ($z = 0$) and the centre of the sorbent bed ($z = S$) between two neighbouring fins. The free gas fills the macropores, whereas the adsorbed gas is disposed on the surface of the micropores. The metal fin (1; disc) with a half-width (δ) is attached to the outer wall (3) of the heat pipe. The annular sorbent bed (2) with radii r_0 and r_1 is adjacent to the fin of the heat pipe. The contact thermal resistance of the heat pipe with the microporous sorbent is characterised by the heat transfer coefficient α_s. The sorbent bed is surrounded by a thin, perforated aluminium tube used for the gas released during its desorption from the sorbent bed. Since the thickness of the sorbent bed is much smaller

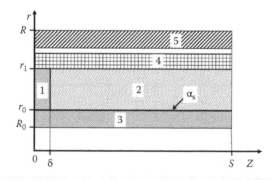

FIGURE 8.6
Scheme of the calculation element of adsorbed gas storage cylindrical vessel: (1) fin; (2) sorbent; (3) shell of the heat pipe; (4) channel for gas formed by the perforated tube and the cylinder body; (5) cylinder body.

than its length, the desorbed gas mostly moves in the radial direction. There is a channel (4) between the perforated tube and the vessel wall through which the gas moves along the cylinder to the pressure regulator. The gas flow rate is maintained constant with the help of a pressure regulator. This factor, as well as the factor of heating/cooling of the sorbent bed, determines the rate and character of pressure change in the vessel. During the period of gas release, the vessel is cooled by air convection (temperature T_{env}) with a coefficient of heat transfer α_{env} and is heated inside with the heat pipe. Depending on the conditions of sorbent bed heating, a constant temperature or heat flow is maintained in the inner surface of the heat pipe.

Boundary conditions were set for four surfaces (Figure 8.6):

$$\left.\frac{\partial T}{\partial z}\right|_{z=0} = 0, \quad \left.\frac{\partial T}{\partial z}\right|_{z=S} = 0; -\lambda\left.\right|_{r=R} = \alpha_{env}\left(T - T_{env}\right) \tag{8.11}$$

$$-\lambda\left.\frac{\partial T}{\partial r}\right|_{r=R_0} = \frac{Q_{HP}}{2\pi R_0 SN} \text{ or } T\left.\right|_{r=R_0} = T_{HP} \tag{8.12}$$

8.3.1.2 Calculated Cells and Boundary Conditions for a Flat Sectional Vessel

Figure 8.7 shows a flat sectional vessel with heat pipes for gas sorption storage at the pressure range 3.5–6 MPa. An efficient system that performs thermal active management of the sorbent bed during its sorption/desorption is the heat pipe heat exchanger. Every separate section has its casing made from aluminium (or reinforced plastics) and filled with a briquette sorbent material where hydrogen is present in adsorbed and compressed states. Heat pipes can be easily implemented inside the sorption storage vessel due to their flexibility, high heat transfer efficiency, cost-effectiveness, reliability, long operating life and simple manufacturing technology. The heat pipe metal fins on the heat pipe surface intensify heat transfer in the sorbent bed. A flat rectangular, rather than cylindrical, form is convenient for gas vessel location in automobiles. The channels for gas supplying and removing were

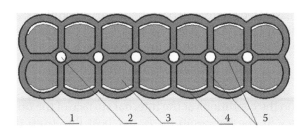

FIGURE 8.7
Thermally controlled sectional vessel for hydrogenous gas sorption storage: (1) vessel casing; (2) heat pipe; (3) sorbent; (4) gas channel; (5) longitudinal fins/partitions.

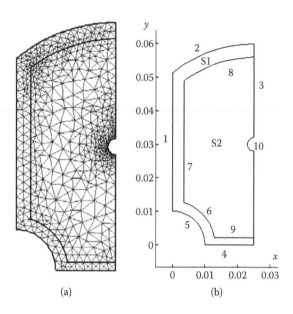

(a)   (b)

FIGURE 8.8

Calculation cell of the sectional vessel: (a) model mesh; (b) model domains and the boundary conditions (S1) metal, (S2) sorbent; (1,3,4) symmetry planes; (2,8) external and internal surfaces of the cylinder casing; (5,6) internal and external surfaces of the heat pipe body; (7,9) boundaries between sorbent and metal fins/partitions.

performed inside of aluminium casing the gas channels can be replaced with cylindrical channels in a sorbent body at the centre of each section parallel to the axis of the heat pipes.

Figure 8.8a illustrates a calculation cell of a sectional vessel with an anisotropic mesh, consisting of 1031 triangular elements for two-dimensional geometries. Figure 8.8b shows cell breakdown to two calculation areas or domains, S1 (the cylinder casing the body of the heat pipe, fins/partitions) and S2 (sorbent), with the indication of boundaries between them. The calculation cell is bounded by the external surface of casing 2, the internal wall of heat pipe (boundary 5), the symmetry planes at the middle of the fins (boundaries 1 and 4) and the middle of the sorbent layer (boundary 3). For the sectional vessel (see Figure 8.8) basic boundary conditions: at the external sides of the casing (boundary 2 in Figure 8.8), the conditions of convective heat exchange with the environment are considered:

$$\mathbf{n} \cdot \mathbf{q} = \alpha_{\text{env}} (T - T_{\text{env}}), \quad \mathbf{q} = -\lambda \nabla T \tag{8.13}$$

where n is the unit vector of the exterior surface normal.

Depending on the conditions at the internal surface of the finned heat pipe body (boundary 5), the temperature, T_{HP}, or heat flux from the fins to the sorbent bed is as follows:

$$\mathbf{n} \cdot \mathbf{q} = q_{\text{HP}}, \quad \mathbf{q} = -\lambda \nabla T, \quad q_{\text{HP}} = \alpha_{\text{HP}} (T_{\text{HP}} - T) \tag{8.14}$$

where αH_p is the heat transfer coefficient from the heat pipe working fluid (vapour) to the internal surface of heat pipe body. The heat flux is equal to zero at symmetry planes.

For boundary 10, where the cylindrical aperture for gas supply is located, the heat flux is determined by the gas flow rate and the temperature T_i:

$$-\mathbf{n}\cdot\mathbf{q} = q_0, \quad q = \lambda\nabla T - \rho_g C_g\, \mathbf{u}, \quad q_0 = -\rho_g C_g T_i(\mathbf{u}\cdot\mathbf{n}) \tag{8.15}$$

In the case of the presence of an axial channel for gas supply at the centre of the cell (boundary 10), the supply pressure is set as follows:

$$P = P_i = P_{ch} = \min\left[\left(P_0 + (P_{stor} - P_0)\tau/\Delta\tau\right), P_{stor}\right] \tag{8.16}$$

The conditions of gas charging were modelled by linearly changing the pressure for $\Delta\tau = 600$ seconds from an initial value, P_0, to a final value, $P_{stor} = 6$ MPa. The gas pressure could be maintained with the help of a reducer. Another variant of the charging process occurs at a constant mass rate outwards from the cylinder at the opening:

$$\mathbf{u} = \mathbf{u}_0 = \frac{\mathbf{n}\cdot G}{NL_b \int \rho_g\, dl} \tag{8.17}$$

8.3.1.3 Solution Algorithm

To solve Equations 8.1 through 8.9, the method of finite elements, realised on COMSOL Multiphysics, was used. Next, we employed interface blocks with some additional programming:

- Conduction and convection application mode:

$$\delta_{ts}\rho_T C_p \frac{\partial T}{\partial\tau} + \nabla\cdot\left(-\lambda\nabla T + \rho_T C_p T\mathbf{u}\right) = Q \tag{8.18}$$

where

$$\rho_T = \rho_g, \quad C_p = C_g, \quad \delta_{ts} = \frac{\varepsilon\rho_g C_g + \rho C + \rho a C_a}{\rho_g C_g}, \quad Q = q_{st}\rho\frac{\partial a}{\partial\tau} \tag{8.19}$$

- Convection and diffusion application mode (for modelling the kinetic equation of sorption):

$$\delta_{ts}\frac{\partial a}{\partial\tau} + \nabla\cdot\left(-D\nabla a + a\mathbf{u}\right) = R \tag{8.20}$$

where

$$\delta_{ts} = 1, \quad R = K_{s0}\exp\left(-\frac{E}{R_\mu T}\right)\left(a_{eq} - a\right) \tag{8.21}$$

- Brinkman equations application mode (for incompressible viscous flow of an isothermal fluid in a porous medium):

$$\rho_B \frac{\partial \mathbf{u}_B}{\partial \tau} + \frac{\eta}{k} \mathbf{u}_B = \nabla \cdot \left[-P\mathbf{I} + \eta \left(\nabla \mathbf{u}_B + \left(\nabla \mathbf{u}_B \right)^T \right) \right] + \mathbf{F} \qquad (8.22)$$

$$\frac{\partial \rho_B}{\partial \tau} + \nabla \cdot \left(\rho_B \mathbf{u}_B \right) = 0 \qquad (8.23)$$

This block does not provide mass transfer, as on the right side of the continuity equation (8.23) the source of sorption

$$\left(-\frac{\partial}{\partial \tau} \left(\rho \frac{a}{\varepsilon} \right) \right) \qquad (8.24)$$

is not present. It is added 'manually' in Equation 8.23, and in the 'feeble' terms it is introduced for the purpose of stabilisation of numerical procedures and overcoming instability. Here, $\mathbf{F} = 0$, $\mathbf{u}_B = \mathbf{u}/\varepsilon$ and $k = K/\varepsilon$. Convergence precision was equal to 10^{-6}. The system of equations with a sorption source is stiff and allows integration only with a minor step. For achieving required accuracy and numerical stability, the integration step did not exceed 2 seconds.

8.3.2 Results of Numerical Modelling of Thermally Regulated Vessels with Sorbents and Hydrogenous Gases

The modelling of the thermodynamic process of hydrogen charging of a thermally controlled vessel was carried out to find the best performance. The main characteristic of the efficiency of a sorption storage system of a hydrogenous gas is volumetric storage density, ρ_v, which represents the ratio of the volume occupied by the gas contained in the vessel at standard temperature 273 K and pressure 0.1 MPa (STP) to the geometric volume of the vessel. The second parameter used in the process analysis is the dynamic coefficient of vessel filling, m (the ratio of current gas amount in the vessel to its value at storage pressure P_{stor} and storage temperature T_{stor}). Other characteristics of the process are the time of gas discharge (or charge), average volumetric temperature and pressure of the sorbent bed. A set of calculations was performed for the activated carbon fibre Busofit-M8 and the wood-based carbon WAC 3-00 as promising gas sorption materials for designing a hydrogen storage system. Table 8.2 presents their empirical coefficients of

TABLE 8.2

Empirical Coefficients of the Dubinin–Radushkevich Equation for Hydrogen Sorption on Carbon Materials

Material	$W_0 \times 10^3$ (m³/kg)	E (kJ/kg)
Busofit-M8	482	1710
WAC 3-00	270	3782

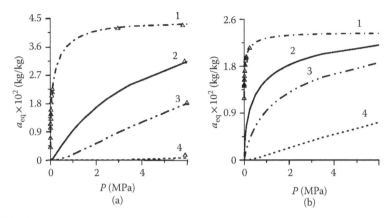

FIGURE 8.9
Hydrogen adsorption isotherms for (a) activated carbon fibre Busofit-M8 and (b) wood-based carbon WAC 3-00 at different temperatures (1) 77, (2) 153, (3) 193, (4) 293 K: points – experimental data; lines – calculated data (Dubinin–Radushkevich equation).

the Dubinin–Radushkevich equation. Figure 8.9 shows the experimental and calculated isotherms of the adsorption of hydrogen on steam-activated Busofit-M8 and WAC 3-00 at different temperatures (77, 153, 193 and 293 K). The large number (volume) of micropores (1.04 and 0.74 mL/g) and advanced surface area (1939 and 1383 m²/g) are the reasons for their good adsorption capability.

To verify the reliability of the mathematical modelling, we compared the results of the calculations with the experimental data (Figure 8.9) on desorption of methane [32], obtained at the Luikov Institute, Minsk, Belarus. A cylinder of volume 12.5 L was tested at an ambient temperature, T_{env} = 290 K, and a constant heat flow, Q_{HP}. As the sorbent material, the commercially available activated carbon fibre 'Busofit-AYTM' was used. In this case, a gas flow rate of 0.2 g/s was considered. Figure 8.10 presents the dynamics of the mean-volume temperature, the pressure and the volume storage density change of the sorbent bed for the heat flow Q_{HP} = 80 W (G = 0.2 g/s). All the basic numerical parameters of the discharge process agree satisfactorily with the experimental data.

8.3.2.1 Results of a Numerical Study on Hydrogen Filling and Extraction for a Cylindrical Vessel

The data of the basic variant of calculation of a cylindrical vessel corresponded with wood-based carbon WAC 3-00; mass flow rate of gas, 0.3 g/s; α_{env} = 3 W/m²·K; α_s = 500 W/m²·K, Q_{HP} = 500 W and T_0 = 195 K. The steel vessel had the following geometric characteristics: a length of 1.82 m, an outside radius of 0.1095 m and a wall thickness of 0.0065 m. The height of the finning discs was in line with the external boundary of the cylindrical sorbent layer. The radius of the heat pipe, R_0, was 90 mm, and the radius of the outer sorbent layer was 91 mm. In calculations, the parameters of the vessel were varied in the following ranges: mass flow rate of gas, 0.15–0.7 g/s; thickness of the aluminium fin, 0.5–3 mm and the step of finning, 20–30 mm. As the results of calculations of two-dimensional fields of temperature, hydrogen adsorption, pressure and velocity were obtained. They are the basis for determining the stored gas amount and charged time. Numerical experiments were carried out for the process of vessel filling through a reducer with a fixed flow rate of hydrogen. In Figure 8.11, the filling performances of a temperature-controlled vessel are shown. During gas filling, there is a rise of pressure and temperature in the volume of the sorbent, with their variation depending on the rate of gas charging. The time of filling is determined by the reaction rate of gas adsorption and heat transfer processes in a

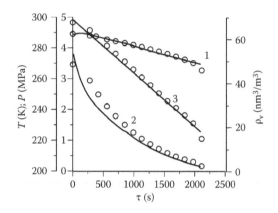

FIGURE 8.10
Experimental points and calculated curves of the values of (1) the mean temperature of sorbent, (2) the pressure and (3) the volume storage density of gas as functions of time in the process of discharging of a cylindrical vessel with sorbent and methane.

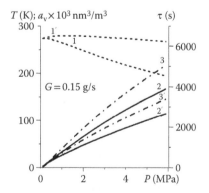

FIGURE 8.11
The mean temperature of the sorbent (1, 1'), the volumetric storage capacity of hydrogen (2, 2') and the time of filling (3, 3') versus the pressure in the process of vessel filling (the heat output $Q_{HP} = 500$ W: 1–3; $Q_{HP} = 0$ W: 1'–3').

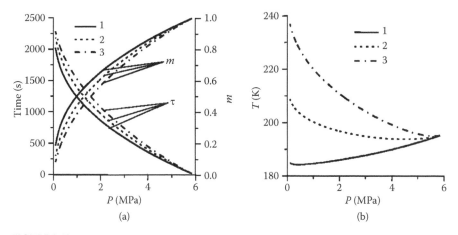

FIGURE 8.12
(a) Discharge time and dynamic coefficient of filling and (b) sorbent mean temperature versus pressure in the vessel for different additional heat flow applied by heat pipe: (1) 0 W; (2) 500 W; (3) 1000 W.

sorbent bed. It is possible to counteract the negative influence of latent heat of adsorption on the gas charge/discharge cycles with the help of a special thermal control system.

Figure 8.12 illustrates change of the dynamic coefficient of the vessel filling, m, the mean temperature and the pressure of the sorbent in the course of a discharging of the vessel as a function of the HP heat flow value. It is an inverse problem to vessel gas charging. The value of m varies from 1 to 0 and shows the fraction of unrecovered mass of gas. We analysed three variants of sorbent bed heating conditions. In one variant, the influence of sorbent material heating is negligible. It was assumed that the heat transfer between the heat pipe, sorbent bed and fins is absent, that is, $\alpha_s = \alpha_f = \alpha_{fs} = 0$. Thus, in the variant where the heat pipe was switched off the sorbent bed thermally interacted only with the environment, whereas at all the remaining boundaries the heat flow was equal to zero. Here, the maximum value of m corresponds to the volumetric storage density of hydrogen, about 170 nm^3/m^3 (see Figure 8.13). The heat flow Q_{HP} increases the gas pressure and temperature inside the vessel. As temperature increases, the pressure in the cylinder during gas desorption decreases more slowly at the same gas flow rate. As a result, at $Q_{HP} = 1000$ W the mass of remaining gas that is not extracted decreases from 20% to 10%, compared with the variant in which the heat pipe is switched off (Figure 8.12a). At Q_{HP} equal to zero, the sorbent layer is cooled by 12 K (the endothermic reaction of gas desorption) below the initial temperature. An increase in the heating power ($Q_{HP} = 500$ W) makes it possible to release the gas from the cylinder more completely and ensure isothermal conditions of gas desorption. In special cases (with increased gas output; see Figure 8.13), to provide maximum hydrogen extraction it

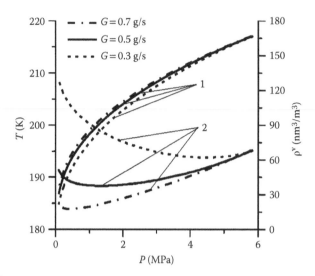

FIGURE 8.13
(1) Volumetric storage density of hydrogen and (2) the mean temperature of the sorbent versus the pressure in the process of discharging of the vessel for different gas flow rates.

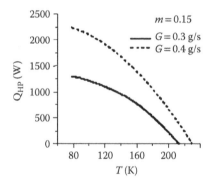

FIGURE 8.14
The heat flow (at which the gas remaining in the vessel accounts for 15%) transferred by a heat pipe in the discharge process of the cylindrical vessel versus an ambient temperature for different gas flow rates.

is necessary to select the optimal heating mode of sorbent at a given environmental temperature and fixed gas rate. The abundant heat flow transferred by the heat pipe in the discharge process to the sorbent bed will result in excessive cylinder heating and an inadmissible increase of pressure. Figure 8.14 shows the results of calculations to optimise the heating conditions of the cylindrical vessel (which ensures about 85% of the use of stored gas). Evidently, at a mean flow rate $G = 0.3$ g/s and a temperature $T_0 = 273$ K, an additional heat flow $Q_{HP} = 300$ W is needed.

8.3.2.2 Results of a Numerical Study on Gas Filling of a Flat Sectional Vessel

In this chapter, the partial optimisation of the operational conditions of a 60-L sectional vessel is executed based on computer analysis. Note that numerical experiments were carried out for one calculation cell and overall characteristics of the vessel operation were obtained with respect to the vessel section number, equal to 14. The two-dimensional surface plots in Figure 8.15 show temperature, adsorption, pressure and velocity as a function of cell position after 600 seconds of hydrogen filling. Figure 8.16 demonstrates the profiles of

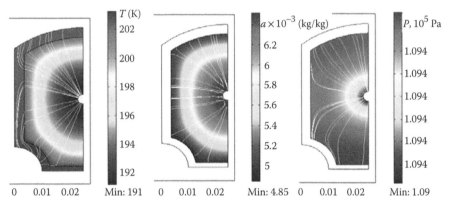

FIGURE 8.15
Stream lines and two-dimensional surfaces of temperature, adsorption and pressure in the cell of a thermally controlled sectional vessel for hydrogen sorption storage after 600 seconds of hydrogen filling.

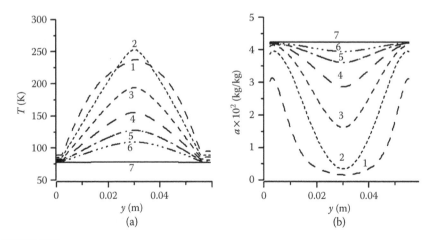

FIGURE 8.16
Profiles of (a) temperature and (b) adsorption in the cell ($x = 0.02$ m; see Figure 8.8) during gas charging (1) 300, (2) 600, (3) 900, (4) 1200, (5) 1500, (6) 800, (7) 4000 seconds by the sorbent bed cooled by a heat pipe at a constant temperature 77 K (at a fixed flow rate of hydrogen of 0.25 g/s).

temperature and adsorption in the cell (see Figure 8.8; cross section $x = 0.02$ m) during gas charging of the hydrogen storage vessel cooled by a heat pipe with a constant temperature of 77 K. The properties of the sorbent material corresponded with the properties of briquette Busofit-M8, having a bulk density of 500 kg/m³. The porosity used in the model is equal to 0.77.

Figure 8.17a shows the evolution of pressure and temperature in a sorbent bed for different variants of gas charging of the sectional vessel, when hydrogen intake occurs with a fixed mass flow rate, the sorbent bed is cooled by the heat pipe down to 77 K, and the pressure increased to 6 MPa. Calculations were carried out for adiabatic conditions on the external surface of the casing $T_{HP} = 77$ K and $\alpha_{env} = 0$ W/m²·K. This version of gas charging corresponds to the situation in which the heat output is not limited and the temperature, T_{HP}, is set. This temperature is maintained at the inner surface of the heat pipe due to the contact of its condensation zone with a massive body of cooled liquid nitrogen. Initially, hydrogen and the sorbent material were in equilibrium state, which was characterised by the pressure $P_0 = 0$ MPa, temperature $T_0 = 273$ K and adsorption $a_0 = f(P_0, T_0)$. The valve was switched on, and the gas moved in the cylinder throughout the channels. The gas pressure was maintained with the help of a reducer equal to the charging pressure. The temperature inside the sorbent is defined by heat pipe heat transfer rate, heat capacity of a design and the latent heat of sorption. The volumetric storage of hydrogen and the dynamic factor of filling are directly proportional to the stored gas amount. These factors of vessel filling are chosen as the ordinates of a graph (Figure 8.9b). The patterns of relationship of modification filling performances are defined by the rate of rise of pressure, which depends on gas flow rate. The advantages of this type of charging are obvious at a gradual increase in pressure and a low gas flow rate. The heat pipe being turned off in

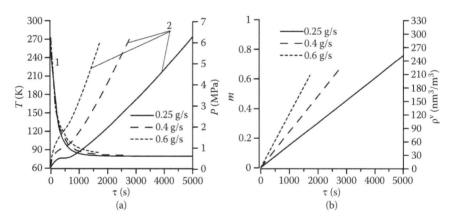

FIGURE 8.17
(a, 1) The mean sorbent temperature, (a, 2) the pressure and (b) the dynamic coefficient of filling and volumetric storage density of hydrogen during gas charging of the sectional vessel cooled by a heat pipe ($T_{HP} = 77$ K, $\alpha_{env} = 0$ W/m²·K) for various flow rates of gas.

a constant-temperature condition (77 K) results in fast cooling and a high gas reserve, equal to 80%, during the first stage of 2400 seconds. It is obvious that the sorbent temperature is decreasing, despite sorption heat release, and the rate of cooling is dependent on heat output. In this case, as in many other cases, the heat pipe serves as an element of the thermal control system. These competing processes determined the resulting course of change in temperature and amount of the accumulated gas in a cylinder. A summary of numerical experiments with various inlet flow rates of hydrogen for the charging process is given in Figure 8.18.

The acceleration of the inlet flow rate of gas from 0.2 to 0.7 g/s has ensured the cutting down of time of filling from 6430 to 1430 seconds, growth of ending temperature of a sorbent from 79 to 90 K and a falling of storage capacity from 5.5 to 4.3 wt% and of the coefficient of dynamic coefficient of filling from 0.8 to 0.6. For reaching the maximum filling (at 6 MPa and 77 K), it is possible to stop the delivery of gas and maintain it at 20%–35% to add to the expense of smoothing the temperatures of the sorbent.

Heating the adsorbed vessel during the charging process is counteracted by cooling the sorbent bed with heat pipes. The filling characteristics of a flat sectional vessel with working heat pipes are shown in Figure 8.19a and b. The gravimetric (mass of adsorbed and compressed gas per unit mass of sorbent bed) and volumetric densities of hydrogen storage are chosen as the ordinates of a graph, plotted in Figure 8.19b. Vessel gas charging was stopped at a mean sorbent temperature of 77 K, providing a great reserve of the compressed and adsorbed hydrogen in the set conditions. This moment characterised the duration of the charging stage. Calculations were carried out for five variants of heat flow, transferred through heat pipe, adiabatic

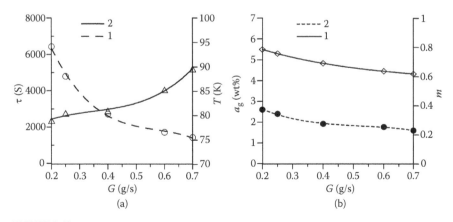

FIGURE 8.18
Influence of inlet flow rates of gas on charging-end characteristics of a sectional vessel ($T_0 = 273$ K) cooled by a heat pipe ($T_{HP} = 77$ K): (a) (1) charged time (up to $P_{st} = 6$ MPa), (2) the mean sorbent temperature (end of filling); (b) (1) gravimetric storage density of gas in the sorbent bed (the adsorbed and compressed hydrogen) and the dynamic coefficient of filling, (2) gravimetric storage density of adsorbed gas in the sorbent bed.

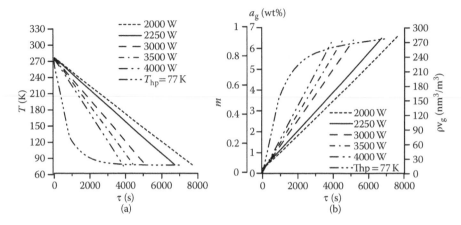

FIGURE 8.19
The mean (a) sorbent temperature and (b) gravimetric and volumetric density of stored hydrogen versus the time of charging process of the sectional vessel ($\alpha_{env} = 0$ W/m²·K) for different values of heat flow, transferred by heat pipes.

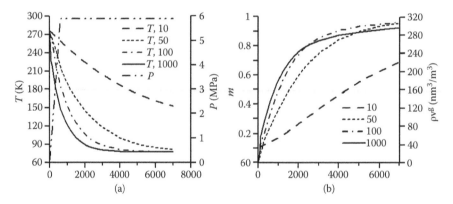

FIGURE 8.20
(a) The mean sorbent temperature and (b) the pressure, the dynamic coefficient of filling, and the volumetric storage density of hydrogen during gas charging with fixed pressure (6 MPa) of the sectional vessel for different heat transfer coefficient values (surrounding temperature = 77 K, $Q_{HP} = 0$ W).

conditions on the external surface of the casing and the activated carbon fibre Busofit-M8. The condition of vessel charging corresponds to the situation in which the heat flow of the heat pipe is known. For comparison, the results of a numerical experiment with different versions of gas charging, when $T_{HP} = 77$ K and $\alpha_{HP} = 1000$ W/m²·K, are given.

Also, we performed the analysis of the vessel gas charging procedure for the case when the heat flow through the heat pipe was equal to zero and the coefficient of heat transfer between the vessel envelope and the surroundings was variable (four different values of the heat transfer coefficient). The data on vessel gas charging are shown in Figure 8.20a and b for $P_{stor} = 6$ MPa and the surrounding temperature 77 K. We observed that for the heat transfer coefficient equal to 100–1000 W/m² · K, the dynamic coefficient of filling is above 1. This is because the heat transfer intensification due to the higher convection coefficient decreases the sorbent bed temperature above the isothermal limit. The convective heat exchange with the surroundings and the heat flow, transferred by the heat pipe, determine the degree and time of the vessel charge/discharge cycle. A peculiarity of the designed storage system is the combination of heat transfer processes and accumulation of hydrogen in one volume. The increase in heat flow, removed by heat pipes from a sorbent bed, from 2000 to 4000 W reduced the filling time of the vessel by approximately three times. Heat pipe turn on in constant-temperature conditions (77 K) has resulted in fast cooling and 80% filling by gas during the 2400 seconds. It is obvious that the sorbent temperature is decreasing, despite sorption heat release, and the rate of cooling is dependent on heat output.

8.3.2.3 Results of a Numerical Study on Gas Extraction from a Flat Sectional Vessel

During hydrogen consumption by a car engine, the pressure in the vessel decreases from 6 MPa down to 0.1 MPa. In the technical gas storage application, only a part of the stored mass of hydrogen can be extracted from the vessel. This fraction of hydrogen is the gas mass reversibly stored/consumed during a charge/discharge cycle between the considered pressure levels. The process of gas extraction from a constant-volume vessel occurs at a variable pressure, temperature and gas mass. The pressure regulator at the outlet of the vessel provides the required gas flow rate, which depends on the velocity of the vehicle and other conditions of its motion. To fulfil the requirements of economical efficiency and safety, the characteristics of the vessel should be optimised. Special features of the thermodynamic processes of gas adsorption and desorption are described by the equation of equilibrium state, $a_{eq} = a_{eq}(P, T)$, and the equation of sorption kinetic that determines the rate of adsorption, a, approaching its equilibrium value, $a_{eq}(P, T)$. According to the equation of state, the lower the temperature of the sorbent, the larger the mass of gas, $M_{dich\_end} \sim a(P_{dich\_end}, T_{dich\_end})$, remaining in the vessel. In the case of intensive extraction of gas, the percentage of its application can be decreased down to 50%–60% due to intensive cooling of the sorbent bed. The heat of exhausted gases or the liquid system of motor cooling and the heat pipe can be used to stimulate gas release from the vessel. It is known that the sorbent bed is cooled during desorption. The cooling of sorbent is the reason for the partial reduction of gas extraction. To increase the mass of released hydrogenous gas, we need to heat the sorbent bed. It could be ensured by sorbent heating with the help of a heat exchanger or by stimulating the heat transfer between the vessel envelope and the surroundings (convective heat transfer). The data of the basic variant of gas extraction calculation of the flat sectional vessel concerned to the sorbent Busofit-M8, (pyrolysis products of cellulose). When stating the problem, it was considered that the reducer holds the discharge rate of gas from the vessel constant. Figure 8.21a and b shows the evolution of pressure (from $P_{st} = 6$ MPa down to $P_{disch\_end} = 0.1$ MPa) and temperature in a sorbent bed during gas discharging of the vessel at a gas flow rate $G = 0.2$ g/s and different heat flows, transferred by heat pipes. The generalised dependences of the final discharging characteristics of the sectional vessel as a function of the heat pipe heat flow value for the final pressure 0.1 MPa are shown in Figure 8.22a and b.

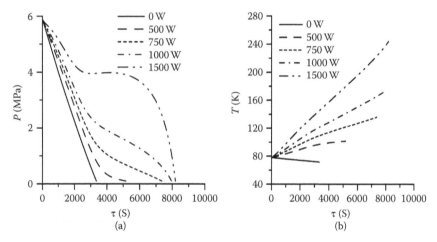

FIGURE 8.21
Variations of the (a) mean pressure and (b) sorbent temperature in the discharge process of a sectional vessel for different heat flows, transferred by heat pipes ($\alpha_{env} = 0$ W/m² · K).

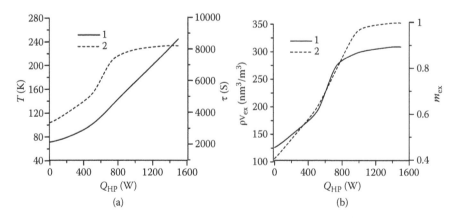

FIGURE 8.22
(a, 1) The mean sorbent temperature, (a, 2) the period of gas release, (b, 1) the volumetric density of released hydrogen and (b, 2) the factor of hydrogen release as a function of heat flow value (through heat pipes). The pressure in the vessel is 0.1 MPa and $\alpha_{env} = 0$ W/m² · K.

Then we performed the analysis of gas discharging from the vessel for the case when the heat flow through the heat pipe was equal to zero and the coefficient of heat transfer between the vessel envelope and the surroundings was variable (five different values of the heat transfer coefficient). The variant of $\alpha_{env} = 0$ W/m²·K corresponds to a thermally insulated vessel. Figure 8.23 shows the changes in averaged pressure and temperature in a sorbent bed during the discharge process for an ambient temperature of 273 K and a gas flow rate $G = 0.2$ g/s. The generalised numerical data of gas discharge from the vessel are shown in Figure 8.24a and b for $P_{disch\_end} = 0.1$ MPa. For

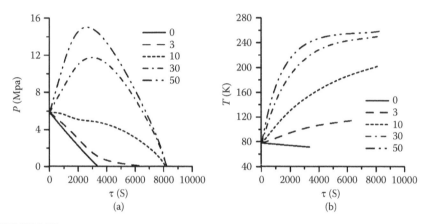

FIGURE 8.23
Changes of (a) pressure and (b) temperature during the discharge of a flat sectional vessel filled with Busofit-M8 for different values of an outside heat transfer coefficient.

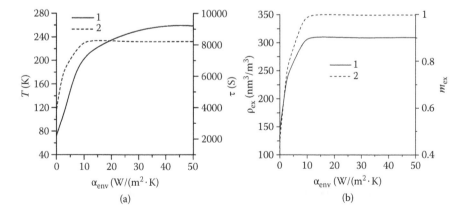

FIGURE 8.24
(a, 1) The mean sorbent temperature, (a, 2) the period of gas release, (b, 1) the volumetric density of released hydrogen and (b, 2) the factor of hydrogen release as a function of the heat transfer coefficient value ($P_{disch\_end} = 0.1$ MPa, $Q_{HP} = 0$ W).

the heat transfer coefficient equal to 20 W/m²·K or the additional heat flow $Q_{HP} = 1000$ W, the factor of gas release (m_{ex}: the ratio of mass of released gas to mass of stored gas at the beginning of the discharge procedure) is above 1. This is because the heat transfer intensification increases the sorbent bed temperature above the isothermal limit. This increase in bed temperature causes an increase in cylinder pressure, maintaining the mass flow rate at the vessel opening constant. This pressure increase can extend the discharge period up to 9000 seconds. Numerical experiments have demonstrated that the heating factor defined by the heat transfer rate of a heat pipe and the condition of convective heat exchange with the environment affects the duration of vessel discharging and helps to extend vehicle mileage at the expense of increasing average sorbent temperature due to the reduction of gas remaining in the vessel.

A set of curves reflects the results of investigating the dependence of vessel behaviour on storage temperature. The variations in mean sorbent temperature and pressure during desorption for three storage temperatures are presented in Figure 8.25. In this case, the boundary thermal conditions of hydrogen discharge of the sectional vessel were $Q_{HP} = 0$ W and $\alpha_{env} = 3$ W/m²·K. The initial temperature of the discharging process or storage temperature defines a gas reserve in the vessel and time of a discharging. The level of mean temperature, directly depending on the storage condition, characterises the isotherm of desorption, which is a function of temperature and pressure. The data presented in Figure 8.26 confirm a reduction in sorption and total capacity of gas storage with the growth of temperature of a sorbent. It is caused by the shape of the sorption isotherm and the perfect gas law for free hydrogen in macropores.

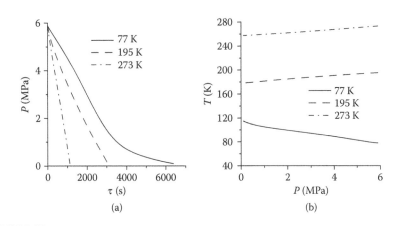

FIGURE 8.25
(a) Mean sorbent pressure and (b) temperature during the discharge process of a sectional vessel with the sorbent WAC 3-00 for various initial temperatures.

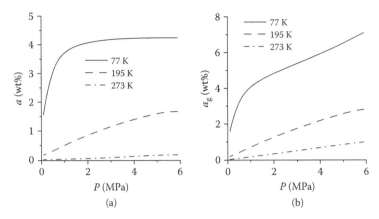

FIGURE 8.26
Influence of initial temperature on the characteristics of discharging of a flat sectional vessel with hydrogen: (a) gravimetric sorption density; (b) gravimetric storage density.

Evidently, the isotherm corresponding to lower temperatures is shown in Figure 8.26a. Transition to the liquid nitrogen temperature ensures the greatest duration of the process of discharge, equal to 6380 seconds, and a maximum gravimetric density of hydrogen storage of 7.1 wt%, which is twice more than the corresponding values at 195 K (Figure 8.26b). Under such conditions, the monotonous growth of mean temperature of the sorbent was observed. The limiting influence of the heat of desorption is weakened because of the reduction of heat of phase transition and the increase of heat gain from a surrounding medium.

8.3.3 Advantages of a Combined System of Hydrogenous Gas Storage

Figures 8.27 and 8.28 illustrate the advantages of a combined system of gas storage (compression and adsorption at the temperature of liquid nitrogen) that is compared with compressed gas storage. Figure 8.27 is drawn for a conventional cylinder based on the p–V–T data of hydrogen. Hydrogen compression at ambient temperature is not effective compared to compression at the temperature of liquid nitrogen (77 K). The pressure of hydrogen must reach 75 MPa at 298 K to store 41 kg of hydrogen in a 1-m³ container, but only 15 MPa is required at 77 K. Storage of hydrogen at as high a pressure as 75 MPa is presently not technically feasible. Adsorption increases the density of hydrogen near the surface of the microporous sorbent and, hence, was applied to enhance the storage of hydrogen or natural gas in the container. As a fundamental feature of physical adsorption, the amount adsorbed depends strongly on temperature. An effective way of enhancing the storage is to reduce the temperature of storage. Liquid nitrogen is cheap and

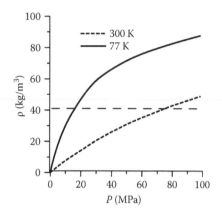

FIGURE 8.27
Storage density of compressed hydrogen versus pressure for the traditional cylinder (1 m³) at temperatures 300 and 77 K.

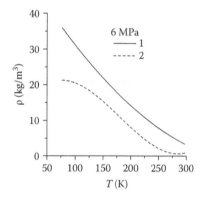

FIGURE 8.28
Storage density of hydrogen on Busofit-M8 versus temperature for a sectional vessel (0.06 m³): (1) the adsorbed and compressed gases; (2) the adsorbed gas.

widely available, and therefore it is a practical cooling media. Figure 8.28 shows the comparison between total storage density and density of the adsorbed phase of hydrogen in a vessel with Busofit-M8. As indicated in the graphs, it is necessary to take into account the presence of compressed gas in the vessel, whose fraction can be measured up to 30%–50% at a pressure of 6 MPa. The best parameters of a vessel filled with carbonaceous sorbent at 36 kg/m³ (for Busofit-M8) correspond to the liquid nitrogen temperature, 77 K, and pressure, 6 MPa. A reduction in storage density down to 3 kg/m³ is observed at 273 K.

8.4 Conclusion

1. The vessels with the heat pipe thermal control and active carbon material and composites on its base were suggested as promising hydrogenous gas storage tank design for vehicles with internal combustion engine and for the vehicles with fuel cells.

2. Heat pipes are proposed as a basic element of the thermal management system for storage vessels. Heat pipes can be easily implemented inside a sorption storage vessel due to their flexibility, high heat transfer efficiency, cost-effectiveness, reliability, long operating life and simple manufacturing technology. The metal fins on the heat pipe surface serve to intensify heat transfer in the sorbent bed.

3. Results of the investigations show that the application of a heat pipe in a gas accumulator enables one to control the temperature of the sorbent bed and to provide optimum charging and discharging. The convective heat exchange with the environment and the heat flow, transferred by heat pipes, determine the degree and time of the vessel charge/discharge cycle.

4. Using the original technology of activation, a new activated carbon fibre material, Busofit-M8, and a wood-based carbon, WAC 3-00, were developed and tested. The materials are efficient for hydrogen reach gases (hydrogen, methane) storage and transportation at low pressures.

5. The composite sorbent material COM 40–50 based on the activated carbon fibre matrix and microparticles of metal hydrides on its surface were suggested and tested for hydrogen storage and transportation. COM 40–50 has a cycle (sorption/regeneration) time close to that of the activated carbon fibre Busofit-M8 and the storage capacity of hydrogen (at room temperature) close to that of the metal hydride $La_{0.5}Ni_5Ce_{0.5}$.

6. The two-dimensional transient and non-equilibrium model of heat, momentum and mass transfer processes in a sorption vessel with finned heat pipes was developed. It takes into account the energy released during sorption.

7. The numerical modelling of the process testifies the possibility of achieving 6 wt% for hydrogen storage on board automobiles (the goal announced by the U.S. DOE). The combination of compressed and adsorbed gas (on activated carbon fibres) at cryogenic temperatures (~77 K) and low pressures (6 MPa) is used for such cases.

8. The vessels developed with finned heat pipes and microporous carbon-base sorbents can be used for power engineering and in the domestic and small business gas industry.

Nomenclature

a	Adsorption (kg/kg)	V_g	Molar gas-phase volume
a_{eq}	Equilibrium adsorption (kg/kg)	V_{DR}	Micropore volume, determined by the Dubinin–Radushkevich method (mL/g)
a_v	Volume of adsorbed hydrogen per unit mass of sorbent bed (mL[STP]/g)	W_0	Maximum microporous specific volume (m³/kg)
C	Specific heat capacity of solid sorbent (J/kg·K)	z_g	Coefficient of gas compressibility
C_a	Specific heat capacity of adsorbed gas (J/kg·K)		
D	Diffusivity (m²/s)		
E	Activation energy (J/kg)	**Greek Letters**	
G	Gas output from a cylinder (kg/s, g/s)	α	Heat transfer coefficient (W/m²·K)
K_{s0}	Pre-exponent constant in the equation of sorption kinetics	β	Thermal coefficient of expansion (1/K)
K	Permeability (m²)	ε	Porosity
L	Length (m)	λ	Effective thermal conductivity of the sorbent bed (W/m·K)
M	Mass of the gas in the cylinder (kg)	η	Dynamic viscosity (kg/m·s)
m	Dynamic coefficient of filling	ρ	Density of the sorbent bed (kg/m³)
N	Number of calculated cells in the cylinder	ρ_g	Density of free gas (kg/m³)
P	Pressure (Pa)	ρ^v	Volumetric storage density of gas (nm³/m³, m³[STP]/m³)
q_{st}	Latent isosteric heat of sorption (J/kg)	τ	Time (second)
Q	Heat flow (W)	**Subscripts and Abbreviations**	
Q	Heat flux (W/m²)	a	Adsorbate
R_{DR}	Size of pore, determined on the Dubinin–Radushkevich method (nm)	ch	Charging
		cr	Critical state
		e	Finite value
R_μ	Gas constant (J/kg·K)	env	Environment
S_{BET}	specific surface area of sorbent (according to the model formulated by Brunauer, Emmert, Teller), determined on nitrogen, m²/g	eq	Equilibrium conditions
		0	Initial value
T	Temperature (K, °C)	sat	Saturation
u	Velocity vector (m/s)		
v	Component of the velocity vector (m/s)	t	Transfer
ν_a	Specific volume of adsorbed medium (m³/kg)	v	Volume
V	Volume (m³)	Standard temperature (273 K) and pressure (0.1 MPa)	
V_a	Partial molar adsorption volume		

References

1. S. Hynek, W. Fuller, J. Bentley, Hydrogen storage by carbon sorption, *Int. J. Hydrogen Energy*, 22, 601–610 (1997).
2. A. Zuttel, Materials for hydrogen storage, *Materials Today*, 6, 24–33 (September 2003).
3. K.M. Thomas, Hydrogen adsorption and storage on porous materials, *Catalysis Today*, 120, 389–398 (2007).
4. G. Sandrock, A parabolic overview of hydrogen storage alloys from a gas reaction point of view, *J. Alloys Compd*, 339, 877–888 (1999).
5. L. Schlapbach, A. Zuttel, Hydrogen-storage materials for mobile applications. *Nature*, 414, 353–358 (2001).
6. E. David, An overview of advanced materials for hydrogen storage, *J. Mater. Process. Technol.*, 162, 169–177 (2005).
7. P.P. Muthukumar, U.A. Kevin Abraham, M. Rajendra Prasad, P. Maiya, S. Srinivasa Murthy, Screening of metal hydrides for engineering applications, *Proc. 16th Int. Conf. on Efficiency, Costs, Optimization, Simulation and Environ. Impact of Energy Syst.*, 2003, Copenhagen, Denmark, 93–96.
8. L. Zhou, Y. Zhou, Y. Sun, Enhanced storage of hydrogen at the temperature of liquid hydrogen, *Int. J. Hydrogen Energy*, 29, 319–322 (2004).
9. L. Zhou, Y. Zhou, Y. Sun, Studies on the mechanism and capacity of hydrogen uptake by physisorption-based materials, *Int. J. Hydrogen Energy*, 31, 259–264 (2006).
10. C. Carpetis, A system consideration of alternative hydrogen storage facilities for estimation of storage costs, *Int. J. Hydrogen Energy*, 5, 423–437 (1980).
11. A.C. Dillon, M.J. Heben, Hydrogen storage using carbon adsorbents: past, present and future, *Appl. Phys. A. Mater. Sci. Process.*, A72, 133–142 (2001).
12. R. Strobel, J. Garche, P.T. Moseley, L. Jorissen, G. Wolf, Hydrogen storage by carbon materials, *J. Power Sources*, 159, 781–801 (2006).
13. M.G. Nijkamp, J.E.M.J. Raaymakers, A.J. van Dillen, K.P. de Jong, Hydrogen storage using physisorption–materials demands, *Appl. Phys. A. Mater. Sci. Process.*, A72, 619–623 (2001).
14. W.-C. Xu, K. Takahashi, Y. Matsuo, Y. Hattori, M. Kumagai, S. Ishiyama, K. Kaneko, S. Iijima, Investigation of hydrogen storage capacity of various carbon materials, *Int. J. Hydrogen Energy*, 32, 2504–2512 (2007).
15. K.J. Chang, O. Talu, Behavior and performance of adsorptive natural gas storage cylinders during discharge, *Appl. Therm. Eng.*, 16, 359–374 (1996).
16. L.G. Lara, P. Couto, R. M. Cotta, D.M.A. Sophia, Gas discharge capacity intensification of adsorbed natural gas reservoirs, *Proc. 11th Brazilian Congress of Thermal Sciences and Eng. – ENCIT*, Braz. Soc. of Mechanical Sciences and Engineering – ABCM, 2006, Curitiba, Brazil, Paper CIT06–0633.
17. L.L. Vasiliev, L.E. Kanonchik, Storage and transportation of hydrogen in the adsorbed state, *Preprint 2* (Luikov Heat & Mass Transfer Institute, National Academy of Sciences, Byelorussia, Minsk) 2006.
18. M. Lamari, A. Aoufi, P. Malbrunot, Thermal effects in dynamic storage of hydrogen by adsorption, *Environ. Energ. Eng.*, 46, 632–641 (2000).
19. A. Delahaye, A. Aoufi, A. Gicquel, Improvement of hydrogen storage by adsorption using 2d modeling of heat effects, *AIChE J.*, 48, 2061–2073 (2002).

20. L.L. Vasiliev, L.E. Kanonchik, A.G, Kulakov, D.A. Mishkinis, A.M. Safonova, N.K. Luneva, Activated carbon fibre composites for ammonia, methane and hydrogen adsorption, *Int. J. Low Carbon Technol.* (Manchester University Press), 2/1, 95–11 (2006).

21. L.L. Vasiliev, Heat pipes in modern heat exchangers, *Appl. Therm. Eng.*, 25, 1–19 (2005).

22. J.P. Barbosa Mota, E. Saatdjian, D. Tondeur, A.E. Rodrigues, A simulation model of a high-capacity methane adsorptive storage system, *Adsorption*, 1, 17–27 (1995).

23. L.Z. Zhang, L. Wang, Performance estimation of an adsorption cooling system for automobile waste heat recovery, *Appl. Therm. Eng.* 17, 1127–1139 (1997).

24. L.L. Vasiliev, L.E. Kanonchik, A.A. Antuch, V.A. Babenko, 'Metal-hydride particles on the fibre' as new sorbents for hydrogen storage, *Proc. Int. Symp., 'Innovative Mater. Processes in Energy Syst.'*, 2007, Kyoto, Japan, IMPRES Paper ID: A046.

25. M.M. Dubinin, The potential theory of adsorption of gases and vapors for sorbents with energetically nonuniform surfaces, *Chem. Rev.*, 60, 235–241 (1960).

26. R.K. Agarwal, High pressure adsorption of pure gases on activated carbon: Analysis of adsorption isotherms by application of potential theory and determination of heats and entropies of adsorption, *Ph. D. Dissertation*, Syracuse, 1988.

27. L.Z. Zhang, Design and testing of an automobile waste heat adsorption cooling system, *Appl. Therm. Eng.*, 20, 103–114 (2000).

28. N.V. Keltsev, *Basics of Adsorption Technology* [in Russian] (Moscow, Chemistry), 1976.

29. F. Gluckauf, Theory of chromatography. Part 10.—Formulæ for diffusion into spheres and their application to chromatography. *Trans. Faraday Soc.*, 51, 1540–1551 (1955).

30. A. Sakoda, M. Suzuki, Fundamental study on solar powered adsorption cooling system, *J. Chem. Eng. Japan*, 17, 52–57 (1984).

31. D.M. Young, A.D. Growell, *Physical Adsorption of Gases* (London: Butterworths), 1962.

32. L.L. Vasiliev, L.E. Kanonchik, D.A. Mishkinis, M.I. Rabetsky, A new method of methane storage and transportation, *Int. J. Environmentally Conscious Design & Manufacturing*, 9, 35–62 (2000).

9

Fundamental Questions of Closed Two-Phase Thermosyphons

M. K. Bezrodny

CONTENTS

9.1 Introduction: General Characteristic of Two-Phase Thermosyphons....320
    9.1.1 Determination and Operation Principle...320
    9.1.2 Modifications of Circuit Designs and Classification of
           Thermosyphons 320
    9.1.3 Basic Practical Tasks That Can Be Solved by Means of
           Two-Phase Thermosyphons..323
9.2 Conditions of Effective Work of Two-Phase Thermosyphons 324
    9.2.1 Thermodynamic Conditions of Effective Work of CTTs..........324
    9.2.2 Hydrodynamic Conditions of Effective Operating Modes
           of Thermosyphons..328
    9.2.3 Choice of Intermediate Heat-Transfer Agent 332
9.3 Characteristics of Transfer Processes...338
    9.3.1 Peculiarities of Transfer Processes 338
    9.3.2 Heat Exchange in the Evaporation (Boiling) Zone of Heat
           Medium..339
    9.3.3 Heat Exchange in the Condensation Zone...................................343
    9.3.4 Limiting Heat Fluxes..346
    9.3.5 Upper Boundary of Heat-Transmitting Capability of
           Thermosyphons 349
    9.3.6 Thermodynamic Similarity of Crisis Phenomena in CTTs 352
9.4 Concluding Remarks...353
9.5 Nomenclature...353
References...354

9.1 Introduction: General Characteristics of Two-Phase Thermosyphons

9.1.1 Determination and Operation Principle

Two-phase thermosyphons (e.g. heat pipes) represent the class of effective autonomic heat-transmitting devices with phase transformation of intermediate heat medium, based on the use of gravity forces as the activator of liquid movement. This accounts for the relative similarity of these devices and their wide range of implementation in apparatus and systems of ground type. Different tasks can be performed with the help of closed two-phase thermosyphons (CTTs).

Closed two-phase thermosyphons work on the following principle: when heat is supplied at the bottom evaporator section, where the liquid pool exists, the medium begins to boil and vapour is created. The vapour rises and moves to the condenser section where it condenses on the walls, delivering the heat of phase transformation to the cooling environment. The generated condensate returns to the evaporator under the action of gravity forces or other mass forces. The processes in the thermosyphon run continuously, which provides the heat transfer from one zone to another. In the transport zone, adiabatic conditions exist, notably heat is not delivered and is not derived through the walls.

9.1.2 Modifications of Circuit Designs and Classification of Thermosyphons

All heat pipes can be divided into two subclasses: wick and wickless heat pipes or thermosyphons (Figure 9.1). Two-phase thermosyphons can be closed and opened. Opened thermosyphons have no condenser, so heat is supplied continuously into the evaporator along the pipe. Such thermosyphons are used, in general, for the boiling process in research. Closed thermosyphons are vacuum processed and not vacuum processed (with non-condensing gases in cavities). There are syphons with side and butt heat supplies. Side supplies are used more often in heat exchanges.

Air syphons are also considered two-phase devices in which the heat flow is passed as a result of forced convection of fluid. During the operation process, the saturated gas is bubbled through the liquid layer, causing its intensive mixing. Heat-transfer coefficients in them can be higher than those in evaporating two-phase thermosyphons. Air syphons, as heat-transmitting devices, are not found in practical use and are used, in general, for studying heat exchange during boiling, because the fluid, bubbled by gas, models this process very well.

Besides two-phase thermosyphons, one-phase thermosyphons are sometimes used based on convection and with heat medium at critical parameters.

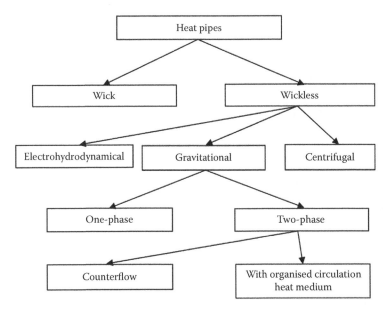

FIGURE 9.1
Classification of heat-transmitting devices, implemented as heat pipes, according to operation principle.

Convectional thermosyphons are nearly fully filled with liquid, and heat transfer is held by free convection and fluid heat exchange. Moreover, they can be opened, notably when connected with external environment, and closed, when mass exchange between thermosyphon cavity and environment is excluded.

There are also one-phase thermosyphons, in which the heat medium at determined conditions exists near the critical point. At the critical point, the difference between fluid and gas disappears, notably there is a one-phase condition, in which the values of several thermophysical properties of fluid (λ,c) are the greatest. In such a thermosyphon, the heat transfer is held at the point of high heat conductivity. Conditional heat conductivity of steel pipe, filled with ammonia at critical parameters, is 20 times greater than that of copper heat conductivity. However, one-phase thermosyphons do not get wide distribution. This is because of the small heat-transmitting capability of convectional thermosyphons and the complexity of thermosyphon exploitation with heat medium at critical parameters because of difficulty of supplying a constant temperature and pressure in the cavity at the change of applied heat flux.

In examined constructions of thermosyphons, the fluid from the condenser returns into the evaporator under the action of gravity forces. But, the cases of fluid flow under the action of other mass forces, for example, centrifugal, are also possible. Such thermosyphon cooling is used in turbine blades, electric motors and other devices. In such thermosyphons, the evaporator is situated on the periphery of the operational wheel (in the most high-heat area) and the condenser is located nearer the rotation axis. Vapour from the evaporator moves to the condenser and the condensate returns to the evaporator under the action of centrifugal forces.

In the last few years, the interest in the research of electrohydrodynamical methods of heat-exchange intensification has increased. An electrohydrodynamical thermosyphon is a grounded pipe, inside of which the coaxial cylinder-electrode is installed, on which constant voltage is applied. The annular gap between electrodes is partly filled with fluid. The dispersion of fluid is held by the interaction of the electrical field with the liquid–vapour mixture, conditioned by induction of the charge on the boundary of liquid–vapour phase division and because of the forces of electrostatic pressure. In such a thermosyphon, the fluid film becomes thinner and the thermal resistance of the thermosyphon decreases as a result of boundary phase rupture in the electrical field.

CTTs are differentiated based on the method of circulation of operational medium, constructive design, conditions of usage and type of intermediate heat medium.

Thermosyphons can be divided into two groups based on circulation method: (1) those with nonorganised circulation and (2) those with organised circulation of intermediate heat medium.

On the basis of constructive indication, heat-transmitting devices of the first group can be divided into two subgroups: thermosyphons of simple geometrical form and thermosyphons with branched system of heat supply and heat removal. The latter have relatively big chambers which facilitate the transfer processes for different parts of the heat-transmitting device. Therefore, the conditions of the transfer processes occurring on different areas of thermosyphons with branched systems of heat supply and heat removal are near to that in place on the respective areas of simple thermosyphons. By taking that into account, the regularities of heat and mass transfer processes of simple thermosyphons can be used for counting and designing complicated thermosyphon systems.

Principle differences from examined thermosyphons are thermosyphons with organised circulation of intermediate heat medium. Different authors have proposed a great deal of constructive variants of thermosyphons, realizing the pointed method of circulation of heat medium. The analysis of similar variants concludes that different constructive schemes of thermosyphons differ only in reciprocal location of channels for sinking and lifting movement of intermediate heat medium. The thermosyphons with organised circulation of intermediate heat medium also can be divided into two subgroups: those with an external sinking channel and those with a sinking channel located inside the evaporating area (arterial thermosyphons).

9.1.3 Basic Practical Tasks That Can Be Solved by Means of Two-Phase Thermosyphons

Heat transfer from one medium to another in varied thermal–technical and thermal–technological processes can be accomplished by one of the following three methods: (1) heat transfer through direct contact of two mediums, (2) heat transfer through the divisional hard wall and (3) heat transfer with the help of intermediate heat-transmitting devices.

Several devices can be used as intermediate heat-transmitting devices (e.g. high-heat conductive metal pivots, one-phase convection thermosyphons, circulation devices with forced movement of intermediate heat medium). But for the implementation of thermal–technological processes in most cases, the most effective method is using autonomic heat-transmitting devices with the look of CTTs. This is because of the high intensity of heat transfer on the areas of boiling and condensation of intermediate heat medium and also high values of maximum passed heat fluxes. With the help of two-phase thermosyphons, the following practical tasks can be accomplished.

One of the tasks that appear in many thermal–technological processes is the necessity of creating 'soft' conditions of heating of different technological products, notably creation of an even temperature field from the side of the heat supply in the heating chamber. This explains why the first industrial implementation of CTTs, called Perkins pipes, was as heating elements in bread ovens. Next, due to such qualities as isothermality on the separated areas and possibility of transformation of heat flux, two-phase thermosyphons permitted improved technological regimens in aggregates for thermal treatment of gas and fluids in different branches of industry. For example, because of these advantages, two-phase thermosyphons are now successfully used in developing devices for regeneration of absorbents for the installation of draining and cleaning natural gas in gas industries [1]. Absorbent fluids such as diethyleneglycol and monoethanolamine are very sensitive to overheating, resulting in their intensive breakdown. The use of two-phase thermosyphons in such apparatus, due to their isothermality, allows for the creation of a 'soft' condition of heat supply and in such way to realise the fire regeneration of absorbent that meet the technological requirements of the heating regimen.

With the help of two-phase thermosyphons, such technological tasks can be decided as a support of heat-exchange surface temperature on a given technological level. With this aim, such a property of two-phase thermosyphons is used as a means of temperature change of intermediate heat medium by the way of changing external conditions of heat exchange and correlation of sizes of heat-exchanging surfaces on the areas of supply and removal of heat. With this in mind, the necessary measured support can be provided in advance for the heat-exchange surface. Similar success with the task was realised with the creation of a topping air preheater with intermediate heat medium, which has higher stability against sulphur corrosion and is intended to protect the main air heater of a steam boiler from corrosive damage [2]. The advantage of such a heater is the possibility of concentration of a corrosive dangerous zone on several rows of pipes and the simple change of corrosion-damaged sections and also the density of heated air through corroded sections.

The most important task of improvement of thermal-technological installation lies in increasing of reliability of their exploitation. In this case, the considerable prospect for using two-phase thermosyphons is opened in connection with research and development of two-circuit systems of cooling of high-heat elements in industrial furnaces [3]. The operation principle of the cooling system in this case is that the heat, detected by constructive elements and knots, is transmitted, with the help of thermosyphons, over the limits of the furnace casing, where it is transferred to the double circuit (fluid or evaporating) of the cooling system. The implementation of CTTs as primary circuit, due to their autonomy, eliminates the disadvantages of the traditional cooling systems of metallurgical furnaces. Thereby the danger of introducing a considerable quantity of water into the operational space of furnace through the damaged areas of pipes is removed.

In many cases, decreasing the temperature of a heat-generating object is an important technological task. With this aim, the properties of two-phase thermosyphons can be used for their capability of transformation (deconcentration) of heat flow for heat removal to the environment. This allows their use as heat sinks for cooling of electrotechnical devices and electronic engineering apparatus [4]. The use of evaporating thermosyphons in the cooling systems of transformative and electronic devices is an effective method of increasing their operation reliability, reducing their weight, and improving their overall characteristics.

A wide range of practical implementation includes heat-exchange recuperators, intended for heat utilisation of waste gases of industrial furnace devices; technological products and low-temperature industrial emissions; and also for receiving technological and extraction energy resources in autonomic heat generators [5,6]. The expediency of use of thermosiphon recuperators is defined by possibility of ribbing of heat-exchange surfaces both with 'hot' and 'the cold' side, small values of hydraulic losses at an external flow of pipes in both streams, possibility of maintenance of temperature of a surface at the set level that in turn promotes the solution of important problems of decrease in specific consumption of materials and power expenses, raise reliability of work of the equipment.

9.2 Conditions of Effective Work of Two-Phase Thermosyphons

9.2.1 Thermodynamic Conditions of Effective Work of CTTs

Operation of CTTs is characterised by the totality of heat-exchanging, hydrodynamic and thermodynamic processes. The complexity in studying heat-transmitting properties of thermosyphons depends on both the variety and interconnectivity of separated process. In connection with this, the rational method of studying the characteristics of thermosyphons is based on the division of these processes and studying them in clean appearance.

While studying the characteristics of CTTs, the processes that should be noticed first are those that are easily controlled by changing external heat-exchange conditions. This is related to the thermodynamic process of state change of intermediate heat medium. Analysis of this process allows the determination of the conditions, which provides effective implementation of other (transfer) processes in thermosyphons [7].

The normal operation regimen of examined devices is one at which the intermediate heat medium is in a two-phase state. The passage of heat medium from two-phase state into one-phase state is accompanied by the sharp degradation of heat-transmitting properties of the thermosyphon. But this passage characterises only the higher boundary of operation ability of CTTs according to thermodynamic parameters.

In real heat-transmitting devices, the condition of reliable and effective operation is in a state of intermediate heat medium, where enough washing of the internal surface by the fluid phase is provided. But in operational conditions with the change of thermodynamic parameters in the subcritical area, the change of heat medium fluid phase quantity is assumed which can lead to the sharp change of heat-transmitting properties of the thermosyphon. For the determination of effective operation properties of the thermosyphon, the analysis of thermodynamic process of state change of intermediate heat medium must be considered, which leads to the installation of the connection between quantity of fluid phase, thermodynamic parameters and basic values.

It is known that the state of operating fluid in a two-phase area is determined by two parameters. In case of a closed device, which is the closed thermosyphon, it is proper to choose a specific volume of contents in the device cavity V_{mix} as one of the parameters because the process of condition change of operation fluid is held isochoric. The average temperature of heat medium T_s can be taken as the second parameter. The inclusion volume fraction of fluid phase of heat medium in the thermosyphon cavity (degree of volume filling) can be taken as the unknown quantity:

$$\varepsilon = \frac{V_l}{V_o} \tag{9.1}$$

Then, the unknown quantity is determined by the equation

$$\varepsilon = f\left(V_{mix}, T_s\right) \tag{9.2}$$

Let us determine the concrete view of these equations. With this aim, the specific volume V_{mix} will be represented as

$$V_{mix} = \frac{V_o}{G_{mix}} = \frac{V_o}{G_v + G_l} \tag{9.3}$$

Determining, according to T_s, the value of specific volume of liquid and vapour phase, on the base of equations 9.1 and 9.3, we get

$$\varepsilon = \frac{v'(v'' - V_{mix})}{V_{mix}(v'' - v')} \tag{9.4}$$

Let us bring into the consideration the degree of mass filling of a thermosyphon, which is characterised by the equation

$$\Omega = \frac{G_{mix}}{G_{cr}} \tag{9.5}$$

where G_{cr} is the mass quantity of heat medium, appropriate to specific volume v_{cr}, in critical point, notably

$$G_{cr} = \frac{V_0}{v_{cr}} \tag{9.6}$$

Then from equations 9.3, 9.5 and 9.6, we get

$$V_{mix} = \frac{v_{cr}}{\Omega} \tag{9.7}$$

With equation 9.7, equation 9.4 takes the view, satisfying the law of appropriate conditions

$$\varepsilon = \frac{v'(v''\Omega - v_{cr})}{v_{cr}(v'' - v')} \tag{9.8}$$

In the limiting cases at $\varepsilon = 0$ and $\varepsilon = 1$ from Equation 9.8, follow the expressions for the determination of mass filling degree values Ω on the boundaries of a two-phase area of intermediate heat medium condition

$$\Omega_{min} = \frac{v_{cr}}{v''} \tag{9.9}$$

$$\Omega_{max} = \frac{v_{cr}}{v'} \tag{9.10}$$

The character of changing of volume filling degree ε from determining parameters inside a two-phase area can be observed on the base of a graphical representation of Equation 9.8. Since the single-valued connection exists between the specific volumes v', v'' and temperatures on the line of saturation, Equation 9.8 can be represented as

$$\varepsilon = f(T/T_{cr}, \Omega) \tag{9.11}$$

The calculating relations of volume filling degree ε from reduced temperature T/T_{cr} and mass filling degree Ω for different operational fluids are depicted in Figure 9.2. From the graphics, it can be seen that at small values of mass filling degree with the reduction of temperature the degree of volume filling decreases and heat medium passes from two-phase into one-phase vaporous condition. At values of Ω near 1, the increase of T/T_{cr} leads firstly to some increase in the value of ε and then to the sharp decrease of value ε to zero. At large values of mass filling degree ($\Omega > 1$), the increase of reduced temperature causes the continuous increase of fluid phase volume in the thermosyphon and as a result all internal cavities of the thermosyphon are filled with heat medium in a fluid condition. The comparison of calculating relations for different operations fluids indicates that these relations have the same characteristic in the limits of the currency for thermodynamic similarity of fluids.

Figure 9.2 allows for the determination of practically important boundaries for correlative characteristics of thermosyphon filling by intermediate heat medium. It is obvious that stable operation of two-phase thermosyphons in the area of small values of filling characteristics can be provided in that case if the change of saturation temperature of heat medium does not lead to the sharp decrease of fluid phase volume. In relation to this, the limit values of filling characteristic on the low boundary of their changes can be determined with the help of the points shown in Figure 9.2, appropriated to the condition $\varepsilon = \varepsilon_{max}$ at $\Omega = $ constant. The relation of limit values ε and Ω from reduced temperatures T/T_{cr} for different operational fluids are shown in Figure 9.3. From the figure, it is seen that for different thermodynamic similar fluids the comparison of settlement dependences for various operating fluids testifies

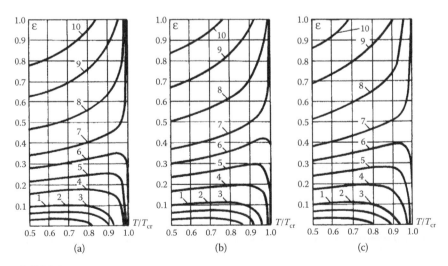

FIGURE 9.2
Relation of volume filling degree ε of closed cavity by fluid phase from reduced temperature and mass filling degree: (a) water; (b) ethanol; (c) R-11; (1) $\Omega = 0.1$; (2) $\Omega = 0.2$; (3) $\Omega = 0.3$; (4) $\Omega = 0.5$; (5) $\Omega = 0.7$; (6) $\Omega = 0.9$; (7) $\Omega = 1.1$; (8) $\Omega = 1.5$; (9) $\Omega = 2.0$; (10) $\Omega = 2.5$.

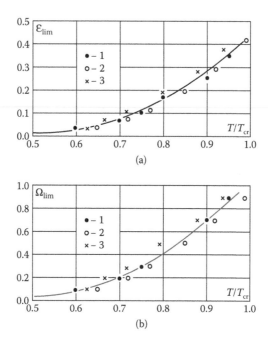

FIGURE 9.3

Relation of limit (minimal) values of filling characteristics from reduced temperature: (a) for degree of volume filling by fluid phase; (b) for degree of mass filling; (1) water; (2) ethanol; (3) R-11.

that in limiting accuracies of thermodynamic similitude of substances these dependences have equal character.

Continuous increase of liquid phase volume with the increase of heat medium temperature in the areas of large characteristics of filling ($\Omega > 1$), as the result of small compressibility of fluid, can lead to the rupture of the thermosyphon frame. In relation to that, the limit value of mass filling degree on the upper boundary of recommended zone of its change can be accepted equal to $\Omega_{lim} = 1$.

The limit characteristics of thermosyphon filling are based on the analysis of thermodynamic processes in the closed thermosyphon and characterise only the boundaries of recommended area of change of average thermodynamic state of intermediate heat medium. The final selection of the filling parameters can be held on the basis of hydrodynamic processes analysis in the thermosyphon with the account of the possible regimens of thermosyphon operation.

9.2.2 Hydrodynamic Conditions of Effective Operating Modes of Thermosyphons

Hydrodynamic operation regimens of closed thermosyphons are connected with the structure of a two-phase system in the device cavity, depending on relative quantity of heat medium liquid phase. Two main regimens are distinguished: the draining film of fluid and the bubble regimen. The preference of one over the other can be given in the dependence on concrete condition of practical usage of thermosyphons.

The regimen of draining film is appropriate to the fluid volume, draining as film along the internal surface of the walls at the condition of full wetting and absence of fluid phase overflow in the low part of the device. The necessary fluid quantity is determined by integration to the height of thermosyphons with a glance at the changing value of fluid film thickness.

The use of draining film of fluid in thermosyphon operation is characterised by different limits, connected with fault of solidity of fluid film on the area of heat supply (thermal instability of film, irregular washing along the perimeter, etc.). The results of these events can be both the degradation of heat-transmitting properties of thermosyphons and the decrease of reliability of their operation. Pointed circumstances have special significance at limit conditions, which are closed to the limit conditions of the second kind. To remove such events, the presence of a fluid overflow in the thermosyphon is needed. Such operation regimen of thermosyphons is called the regimen of two-phase mixture (bubble regimen). In this regimen, it is established that crisis events (the event of drying) happen both with a decrease of fluid level in the thermosyphons and also with an increase of operational temperature of heat medium (or the saturation pressure). The analysis of known data allows the conclusion that crisis events happen when the level of the two-phase mixture in the thermosyphon drops below the upper boundary of the heated area. Thereby the condition of normal operation of the thermosyphon in the bubble regimen is the full washing of the heat supply area by two-phase mixture.

The needed degree of volume for filling the thermosyphon by liquid phase can be determined in this case from the condition $V_h \le V_{mix} \le V_h + V_a$. Neglecting the fluid volume, which is the film on the condensed area, this can be written as

$$V_1 = V_{mix} - V_v = V_{mix}(1 - \bar{\varphi}) \tag{9.12}$$

where $\bar{\varphi}$ is the average volume of vapour content of two-phase layer.

Then, the expression for volume filling degree of the thermosyphon by fluid phase has the following view:

$$\varepsilon = \frac{V_1}{V_o} = \frac{V_{mix}}{V_o}(1 - \bar{\varphi}) \tag{9.13}$$

Full washing of the heating surface in the bubble regimen of thermosyphon operation is reached at $V_{mix} \ge V_h$. Then the minimal value of volume filling degree of the thermosyphon by liquid will be determined by the equation

$$\varepsilon_{min} = \frac{V_h}{V_o}(1 - \bar{\varphi}) \tag{9.14}$$

In this way, for the given geometric parameters of the system, the minimal value for the degree of filling by volume with the liquid phase in the bubbling operation is directly related to the value of the average volume vapour content in the two-phase layer of the working fluid. Analysis of well-known works has allowed us to conclude that the mechanism of the change in values $\bar{\varphi}$ in thermosyphon conditions is not entirely known. The specific character of the conditions of flow of hydrodynamic processes in thermosyphons does not allow the use of known dependencies for the determination of $\bar{\varphi}$, which is justified for other gas–liquid systems. In relation to this, specialised experiments were carried out and the results are given below. In the course of experiments it was established that in the working conditions of two-phase mixtures, depending on the diameter of the thermosyphon and the characteristics of the working fluid, two basic patterns of vapour flow may occur:

slug flow where relative diameter $d/\delta < 18$ and

bubble flow where $d/\delta > 30$ (here, $\delta = \sqrt{\sigma/g(\rho' - \rho'')}$ is value of Laplace).

In area $18 < d/\delta < 30$ we have an intermediate regimen. The generalisation of experimental results for average vapour contents of a dynamic two-phase layer is carried out on the basis of cinematic models of two-phase flow, which in the case of a bubbling system leads to the following relation [7]:

$$\bar{\varphi} = \frac{1}{1 + (W^\circ/W''_o)\psi} \tag{9.15}$$

where W''_o is the reduced velocity of steam at the outlet of the heating unit; W° is the velocity at which single vapour bubbles rise to the surface and ψ is the coefficient of bubble interaction.

For the surfacing velocity of single vapour bubbles, the critical surfacing velocity of vapour bubbles is taken as a standard, which may be determined by the equation [7]

$$K = CK_p^n \tag{9.16}$$

or

$$\frac{W^\circ \sqrt{\rho'}}{\sqrt[4]{\sigma g(\rho' - \rho'')}} = C\left(\frac{P\sigma}{\delta}\right)^n$$

where $C = 8.2$, $n = -0.17$ for $K_p \leq 4 \times 10^4$ and $C = 1.35$, $n = 0.0$ for $K_p \geq 4 \times 10^4$

Equation 9.15 has been compared with experimental results by φ for various working fluids (Figures 9.4 and 9.5). According to these relations, the following relation was obtained for the coefficient of vapour–bubble interaction:

$$\psi = 110 \, Ar^{-0.25} \tag{9.17}$$

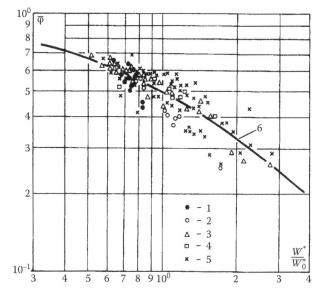

FIGURE 9.4
Generalisation of experimental data of the average volume vapour content in the two-phase layer for slug-flow regime: (1) R-11, $d_{in} = 14\,mm$; (2) ethanol, $d_{in} = 14\,mm$; (3) water, $d_{in} = 36\,mm$; (4) water, $d_{in} = 20\,mm$; (5) R-11, $d_{in} = 20\,mm$, according to Equation 9.15 for ψ in Ref. (17).

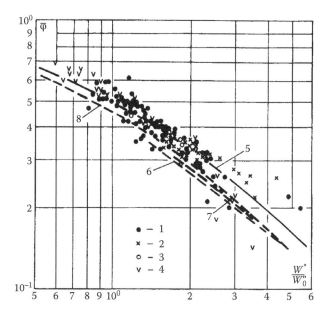

FIGURE 9.5
Generalisation of experimental data on $\bar{\varphi}$ for the bubble-flow regime: (1) R-11, $d_{in} = 36\,mm$; (2) R-11, $d_{in} = 66\,mm$; (3) water, $d_{in} = 66\,mm$; (4) ethanol, $d_{in} = 36\,mm$; (5) according to Equation 9.15 for $\psi = 3.78$.

for the slug pattern and

$$\psi = 3.78 \tag{9.18}$$

for the bubble pattern in vapour-phase movement.

The introduction of these relations allows us to calculate the minimum values for the degree of filling by volume of thermosyphons required for the prevention of crises of heat exchange associated with interruptions in the constant washing of the heating surfaces with the liquid phase. These crises are connected with hydrodynamic processes, leading to redistribution of the liquid in the cavities of the thermosyphon. The characteristic curve selected in practice for the filling of the thermosyphon must satisfy the condition $\varepsilon \geq \varepsilon_{min}$ for the determination ε_{min}, proceeding from both hydrodynamic and thermodynamic processes.

A guarantee of the necessary operating characteristics for the filling of thermosyphons eliminates the incidence of heat-exchange crises connected with the change in volume of the liquid phase in the thermosyphon and provides methodological basis research upper-limit heat-flow as a characterisation of the conducting properties of two-phase system of the thermosyphons.

9.2.3 Choice of Intermediate Heat-Transfer Agent

The operation term and metal intensity of a heat-transmitting device depends on the right selection of heat medium and frame materials. During the selection of heat medium, the following conditions must be taken into account: operation diapason of temperatures and pressures, maximal heat flow in evaporator of thermosyphon, thermal physical properties (such as boiling temperature, fusion, critical parameters, heat of vapourisation and vapour density), toxicity, fire and explosion risks and device operation condition during long periods of time (possibility of fluid freezing during the time of stopping, startup regimen, emergencies—consequences at the burnout of evaporator or condenser depressurisation, etc.). Theoretically, any chemical compounds and substances that have fluid and vapour phases in the operational diapason of temperatures and pressures can be used as heat medium. In practice, a very limited quantity of fluids is used.

Critical parameters of heat medium must be higher than operational temperature of the thermosyphon. Vapour pressure of heat medium at operational temperatures must be high to provide the transfer of a great amount of heat. But, it should be noted that higher the pressure is, the more metal intensive construction is and also that at $P > (0.2/0.3)P_{cr}$, the limited heat flux at fluid boiling begins to decrease. At low pressures, the sizes of vapour bubbles sharply increase. This can lead to the overlap of a cross section of the evaporator at the startup and leads to damage (burnout) of its wall.

The low limit of operational temperature must be higher than the temperature of phase passage from the fluid condition into the solid. It is very important for the heat mediums because at phase passage the volume of solid phase is greater than fluid, which can lead to damage of the thermosyphon frame. Special attention must be paid to the selection of heat medium for thermosyphons which operate at high temperatures and the radiation method of the heat supply. In these conditions, the heat mediums with complicated molecular structure can change thermophysical properties.

Analogical to the heat medium classification due to temperature, the thermosyphons can be divided into cryogenic, low-, average- and high-temperature thermosyphons. In practice, low-temperature thermosyphons and heat mediums are used.

Due to several parameters, water is the best heat medium. It supports the most heat transfer from all known heat mediums, except liquid metals. Water is easily accessible and fire and explosion secure. Despite its advantages, water as heat medium has the following disadvantages: high melting point (freezing), reaction with some substances (alkaline metals and other), and hydrogen separation, which can lead to explosion at determined conditions.

Besides water, alcohols, ethers and Freons also can be used as low-temperature heat mediums. Such heat mediums are harmful to humans.

There is a special interest in the use of two-component or multi-component mixtures as heat mediums. The use of multi-component mixtures allows for extended operational diapason of temperatures. One widely used multi-component mixture is water–glycol. Glycols are added to the water to reduce freezing temperature. Such mixtures are called antifreezes. By mixing water with glycols, we can get mixtures with boiling temperatures from 100 to 200°C at atmospheric pressure and temperatures of freezing from 0 to –75°C (see Table 9.1).

To prevent water freezing at low temperatures, the water–alcohol mixture is used with low alcohol content (1%–2% by volume). Such a mixture excludes frame damage at temperatures higher than –40°C and has the same heat-transmitting characteristics as water.

TABLE 9.1

Crystallisation Temperatures of Water Solution of Some Compounds (°C)

Compound	Content in Water Solution (% Due to Mass)							
	10	20	30	40	50	60	70	80
Acetone	−2	−5	−9	−12	−20	−25	−32.5	−45
Methanol	−6	−15	−25	−40	−53.5	−60	–	–
Isopropyl alcohol	−3	−5	−8	−9	−10	−10.5	−12	−16
Ethanol	−4	−10	−22	−31	−37	−45	−55	−60
Ethylene glycol	−2.8	−8.3	−16	−26	−37	–	–	–

The two-component heat medium are referred to as azeotropic solutions, which are binary solutions and the content of which is identical to the content of equilibrium with the vapour. Azeotropic solutions boil at certain temperature (for the given pressure), despite the big difference of partial pressures of making pure components (Table 9.2). These solutions exist only at defined correlations of binary mixture components, and they cannot be divided into clean components through distillation.

In the temperature range of 200–400°C, a dowtherm A and diphenyl mixture is used. Also, a silicon–organic fluid mixture can be used as an average-temperature heat medium. They have relatively low critical pressures, a high critical temperature and a low melting temperature. There are data in the literature about the testing of heat pipes with such heat mediums, but long term resource tests were not held.

Recently, naphthalene was used as an average-temperature intermediate heat medium. In the literature there are data that thermosyphons, produced from steel 20 and titanium, filled with naphthalene have stable heat-transmitting characteristics and they can be used to temperatures of 350–400°C. But it should be noted that the operation of thermosyphons with naphthalene has peculiarities, which can lead to the rupture of the heat-transmitting device. Under normal conditions, the naphthalene is in the hard phase and its heat of evaporation and crystallisation are small so that it easily settles on the cold surfaces, which can lead to the creation of the corks out of the heating zone. At startup of such thermosyphons as with thermosyphons using any other heat medium, the medium is in the solid condition, and careful attention should be paid because frame rupture is possible, which could include the emission of hot, and more often, poisonous, vapours and fluid heat medium.

High-temperature heat mediums are used at temperatures of 400–1000°C. The most common heat mediums are alkaline metals. Lithium and sodium are used for the best thermal-physical properties in thermosyphons and heat pipes. For temperatures lower than 600°C, potassium or cesium is used.

TABLE 9.2

Boiling Temperatures of Some Azeotropic Solutions

Component		Component Content I in Solution (%)	t_s (°C)		
I	II		I	II	Solution
Water	Ethyl alcohol	4.43	100.00	78.30	78.13
Water	Isopropyl alcohol	12.10	100.00	82.44	80.37
Methanol	Benzene	39.55	64.70	80.20	58.34
Ethanol	Benzene	32.37	78.30	80.20	68.24
Ethanol	Chloroform	7.00	78.30	80.20	68.24
Toluol	Tin chloride	48.00	110.70	113.85	108.15
Carbon bisulphide	Acetone	66.00	46.25	56.25	39.25

Operational ability CTT in concrete conditions of practical usage depends on thermodynamic and thermophysical properties of the intermediate heat medium. Here, without factoring in that the operation CTT is principally possible in the diapason of temperature from the triple to critical point of heat medium, the operational interval of temperatures must be the average of pointed diapason, excluding the condition of deep vacuum and pressures that are near critical.

Concrete recommendations are followed from joint analysis of thermodynamic and thermal physical properties of operational substances. The maximum heat fluxes in CTT are restricted by crises of heat transfer, which are characterised by some values of stability criterion

$$K = \frac{q}{r\rho_v^{0.5}} (\sigma g(\rho - \rho_v))^{0.25} \tag{9.19}$$

and we gain a complex thermophysical properties of heat-transfer agent as

$$q_* = r\rho_s^{0.5} (\sigma g(\rho - \rho_s))^{0.25} \tag{9.20}$$

determining the heat-transmitting capabilities of thermosyphons. Relations of quantity q_* for different heat mediums from saturation temperature are shown in Figure 9.6. Restricting these relations from one side by creating a vacuum

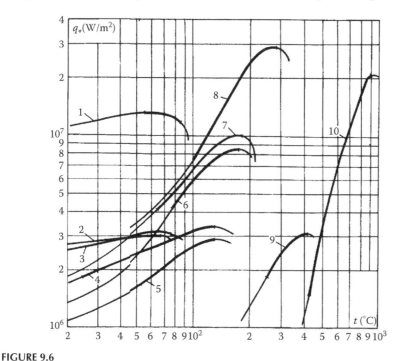

FIGURE 9.6
Selection of working fluid according to the temperature dependence of thermophysical characteristics (1) ammonia; (2) R-12; (3) R-42; (4) R-11; (5) R-113; (6) methanol; (7) ethanol; (8) water; (9) dowtherm A; (10) sulphur.

in CTT and from the other side by extreme values q_* at reaching the critical parameters, we get recommended diapasons of operational temperatures in which different fluids can be used. These diapasons correspond to the thick areas of the curves $q_* = f(t_s)$, $P = f(t_s)$, represented in Figures 9.6 and 9.7.

For the selection of heat medium, its compatibility with constructive material of the thermosyphon frame must be considered. Material must be selected in a way that it does not corrode in heat medium and in the heating or cooling environment. It should be noted that at interaction of separated pairs of constructive material—heat medium—the noncondensable gases can escape. Noncondensable gases do not substantially influence the heat-transmitting characteristics. But during accumulation in a closed cavity they can create substantial heat emission in the condenser and thus increase the primary pressure in the thermosyphon. The compatibility of different combinations of constructive materials and heat mediums for thermosyphons is shown in Table 9.3.

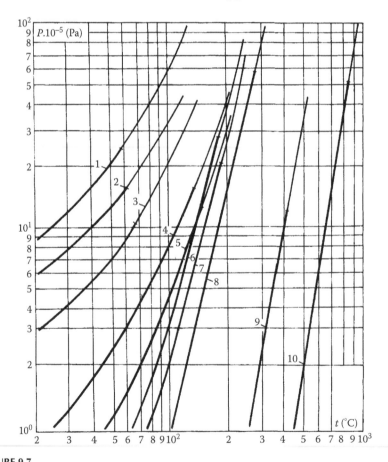

FIGURE 9.7

Selection of working fluid according to the admissible working pressure in thermosyphons (see Figure 9.6 for designations).

TABLE 9.3

Compatibility of Heat Mediums and Constructive Materials

Heat medium	Aluminium	Wolfram	Iron	Copper	Molybdenum	Nickel	Nioubium	Stainless Steel	Carbon	Tantalum	Titan
Nitrogen	C	–	C	C	–	C	–	C	–	–	–
Ammonia	C	–	C	H	–	C	C	C	H	–	C
Acetone	C	C	–	C	C	C	C	C	C	C	–
Water	R	C	–	C	C	C	C	C[a]	H	C	C
Dowtherm A	–	C	–	C	C	C	C	C	C	C	–
Potassium	H	C	–	C to 600 K	C	C	C	C	H	C to 900 K	H
Lithium	–	–	–	–	C	C	C	C	H	–	–
Methane	C	C	–	C	–	–	–	C	–	–	–
Methanol	H	C	C	C	–	C	C	C	H	–	C
Sodium	H	C	–	C to 600 K	C	C	C	C	H	C to 900 K	H
Naphthalene	–	–	–	–	–	–	–	C	–	–	C
H-Octane	–	–	–	–	–	–	–	C	–	–	–
Toluol	–	–	–	–	–	–	–	C	–	–	–
Freons	R[b]	C	–	C	C	C	C	C	C	C	–
Ethanol	R	–	–	–	–	C	–	C	–	–	–

Note: C, compatible; H, incompatible; R, restricted compatible.
[a] Possible abstraction of hydrogen.
[b] Possible corrosion.

9.3 Characteristics of Transfer Processes

9.3.1 Peculiarities of Transfer Processes

Peculiarities of transfer processes in two-phase thermosyphons are determined by peculiarities of the two-phase system, which can take place in thermosyphons of different constructions. In turn the characteristics of these two-phase systems in any case is determined by the influence of field of gravity mass forces, forming corresponding distribution of vapour and fluid phase on different areas of thermosyphon devices. For determining the character of two-phase structure in thermosyphons, the following should be referred to: degree of thermosyphon internal cavity filling by operational fluid (heat medium), geometrical sizes of thermosyphons and their correlation on different areas, correlation of conditions of heat supply and removal, kind of operational fluid and presence of internal devices that influence the circulation of operational medium.

As shown above, the principle operation condition of device is such correlation of conditions of heat supply and removal, which provides current operational fluid for its thermodynamic state, corresponding to two-phase area of substance state. In implementing this condition in relation to the range of other factors for thermosyphons of classical construction (without internal devices) the following characteristics of two-phase systems must be singled out:

two-phase system with bubble layer in the condition of large volume

two-phase system with undivided character of phase distribution in restricted space

two-phase system with enough accurate surface of phase division, dividing the liquid and vapour flows in the thermosyphon

In practicality, the internal cavity of the thermosyphon can be completely occupied by these systems separately or together.

The basic peculiarities of transfer processes in comparison with transfer processes in similar-known canonical systems are in correlative influence of these systems, determined by completeness of internal cavity of the thermosyphon; in mutual influence of geometrical parameters and physical properties of operational fluids; in straightened character of conditions for the development of vapour creation process and hydrodynamic phase flow and in specificity of formation of two-phase structure in relation to conditions of heat supply and location of thermosyphons in the space. The presence of pointer peculiarities needs corresponding research organisation and study of heat-transmitting processes as in separated above pointed two-phase systems, singling them out as independent objects of research, and also in the conditions of their joint existent in the thermosyphon cavity.

For thermosyphons with organised circulations of heat medium, the characteristic peculiarity of the system is in the presence of a free-convection circulation frame with separated channels for ascending and descending flows of heat medium. It is known that transfer processes in the evaporating circulation frame depend on constructive formation of the frame, conditions of heat supply and are complicated with the appearance of contour instability at the increasing of heat flows. Insufficient study of these processes in traditional evaporating contours of industrial devices and the variety of peculiarities of operation regimens and constructive design of circulation contour in two-phase thermosyphons need the special consideration of systematic research of pointed processes with the aim of studying physical events of transfer and receiving of generic calculating correlations.

9.3.2 Heat Exchange in the Evaporation (Boiling) Zone of Heat Medium

In relation to the degree of thermosyphon cavity filling of heat medium in the area of heat supply, different regimens of heat flowing and heat exchange can take place, determined by the structure of two-phase medium. Most frequently, thermosyphons with partly and full filling of heating zone by liquid are used. Here, a regimen of heat exchange at boiling takes place, followed by the bubbling of vapour bubbles through the liquid layer.

At boiling, it is considered to be a measure of intensity of heat exchange magnitude of factor of a convective heat exchange $\alpha = q/\Delta T$. Notably the density of heat flux on the surface division of the hard body fluid q, referring to the temperature drop between heating surface and fluid, determined as $\Delta T = T_W - T_s$. For practical aims of calculations and designing of technical devices, the most important values are the heat-exchange intensity at bubble boiling (heat-exchange coefficient α) and its crisis (notably, the critical density of heat flux q_{cr}).

The results of visual examinations conducted by many authors showed that at partial filling of the heating zone by fluid at densities of heat fluxes in the diapason 160×10^4 W/m^2, the liquid evaporates from the surface of phase division. At this value the steady film of condensate moves toward adiabatic and condensation zones. With the increase of heat generation in the evaporator of the thermosyphon, there is observed contraction of film in the separated streamlets on the input of the heating zone. The intensive evaporating of film begins in the heating zone. The influence of geometrical sizes in this case is explained by the change of ratio of surface area, washing of film, moving as separated streamlets, to the drained surface.

By reaching some wall overheating in relation to temperature of film saturation, the centres of vapour creation appear. Bubble boiling in this case is followed by intensive spray of fluid onto drained areas. The boiling leads to a temperature decrease of the wall in comparison to an evaporating regimen. At developed bubbled boiling, all drained surface of the evaporator is evenly covered by evaporating liquid drops.

By decreasing the angle of inclination of the thermosyphon relatively, the horizontal flatness of the intensity of heat exchange is decreased. This is explained by increasing of area of drained surface and of part of heat flux, transmitted to streamlets of film by heat conductivity, which leads to the increase of thermal resistance in the evaporator.

The maintenance of the pointed operation regimens of thermosyphons under real conditions is very complicated because at the changes of regimen parameters, the draining of the evaporator is possible through the liquid phase shortage. That is why there is practical interest to operate in the presence of some volume of boiling liquid in the evaporator of the thermosyphon.

The regularity of the heat-exchange process at developed boiling in the thermosyphon differs from that of the boiling process in a large volume because of restricted sizes of evaporators and surface of phase division. The difference is that the vapour bubbles continuously collide with each other, changing the direction of their movement, which leads to a decrease in surfacing speed and complicates their output through the surface of phase division. Restricted boiling conditions enable the increase of bubble concentration and degree of their correlative influence at the passing to the surface of phase division of liquid and vapour. With the increase of heat load, the chaotic movement of bubbles and layer thickness increase and gets saturated by the vapour phase. In the evaporator intensive turbulisation of the wall layer and all volume of fluid by the chaotic moving bubbles are observed. In the range of works it is noticed that at developed boiling the vapour bubbles push the liquid back from the heating surface, leaving on it the thin film which is able to increase the heat-exchange intensity.

Pointed circumstances indicate significant changes of heat-exchange conditions at boiling in the thermosyphon compared with boiling in large volume. That is why it is reasonable to indicate the regularities of heat exchange at boiling in thermosyphons on the basis of comparison with such large volumes. But the analysis of known experimental data about heat exchange at fluids boiling on metallic surfaces in large volume show that because of the difference in the surface conditions, the values of coefficients of heat exchange at the same heat fluxes and pressures can considerably deviate from some statistically average current fluid level. In relation to that, rating-test experiments were held to explore the intensity of heat transfer at fluids boiling in large volumes. They gave the chance to detect further peculiarities of heat exchange at boiling in thermosyphons in comparison with boiling in large volumes by way of comparison of appropriate data, received on surfaces with the same roughness. The results of corresponding researches were shown in Refs. [8] and [9].

Also it was discovered that at fluids boiling in the thermosyphon (as in large volume) by changing the heating flux in the wide diapason, two characteristic areas of heat exchange are observed: undeveloped boiling and developed boiling. The degree of influence of heat flux in both areas is approximately the same as that in the conditions of large volume. But, absolute values of heat-exchange coefficients in the thermosyphons (at $P = 0.1$ MPa) are higher than in the conditions of large volume.

Visual observations showed that in the area of undeveloped boiling the process of heat exchange is followed by surfacing of separate bubbles, which in their movement turbulisate the wall layer of liquid on all areas of heating part. This leads to the considerable intensification of the heat-exchange process.

In the area of developed boiling, vapour bubbles appear in the overheated layer of the liquid on all the heated surface. So the main contribution to the heat-exchange intensity is introduced by turbulisation of liquid by vapour bubbles, which appear on the heated surface at about the degree of influence of heat flux on heat-exchange coefficient, which is equal to 0.75. But the axial flow of vapour in the central part of the channel in general influences the structure of wall layer, leading to the creation of a thin film of liquid on the internal wall of thermosyphon and leading to premature breakage of appearing vapour bubbles which leads to the intensification of the heat-exchange process.

The experiments showed that the intensification of heat exchange at boiling in the thermosyphon in comparison with large volume is different for different heat mediums. It is connected with the structure of axial vapour flow, depending on fluid properties, and in considerable measure can explain the discrepancy of literature data. In turn the structure of vapour flow depends on pressure in the internal cavity of the thermosyphon, the influence of which is in the same range of its change for different liquids is not univocal.

The generalisation of experimental data in the area of the developed boiling of liquids in thermosyphons is presented in Figure 9.8. As degree of influence of a heat flux on intensity of heat exchange in this area practically remains the same, as in the conditions of large volume, the basic attention at generalisation of test data is given to the degree of influence of pressure of the intermediate heat-transfer medium. The dependence presented in Figure 9.8

$$\frac{\alpha}{\alpha_{B.V.}} = f\left(\frac{P}{P_{cr}}\right) \qquad (9.21)$$

shows that the influence of restricted conditions on heat exchange in the thermosyphons is observed only in the defined diapason of pressure changes of intermediate heat medium. It should be noted that in accepted coordinates the character of equation (9.21) is the same for different heat mediums and changes by only changing the regimen of vapour phase movement. The area of maximum values of relation $\alpha/\alpha_{B.V.}$ is observed at values of reduced pressure $P/P_{cr} = 0.01\ldots0.1$. Increasing and decreasing pressure of relatively pointed values leads to the decrease of relative heat-exchange coefficient. Characteristics of the gained dependence confirm the assumption resulted above that the heat exchange intensification in ЗДТ occurs at the expense of agency of an axial flow of steam on a liquid wall layer.

At low pressure, as a result of large volumes of vapour bubbles, the boiling process in CTT has instable characteristics and is followed by pulsation, fluid

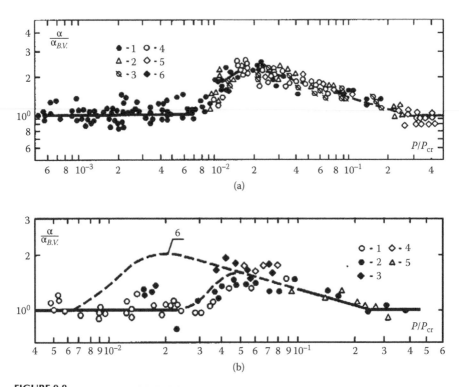

FIGURE 9.8
Generalisation of test data on heat exchange in the field of the developed boil: (a) slug regime: (1) water, (2) ethanol, (3) R-11, (4) R-113, (5) R-12, (6) methanol; (b) bubble regime: (1) water, (2) ethanol, (3) R-113, (4) R-11, (5) R-12, (6) a line for slug regime.

ejections from the heating zone and noise effects. In this area, the experimental data are the same as corresponding data for large volume. Analogical results were obtained in other works of water boiling in thermosyphons in areas of decreased pressure (from 0.01 to 0.25 MPa).

With the increasing of pressure, the process stabilisation of boiling process and heat exchange takes place and as a result of this the heat-exchange coefficient increases. Further increasing of pressure leads to the decreasing of vapour speed and its influence on heat-exchange intensity. At $P/P_{cr} \geq 0.3$, additional turbulisation of the wall layer of liquid at boiling in the restricted conditions does not influence the heat-exchange coefficients and they can be determined due to known relations for large volume.

The shown data allow for some practical recommendations concerning to the selection of operation regimens of CTT or kind of heat medium. At a selected heat medium, the best regimen of thermosyphon operation is achieved at values of reduced pressure $P/P_{cr} = 0.016\ldots0.1$. On the other hand, at given value of absolute pressure the heat medium can be selected, which provides the pointed value of reduced pressure.

9.3.3 Heat Exchange in the Condensation Zone

Condensation or phase transition of the first kind represents the transition of vapour into liquid state. The condensation process, also known as the process of evaporating or boiling, is the main process for CTT. It runs at subcritical conditions of vapour and is held by the way of its cooling in the condenser of the thermosyphon. In relation to regimen parameters and constructive peculiarities of the condenser, condensation can happen in the fluid volume (if the degree of filling of heating zone is 100% and more, the length of the transport zone is insignificant, notably the carrying out of fluid in the condenser takes place), in the vapour volume (if the vapour is considerably overcooled, relative to the temperature of saturation) or on the cold surfaces (if the surface temperature is less than the temperature of saturation at current pressure). During condensation of the vapour on hard surfaces, the creation of condensate film is possible with a thickness which exceeds the distance of effective action of intermolecular forces (film condensation) or during condensation of the vapour on the surfaces, the separated condensate drops are formed (drop condensation).

Film condensation happens if the fluid wets the surface. It is common that in the thermosyphons film condensation takes place on clean metallic surfaces. If the fluid does not wet the surface, which is observed in the presence of contamination on the surface, drop condensation happens. In general, the heat-exchange coefficients during drop condensation can be considerably higher than that during film condensation. This is connected with the fact that the condensate film has high thermal resistance during heat transmission to the wall. The resistance depends on film-flow regimen. During laminar movement of the film, the heat is transmitted by heat conductivity and during turbulent movement of the film, the heat is transmitted by heat conductivity and convection. The passage from one film-flow regimen to another is determined by critical Reynolds number $Re_{cr} = W\delta/\nu$, where W is the flow speed of the film in cross section and δ is film thickness. The value Re_{cr} can be in the limits of 60–500.

For calculating the heat-exchange coefficient during condensation of clean vapour inside vertical thermosyphons, the range of regularities were obtained by different authors, which are based on the use of the theory of film condensation of Nusselt. For example, in Ref. [10] more than 16 publications of different authors are analysed which contain primary data about vapour condensation of different heat carriers in two-phase thermosyphons and the effort to get one universal regulation in wide diapason of changes of regimen and geometrical parameters was made. It should be noted that similar regulations have artificial character and are of little use for practical calculation because they do not take into account the operation regimens and peculiarities of process development in the conditions of thermosyphons.

The following points of characteristic peculiarities of the vapour condensation process in the conditions of the closed cavity of the thermosyphon must be considered: the influence on the regularity of the heat exchange of residual and escaping noncondensate gases in the operation process, possible interaction of phase on the boundaries of division in the conditions of counterflow of the liquid film and vapour flow, changes of physical level of two-phase mixture and possible submergence by two-phase mixture of the condensation surface and correlative influence of hydrodynamic processes on the zones of boiling and condensation, followed by drop fluid occurring on the condensation area.

The analysis of such peculiarities allows us to conclude that the obtained results concerning intensity of heat exchange at condensation in the thermosyphons can be achieved only under the condition of controlled character operation regimens of the condenser, which can be singled out on the basis of preliminary study of hydrodynamic characteristics of two-phase systems of thermosyphons. The main (limit) regimens of operation of the condensation area of the thermosyphon can be singled out as (1) film condensation of clean vapours of intermediate heat carrier in the conditions of counterflow of descending film of fluid and ascending vapour flow, and (2) condensation of vapours from the dynamic two-phase layer of heat transfer at conditions of full submergence of the condenser surface by two-phase mixture.

For singling out the pointed regimens of heat exchange, the results of research of the physical level of two-phase mixture in the thermosyphon [11] can be used, on the basis of which the boundary between areas of undivided (dynamic two-phase layer) and divided (liquid film–vapour kernel) flowing of heat-carrier phases in the condenser were implemented. Results received provided the methodical base for holding the research of heat exchange in the above operation regimens of condensation areas of thermosyphons.

Research conducted in Ref. [12] showed that the regimen of film condensation of clean vapour can be provided with a consideration of selection of appropriate length of adiabatic zone and degree of thermosyphon filling by heat carrier, which guarantee the conditions of not flooding of the condenser surface by two-phase mixture, the removal of noncondensable gases through the drainage valve in the upper part of the thermosyphon. and the removal of the carrying out of the liquid phase of heat carrier with the help of separating inserts. In this case, the average due to the condenser height value of heat-exchange coefficients are interfaced with known regulations [13] at appropriate values Re_{cond} up to the crisis of heat transfer, determined by the upper boundary of the flooding regimen of two-phase flow in the thermosyphon conditions [14–16].

The research results of heat-exchange intensity in the conditions of flow of the condensation area of thermosyphon by two-phase mixture showed that the regularities of the heat exchange at condensation depends on the regimen of the vapour phase movement. The generalisation of experimental data, held for slug and bubble regimens in accordance with given boundaries (Bo < 18—slug regimen, Bo < 30—bubble regimen), is shown in Figure 9.9.

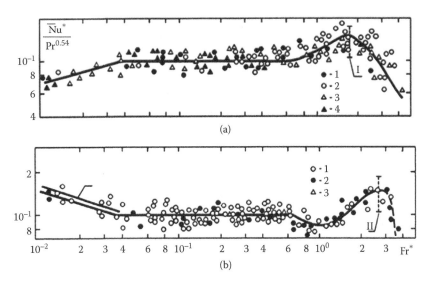

FIGURE 9.9

Generalisation of the data due to vapour condensation from dynamic two-phase layer: (a) slug-flow regime: (1,2,3) $d_{in} = 18$ mm; (4) 36; (1) R-11; (2) ethanol; (3,4) water; (b) bubble-flow regime: (1,2) $d_{in} = 36$ mm; (3) 66; (1) ethanol; (2) R-11; (3) water; (4) due to Nusselt formula for immovable vapour.

As follows from the shown generalisation of $Fr^* < 4 \times 10^{-2}$, regularities of heat exchange in separated regimens of the vapour flows have different characteristics, which are conditioned by different hydrodynamic conditions in the cooling zone of the slug and bubble regimens. In the first case, the condensation zone is filled nearly full by alternate vapour or liquid slugs, whereas in the second case, the condenser is filled by a two-phase mixture. As the result of pointed peculiarities the heat exchange coefficients in the bubble regime in the pointed diapason of parameter Fr* practically are submitted the relation due to equation Nusselt. In the slug regimen, a considerable decrease in the heat-exchange intensity is observed. The average values of heat-exchange coefficients corresponded to the equation

$$Nu^* = 0.21 \, Pr^{0.54} \, Fr^{*0.24} \tag{9.22}$$

where

$$Nu^* = \frac{\alpha_c}{\lambda} \left[\frac{v'^2}{g\left(1-(\rho''/\rho')\right)} \right], \quad Fr^* = \frac{W''^2}{g\delta} \cdot \frac{\rho''}{\rho'}$$

In the diapason $Fr^* = 5 \times 10^{-2}/6 \times 10^{-1}$, regularities of heat exchange at condensation both in the slug and bubble regimens have close characteristcs and can be calculated by the relation

$$Nu^* = 0.1 \, Pr^{0.54}. \tag{9.23}$$

Some intensification of heat exchange, observed at values of complex $Fr^* > 0.6$, is connected with active interaction of ascending (in the flow kernel) and descending (near the wall) flows of heat carrier. The comparison of received results with data of hydrodynamic research of counterflow phase movement in the conditions of CTT [14–16] show that the beginning of the increased heat-exchange intensity corresponds to the beginning of the flooding regimen, which is characterised by the beginning of the sharp increase of hydraulic coefficient of resistance to the vapour flow. The maximum of relation $Nu^* = f(Fr^*)$ is observed at complex values Fr^*, which corresponds to the reduced speed of vapour flow on the input into condensation area in the moment of hanging of the liquid phase of the heat carrier. Further increase of reduced vapour speed promotes the accumulation of liquid phase on the condensation area, which leads to the sharp decrease in intensity of heat exchange and stopping of normal work of the thermosyphon.

During the operation of vacuum thermosyphons, the emission of noncondensable gases from the heat carrier or casing is possible. The presence of a relatively small admixture of gases can sharply decrease the intensity of heat exchange at condensation. This is in keeping with the notion that the noncondensable gases in the current temperature interval create the additional diffusion resistance to the vapour transfer to the heat-exchange surface. But at the defined speed, the vapour flow influences noncondensable gases dynamically, which leads to the gas edging into the upper part of the condenser. It is known that the passage area between vapour and air zones is too narrow which is why the clean vapour condensation is possible in the biggest area of the condenser. At large degrees of heating zone filling by the heat carrier and at defined heat load, the vapour flow carries part of the liquid into the upper part of condenser, which separates the noncondensable gases from vapour and so clean vapour will be located under the liquid plug.

9.3.4 Limiting Heat Fluxes

To determine the maximum heat-transfer capabilities of two-phase thermosyphons, systematic experimental research was conducted into the intensity of heat flow in a cross section of the apparatus limited by crises of heat exchange. As a result of the experiments, the influence of each of the following determining parameters was indicated: the degree of liquid filling of the cavities, the geometric dimensions of individual parts of the thermosyphon, the pressure, angle slope of the thermosyphon to horizon and type of working fluid.

The performance of systematic research into maximal heat flow in CTTs bears evidence of the fact that the heat-transfer capabilities of these structures may be limited by crisis occurrences of different physical natures.

Analysis of the specifics of these crisis occurrences allows us to place them in two groups:

1. crises connected with the amount and distribution of the liquid phase in the cavities of the thermosyphon
2. crises connected with the hydrodynamic processes of mutual inter-action of the liquid and steam phases of the working fluid

In the first of these groups, we include crises of heat exchange as the work-ing fluid transfers to single-phase state and similarly crises arising from a reduced level of the dynamic two-phase layer and evaporation of the liquid film in the lower part of the heating zones. Quantitative characteristics of these crises in heat transfer may be calculated with reasonable accuracy on the basis of a generalised thermodynamic analysis of the state of the working fluid and, in the same way, of the mechanism of the change in level of the dynamic two-phase layer represented above. Together with this, the analysis of conditions at the onset of crisis phenomena of this group allows us to conclude that it might be eliminated by simple organisation of heat-exchange processes through selecting suitable values for operational parameters (degree of filling, conditions of heat exchange in the cooling section, etc.).

Under these conditions, the maximum heat-transmitting capacity of two-phase thermosyphons may be limited only by crisis phenomena of the second group (Figure 9.10), strictly associated with the principle feature of normal thermosyphons, determining the reciprocal action of the phase in the counterflow of liquid and steam in various parts of the two-phase system. In this, as shown by analysis of the experimental

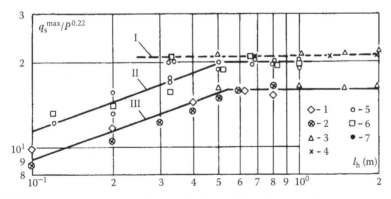

FIGURE 9.10
Comparison of maximum heat fluxes for different working regimes of thermosyphon (water is working fluid): (I) liquid film flow regime; (II,III) two-phase flow regimes ([II] slug-flow regime; [III] bubble-flow regime); (1) $d = 8$ mm; (2) 12 mm; (3) 20 mm; (4) 27 mm; (5) 25 mm; (6) 28.3 mm; (7) 33.2 mm.

results, the regularity of crises of heat exchange in general depends on the geometric dimensions of the thermosyphon and on its mode of operation (liquid film flow mode or bubbling mode). At the same time, as seen in Figure 9.10, in the range of change in geometric measurements important in practice for industrial applications the values of maximum heat flow are self-similar, both with regard to determining the measurements and with regard to the mode of operation of the thermosyphon. In this range, the maximum heat flow q_s^{max} depends only on the pressure in the thermosyphon and on the type of working fluid (Figure 9.11).

Under these conditions, the maximum heat-transmitting capability of a thermosyphon can be restricted only by the processes, which run in the adiabatic area in the conditions of countermovement of the liquid film and vapour flow. At use (as an intermediate heat carrier) of low-friction fluids, such conditions

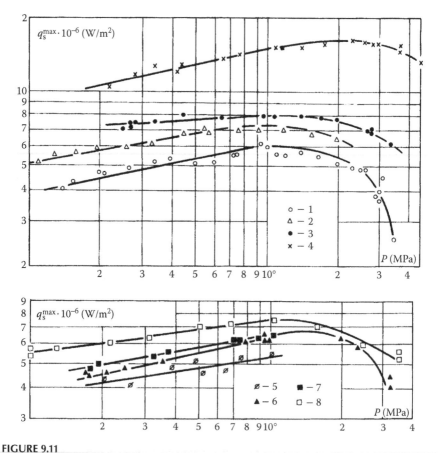

FIGURE 9.11

Experimental results of q_s^{max} for the thermosyphons in the field of a self-similarity with regard to geometrical measurements: (1) R-11; (2) toluene; (3) acetone; (4) methanol; (5) R-113; (6) tetrachloride carbon; (7) ether; (8) benzene.

take place in the thermosyphons with length of heating zone $l_h > 0.5$ m. The corresponding to these limiting heat flows are the upper boundary of maximum heat-transmitting capability of classic thermosyphons.

9.3.5 Upper Boundary of Heat-Transmitting Capability of Thermosyphons

The analysis shows that the corresponding to the upper boundary of the heat-transmitting capability of thermosyphons limiting heat flows are conditioned by crisis events 'flooding' in the counterflow movement of liquid and vapour in the vertical channels. These events are well studied and elucidated in the literature concerning gas–fluid systems of such type as water–air in the opened vertical channels. The peculiarities of the development of corresponding hydrodynamic processes in closed systems, which the closed thermosyphons are, are shown in detail in Ref. [17]. In this work, it is shown that the event flooding is followed by crisis events, characterised by different stages of the process of interaction of counterflows of vapour and liquid phases (Break-down of drops frustration from the film surface, 'hanging' of the liquid film and ejection of the fluid from the flow kernel into the upper part of the channel). There is an interest from the researched crisis events in the counterflow phase movement that are considerable for heat-transmitting processes in the concrete constructions of the CTTs from the point of view of restriction of their maximum heat-transmitting capability.

It is obvious that the condition of normal functioning of thermosyphons as a heat-transmitting device is the equality of axial mass rates of vapour and liquid phases. The disturbance of this equality must lead to the sharp change of heat-transmitting properties of the thermosyphon. In this connection, there are essential processes of heat transfer those from the crisis phenomena in processes of interaction of phases, which can lead to infringement of balance of mass flow rates of steam and liquid. From this point of view, it can be predicted that crisis events, connected with break down of drops of a liquid from a film surface are considerable for processes of limit heat transfer only in such construction of thermosyphons in which the possibility of eviction from the heating zone of excess fluid has place more than quantity, determined by equilibrium of the vapour and liquid phases in counterflow movement.

Such conditions can take place in the presence of the upper part of the thermosyphon of a big tank. In closed thermosyphons of the simplest construction (with continuous cross section on different areas), the frustration and carry-over of the fluid as drops leads only to the additional circulation of fluid in the closed space of thermosyphon. Fundamental failure of the structure of two-phase flow is connected with the beginning of hanging of the fluid film, which results in the sharp increase of hydraulic resistance to the vapour flow. As shown in the Ref. [14], in relation to pipe diameter (or number Bo), different regimens of movement of two-phase flow can take place: slug, transitional (emulsion) and annular.

The analysis of experimental data showed the presence of different regimens of movement in the moment of sharp change of the structure of two-phase flow leads to the different values of maximal heat flows in the thermosyphon.

The generalisation of experimental data due to maximum heat fluxes in the area of slug regimen of movements of two-phase flow is shown in Figure 9.12 (at Bo < 10). From Figure 9.12, we can see that the experimental data for different intermediate heat carriers are well generalised in the coordinates $K \rightarrow K_p$ and are approximated by the equation

$$K = CK_p^n \tag{9.24}$$

where

$C = 8.2;$ $n = -0.17$ for $K_p \leq 4 \times 10^4$
$C = 1.35;$ $n = 0$ for $K_p \geq 4 \times 10^4$

The comparison of Equation 9.24, which characterises the crisis of heat transfer into the area of slug regimen, with gained in relationships [17] for crises of a current two-phase flow shows that it fully coincides with the equation, describing the experimental data due to crisis of flow, connected with hanging of fluid film in vapour flow. This coincidence points to the physical nature of heat-transfer crises, which appears in sharp failure of conditions of heat exchange on the area of heat supply of the thermosyphon. During this, the crises events in the processes of heat and mass transfer are developed in the following way.

While reaching the critical speed of vapour phase, corresponding to the event hanging of fluid film, the hydraulic resistance to the vapour flow on the adiabatic area of thermosyphon increases sharply. In turn, the increasing of hydraulic resistance holds to the decreasing of outlay of vapour phase and

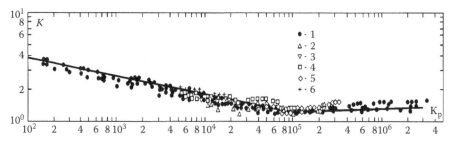

FIGURE 9.12
Generalisation of experimental data according to the crises of heat and mass transfer in closed two-phase thermosyphon in the flooding regime (slug-flow regime, Bo < 10); (1) water; (2) ethanol; (3) methanol; (4) R-11; (5) R-12; (6) R-113.

corresponding increasing of vapour slugs in the area of heat supply. As the result of this event, the evaporating and frustration of wall liquid film occur, which cause sharp degradation of heat-exchange conditions on the surfaces of heating.

At the values of Bo > 10 in the second stage of the regimen, flooding is observed to be transitional and further at Bo $\gtrless$ 30 annular regimen, of running of two-phase flow is observed. Generalisation of experimental data due to maximum heat fluxes q_s^{max} in the areas of pointed regimens is shown in Figure 9.13. The experimental data for different operational fluids is described by the equation

$$K = CK_p^n \qquad (9.25)$$

where
$$C = 10.2; \quad n = -0.17 \quad \text{for} \quad K_p \leq 4 \times 10^4$$
$$C = 1.7; \quad n = 0 \quad \quad \text{for} \quad K_p \geq 4 \times 10^4$$

Equation 9.25 fully coincides with the equation in Ref. [15], characterising crisis events and connected with sharp ejection of the fluid from the flow kernel (on adiabatic area) into the upper part of thermosyphon.

Hereby, the crisis of heat transfer in thermosyphons in relatively big diameters (Bo > 10) corresponds to the critical speed of vapour phase in the moment of transfer from foamed to the annular co-current regimen of the two-phase running. Here, the crisis events in the processes of flow and heat exchange are connected by the following method. With the increasing of the vapour speed on the adiabatic area of the thermosyphon to the value corresponding to hanging of fluid film, the two-phase system of the thermosyphon (due to pulsations of the phases rate) is in such quasi-stationary condition at which aggregated balance of mass rates of vapour and liquid is not disturbed. The wall surface in the heating zone continues to be washed by a two-phase mixture of intermediate heat medium without decreasing of heat-exchange

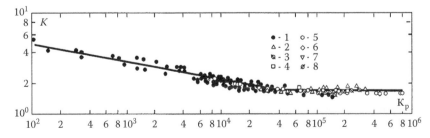

FIGURE 9.13
Dependence $K = f(K_p)60°$ for emulsion and annular flow regime (Bo > 10): (1) water; (2) ethanol; (3) R-12; (4) R-11; (5) methanol; (6) benzene; (7) tetrachloride carbon; (8) R-113.

intensity. By further increasing the vapour speed, the flow crisis happens, which disturbs the balance of mass rates of liquid and vapour. The specified crisis of current is accompanied by pipe unwatering in the heating zone, leading to a sharp decline of conditions in heat exchange.

9.3.6 Thermodynamic Similarity of Crisis Phenomena in CTTs

The generalisation of experimental results shown above may be used for working fluids with known thermophysical characteristics. For a range of working fluids, these characteristics are unknown. In these cases, a general functional dependency for the calculation of maximum heat fluxes may be similarly constructed on the basis of knowledge of only the major thermodynamic characteristics of a substance [18]. Presupposing a self-similarity of impulses of critical heat flow with regard to properties and the composition of heating surfaces, this dependency appears as

$$\frac{q_{cr}}{P_{cr}}\sqrt{RT_{cr}} = f\left(\frac{P}{P_{cr}}\right)\frac{C_{V_o}}{R_\mu} \tag{9.26}$$

where P_{cr} and T_{cr} are critical working parameters of the liquid; R_μ is the universal gas constant; and C_{V_o} is the molar heat capacity of the substance in ideal gas state.

The results of the generalisation of experimental data for maximum heat flow in thermosyphons on the basis of functional dependence in Equation 9.26 are presented in Figure 9.14. It follows to note that liquids used in experiments have substantially varying critical coefficient values, which characterise the degree of thermodynamic similarity of these substances. Nevertheless, the generalisation obtained may be considered fully satisfactory. This supports the known assertion of the approximate thermodynamic similarity of all liquids and allows us to use the general dependence relation obtained to evaluate the maximum heat flow in thermosyphons for other working fluids [19].

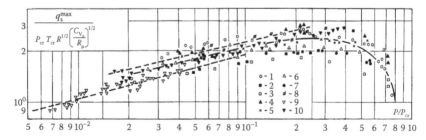

FIGURE 9.14

Generalisation of the maximum heat flux experimental results by the method of thermodynamic similarity: (1) R-11; (2) ether; (3) benzene; (4) tetrachloride carbon; (5) methanol; (6) toluene; (7) acetone; (8) R-113; (9) water; (10) ethanol.

9.4 Concluding Remarks

In summary, it is necessary to note the following. The fundamental questions of an effective operation of the CTTs include the features of behaviour and development of interconnected thermodynamic, hydrodynamic and heat-transfer processes in the closed cavity of a thermosyphon. As appears from the material set forth above, effective work of the CTT depends on both external conditions of heat supply and heat removal defining temperature level of behaviour of internal processes and internal conditions of their implementation (geometrical characteristics of the closed cavity, physical properties of intermediate heat-transfer agent and relative quantity of working fluid). Under the set and adjustable external working conditions, efficiency of the behaviour of internal heat-transfer processes is defined by internal parameters of the device. In the given work, the summary of the basic characteristics of these processes for thermosiphons of the classical form is described. Questions of an intensification of these processes, work of thermosyphons in special conditions and also questions of increase in heat transport ability of thermosyphons with the organised circulation of intermediate heat-transfer agent are stated in the monograph in detail [20].

9.5 Nomenclature

d_{in}	internal diameter, m	Ar	Archimedes number
l_h	length of heating zone, m	Fr*	modified Frud's number
q_s^{max}, q_*	heat flux density, W/m²	K	Kutateladze number
r	heat of vapourisation, J/kg	Nu	Nusselt number
t_s	temperature of saturation, °C	Pr	Prandtl number
P, P_{cr}	pressure, critical pressure, respectively, N/m²	Re	Reynolds number
v'', v', v_{cr}	specific volume of vapour, liquid and critical, respectively, m³/kg	K_p	pressure criteria
W	velocity, m/s	**Greek Letters**	
T_W	temperature of wall, K	α	heat-transfer coefficient, W/(m² · K)
T_s	temperature of saturation, K	δ	value of Laplace, thickness, m
T/T_{cr}	reduced temperature	λ	heat conductivity, W/(m · K)
G_1, G_v, G_{cr}	mass of liquid, vapour and critical mass of heat medium, respectively, kg	ρ'', ρ'	densities of vapour and liquid, respectively, kg/m³
V_1, V_v, V_{mix}	the volume of liquid, vapour, mixture, the heat zone and full thermosyphon, m³	σ	coefficient of the surface tension, N/m

(Continued)

V_h, V_a, V_o	the volume of heat zone, adiabatic zone and full thermosyphon, m^3	ν	kinematic viscosity coefficient, m^2/s
C_{V_0}	molar heat capacity, $J/(mol \cdot K)$	ε	degree of filling by volume
R_μ	universal gas constant, $J/(mol \cdot K)$	Ω	coefficient of the surface tension, N/m
R	gas constant, $J/(kg \cdot K)$	$\bar{\varphi}$	the average volume vapour contents

References

1. Sedelkin V.M., Bezrodny M.K. The investigation and method of the calculation of the two-phase thermosyphon for fire heaters of the absorbers. In: *The techniques and technology of the gas extraction and the exploitation of underground gas warehouses*. Moskow: WNIIEGAZPROM, 1/7, 1979, pp. 14–25.
2. Gavrilov A.F., Lyah V.Y. Air heaters with intermediate heat carrier. *Teploenergetika* [Thermal Engineering], 1965, No 3, pp. 11–17.
3. Bezrodny M.K., Volkov S.S., Ivanov V.B., Petrov V.N. New energotechnological system of heat utilization in non-ferrous pyrometallurgical plants. *Journal of Heat Recovery Systems*, 1985, Vol. 5, No 2, pp. 127–131.
4. Isakeev A.I., Kiselev I.G., Filatov V.V. *Effective ways of cooling the power semiconductor devices*. Leningrad: Energoizdat, Leningrad Department, 1982.
5. Bezrodny M.K., Volkov S.S., Ivanov V.B. Thermosyphon waste heat boilers for exhaust gases from furnaces in non-ferrous metallurgy. *Heat Recovery Systems and CHP*, 1990, Vol. 10, No. 2, pp. 99–105.
6. Vasiliev L.L., Kiselev V.G., Matveev Y.N., Molodkin F.F. Heat exchangers based on heat pipes for heat recovery. Minsk: Nauka I Technika, 1987.
7. Bezrodny M.K., Kondrusik E. Thermosyphons for industrial heat transfer: Fundamental questions for effective applications. *Archives of Thermodynamics*, 1994, No. 1–4, pp. 21–39.
8. Bezrodny M.K., Alekseenko D.V. Heat exchange research at boiling of liquids in the closed two-phase thermosyphons. *Izvestija VUZov USSR Energetika*, 1976, No. 12, pp. 96–101.
9. Bezrodny M.K., Alekseenko D.V. Intensity of heat exchange on a boiling section evaporative thermosyphons. *Teploenergetika* [Thermal Engineering], 1977, No 7, pp. 83–85.
10. Gross V., Hahne E. Condensation heat transfer inside a closed thermosyphon—Generalized correlation of experimental data. Proceedings of the 4th International Heat Pipe Conference, London, September, 1981, pp. 466–471.
11. Bezrodny M.K., Alabovskij A.P., Mokljak V.F. Research of average volume steam content of a dynamic two-phase layer in the closed thermosyphons. *Izvestija VUZov USSR Energetika*, 1981, No. 9, pp. 58–63.
12. Bezrodny M.K., Mokljak V.F. The heat exchange by the condensation in vertical closed thermosyphons. *Inzhenerno-Fizicheskii Zhurnal* [Journal of Engineering Physics and Thermophysics], 1986, Vol. 51, No. 1, pp. 9–16.
13. Kutatelsdze S.S. *Fundamentals of heat transfer theory*. Novosibirsk: Nauka, 1970.

14. Bezrodny M.K., Alabovskij A.N., Volkov S.S. Research of hydrodynamic characteristics of a two-phase stream in the conditions of the closed thermosyphon. *Izvestija VUZov USSR Energetika*, 1980, No. 2, pp. 116–121.

15. Bezrodny M.K. Flooding regime in a counterflow current of a film of a liquid and a gas stream in riser pipes. *Chim. i neft. Mashinostrojenije* [Chemical and Petroleum Engineering], 1980, No. 5, pp. 30–32.

16. Bezrodny M.K., Volkov S.S. Research of hydrodynamic characteristics of two-phase stream in a closed thermosyphons. *Air, fluid-flow and vapour cooling of power semiconductor devices and converting units on their basis.* Tallinn: Souzpreobrazowatel, 1982, pp. 119–125.

17. Bezrodny M.K. Hydrodynamic theory of heat transfer in crises in free-convective two-phase systems. *Promyshlennaja teplotechnika* [Industrial Heat Engineering], 2000, No. 5–6, pp. 41–49.

18. Mostinskj J.L. The application of the law of respective conditions for the calculation of the heat transfer and critical heat fluxes by boiling of the liquid. *Teploenergetika* [Thermal Engineering], 1963, No. 4, pp. 66–71.

19. Bezrodny M.K., Alekseenko D.V., Kazhdan A.Z., Volkov S.S. The generalization of experimental data on limit heat transfer in two-phase thermosyphons by method of thermodynamical similarity. *Izvestija VUZov USSR Energetika*, 1980, No. 7, pp. 121–124.

20. Bezrodny M.K., Pioro I.L., Kostyuk T.O. *Transfer processes in two-phase thermosyphon systems. Theory and practice.* Kiev: Fact, 2005.

10

Thermal Control Systems with Variable Conductance Heat Pipes for Space Application: Theory and Practice

V. M. Baturkin

CONTENTS

10.1 Introduction 357
10.2 Survey of Heat and Mass Transfer Research of Variable
 Conductance Heat Pipes 358
10.3 Main Heat Balance Correlations in Passive TCSs for
 Thermostating Autonomous Devices 373
    10.3.1 Principles of Operation of TCSs on GLHPs for
 Autonomous Devices 376
10.4 Integration of Heat Pipe TCS in PCB Design...............................386
10.5 Thermal Control of Devices without Heat Generation 393
10.6 Thermal Control of Small Satellite Compartments 398
10.7 Conclusions...399
Acknowledgements 405
Nomenclature 406
References..406

10.1 Introduction

This chapter describes the scientific and applied problems of creation of high-performance thermal control systems (TCSs) for equipment operating in non-hermetic compartments of a spacecraft. In such systems, heat pipes with variable thermal conductance perform the heat transport function together with controlling the device temperature, keeping the heat balance in the 'mounting place of device–device–space environment' system at the required device temperature level. Heat balance equations are the basis for defining the thermal parameters of constituent elements of the system: heat pipe, radiator, thermal insulation, flexible elements, low conductance stand-offs, cables and contact resistances, providing the function of passive temperature regulation. Proposed conceptions of systems with heat pipes do not foresee subsidiary electric power consumption of a spacecraft (SC), and they are based on the use of own inherent generation of scientific devices or heat from available sources (e.g. the sun), providing stabilisation of device temperature on the level of 20°C. Elaborated thermal schemes are experimentally proofed for groups of scientific devices, an individual electronic card (with a mass of 0.3 kg), an autonomous electronic unit (mass up to 5 kg) and a device panel of a spacecraft compartment (mass of 60 kg), at the change of inherent heat generation by 10 times, spacecraft temperature from −20°C to +50°C and illumination with external sources (the sun, the earth) to 270 W/m².

10.2 Survey of Heat and Mass Transfer Research of Variable Conductance Heat Pipes

This section summarises the achievements in the field of heat and mass transfer processes for heat pipes with variable properties, such as gas-loaded heat pipes (GLHPs). The main outputs in this field are generalised to define basic features important to the design of variable conductance heat pipes (VCHPs).

The development of space thermal control techniques is tightly connected to the achievements of heat pipe science and technology during the past 40 years (Bienert et al., 1971, Marcus 1972, Marshburn and McIntosh 1978, McIntosh and Ollendorf 1978, Delil and van der Vooren 1981, Antoniuk 1987, Bobco 1989, Faghri 1995, Edom 2001, Goncharov et al., 2007). As the main elements of a TCS (other names are thermal management system, cooling system, thermostating system and thermoregulating system), different modifications of heat pipes are used, which may differ in the type of capillary structure (CS) used, in operation principles and in temperature-regulating mechanisms. Conditionally, in terms of heat pipe application it is possible to mark out the following groups: conventional heat pipes; VCHPs and thermal diodes (TDs), which can be characterised by their thermal conductance K (W/K) = $Q/(T_{ev} - T_{con})$ or thermal resistance R(K/W) = $(T_{ev} - T_{con})/Q$, where W is Watt, unit of transferred power, K is degree Kelvin. (Figure 10.1).

Conventional heat pipes are usually intended for heat transport from a heat source to a heat sink in one direction. In certain regimes, they can operate in both directions and have several heat sources and heat sinks along the heat pipe length. Heat pipe thermal conductance, K_{HP}, should be high (up to 100 W/K) in order to transfer heat power Q at the minimal temperature difference between T_{ev} and T_{con}.

VCHPs may regulate their thermal conductance passively (without inputting additional electrical power). The most usable type of VCHP is GLHP with non-condensable gas (NCG), which essentially reduces the intensity of vapour condensation in a certain part of the GLHP. Such heat pipes transfer heat from heat source to heat sink in one direction. Variation of conductance,

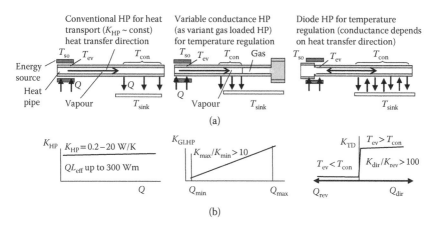

FIGURE 10.1
Schemes of conventional, variable conductance and diode heat pipes (a) and heat conductance dependencies on power transferred (b): T_{ev} – temperature of heat input zone (evaporator); T_{con} – temperature of heat output zone (condenser); T_{so} – temperature of heat source; T_{sink} – temperature of heat sink; L_{eff} – effective length of heat transfer; Q – heat power.

K_{GLHP}, can reach more than one at change of power and temperature difference between T_{so} and T_{sink}.

TDs have high conductance at direct heat transfer, K_{dir} ($T_{so} > T_{sink}$), and very small ones when $T_{so} < T_{sink}$ (reverse regime, K_{rev}). The ratio K_{dir}/K_{rev} reaches 100 and more. VCHPs and TDs typically have less heat transfer capability in comparison with conventional heat pipes.

Intensive investigations of heat and mass transfer processes in GLHPs are being conducted in many scientific centres all over the world, as GLHPs are very promising. The combination of different mechanisms of heat and mass transfer such as evaporation, condensation on the CS surface, conductivity in solid shell and diffusion vapour–gas along the heat pipe length, thermodynamics of gas, filtration of a liquid in porous medium in the presence of solute gas and rectification of vapour space from gas have been studied on a definite level of understanding in order to obtain an engineering model of GLHP operation. This successful activity has created a background for subsequent introduction of this self-regulating heat transfer instrument into complex TCSs for space applications.

The main feature of a GLHP is its ability to change the thermal conductance between the source of energy and the heat sink according to the variation of thermal parameters of the thermal network, keeping the heat balance at the prescribed temperature level. This mechanism is based on variation of condensation area in the heat removal zone. The distinctive character of a typical regulating function, namely the temperature of a vapour (or the temperature of a heat source), of transferred power and sink temperature allows the creation of manifold TCSs, which have a large track of successful exploitation in space (Kirkpatrick and Marcus 1972, Marcus 1972, Anderson et al.,

1974, Groll and Hage 1974, Mock et al., 1974, Edelstein and Flieger 1978, Lanteri and Henning 1988, Dvirnyi et al., 1999). The principle of GLHP operation and main correlations are described in plentiful detail in fundamental literature sources (Marcus 1972, Bienert 1972, Vasiliev et al., 1973, Vasiliev and Konev 1977, Shimoji and Kimura 1978, Brennan and Kroliczek 1979, Delil and van der Vooren 1981, Furukawa 1983, Antoniuk 1987, Bobco 1989, Faghri 1995). The most important results of GLHP investigation are summarised here:

1. A heat pipe has the following parts (Figure 10.2): evaporator (length = L_{ev}), adiabatic zone (T_{ad}, L_{ad}), condenser (L_{con}), connecting tube (optional; T_t, L_t) and reservoir (T_{res}, L_{res}). All these elements have a common vapour space and CS for liquid interconnection. In the inner vapour space, a certain mass of an NCG is added (m_g). The regulating mechanism of GLHP is based on the possibility of collecting the NCG in the coldest inner volume of the heat pipe (in adiabatic zone, condenser, tube and reservoir) that extremely reduces the intensity of vapour condensation. Due to the movement of vapour from the heat pipe's hottest part (evaporator) to the condenser a pure vapour mixture (vapour + gas) is formed at the boundary, and this boundary may move along the heat pipe length. This area, which is free from gas, is available for effective vapour condensation (as in conventional heat pipes), and it may be enlarged or reduced with the increase or decrease of vapour temperature. A flat 'vapour–gas' front assumption foresees a sharp change in gas concentration and heat exchange intensity in the passage from the gas-free zone to the vapour + gas mixture. Conditionally, the volume of the vapour + gas mixture has three distinctive lengths, L_g, which coincide with $L_g' = L_{ad} + L_{con}$; $L_g'' = L_{con}$ and $L_g''' = (0.1–0.2) L_{con}$ (Figure 10.2a). The regulation function 'vapour temperature in evaporator of applied heat power' has the following sectors (Figure 10.2b and c):

   Sector A: $T_{sink,min} < T_v < T_v'$ at $Q < Q_{min}$; the vapour–gas front has not been formed, and an increase in partial vapour pressure occurs in the evaporator, $P_v(T_{ev})$, and it becomes equal to the total pressure, P.

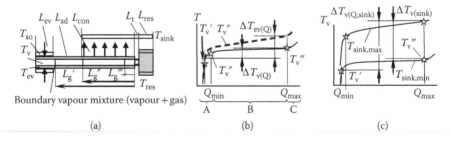

FIGURE 10.2
Scheme of vapour–gas front movement in (a) gas-loaded heat pipe (GLHP) and (b) GLHP regulating function evaporator and vapour temperature of transferred power and of (c) T_{sink}.

Sector B: $T'_v < T_v < T'''_v$ at $Q_{min} < Q < Q_{max}$, regulating temperature range, $T'''_v - T'_v = \Delta T_v$, within power $Q_{min} - Q_{max}$; this is nominal regulation diapason, and the vapour–gas front moves into the adiabatic and condenser zones (Figure 10.3a and b).

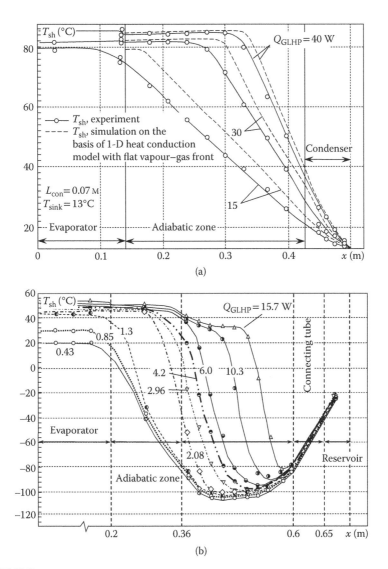

(a)

(b)

FIGURE 10.3

Shell temperature (T_{sh}) distribution in a heat pipe: (a) copper–water–argon heat pipe with forced liquid cooling when the vapour–gas front moves into adiabatic zone; (b) stainless steel–acetone–argon heat pipe with radiation cooling, when front is not formed ($Q_{GLHP} = 0.43$ and 0.85 W), moves into adiabatic zone ($Q_{GLHP} = 1.3$ and 2.96 W) and moves into condenser ($Q_{GLHP} = 6$–15.7 W).

Sector C: $T_v > T_v'''$; the vapour–gas front moves to the condenser end and even enters the reservoir, and the regulation of vapour temperature ceases.

The accuracy of GLHP regulation is characterised by a rise in vapour temperature, $\Delta T_{v(Q)}$, and evaporator temperature, $\Delta T_{ev(Q)}$, at a heat flux variation of $Q_{min} \rightarrow Q_{max}$ (Figure 10.2b); by a change of vapour temperature, $\Delta T_{v(sink)}$, at a sink temperature change of $T_{sink,min} \rightarrow T_{sink,max}$ and by $\Delta T_{v(Q,sink)}$, when Q and T_{sink} are changing (Figure 10.2c).

The length of the gas-filled zone L_g in the flat vapour–gas front theory depends on the following parameters:

$$\frac{m_g}{M} = \frac{[P - P_v(T_{res})]V_{res}}{R_\mu T_{res}} + \frac{[P - P_v(T_{g,con})]F_{con}L_{con}}{R_\mu T_{g,con}}$$

$$+ \frac{[P - P_v(T_{g,ad})]F_{ad}\left(L_g - L_{con}\right)}{R_\mu T_{g,ad}}; \ L_g' > L > L_g'' \tag{10.1}$$

$$\frac{m_g}{M} = \frac{[P - P_v(T_{res})]V_{res}}{R_\mu T_{res}} + \frac{[P - P_v(T_{g,con})]F_{con}L_g}{R_\mu T_{g,con}}; \ L_g'' > L > L_g''',$$

where F_{con} and F_{ad} are the cross sections of vapour channel in the condenser and adiabatic zones respectively; $T_{g,ad}$ and $T_{g,con}$ are average temperatures in the zone filled with gas in the adiabatic and condenser zones respectively; P is pressure in the vapour space of GLHP, which is constant along its length; $P_v(T_g)$ is partial vapour pressure at the temperature T_g; M is the gas molecular mass; m_g is the mass of gas R_μ is universal gas constant V_{res}, $V_{con} = F_{con}*L_{con}$— inner vapour/gas volume of reservoir and condenser, respectively.

2. The minimal heat flux, Q_{min}, is defined by two components: heat transfer by conductance in shell and CS (Q_{cond}) and heat transfer by vapour diffusion (Q_{dif}), that is, $Q_{cond} \sim 1$ sh$*F_{sh}*$grad (Tsh) and $Q_{dif} \sim D_{v-g}*F_v*$grad ($T_v$) where F is cross section, D_{v-g} is coefficient of diffusion of vapour into gas and grad is gradient of temperature. Some recommendations on the calculation of Q_{cond} and Q_{dif} are given by Konev (1976), Baturkin (1979), Barsukov (1981), and Edom (2001). The value of Q_{max} is defined by the ability of the condenser to remove the heat from the condenser shell to the sink at the active length $L_a = L_{con} (1 - L_g'''/L_{con})$ and at vapour temperature T_v'''. This value $Q_{max} \sim L_a (T_v''' - T_{sink})$ TETA, where TETA – thermal efficiency of heat removing elements. The used value of L_g''' should guarantee that the reservoir temperature does not exceed the assumed value of $T_{res,max}$.

3. There are several modelling approaches for GLHP temperature field prediction: (1) flat front vapour mixture, (2) flat front vapour mixture with axial conductance in shell, (3) one-dimensional conductance

in shell and gas diffusion, (4) one-dimensional diffusion with initial effects, and (5) two-dimensional diffusion. The chronological survey of GLHP models is presented by Edom (2001). The models (1), (2) and (3) are the most usable for engineering calculations and can be realised in commercial codes, for example, the software products Systema 4.5.0 and NX7.5. The applicability of the simplification used in GLHP models may be estimated by elaborate criteria, and they are collected in the work of Edom (2001).

4. The minimisation of regulating temperature range $T_v'-T_v'''$ (in some works, $T_v''-T_v'''$) requires the use of a reservoir, which reduces the variation in vapour pressure, P, inside the heat pipe. The following are the main schemes of GLHP, which differ in design and temperature regime of the reservoir (Figure 10.4):

- Reservoir has CS (schemes 1–3): it may accept the temperature of sink or have a higher temperature (scheme 1), have a constant temperature (scheme 2) or may undergo reduction in temperature with rise in power. Partial gas pressure in the reservoir is $P_{g,res} = P - P_{v,res}(T_{res})$, where $P_{v,res}(T_{res})$ is the partial vapour pressure at the temperature T_{res}. This value depends on heat sink temperature variation and on heat balance for the reservoir.

- Reservoir does not have CS (schemes 4–6): it may accept the temperature of sink or have a higher temperature; gas pressure in the reservoir is $P_{g,res} = P - P_{v,res}$, where $P_{v,res}$ is partial vapour pressure in the reservoir, which can be the function of the condenser end temperature (scheme 4), can be omitted or may have some undefined value (scheme 5). Reservoir temperature, T_{res}, is dependent on heat sink temperature variation and on heat balance for reservoir.

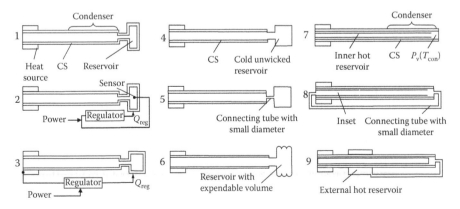

FIGURE 10.4
Gas-loaded heat pipe configurations: (1,2,3) with capillary structure in reservoir; (4,5,6) with cold reservoir without capillary structure; (7,8,9) with hot reservoir without capillary structure.

- Reservoir does not have CS (schemes 7–9): it accepts the temperature of a vapour. Partial gas pressure in the reservoir is $P_{g,res} = P - P_v(T_{con})$, where $P_v(T_{con})$ is partial vapour pressure at the condenser end (scheme 7), which is the function of the condenser end temperature, T_{con}; partial pressure P_v can be omitted (schemes 8 and 9) or may have an undefined value.

5. During operation, the length of the gas-filled zone changes with the variation of any thermal parameter, which is inside the thermal line 'vapour temperature–heat sink'. The typical disturbances are the following: transmitted heat power temperature of the sink, heat exchange intensity between vapour and the condenser shell and heat exchange intensity between the condenser shell and sink. The total differential, dT_v, of these values was obtained by Bienert (1972). Another approach to analyse the temperature disturbances is based on the definition of the ratio of vapour volume of reservoir to vapour volume of condenser, V_{res}/V_{con}, or vapour volume of reservoir to gas-filled zone change, $V_{res}/\Delta V_g$, as the function of vapour and heat sink temperature variation (Marcus 1972).

In the studies by Marcus (1972), Vasiliev et al. (1973), Brennan and Kroliczek (1979) and Baturkin (1979), the comparative analysis of different schemes was conducted. It was shown that at constant sink temperature the basic schemes 1, 4 and 7 (Figure 10.4) are nearly identical at the value $V_{res}/V_{con} < 20$ used in practice. The value $\Delta T_v = 1°C - 5°C$ can be reached, and this value depends on parameters of the saturation curve, $p = f(T_v)$, for the working fluid used (Figure 10.5a and b). The recommended criterion for working fluid selection is the highest value of $d(\ln(P)/dT_v)$ (Marcus 1972).

When T_{sink} varies from $T_{sink,min}$ to $T_{sink,max}$, all heat carriers have the minimal reachable value of ΔT_v (best theoretical stabilisation) that is obtained at $V_{res}/V_{con} \rightarrow \infty$ (Figure 10.6a and b). For ethanol and methanol, the scheme with wicked reservoir can reach $\Delta T_v = 8°C–10°C$; for ammonia, $\Delta T_v = 18°C$ for $Q = var$, $T_{sink} = var = -40°C–0°C$, $T_{v,nom} = 20°C$ and $T_{v,nom} = (T_{v,max} + T_{v,min}) / 2$ is nominal or average vapour temperature.

The comparison of schemes 1, 2, 5 and 7 (Figure 10.4) for $Q = var$, $T_{sink} = var = -40°C–0°C$ leads to the following conclusions (Figure 10.7):

Maintaining reservoir temperature at the heat sink maximal level ($0°C \pm 1°C$) enables the possibility of achieving $\pm(1–2 \text{ grad})$ heat pipe vapour temperature stability. The scheme with a cold unwicked reservoir has a stability of $\pm(2–3 \text{ grad})$, but it may suffer a loss of accuracy if liquid enters the reservoir. The scheme with a hot unwicked reservoir provides a temperature stability of $\pm(3–4 \text{ grad})$, but it may suffer high temperature excursions if liquid imbues into the reservoir. The scheme with wicked reservoir is characterised

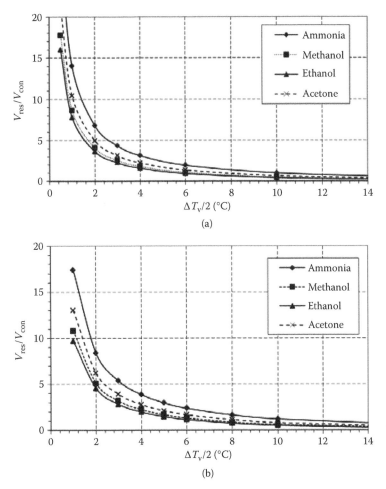

FIGURE 10.5
Requirements for the ratio V_{res}/V_{con} under the conditions Q = var, T_{sink} = const = $-40°C$ and $T_{v,nom}$ = $20°C$: (a) scheme with cold wicked reservoir and (b) scheme with hot unwicked reservoir.

by an accuracy of $\pm(5-6$ grad$)$ for methanol and $\pm(9-10$ grad$)$ for ammonia, but it gives stable results. At a greater difference between T_v and $T_{sink,max}$, the regulation diapason ΔT_v becomes smaller. A comparative analysis of technical characteristics of GLHP with different reservoir designs is summarised by Marcus (1972), Brennan and Kroliczek (1979) and Dunn and Reay (1976).

6. The heat source temperature T_{so}, the vapour temperature T_v and T_{ev} are connected by the following equation:

$$T_{so} = T_v + \frac{Q}{K_{so-v}} = T_v + \left(\frac{1}{K_{ev-v}} + \frac{1}{K_{cont,\,so-ev}} + \frac{1}{K_{so}} \right) Q = T_{ev} + \left(\frac{1}{K_{cont,\,so-ev}} + \frac{1}{K_{so}} \right) Q \quad (10.2)$$

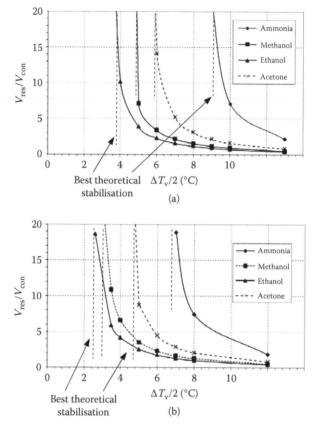

FIGURE 10.6
Requirements for the ratio V_{res}/V_{con} under the conditions Q = var, T_{sink} = var = $-40°C-0°C$ and $T_{v,nom}$ = 20°C: (a) scheme with cold wicked reservoir and (b) scheme with hot unwicked reservoir.

where K_{so-v} is the conductivity of line heat source–vapour, K_{ev-v} = $k_{ev}F_{ev}$ is the conductivity of phase change processes in heat pipe evaporator, k_{ev} is the intensity of heat exchange process in evaporator, F_{ev} is the area of heat exchange in evaporator, $K_{cont,so-ev}$ is the conductivity of contact between heat pipe shell and heat source and K_{so} is the conductivity of heat transfer line inside the thermostating object. If heat input to GLHP, Q, changes within the range $Q_{min} - Q_{max}$ and it can be assumed that $[(1/K_{ev-v}) + (1/K_{cont,so-ev}) + (1/K_{so})]$ = const, the variation of T_{so} can be defined as follows:

$$\Delta T_{so} = \left(\frac{1}{K_{ev-v}} + \frac{1}{K_{cont,so-ev}} + \frac{1}{K_{so}} \right) (Q_{max} - Q_{min}) + \Delta T_v. \qquad (10.3)$$

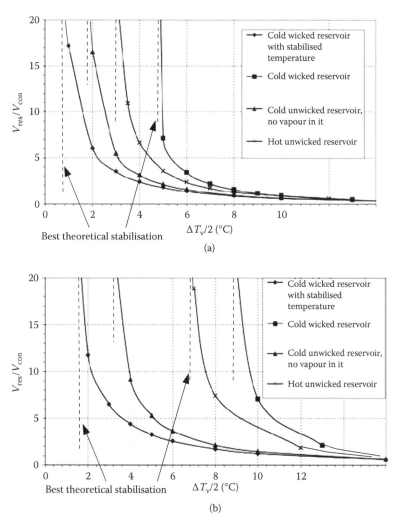

FIGURE 10.7
Ratio V_{res}/V_{con} under the conditions Q = var, T_{sink} = var = −40°C–0°C and $T_{v,nom}$ = 20°C: (a) for methanol and (b) for ammonia.

The complex $[(1/K_{ev\text{-}v}) + (1/K_{cont,so\text{-}ev}) + (1/K_{so})](Q_{max} - Q_{min})$ should be minimised. Otherwise, at low values of ΔT_v a wide variation of ΔT_{so} may take place and the effect of GLHP application will be diminished. A classification of methods of $K_{ev\text{-}v}$ increase for different types of CSs is presented by Smirnov and Tsoy (1999), and the methods of contact resistance improvement are given by numerous researchers (Gilmore 2002, Donabedian 2004, Electronics-cooling 2004).

7. For the device thermostating nominal temperature range of 10°C–70°C, the practice of GLHP usage in space missions recommends the following combinations of 'shell/CS–working fluid–gas':

    - Stainless steel–methanol–nitrogen (argon)
    - Stainless steel–ammonia–nitrogen
    - Stainless steel–ethanol–nitrogen (argon)
    - Stainless steel–acetone–nitrogen (argon)
    - Aluminium alloy–ammonia–nitrogen
    - Aluminium alloy–acetone–argon

    Technological aspects of GLHP fabrication are summarised by Anderson et al. (1974) and Brennan and Kroliczek (1979).

8. Constructive features: the design of GLHPs is more complicated than that of conventional heat pipes. New properties, which should be kept in mind when designing GLHPs, are the following:

    - The adiabatic zone should have low axial conductance to diminish the minimal heat flux, Q_{min}.
    - To have a more sharp temperature profile in the area active condensation zone–zone filled with gas, the radiator/condenser should have the minimal axial conductance that diminishes the unregulated heat flux, transferred by conductance along the heat pipe axis. In practice, this is realised by slitting and sectioning the radiator.
    - Designs of a high axial conductive radiator (non-slitted radiator) are in practice (Mock 1974, Dvirnyi et al., 1999, Edom 2001).
    - The cold area (trap) should be arranged at the end of the condenser to minimise condenser temperature and reduce vapour partial pressure at the entrance of the reservoir in the regime with maximal power.
    - At GLHP ground testing, free convection in the vapour–gas zone may increase the heat exchange between cold and hot parts of condenser and reservoir. To eliminate this undesirable heat exchange, the NCG for ammonia should be a neon–helium mixture (Cima and Abrose 1995).
    - The ratio V_{res}/V_{con} can be increased by preserving of condenser and reservoir external geometry, if an inset in adiabatic (transport) and condenser zones is installed (schemes 7–9, Figure 10.4). The longitudinal porous artery for liquid movement can be used as such an inset.

9. Software used for GLHP design and for GLHP model integration into space TCSs. The following are commercially distributed codes for GLHP modelling in itself and as a part of more complicated TCSs:

- One-dimensional axial conduction in heat pipe shell and vapour diffusion, heat rejection from condenser by convection and radiation (Edwards et al., 1973)
- Flat vapour–gas front assumption, one-dimensional axial conduction, heat rejection from condenser by convection and radiation; subroutine VCGHPDA in the SINDA/G thermal analyser (Antoniuk 1987, SINDA/G User's Guide 2003)
- Flat vapour–gas front assumption, one-dimensional axial conduction in heat pipe shell; subroutine named VCHP should be part of SINDA/FLUINT subroutine library (Matonak and Peabody 2006)
- Flat vapour–gas front assumption, one-dimensional axial conduction in heat pipe shell model integrated into the thermal analyser TMS, subroutines VCHP and VCHPD (ESATAN-TMS Thermal User Manual 2011)
- Flat vapour–gas front assumption, one-dimensional axial shell conduction model of GLHP; subroutine VCHPDA has been incorporated into the MSC Sinda thermal analyser (User's Guide MSC Sinda 2009)

GLHP models (except the ones discussed by Edwards et al. [1973]) define the axial one-dimensional temperature distribution in a heat pipe shell, alternating the position of the flat vapour–gas front to achieve a balance of power and NCG mass in the inner heat pipe volume. So the assumption of a flat vapour–gas front and neglecting the diffusion is the current status of software for GLHP simulation in commercial codes.

The main balance equations used are the following (Edwards et al., 1973, User's Guide MSC Sinda 2009):

- The total heat load, Q, on a GLHP is defined as

$$Q = \int_0^{L_{\text{GLHP}}} k_{\text{sh-v}}(z) \Pi_{\text{in}}(T_v, z)(T_v - T_{\text{sh}}(z)) \, dz \qquad (10.4)$$

where z is the coordinate starting from the end of condenser towards evaporator, T_v is vapour temperature in the active section of the heat pipe, $k_{\text{sh-v}}(z)$ is the coefficient of heat transfer between the vapour and the heat pipe shell, Π_{in} is the perimeter of heat exchange inside the heat pipe and $T_{\text{sh}}(z)$ is the temperature of the heat pipe shell along the heat pipe.

- For the definition of length of the gas-filled zone, L_g, the mass balance of NCG is determined by integrating the equation for distribution of gas partial pressure:

$$\frac{m_g}{M} = \frac{[P - P(T_{\text{res}})]V_{\text{res}}}{R_\mu T_{\text{res}}} + \int_0^{L_g} \frac{[P - P_v(z)]F(z)}{R_\mu T_{\text{sh}}(z)} \, dz \qquad (10.5)$$

where $F(z)$ is vapour channel cross-sectional area; $P_{v,res}$ and $P_v(z)$ are the partial vapour pressures of working fluid at T_{res} and $T_{sh}(z)$ in zone $z < L_g$ respectively and P is total pressure in the GLHP. The temperature of a vapour in a zone filled with gas, $z < L_g$, is equal to $T_{sh}(z)$. The saturated vapour pressure in the vapour space is defined as the functional $P_v = f(T_v)$, and temperature is in kelvin.

- The temperature profile in the heat pipe shell, $T_{sh}(z)$, is defined on the basis of the one-dimensional conductivity equation:

$$\left(\sum_{n=1}^{N} \lambda_n F_n \right) \frac{\partial^2 T_{sh}}{\partial z^2} - k_{sh\text{-}v}(z) \Pi_{in} (T_{sh} - T_v) - k_{ex} \Pi_{ex} \eta (T_{sh} - T_{sink})$$

$$- \varepsilon \sigma \Pi_{ex} \eta (T_{sh}^4 - T_{sink}^4) + q_{ab} \Pi_{ex} = \left(\sum_{n=1}^{N} c_{p,n} \rho_n F_n \right) \frac{\partial T_{sh}}{\partial \tau}, \tag{10.6}$$

where $\lambda_n F_n$ is the product of axial conductivity and area of the nth element (CS, heat pipe shell, flange); Π_{ex} and Π_{in} are the external and internal perimeters of heat exchange respectively; η and ε are efficiency and emittance of the external rib system respectively; σ is the Boltzmann constant; k_{ex} is the external convective heat transfer coefficient; $c_{p,n} \rho_n F_n$ is the product of specific heat capacity, density and area of the nth element (CS, heat pipe shell, flange). This equation should be transformed into a lump discrete scheme that enables the possibility of solving it using a thermal analyser. A typical example of GLHP simulation (User's Guide MSC Sinda 2009) includes 13 nodes, representing the heat pipe shell (1 node for evaporator, 1 for adiabatic zone, 1 for vapour in the zone free from gas, 1 for connecting tube, 1 for reservoir and 8 for condenser), and 9 nodes for the external radiator.

Several in-laboratory developed codes exist as well. They are used for detailed studies of GLHP, prediction of characteristics, solution of dedicated tasks and verification of scientific hypotheses. Practically every scientific centre and heat pipe manufacturer has developed their own GLHP software, which can model the specific features implemented in the GLHP design used.

10. Unsteady regime of GLHP operation: The calculation of the function $T_{sh} = f(\tau)$ is based on the solution of a system of differential balance equations of heat flux for heat source, evaporator and adiabatic zone, composed for three GLHP working periods with the following assumptions: (1) the vapour–gas is not formed and does not enter the adiabatic zone, $T_{sink} < T_v < T_v'$; (2) the front is moving into the adiabatic zone, $T_v' < T_v < T_v''$; and (3) the front is located in the

condensation zone, $T_v'' < T_v < T_v'''$. The model foresees that GLHP zones are objects with concentrated heat capacities and temperatures, thermal resistances are calculated according to the equations for stationary modes, and the main simplifications of the flat front model have been accepted. Heat flux is transmitted by thermal conductivity to the condenser in the first period; by vapour flow, and then by heat conductivity along the shell in adiabatic zone, in the second; and by vapour flow in the third. Heat balance equations are formulated for the vapour temperature, T_v, for each of these periods respectively:

$$Q = \frac{dT_v}{d\tau}(C_{ev}\chi_{ev} + C_{ad}\chi_{ad}) + \frac{T_v - T_{sink}}{\Omega'_{ad}}, T_{sink} < T_v < T_v';$$

$$Q = \frac{dT_v}{d\tau}\left[C_{ev}\chi_{ev} + C_{ad}\chi_{ad}\left(1 + \frac{L_{con}}{L_{ad}}\{1 - \chi_{ad}\}\{(1-\upsilon)\}\right)\right]$$

$$+ \frac{T_v - T_{sink}}{\Omega_{ad}(T_v)}, T_v' < T_v < T_v''; \qquad (10.7)$$

$$Q = \frac{dT_v}{d\tau}[C_{ev}\chi_{ev} + C_{ad}\chi_{ad} + C_{con}\chi_{con}(1-\upsilon)]$$

$$+ \frac{T_v - T_{sink}}{\Omega_{v\text{-sink}}}(1-\upsilon), T_v'' < T_v < T_v''';$$

where $C = \sum c_{p,i}\rho_i F_i L$ is the total heat capacity of the considered zone; χ_{ev}, χ_{ad} and χ_{con} are coefficients that take into account the distinction in temperatures of evaporator, adiabatic zone and condenser, respectively, compared with vapour temperature; $\Omega'_{ad} = L_{ad}/A_{ad} + [A_{con}f_{con}\text{th}(f_{con}L_{con})]^{-1}$ is the thermal resistance caused by thermal conductivity in adiabatic and condenser zones in the axial direction; A is the product of conductivity and cross-sectional area; $f_{con} = [A_{con}/(k_{ex}\Pi)]^{-0.5}$; $\Omega_{ad}(T_v) = L_{con}(\upsilon - 1)/A_{ad} + [A_{con}f_{con}\text{th}(f_{con}L_{con})]^{-1}$ is the thermal resistance caused by thermal conductivity in the adiabatic zone in axial direction, when the front moves to the adiabatic zone; $\upsilon = L_g/L_{con}$ is the dimensionless length of the gas-filled zone; and $\Omega_{v\text{-sink}} = (k_{ex}\Pi_{ex}L_{con})^{-1} + (k_{sh\text{-}v}\Pi_{in}L_{con})^{-1}$ is the thermal resistance of condenser between vapour and heat sink in the radial direction. These equations are solved together with a mass gas balance equation, in which it is assumed that inactive parts of condenser and reservoir have a constant temperature.

Duration of the first period, τ', exceeds the durations of the second and third periods due to greater temperature difference $(T_v' - T_{sink})$ than the small value of $T_v''' - T_v'$). The time to reach temperature T_v' can be defined by the following simplified equation:

$$\tau_{start} \approx \tau' = (C_{so}\chi_{so} + C_{ev}\chi_{ev} + C_{ad}\chi_{ad})\Omega'_{ad} \ln[1 - (T_v' - T_{sink})/Q\Omega'_{ad}]. \quad (10.8)$$

The collection of starting curves, $T_v = f(\tau)$, is presented in Figure 10.8a, and the summary of data for the duration of the first period is given in Figure 10.8b. For GLHPs with copper shell and forced liquid cooling of condenser, Equation 10.8 gives the prediction of τ' with less than +20% deviation relative to experiments.

A detailed study on GLHP unsteady regimes and proper models for start-up and transient performance is quoted in the literature (Bienert et al., 1971, Edwards et al., 1972, Schlitt 1973, Groll and Hage 1974, Baturkin 1979, Barsukov 1981, Dulnev and Belyakov 1985).

Important questions concerning unsteady GLHP operation are as follows:

Feedback control of reservoir (schemes 1 and 2 in Figure 10.4) (Bienert et al., 1971, Schlitt 1973, Groll and Hage 1974, Vasiliev and Konev 1977)

Gas diffusion into reservoir (schemes 4–9 in Figure 10.4) (Marcus and Fleischman 1970, Hinderman et al., 1971)

Gas movement in capillary, connecting balloon and condenser (schemes 5, 8 and 9 in Figure 10.4) (Baturkin 1979, Edom 2001)

Operation of GLHP during periodic change of power and heat sink temperature (Baturkin 1979)

Gas bubbles dissolving in a liquid (Antoniuk and Edwards 1980)

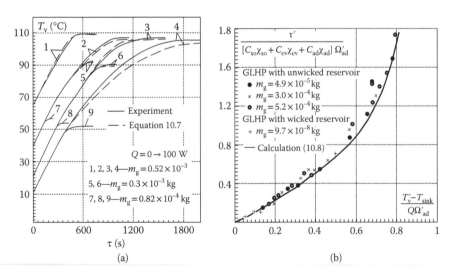

FIGURE 10.8
The start-up performance of a gas-loaded heat pipe: (a) function dependence of vapour temperature on time and (b) summary of data on duration of start-up regime.

Start-up of GLHP with a frozen heat carrier (Edwards et al., 1972, Gnilichenko 1982, Antoniuk 1987)

Pressure and temperature fluctuation in GLHP (Anderson et al., 1974, Edom 2001).

10.3 Main Heat Balance Correlations in Passive TCSs for Thermostating Autonomous Devices

In this section, the balance equations (heat energy and mass of NCG) are formulated and analysed on system and component levels to define the operation limitations of TCSs on GLHPs. The following are among the possible disturbances to be considered: variable power generation of a device, external heat fluxes from the sun and the earth, mounting place temperature changes and variation of thermal and optical properties of TCS elements. The practical application of the proposed passive principles for cooling and temperature regulation concerns: individual elements of equipment, autonomous device cases, electronic elements installed on printed cards, devices that do not generate their inherent power and isothermal mounting surfaces (places) for device montage.

Examples of heat pipe (and GLHP) implementation in space equipment have shown the successful realisation of this idea of thermal control. The mechanisms of temperature regulation of space devices by means of heat pipes are summarised in Figure 10.9.

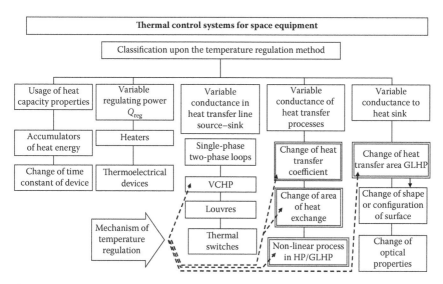

FIGURE 10.9
Classification of temperature regulation principles.

A GLHP acts as the heat conductive line between heat source and heat sink with variable thermal properties. The change of condensation intensity in the presence of an NCG and a vapour condensation area takes place. The non-linear processes on boiling in CSs may have a stabilisation effect as well. Variable radiation thermal coupling between condenser and heat sink is defined by a changeable area of radiation (in case of low axial condenser heat conductance). GLHPs are used as the effective means for device temperature stabilisation at its variable power generation in changeable external illumination conditions, which take place at winter and summer solstice and during satellite movement over the orbit, being illuminated with the sun and the earth and in shadow. The possible deviation in optical properties of coating and multi-layer insulation (MLI) due to impact of space factors or variation of thermal conductance (contacts) is likely to be considered as the reason for the change in device temperature. When device power is dramatically reduced (e.g. satellite safe mode, silent mode, calibration), GLHP increases its thermal resistance to heat sink, diminishing heat leaks outside the system. Another important feature of such systems is that device attachment to GLHP is characterised by a uniform temperature in the evaporator zone, as more than 95% of the heat generated is transferred to a vapour having a very homogeneous temperature.

GLHPs may prevent the overcooling of electronic components, being the alternative to automatically controlled heaters. A TCS with a GLHP creates the design basis for large-sized power-rejected systems (Savage and Aalders 1978, Savage et al., 1979), SIGMA x-ray telescope (International Astrophysical Observatory 'GRANAT' 2010) and communication satellites with a power range of 200–250 W. Further, they appear to be useful in autonomous optoelectronic devices with a power generation of 20 W. A summary of information on the temperature regulation of devices achieved by using TCSs on GLHPs is given in Figure 10.10. Some of these data are obtained at variable

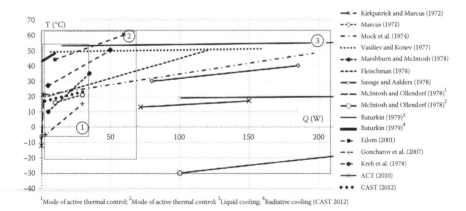

[1]Mode of active thermal control; [2]Mode of active thermal control; [3]Liquid cooling; [4]Radiative cooling (CAST 2012)

FIGURE 10.10
Temperature of a thermostating object as the function of device power generation.

heat sink temperatures. One can find that the minimal power of TCS operation is several watts and the ratio of 'power open/power closed' reaches 10–40. Three levels of regulated power (Figure 10.10) could be conditionally defined as follows: (1) up to 15–20 W, (2) 50–70 W, and (3) 200 W.

Some studies, for example, the one by Anderson et al. (2008), present the design of a GLHP radiator for 30 kW of rejected power at exploitation on lunar surface with a variable heat sink temperature between −63°C and 41°C (for lunar equator) and between −143°C and −60°C (for Shackleton Crater) to keep liquid temperature in the heat exchanger at the level of 100°C–130°C.

The integration of a GLHP into a passive TCS for space electronics foresees the solution of the following tasks: joint design of heat pipe and electronic unit, thermal and mechanical design of heat pipe, and thermal and mechanical design of additional elements of thermal schemes (as MLI, heat conductive lines, stand-offs, cables). As a result of successful design work, a TCS always operates at positive thermal balance of the system—the quantity of generated and incoming heat is enough to compensate all types of heat leaks. The boundary conditions (mounting places, spacecraft environment) have a dominant influence on components of device heat balance, and under certain conditions the positive effect of TCS operation can be lost.

The hierarchy of VCHP integration into optoelectronic equipment, described in literature, is summarised by Baturkin (2011) and illustrated in Figure 10.11.

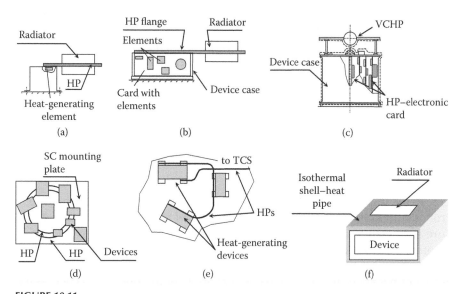

FIGURE 10.11
(a) Integration of a heat pipe for cooling and temperature regulation of individual elements; (b) thermal interface surface of device; (c) elements located on electronic plate and (d) group of devices with heat exchange via mounting plate, (e) with direct heat exchange and (f) with creation of isothermal shell.

The application of every scheme is defined by concrete exploitation conditions and usually requires complex preliminary analysis. Some remarks on the characterisation of these schemes are as follows:

a. Heat pipes provide the thermal regime of an individual element or a part of a device. Heat pipes have limited sizes of surface thermal interface with heat-generating elements or parts of devices. The thermal regime of the whole device is provided by another TCS.

b. Heat pipes provide the thermal regime of an autonomous device. Heat pipes have limited sizes of surface thermal interface with a device. A radiator removes heat from the device to the environment.

c. Heat pipes, being a part of an electronic device construction or an electronic plate, are assembled as a component of electronic equipment. The TCS on heat pipes provides the thermal regime of the device.

d. Heat pipes, which are built in a mounting plate, create an isothermal surface for setting devices and equipment. The thermal regime of a device is provided by an additional TCS or a system on heat pipes.

e. Heat pipes form a thermal network for direct exchange of thermal energy between parts of objects or devices. The thermal regime of devices is provided jointly by the TCS of each device or a centralised system on the basis of heat pipes.

f. Heat pipes keep the isothermality of the container (shell, surface) in which a device is in. The thermal regime of the device is provided by a system on heat pipes or in combination with active control (heater or thermoelectric cooler, regulator, source of electrical energy).

10.3.1 Principles of Operation of TCSs on GLHPs for Autonomous Devices

One of the important practical applications of GLHP systems is thermal control of autonomous devices. At the second level of integration for temperature stabilisation of a device case (Figure 10.11b), in which electronic cards or other sensitive equipment are installed, a TCS on a GLHP may function, using only heat generated by the device, Q_{dev}. In the event that Q_{dev} is insufficient or negative, an additional energy supply with power Q_{ad} is necessary (Figure 10.12).

In general, the device is installed on the mounting surface of a spacecraft and is impacted by the variable temperature of SC (T_{mp}), external environment temperature (T_o) and probably some other temperature sources.

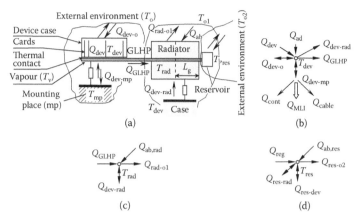

FIGURE 10.12
Structure of heat fluxes in thermal control system (TCS) with gas-loaded heat pipe: (a) scheme of TCS with device; (b) balance of heat fluxes for device case; (c) balance of heat fluxes for radiator and (d) balance of heat fluxes for external reservoir.

So in the steady-state regime, the heat generated by a device, Q_{dev}, or Q_{dev} + Q_{ad}, is distributed between the following components:

$Q_{dev-rad}$: heat exchange between device case and radiator via radiative and conductive elements. It is true when the radiator is installed on the device case.

Q_{dev-o}: heat exchange between device case and external environment (space, the earth and elements of SC can be considered). The heat exchange is mainly radiative, although conductive heat transfer can take place as well. Typical heat exchange is via MLI blankets, which cover the external surface of the device, via optical surface finishing and optical windows for free from MLI parts.

Q_{dev-mp}: heat exchange between device case and mounting place. It includes the following heat transfer components: Q_{cont}, via stand-offs, direct contact of device case and mounting place; Q_{MLI}, radiative heat exchange via MLI or low emissive finishing and Q_{cable}, via electrical cables.

Q_{GLHP}: heat is transferred by GLHP to radiator.

Components of heat balance for radiator and reservoir are the following:

Q_{rad-ol}: heat exchange between radiator and heat sink.

$Q_{ab,rad}$: heat from external light sources (the earth, the sun, reflecting surfaces), absorbed by radiator.

Q_{GLHP}: heat transferred from evaporator by vapour to active part of radiator. For the part filled with NCG, $Q_{GLHP} = 0$.

Q_{reg}: heat flux from electronic regulator.

$Q_{ab,res}$: heat from external light sources (the earth, the sun, reflecting surfaces), absorbed by reservoir.

$Q_{res-dev}$: heat exchange between reservoir and device case (if reservoir requires mechanical fastening).

On the basis of joint analysis of heat exchange in the system 'SC–device–TCS–space environment' and specific features of GLHP regulating characteristics, the following conditions are proposed for consideration:

1. Heat flux transferred by a GLHP, Q_{GLHP}, should be within the limit $Q_{min} - Q_{max}$, defined by the physical processes of heat transfer in the GLHP, based on its regulating characteristics (see Figure 10.2b and c).

2. Components of device thermal balance, such as heat exchange with the mounting place (Q_{dev-mp}) and external environment (Q_{dev-o}), at the minimal values of T_{dev}, T_{mp} and T_o are within the limit $Q_{min} < (Q_{dev,min} - Q_{dev-mp} - Q_{dev-o})$ and at the maximal values of T_{dev}, T_{mp} and T_o should ensure $Q_{max} > (Q_{dev,max} + |Q_{dev-mp}| + |Q_{dev-o}|)$.

3. The value of radiator area, F_{rad}, its emittance, ε_{rad} and thermal efficiency, η_{rad}, should fit the following condition:

$$\varepsilon_{rad} \eta_{rad} F_{rad} \sigma \left(T_{v,max}{}^4 - T_{o1,max}{}^4 \right) \geq \left(Q_{dev,max} + |Q_{dev-mp}| + |Q_{dev-o}| + |Q_{dev-rad}| + Q_{ab,max} \right)$$

4. The change in vapour temperature, T_v, is conditioned by thermodynamics of gas plug movement in the ranges of $T_{v,min} \geq (T_{dev,min} - Q_{dev,min} R_{dev-v})$ and $T_{v,max} \leq (T_{dev,max} - Q_{dev,max} R_{dev-v})$, where R_{dev-v} is the thermal resistance of line device–vapour in the GLHP evaporation zone.

5. The minimal and maximal temperatures of radiator (condenser) in the gas-filled zone and of reservoir are defined by external thermal boundary conditions and by mutual heat exchange. These temperatures define the partial pressure of gas in 'gas–mass' balance and influence on the accuracy of vapour temperature regulation. Neglecting axial conductivity in radiator and between radiator and reservoir, and assuming that the geometrical view factor 'radiator–environment', 'reservoir–environment' equals 1, these temperatures are expressed as follows:

$$T_{g,rad} = \left[\frac{q_{ab,rad} F_{ab,rad}}{(\varepsilon_{rad} F_{rad} \sigma)} + T_{o1}^4 \right]^{0.25} , \quad T_{res} = \left[\frac{q_{ab,res} F_{ab,res}}{(\varepsilon_{res} F_{res} \sigma)} + T_{o2}^4 \right]^{0.25}$$

where $q_{ab,rad}$ and $q_{ab,res}$ are absorbed heat fluxes (from all possible sources), $F_{ab,rad}$ and $F_{ab,res}$ are corresponding areas of radiator or reservoir that accept these fluxes.

The components of heat flux Q_{dev-mp} are proportional to $T_{dev} - T_{mp}$ and $T^4_{dev} - T^4_{mp}$, and the components of Q_{dev-o} are proportional to $T_{dev} - T_o$ and $T^4_{dev} - T^4_o$. They depend on device mass and device side surface area. Quantitative estimation of thermal balance components for the device may cause essential difficulties and can be performed with some margins. For their definition, as usual, commercial thermal analysers are recommended (SINDA/G User's Guide 2003, User's Guide MSC Sinda 2009, ESATAN-TMS Thermal User Manual 2011, software products Systema 4.5.0 and NX7.5), which solve conduction and radiation heat transfer problems for the complicated geometry of spacecraft radiator and reservoir at the non-uniform temperature of their surroundings. The optical properties of components used (MLI, optical finishing) are summarised in handbooks and numerous studies (Favorskiy and Kadaner 1972, Andreanov et al., 1973, Kaganer 1979, Karam 1998, Gilmore 2002, Donabedian 2004, Material Properties 2012). Low conductance stand-offs and contact conductance are reflected in studies by Kaganer (1979), Gilmore (2002), and Donabedian (2004). The prediction of cable (harness) thermal behaviour usually foresees the uncertainty of heat fluxes going out of and into a device via cable. The simplified calculation of heat leak via cable for the considered device can be presented as follows:

$$Q_{cable} = \left(\sum_{n=1}^{N} \lambda_n F_n \right)(T_{dev} - T_{mp})/L_{cable}, \quad \text{if the cable has perfect thermal}$$

contact with the mounting place

$$Q_{cable} = \left(\sum_{n=1}^{N} \lambda_n F_n \right)(T_{dev} - T_{dev2})/L_{cable}, \quad \text{if the cable has perfect thermal}$$

contact with the nearest device

where $\lambda_n F_n$ is the product of axial conductivity and area of the nth element (metallic conductors, electrical insulation, shielding, mechanical protection etc.). It is assumed that ohmic power generation in cables can be neglected compared with the heat transferred by conductivity. More complicated thermal models of cables are discussed by Van Benthem et al. (2011) and Doctor and van Benthem (2011).

The offered principle of temperature regulation was experimentally checked for mock-ups of autonomous devices for the following characteristics:

Device case dimension: $0.36 \times 0.24 \times 0.155$ m^3

Device mass: <4 kg

Power generation: 1.5–20 W

Thermostating temperature of device case: 25°C

Temperature of mounting place: −20°C to +50°C

Heat fluxes absorbed by a flat surface: $q_{ab} = Q_{ab}/F_{rad} = 120–270 \ \text{W/m}^2$

The harness consists of 50 copper wires with an effective cross section of 0.12 mm² and 25 copper wires with a cross section of 0.45 mm², and it has a length of 1 m. Such a cable design was used in practice.

Design features of TCS: TCS should be autonomous, have sizes less than 0.36 × 0.24 × 0.055 m³, and have mass less than 0.8 kg.

The working diagram in Figure 10.13a shows that under cold conditions heat generation, Q_{dev}, should compensate the leak via GLHP to the radiator

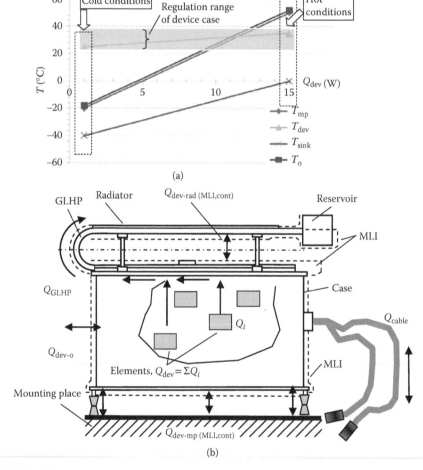

FIGURE 10.13
(a) Map of temperatures of thermal control system (TCS) elements for cold and hot conditions, and (b) scheme of assembly of device and TCS.

(which reaches the temperature T_{sink}) via cables, insulation, stand-offs to mounting place and insulation to external environment. Under hot conditions, besides device power generation, Q_{dev}, there is an additional heat input to the device from the mounting place and external environment. A TCS has a modular principle of design. It consists of GLHP (heat pipe, flange, radiator) and is attached to a device by flanges and stand-offs (Figure 10.13b).

An analysis of heat balance components for TCS operation under cold conditions has been carried out at constant thermal conductance of cables ($GL_{cont} = 1/145$ W/K) and minimal value of $Q_{min,GLHP}$ (0.5 W) (Figure 10.14). The functional dependence of minimal power, $Q_{dev,min}$, on conductivity of stand-offs shows its inessential effect at high values of $\varepsilon_{eff,MLI} > 0.05$ and requires that $R_{cont} > 200$ K/W at $\varepsilon_{eff,MLI} < 0.01$, which can be reached in practice. To reach $Q_{dev,min} \sim 1.5$ W, the effective emittance of MLI blankets should be $\varepsilon_{eff,MLI} < 0.01$.

At selected parameters of $R_{cable} = 145$ K/W, $\varepsilon_{eff,MLI} = 0.01$ and $R_{cont} = 400$ K/W, the heat balance for a device with TCS looks as shown in Figure 10.15. These calculations show the possibility of reaching thermal stabilisation at device heat generation of 1.5 W.

The design features of proposed TCS are as follows:

- GLHP with a wicked reservoir. Regimes of operation include stabilised reservoir temperature (supported by a miniature electronic controller) and variable reservoir temperature (self-installation on the basis of reservoir heat balance).

- Heat pipe shell and reservoir shell were made from a stainless steel tube of thickness 0.0004 m.

- The CS is made of stainless steel discrete fibre felt, and diameter and length of the fibre are 30×10^{-6} and 3×10^{-3} m respectively. The structure has been formed to obtain the configurations in evaporator, adiabatic and condenser zones shown in Figure 10.16. The distinctive

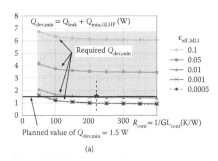 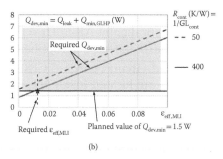

(a)   (b)

FIGURE 10.14

Required minimal device power as a function of parameters of conductive heat exchange via stand-offs, R_{cont} (a), and parameters of radiative heat exchange via multi-layer insulation, $\varepsilon_{eff,MLI}$ (b), for cold conditions: the cable conductance is constant, and $T_{dev} = 20°C$ and $Q_{leak} = Q_{dev-o} + Q_{dev-mp} + Q_{dev-rad}$.

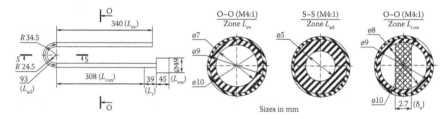

FIGURE 10.15
Quantitative content of heat balance in cold conditions (a) and relative content (b) at $Q_{dev,min}$ = 1.6 W, $T_{dev,min}$ = 20°C, T_o = −20°C, T_{rad} = $T_{sink,min}$ = −40°C and T_{mp} = −20°C.

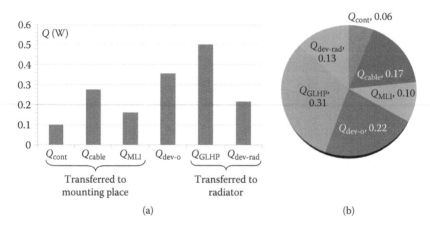

FIGURE 10.16
Scheme of gas-loaded heat pipe and capillary structure configurations.

feature of CS design is the rectangular artery in condenser that reduces vapour cross-sectional area and enlarges the ratio V_{res}/V_{con} to 8.5.

- Radiator has slitted design and contains two zones (A and B) with different widths of ribs. The ribs are made of aluminium and have been brazed to condenser tube, contacting with 90% of the tube perimeter. The optical paint AK-573 (ε = 0.85, α_s = 0.24 at the beginning of coating life (BOL)) has been applied as the radiator and reservoir coating.
- Printed circuit boards (PCBs) with electronic heat-generating elements are located inside the device case and have thermal contact with a lid and the device case.

The appearance of the 'TCS + device' assembly without an MLI jacket is presented in Figure 10.17a, and the assembly with MLI and radiator is shown in Figure 10.17b and c.

Investigations on TCSs have been conducted in thermovacuum chamber with heat rejection from the radiator to a large-sized flat liquid nitrogen screen. The thermal disturbances have been imitated by the following equipment:

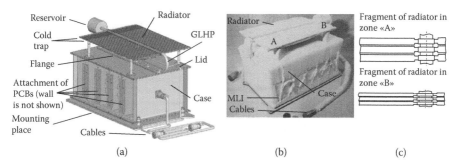

(a)   (b)   (c)

FIGURE 10.17
Autonomous device with thermal control system on gas-loaded heat pipe: (a) without multi-layer insulation jacket; (b) before thermal vacuum test and (c) scheme of radiator ribs.

temperature of the mounting place was kept between −20°C and +50°C with an accuracy of ±1 grad by regulated heaters with a temperature non-uniformity less than 0.5°C. Heat fluxes dropped to radiator and reservoir from the sun and the earth were imitated by electric heaters, placed on the radiator side, undirected into the space, and on reservoir under the layer of optical coating. The used method of absorbed flux imitation is characterised by its relative simplicity of realisation; it has a short reaction time due to the low thermal mass of heaters in comparison with radiator and reservoir, it allows (in the presence of heat insulation) exact control of heat flux, and it enables the simulation of coating behaviour with different solar absorptance, α_s, values. A weakness of this experimental technique is the impossibility of confirming in thermovacuum experiments the real value of α_s, which should be measured additionally.

PCB heat generation has been imitated by fitted cylindrical heaters. Copper–constantan thermocouples (T type, ⌀ 0.2 mm) have been installed in the evaporator zone of GLHPs (2 units), in adiabatic zones (2 units), on discrete elements of radiators in zone A (7 units), in zone B (3 units) and on reservoirs (2 units). Also, sensors have been placed in PCBs and in device cases. The tests studied the reaction of TCS to the change in device power generation and external absorbed flux (Figure 10.18).

The change in evaporator zone temperature characterises the thermostating properties of GLHPs. The isothermality along evaporator length is less than 1°C for Q_{dev} = 1–15 W. At small power Q_{dev} < 1–2 W, the evaporator self-purges refines from gas. With an increase in generated power the gas front moves towards reservoir, opening for condensation the larger area of radiator. As a result, the thermal resistance between device case and heat sink changes from 43 to 5.5 K/W. In the zone filled with gas, radiator temperature falls to the level of −40°C, which is kept by the radiator heater on the condition that q_{ab} = const. Reservoir temperature remains constant at 0°C (provided by the miniature electronic regulator, which supplies power to the heater, located on the reservoir). At powers higher than 13 W, the gas front reaches the condenser end and the temperature of radiator becomes higher than −40°C. At variable external heat

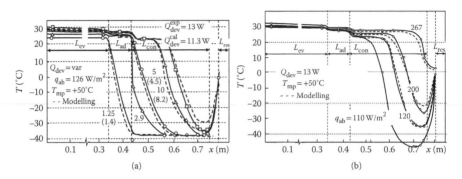

FIGURE 10.18
Temperature distribution in gas-loaded heat pipe shell at change of heat power generation (a), and external absorbed heat flux (b) in steady-state conditions.

fluxes within the range of 120–267 W/m² and Q_{dev} = const, the vapour–gas front moves to the end of the condenser and begins to increase the reservoir temperature. The thermal resistance between device case and heat sink changes from 5.5 to 2.3 K/W. At $q_{ab} > 270$ W/m², the vapour–gas front enters the reservoir, causing it to heat to temperatures higher than 0°C. As a result of the rise in T_{res}, partial pressure in the reservoir becomes higher, causing an increase of vapour pressure in the GLHP and the subsequent rise of device case temperature.

The simulation of temperature profile in GLHP shell has been conducted with the use of a method discussed by Semena et al. (1990), which allows the modelling of the temperature field with stepwise longitudinal heat conductivity, typical for a GLHP slitted radiator. The location of the vapour–gas front (which is one input to the model) was defined experimentally. The comparison of experiments and calculations testifies the similarity of profiles. The difference in heat flux values is caused by accuracy of definition of heat balance components. The impact of mounting place temperature is summarised in Figure 10.19 in the form of dependencies $T_{dev} = f(Q_{dev})$, $T_v = f(Q_{dev})$ at similar q_{ab}. At variation of Q_{dev} from 3 to 13 W and variation of T_{mp} from −20°C to +50°C, T_{dev} changes from 26°C to 32°C (set N I in Figure 10.19). As shown in Figure 10.14b, MLI design has an essential impact on the value of $Q_{dev,min}$. Improvement of MLI design allows the reduction of minimal device heat power $Q_{dev,min}$ to 1.5 W (set N II in Figure 10.19). The regulating characteristics are kept on a lower temperature regulation level (set N II in Figure 10.19). For both devices, the value of $Q_{dev,max}$ is approximately 13 W. At variable T_{mp}, Q_{dev} and q_{ab}, the level of stabilisation of device case temperature, T_{dev}, is 20°C ± 10°C in the passive mode [$T_{res} = f(q_{ab})$] and 29°C ± 3°C in the semi-passive mode (T_{res} = const).

The optimisation of heat exchange in TCS elements (low conductance standoffs, MLI, cables and GLHP shell) enables the regulation of device temperature at $Q_{dev,min} = 1.5$–2.0 W. Under identical conditions, the traditional passive system could control device temperature in the range of −20°C to 50°C and the active system (heater-controlled system) provided temperature stabilisation on the level of 20°C ± 10°C at an additional energy of regulated heater, Q_{reg}, of 15 W.

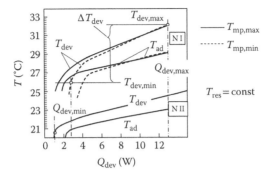

FIGURE 10.19

Thermal control system (TCS) behaviour at $T_{mp,min} = -20°C$ and $T_{mp,max} = 50°C$ in the mode with constant reservoir temperature for TCS modifications N I and N II (TCS set numbers).

At TCS exploitation in the earth orbit, the radiator and the reservoir will be periodically illuminated with an external light flux. Imitation of satellite movement over an orbit has been conducted by changing the values of heat flux absorbed by radiator, $q_{ab,rad}$ (and finally by changing total absorbed flux $Q_{ab,rad}$), and by reservoir, $q_{ab,res}$ (total absorbed flux, $Q_{ab,res}$). These values have been calculated for a circular orbit (altitude = 300 km, inclination = 90 grad), assuming the use of coating AK-573 with $\varepsilon = 0.85$ and $\alpha_s = 0.4$ (values are for end of life, EOL), which is shown in Figure 10.20a. The device case temperature, T_{dev}, radiator temperature of the joint with adiabatic zone, T_{rad1}, and end of radiator, T_{rad2} and reservoir temperature, T_{res}, are cited in this figure. At maximum external irradiation, temperatures of the device and the radiator coincide and the vapour–gas front moves to the reservoir, increasing the device temperature (ΔT_{dev}) by 13°C and T_{res} from 0°C to 25°C. After the reduction of q_{ab}, temperatures T_{rad} and T_{res} return to their initial values. In this regime, due to gas front displacement in the reservoir under the condition $T_{rad2} > T_{res}$, the condensation and accumulation of heat carrier take place in the reservoir. For the part of orbit with smaller values of q_{ab}, the heat carrier should be returned from the reservoir to the CS of condenser and evaporator zones. At considerable hydraulic resistance of reservoir–evaporator liquid path and essential height of liquid lifting in reservoir, infringement of liquid feeding may occur. This was observed in our tests in an increase of device case temperature after a run of several orbits (not shown in Figure 10.20a). With the application of modern optical paints and coatings, for example, silver coated FEP (Teflon) tape (The Red Book 2010), which provides $\alpha_s < 0.2$ and $\varepsilon = 0.85$, TCS may function without falling outside regulation limits. Reaction of TCS to external light disturbances at these coating parameters had the character of change in T_{dev}, T_v and radiator temperature (T_{rad1}, T_{rad3}, $T_{rad,middle}$ – temperatures of extreme points and middle of radiator) demonstrated in Figure 10.20b. The device temperature was within regulation limits. The front moved along the radiator length without reaching the reservoir ($L_g > 0$). The reservoir temperature was maintained during all cycles at the regulation level.

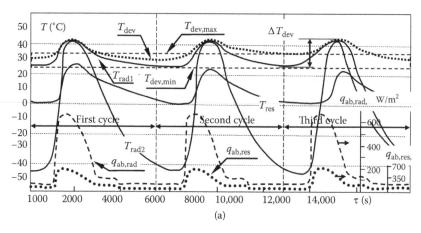

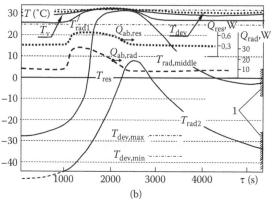

FIGURE 10.20
Thermal control system operation in an unsteady regime at imitation of external heat flux, appearing due to satellite movement over the orbit (--: heat absorbed by radiator; –: heat absorbed by reservoir), $Q_{dev} = 13$ W, and $T_{mp} = 323$ K: (a) radiator coating AK-573 ($\alpha_s = 0.4$) and (b) silver-coated Teflon ($\alpha_s = 0.2$); 1 – diapason of steady-state values of T_{rad2} and $T_{rad,middle}$.

The design of PCB with TCS has passed complete mechanical tests, imitating transport and start of launcher mechanical loadings that have confirmed the rightness of the chosen structural and technological decisions for TCS thermal elements.

10.4 Integration of Heat Pipe TCS in PCB Design

The subsequent level of heat pipe integration foresees two-level TCS. This assumes direct setting of heat-generating and passive electronic elements of PCBs on an isothermal substrate (IS) that can be realised by flat HP. This

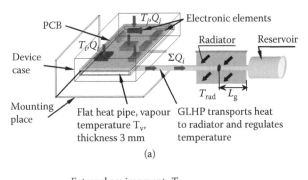

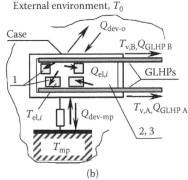

(b)

FIGURE 10.21
Principle of electronic elements' cooling in autonomous electronic printed circuit board (PCB) with isothermal substrate (a), and components of heat balance for device case (b): (1) heat-generating elements; (2) electronic PCB; (3) isothermal substrate–heat pipe.

approach has been successfully tested in the work of Suntsov et al. (2009) and others. In current designs, the temperature of a flat heat pipe is regulated by GLHP (Figure 10.21a). The heat balance for a device is similar to the previous TCS scheme (see Section 10.3). The difference is in appearance in the thermal scheme of additional temperature – vapour temperature T_v in flat heat pipe, which, in turn, is defined by vapour temperatures $T_{v,A}$, and $T_{v,B}$ in GLHP A and GLHP B in Figure 10.21b.

At the elaboration of TCS thermal and mechanical design, the following initial data are accepted:

- An electronic PCB has unilateral elements' installation. The PCB size is 180×130 mm^2, and thickness is 0.5–1.5 mm; it can be rigid or flexible.

- Under operating conditions, the temperature level of electronic elements should not exceed +70°C. The total heat generation of a PCB changes from 1 to 7 W. The heat-generating elements are 25 units of cylindrical shape with sizes of $\varnothing 2 \times 6$ mm to $\varnothing 5.5 \times 12$ mm and a ratio of maximal to minimal heat power generation of 10:1.

- The temperature of device mounting location changes from −20°C to +50°C.
- Heat removal is from device to space.
- TCS should provide stabilisation of IS temperature on the influence of internal and external disturbances, in particular, change of external radiant heat fluxes, absorbed by TCS radiator and reservoir, within the range of 100–300 W/m².
- TCS can function in two modes: with feedback electric control (using as reference the element temperature or the substrate temperature) and in passive mode without electric regulator.
- Device design should allow the detachment of a substrate from the TCS.

As a variant of realisation of these technical requirements, the technical concept shown in Figure 10.22 is proposed.

The IS–flat HP is located inside the device case. It has thermal contact with two GLHPs. Zones of condensation of the GLHPs are connected to a slitted radiator. At the end of the condensation zone, two wicked reservoirs are located. For stress relief of a bending zone, the radiator end is supported with two fibreglass stand-offs (7 in Figure 10.22). The device case is mounted on SC seats by four fibreglass stand-offs (8 in Figure 10.22). The device case and the inner radiator surface are covered with MLI. The IS has a size of 0.180 × 0.130 × 0.0038 m³. To one of its surfaces, a dielectric flexible layer with the resistors simulating electronic elements is pasted. The resistors are attached to the dielectric layer by epoxy glue. GLHPs are fabricated from stainless steel tube ⌀8 with a wall thickness of 0.3 mm. A CS is made of discrete metal fibre felt with a porosity of 0.8 and a thickness of 0.85 mm. Along the heat

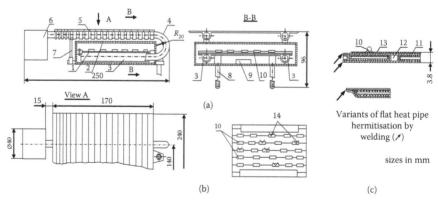

(a)

(b)   (c)

FIGURE 10.22
Design of an autonomous electronic device with isothermal substrate and gas-loaded heat pipe (GLHP) (a), location of heat-generating elements and sensors (b), design of isothermal substrate (c): (1) isothermal substrate; (2) device case; (3) flanges; (4) GLHP; (5) section radiator; (6) reservoir; (7,8) stand-offs; (9) temperature regulator (optional); (10) elements of PCB; (11) capillary structure; (12) stand-offs; (13) PCB; (14) thermocouples.

pipe length, there is a rectangular artery of 2 mm thickness and porosity 0.8. The artery has holes of $\varnothing 2$ mm every 15 mm. The reservoir has a CS of thickness 1 mm, porosity 0.8 and artery thickness 2 mm. The condenser and reservoir CSs are hydraulically connected via arteries. The lengths of GLHP zones are as follows: $L_{ev} = 115$ mm, $L_{ad} = 68$ mm, $L_{con} = 170$ mm, $L_t = 15$ mm and $L_{res} = 40$ mm. The ratio V_{res}/V_{con} is 15.1 for methanol as working fluid. The flat heat pipe was fabricated from stainless steel sheets of thickness 0.3 mm. The CS has a thickness of 0.5 mm and porosity of 0.8 and has been fabricated from sintered discrete stainless steel fibres with a diameter of 30×10^{-6} m and length of 3×10^{-3} m. Hydraulic connection of both sides of the flat heat pipe is realised by point-type supports (5×5 mm^2, 15 points), which prevent the deformation of flat sheets due to pressure change, and by longitudinal arteries of 2 mm width.

The scheme of thermal connections in the proposed design of thermostating PCBs with GLHPs includes the nodes and junction shown in Figure 10.23.

The heat power generated by every ith element ($i = 1, 2, ..., n$), $Q_1, Q_2, ...,$ Q_n, through element–substrate transition thermal resistances, $R_1, R_2, ...,$ R_n, is accepted by a shell of IS–flat heat pipe with substrate temperatures under every element designated as $T_{sub,1}, T_{sub,2}, ..., T_{sub,n}$. Then the heat power is transferred to the vapour via the thermal resistance of the phase change

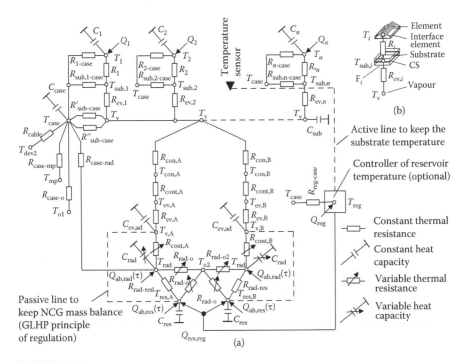

FIGURE 10.23

(a) Scheme of thermal connections in the proposed design of thermostating printed circuit board and (b) heat transfer scheme from ith element to a vapour of substrate.

process designated as $R_{ev,1}$, $R_{ev,2}$, ..., $R_{ev,n}$. The temperature of vapour space, T_v, under every element is constant, which promotes the creation of an isothermal base for the installation of elements. Afterwards, the heat is transferred by vapour to two condensation zones, 'A' and 'B'; the resistance at heat condensation is designated as $R_{con,A}$ and $R_{con,B}$. The minimal temperatures of a substrate in these zones of connection with GLHPs are $T_{con,A}$ and $T_{con,B}$. The vapour temperature of the substrate is defined by the vapour temperatures in GLHPs, $T_{v,A}$ and $T_{v,B}$; thermal contacts between IS and GLHP evaporators, $R_{cont,A}$ and $R_{cont,B}$; and thermal resistances of evaporation zones, $R_{ev,A}$ and $R_{ev,B}$.

The heat output by GLHPs, $Q_{GLHP\,A}$ and $Q_{GLHP\,B}$, depends on the influence of device case temperature, T_{case}. A certain quantity of total heat generation, $Q_{sub} = \sum Q_i$, may be absorbed by the case via radiation from elements (resistances $R_{1\text{-case}}$, $R_{2\text{-case}}$, ..., $R_{n\text{-case}}$) from substrate surface without elements $(R'_{sub\text{-case}})$ and due to the contact of substrate with case $(R'_{sub\text{-case}})$. The case temperature, T_{case}, is connected to the mounting place temperature, T_{mp} and environment temperature, T_{o1}, via combined resistances, $R_{case\text{-mp}}$ and $R_{case\text{-o}}$ respectively. The cable thermal resistance, R_{cable} (and $R_{case\text{-mp}}$, $R_{case\text{-o}}$), plays an essential role in defining the minimal required IS power generation. Heat rejection from radiator into the environment, T_{o2}, is realised through variable resistance $R_{rad\text{-o}}$, which is changing due to the variation of the vapour–gas front location and, correspondingly, radiation area. External heat inputs to radiator and reservoir are defined by absorbed fluxes $Q_{ab,rad}(\tau)$ and $Q_{ab,res}(\tau)$. During the operation of electrical feedback controllers, the regulating power $Q_{res,reg} = Q_{res\,A,reg} + Q_{res\,B,reg}$ is added depending on the IS temperature.

Assuming that the main heat transfer from the element is realised to a substrate and further to a vapour, the temperature of the ith element is defined as $T_i = T_v + Q_i (R_{ev,i} + R_i)$. In order to reach the close temperature for all elements $T_i = $ const, it should be $Q_i (R_{ev,i} + R_i) = $ const. This requirement may assist in determining the functions $R_{ev,i}(Q_i)$ and $R_i(Q_i)$ for concrete types of elements. If $R_i = $ const, as the single stabilisation factor could be an appropriate function $R_{ev,i}(Q_i)$, which decreases with increasing Q_i. Such a behaviour of this function may take place at phase change processes (evaporation, boiling) in certain CSs, for example, in metal felts (Semena et al., 1984). For evaporation $R_{ev,i} \neq f(Q_i)$, and for boiling $R_{ev,i} \sim Q_i^{-0.6} F_i^{-0.4}$, where F_i is the contact area of the ith element (or interface element) with IS.

If $R_{ev,i} = $ const, the smoothing of element temperature can be achieved by varying element interface resistance R_i, which includes contact and conductance resistances of this element. The common requirement for R_i and $R_{ev,i}$ is their small absolute value. The accuracy of ith element temperature stabilisation, ΔT_i, is defined as $\Delta T_i = \Delta T_v + \Delta Q_i (R_{sub,i} + R_i)$, where ΔT_v is the variation of substrate vapour temperature and $\Delta Q_i = Q_{i,max} - Q_{i,min}$. The goal of design is to reach $\Delta T_v > \Delta Q_i (R_{sub,i} + R_i)$. However, the elaboration of the constructive and thermophysical solutions, allowing equalising the temperatures of heat-generating elements in a wide range of generation and individual elements' configurations is rather topical problem.

The value of ΔT_v is defined by GLHP characteristics. At active regulation, $\Delta T_v < 0$ (Bienert et al., 1971, Schlitt 1973, Groll and Hage 1974), which allows the reduction of ΔT_i. At passive regulation, $\Delta T_v > 0$, and for reducing ΔT_v, it is necessary to minimise resistances in the condensation zone of substrate, $R_{con,A}$ and $R_{con,B}$; contact resistances between substrate and GLHPs, $R_{cont,A}$ and $R_{cont,B}$; resistances of GLHPs' evaporation zones, $R_{ev,A}$ and $R_{ev,B}$; and variation of GLHPs' vapour temperatures, $\Delta T_{v,A}$ and $\Delta T_{v,B}$. Vapour temperature T_v can be expressed as follows:

$$T_v = \frac{\left(\sum Q_{el,i} + T_{v1}/R_{\text{IS-GLHP A}} + T_{v2}/R_{\text{IS-GLHP B}}\right)}{\left(1/R_{\text{IS-GLHP A}} + 1/R_{\text{IS-GLHP B}}\right)} \tag{10.9}$$

where $R_{\text{IS-GLHP A}} = R_{con,A} + R_{cont,A} + R_{ev,A}$, $R_{\text{IS-GLHP B}} = R_{con,B} + R_{cont,B} + R_{ev,B}$.

The temperature $T_{v,IS}$ is stabilised due to constancy or a small rise of T_{v1} and T_{v2} at passive regulation, or their reduction with the increase of $Q_{\text{GLHP A}} + Q_{\text{GLHP B}}$ at active regulation of reservoir temperatures, $T_{res,A}$ and $T_{res,B}$.

For the execution of thermal vacuum tests (TVTs) of a device with TCS (Figure 10.24a and b) and in order to have the freedom of elements layout on a PCB, there is an additional requirement for the CS of IS–flat heat pipe, namely its ability to operate at any orientation relative to the gravity vector.

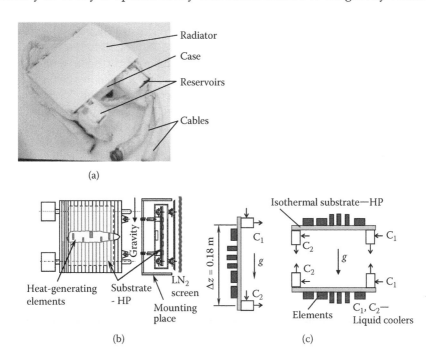

FIGURE 10.24
(a) Assembly of an autonomous device with thermal control system, (b) scheme of assembly layout in a thermal vacuum test and (c) substrate orientation at its autonomous testing.

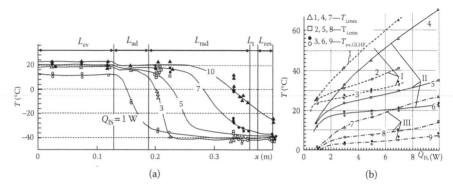

FIGURE 10.25
(a) Temperature distribution along the gas-loaded heat pipe (GLHP) length at variable device power generation in passive mode and (b) dependence of regulating characteristics for printed circuit board elements, T_i, and for the GLHP evaporation zone, T_{ev}, on power generation, Q_{IS}, on three levels of external heat disturbances: $T_{sink} = 0°C$ (I), $-40°C$ (II) and $-60°C$ (III).

The worst variant is when the vector of gravity lies in the plane of IS, and geometrical exceeding of heating zone above condensation zone Δz equals 0.18 m. In autonomous experiments (Figure 10.24c), in this position of IS the temperatures of elements $T_{el,i}$ and T_v are kept at the same level as in the horizontal position ($\Delta z = 0$), and the change of orientation of elements from above or below in the horizontal plane of IS does not influence their temperatures.

Experimental investigations of the device + TCS assembly (Figure 10.25a) in a TVT with radiation cooling on a cold flat liquid nitrogen screen have shown that the range of regulation temperatures of PCB elements in the passive mode makes $23°C \pm 10°C$ for non-heat-generating elements (corresponds to $T_{ev,GLHP}$), $40°C \pm 25°C$ for elements with heat generation of $0.1 \times Q_{IS}$ and flux density of element $q = 12 \times 10^4$ W/m² ($T_{i,max}$), and $28°C \pm 13°C$ for elements with heat generation of $0.011 \times Q_{IS}$ and heat density up to $q = 6 \times 10^4$ W/m² ($T_{i,min}$), which is demonstrated in Figure 10.25b. These data were obtained at a T_{mp} over the range of $-20°C$ to $+50°C$; the total heat generation of the IS is $Q_{IS} = 1–8$ W, and at $q_{ab} = 140–280$ W/m². These flux densities correspond to T_{sink} for radiator and reservoir within the limits of $-40°C$ to $0°C$.

The autonomous device with installed TCS has passed complete mechanical tests, imitating transport and launcher starting loadings that confirm the rightness of the chosen structural and technological decisions for the design of TCS elements.

Thus, the following conclusions can be made:

1. The developed approach is an example of unified designs of space scientific optoelectronic devices and TCS with adjustable temperatures of elements. The flat heat pipe is the base for installation of elements, heat accumulation and sequent heat transfer to the external TCS. The external TCS regulates the temperature of interface with

a flat heat pipe at variable mounting place temperature, power generation of elements, external illumination conditions and change of optical properties of coatings. TVTs of TCS and devices have shown a moderate impact of mounting place temperature within the range of −20°C to +50°C on minimal regulation power that is correct at the level $Q_{IS, min} < 1.0$ W.

2. The temperatures of the majority of heat-generating elements exceed the IS temperature by 30–40 grad.

3. The design proved itself steady against influence of a complex of mechanical loadings, simulating transportation and launch conditions, and kept the thermal characteristics after mechanical tests.

4. The advantages of substrate–electronic PCB, executed in the form of a flat heat pipe, become more significant in comparison with aluminium substrates at increasing PCB sizes (more than 0.13 × 0.18 m^2).

5. Equalisations of the temperatures of electronic elements may be achieved only with the improvement of the elements' contact with the substrate or if the individual heat generation of an element, Q_i, is less than 0.2 W. At larger heat generation values, more efficient ways of connecting elements to a substrate are needed.

10.5 Thermal Control of Devices without Heat Generation

For scientific devices that do not have heat generation ($Q_{dev} = 0$) but require the narrowing of exploitation temperature range (typically T_{dev} is in the range of 10°C–30°C), the use of classic passive TCSs and their combinations is problematic if devices are exploited for a wide range of solar constants, for example, during the Earth–Mars cruise phase under conditions of SC energy deficit (Baturkin 2011). For a device in an unpressurised module, the temperature of the device, T_{dev}, can be 'adhered' to the temperature of a mounting place, T_{mp}, or to the equilibrium temperature, which is defined by solar constant. The heat balance of the isothermal device case includes the heat exchange with SC through stand-offs; thermal insulation; cables and heat exchange device–space, consisting of heat supply from the sun and heat radiation to the space (Figure 10.26). Steady-state heat balance for a system with one radiator could be described by the following equation:

$$Q_{dev} + \alpha_{s,abs} q_s F_{abs} \cos \varphi + \frac{(T_{mp} - T_{dev})}{R_{dev-mp}} = \varepsilon_{rad} F_{rad} \sigma (T_{dev}^4 - T_o^4) \qquad (10.10)$$

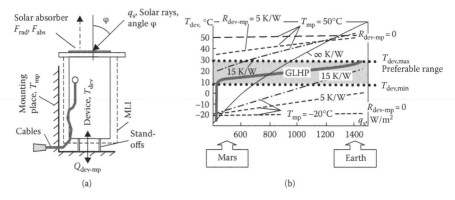

FIGURE 10.26
Passive concept of a device thermostating at $Q_{dev} = 0$: (a) scheme of a thermal control system and (b) estimation of the influence of parameters T_{mp}, q_s and $R_{mp\text{-}dev}$ on device temperature T_{dev}.

At $Q_{dev} = 0$ W and in cases where heater control is impossible (this could be at device operation and transportation in far space missions in a direction opposite to the sun), potential ways of thermal control are the following:

Minimisation of thermal resistance, R_{mp}: at $R_{dev\text{-}mp} = 0$ K/W, $T_{dev} \equiv T_{mp}$.

If $R_{mp} \to \infty$ is maximised, then $T_{dev} = [\alpha_s q_s F_{abs}\cos(\varphi)/(\varepsilon F_{abs}\sigma) + T_o^4]^{0.25}$.

At the value $R_{dev\text{-}mp} = 5$–15 K/W, which can be reached for a device with linear size 0.2–0.4 m and mass 5–10 kg at $T_{mp,min} = -20°C$ and $T_{mp,max} = 50°C$, the variation $|T_{dev} - T_{mp}| < 20°C$–25°C for every T_{mp} to keep the required device temperature diapason of 10°C–30°C is impossible without changing ε/α_s, F_{abs} and F_{rad} during the mission (increasing α_s and F_{abs} at reducing q_s), which is shown in Figure 10.26b. Such regulation is possible by means of louvres. However, there is another approach to solve this task. For such conditions, the concept of integration of passive heating element–absorber of the sun's radiation heat for the function of TCS (heat input into the system, $Q_{s,abs}$) and GLHPs, which compensates the changes in heat balance on selected temperature level (Figure 10.27), is offered by Baturkin (2011).

Device and radiator heat balances at the device temperature level of $T_{dev} = 20°C$ must be maintained in the whole range of initial parameters: q_s ($q_{s,max}/q_{s,min} = 1400/500 = 2.8$), angle $\varphi = \pm20°$ at the manoeuvres of SC, $T_{mp} = -20°C$–$+50°C$, mass of device = 5–10 kg, sizes up to $0.3 \times 0.2 \times 0.5$ m³ and thermal resistance of $R_{dev\text{-}mp} > 15$ K/W:

$$Q_{GLHP} = -\varepsilon_{abs}F_{abs}\sigma\left(T_{abs}^4 - T_o^4\right) + \alpha_{s,abs}q_s F_{abs}\cos(\varphi) + \frac{\left(T_{mp} - T_{dev}\right)}{R_{mp\text{-}dev}} \geq 2Q_{min} \quad (10.11)$$

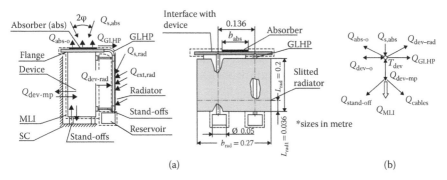

(a)   (b)

FIGURE 10.27
(a) Appearance of a thermal control system with a gas-loaded heat pipe and (b) structure of heat balance for the device.

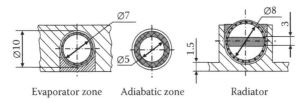

Evaporator zone   Adiabatic zone   Radiator

FIGURE 10.28
Configurations of capillary structure in gas-loaded heat pipe zones (sizes in millimetres).

$$Q_{GLHP} + \frac{(T_{dev} - T_{rad})}{R_{dev\text{-}rad}} + \alpha_{s,rad} q_s F_{rad} \sin(\varphi) + Q_{ext,rad} < \varepsilon_{rad} F_{rad} \eta_{rad} \sigma (T_{rad}^4 - T_o^4) \qquad (10.12)$$

where F_{abs} and F_{rad} are the areas of absorber and radiator, respectively, of the GLHP; $R_{dev\text{-}rad}$ is the thermal resistance of device–GLHP radiator; $Q_{ext,rad}$ is the absorbed additional heat flux; η_{rad} is the thermal efficiency of radiator; and $T_{abs} = T_{dev} \approx T_v$.

On the basis of Equations 10.11 through 10.12, the heat flux, passed via GLHP, Q_{GLHP}, is about 1–14.5 W. The density of flux absorbed by radiator, $q_{ab,rad} = 18$–240 W/m², is used to define the variation of temperatures of the gas-filled zones of radiator and reservoir. To keep the device temperature range of 10°C–30°C, the reservoir with CS should have the ratio $V_{res}/V_{con} > 12$ and methanol as working fluid. The CS, made of metal felt with discrete fibres of diameter $\varnothing 30 \times 10^{-6}$ m, length 3×10^{-3} m and porosity 0.8 ± 0.01 in all zones, has the configurations shown in Figure 10.28.

For elaboration of engineering model the next technological tasks should be solved: development of absorber with $\alpha_{s,abs} > 0.9$ and $\varepsilon_{abs} < 0.18$, background and verification of methods of diminishing the sizes of GLHP reservoirs, minimisation of heat exchange through stand-offs and via MLI, and integration of two GLHPs in one common radiator.

As a variant of the material for fabrication of solar absorbers, the low vacuum condensate of aluminium oxide on polished aluminium has been tested (Palatnik et al., 1972), and it has shown the predicted characteristics, $\alpha_{s,abs} = 0.94$ and $\varepsilon_{abs} = 0.15$. The other selected materials are anodised titanium foil with $\alpha_{s,abs} = 0.7$ and $\varepsilon_{abs} = 0.1$ and Inconel X foil with $\alpha_{s,abs} = 0.52$ and $\varepsilon_{abs} = 0.1$ (Gilmore 2002). The use of these coatings requires a larger area of absorber and a bigger mass.

The proposed thermal conception of device temperature control was realised in the engineering model of TCS (Figure 10.29a). The TVTs of this model have been conducted in a chamber with a closed cylindrical screen cooled by liquid nitrogen. These tests have shown that for its operation in the Earth orbit at $\varphi = 0°$ and variation of T_{mp} from $-20°C$ to $50°C$, the passive TCS provides $T_{dev} = 23°C–25°C$, and in the Mars orbit T_{dev} ranges from $16°C$ to $19.5°C$ (Figure 10.29b). As angle φ changes to $20°$ (at SC manoeuvres), the range of T_{dev} for the Earth orbit is increased to $26°C–32°C$ and for the Mars orbit to $22°C–23.5°C$.

The structure of heat balance for a device with a TCS at operation near Mars at the minimal mounting place temperature shows that the minimal heat transfer by GLHP is nearly $0.7–1$ W (Figure 10.30a). This value should be increased to guarantee positive heat balance for GLHP. At a larger absorber area (1.5 of nominal) under the cold conditions ($T_{mp} = -20°C$) for the Mars orbit, the flux via GLHP may rise to 2.5 W, which will be sufficient for GLHP operation (Figure 10.30b). The increase of absorber area may be realised once at $q_s < 700$ W/m^2. This enhancement could be implemented in the absorber design with an opening lid. Opening of the lid changes the value of F_{abs} and,

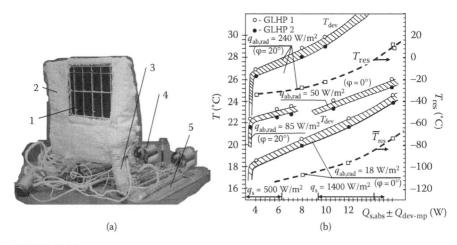

(a)   (b)

FIGURE 10.29
(a) Passive thermal control system for a device without inherent power generation and (b) regulating characteristics of the system as a function of the temperatures of device, adiabatic zone and reservoir of power transferred: (1) solar absorber with coating CΠ-P3; (2) flange with MLI; (3) gas-loaded heat pipe; (4) reservoir; (5) radiator.

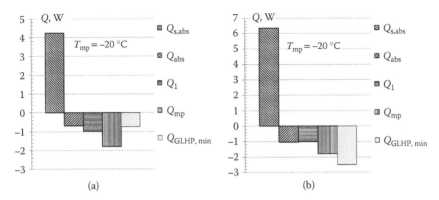

FIGURE 10.30
Heat power budget for the assembly device + TCS in the vicinity of Mars at $T_{mp} = -20°C$: (a) variant with $F_{abs} = 0.01$ m²; (b) variant with $F_{abs} = 0.015$ m². It is noted that $Q_{s,abs}$ is absorbed solar flux; Q_{abs} is heat radiated by absorber; Q_l is leak via MLI to space; Q_{mp} is leak to mounting places; and $Q_{GLHP,min}$ is minimal flux transferred by gas-loaded heat pipe. Positive values signify input to device and negative signify removal.

consequently, increases the quantity of heat, $Q_{s,abs}$, introduced into the system. The lid operation should be unitary to simplify the lock construction and the drive mechanism and be performed by command from the earth or from an SC computer. Such a mechanism has been elaborated and introduced into TCS design.

Thus, the following conclusions can be made:

1. The principle of passive TCS design is proposed and proved for devices that do not have their inherent heat generation but require temperature stabilisation on the level of 20°C. The deficit of thermal energy is compensated by a solar energy absorber, and GLHP keeps the heat balance of the system at a pre-defined temperature level. The given principle could be applied for instruments having solar orientations, in the use of an additional heat pipe, thermally connecting a device with an absorber and for devices located arbitrarily.

2. A device temperature regulation range of 16°C–32°C is provided in the following conditions: solar constant in the range of 500–1400 W/m², temperature of mounting place between −20°C and +50°C, and direction of the device to the sun owing to manoeuvres in an angle of ±20 grad.

3. The main difficulty of keeping device heat balance is in the minimal value of solar constant and minimal temperature of mounting place. To overcome this problem, the use of an additional element—a lid with a lock—and the optimisation of characteristics of insulation elements (low conductance stand-offs, cables and MLI) have been proposed. The mass of the TCS is 12% of the device mass.

10.6 Thermal Control of Small Satellite Compartments

The next stage of VCHP implementation in spacecraft design deals with an elaboration of the thermal and mechanical conception of isothermal compartment for device layout, which combines schemes (d) and (e) of Figure 10.11. Realisation of this idea and its successful example for a large space research canister ($1 \times 1 \times 3$ m^3, 340 kg of equipment, 100–400 W of power) is presented by McIntosh and Ollendorf (1978) for shuttle missions.

A two-level TCS system is proposed for small satellites. Four panels for device montage form the compartment. Each panel has a distributed network of axially grooved aluminium heat pipes with thermal interface to a GLHP. The last unit realises temperature regulation of the device's mounting places (Figure 10.31a and b).

A mounting panel is an all-metal structure, made of alloy Д16 (analog of alloy 2024) without an inner honeycomb structure. It was fabricated by milling (without pressing machinery), which gives a substantially higher mobility at fabrication of the whole compartment, keeping the mass and thermal parameters of panels close to those of honeycomb panels. The thermal network in panels and the method of integration of heat pipes into panels enable heating or cooling of arbitrarily located devices and transportation of heat in the zone of heat exchange with a GLHP. Each device panel had 0.8 m^2 surfaces for two-sided device mounting with a heat generation of up to 60 W. The mass of a panel with heat pipes was 7 kg. In the design, an ancillary thermal energy source (with a mass of 2.5 kg) was foreseen, which allowed the TCS to be independent of the device's heat generation in different satellite operating modes. The range

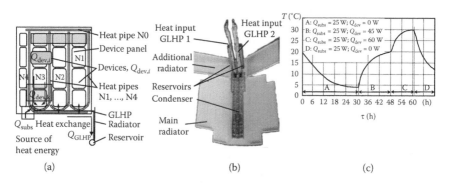

FIGURE 10.31

(a) Scheme of an isothermal panel for device thermal control with regulating temperature of a mounting place, (b) appearance of a gas-loaded heat pipe with a hot reservoir and (c) the regulating characteristic of a thermal control system.

of temperature control of a panel was 16°C ± 14°C at the variation of heat generation of all devices, $Q_{dev} = 0$–60 W (Figure 10.31c). At a density of heat generation of 5 kW/m², non-isothermality of the panel along a length of 0.75 m did not exceed 2°C. This principle has been applied in designing the device compartment of space laboratory Regata (developer of the laboratory is Space Research Institute of the Russian Academy of Sciences, Moscow, Russia), in space by the sun light pressure of light from the sun (Baturkin et al., 1992).

10.7 Conclusions

1. VCHPs, particularly GLHPs, have been implemented into a thermal and structural design of four types of scientific instruments: autonomous devices with and without heat generation, electronic cards and satellite device compartments. They differ in linear size and mass in one order and operate in the temperature level of 10°C–30°C with a temperature control zone narrower than ±15°C at variable power generation and under changeable external thermal conditions.

2. For autonomous devices with a ratio of heat generation, Q_{dev}, to mass of up to 3.3 W/kg (950 W/m³), TCSs reach temperature stabilisation at minimal device heat generation on the level of 1.5 W. The temperature of the device case, on which the electronic PCB is installed, is 25°C–32°C in semi-passive mode (at constant temperature of reservoir with NCG) and 10°C–30°C in a passive mode (without any additional energy sources) at the following thermal disturbances: device heat generation ($Q_{dev,min}/Q_{dev,max} = 1.5$–13 W), external heat absorption flux ($q_{ab,min}/q_{ab,max} = 120$–270 W/m²) and the temperature of mounting places ($T_{mp,min}/T_{mp,max} = -20°C$–+50°C).

3. The concept of a two-level TCS, regulating the temperature of electronic elements installed on a printed board circuit, is proposed. A flat heat pipe (0.13 × 0.18 × 0.003 m³), which can operate at any orientation relative to gravity, is the basis for the PCB layout. This flat heat pipe is attached to GLHPs, which control its temperature. The proposed PCB thermal design significantly narrows the range of temperatures of the elements: for non-heat-generating elements the range is up to 23°C ± 10°C and for heat-generating elements up to 40°C ± 25°C, with a variable heat load ($Q_{max}/Q_{min} = 8:1$), temperature of mounting place (−20°C to +50°C) and external thermal disturbances (140–280 W/m²).

4. For devices having limited or no own heat generation, the principle of temperature stabilisation has been reformulated to use solar light as the source of energy and GLHP for keeping the heat balance at the required temperature level. The experimentally obtained device temperature stabilisation is 16°C–32°C (at passive control) at external disturbances, which include change of the solar constant from the Earth orbit to the Mars orbit, change of the temperature of device mounting location from −20°C to 50°C, and change of spacecraft orientation relative to the sun as a result of manoeuvres in angles of ±20°.

5. The design of an isothermal panel for a device compartment combines the functions of mechanical fastening of devices and their cooling/heating and redistribution of thermal energy, providing isothermality of mounting seats for devices less than 5°C at temperatures of device mounting places of 16°C ± 14°C, when heat generation is from 0 to 60 W. Four such panels create the frame of a payload compartment of a small space laboratory.

6. All the proposed TCS designs integrated in four types of scientific instruments have passed complete mechanical tests and TVTs, which confirmed the rightness of the structural and technological decisions that were made.

7. TCSs on GLHPs are reliable instruments of space thermal technology, which together with other thermal control means create the basis for successfully solving complicated thermal tasks of near-earth, interplanetary and ground missions (Table 10.1). Comparison of different thermal control means and GLHPs outlines the following conditions of GLHP application:

Power transfer range: 2–100 W.

Temperature regulation: level of 10°C–30°C; level of 50°C–70°C.

Accuracy of temperature regulation: in passive mode 1°C–20°C, in semi-active mode 1°C–5°C (reservoir temperature stabilisation) and in active mode 0.5°C–2°C (reservoir temperature variation).

Thermal disturbances to be compensated: variable power transferred, variable heat sink temperature (or variable external heat flux, absorbed by radiator), changes in radiator optical properties due to environmental factors, SC temperature variation and mounting place temperature variation.

Distance 'heat source–radiator': less than 1 m.

Restriction: the orientation relative to gravity or other mass forces should not prevent the circulation of heat carriers.

Space approved technology: yes.

TABLE 10.1

Comparison of Different Space Thermal Control Means (Heat Transfer of 20–100 W)

Thermal Control Means	Ability to Transport Heat over Distances	Control Temperature in Passive Mode	Control Temperature in Active Mode	Technical Characteristics	Ability to Operate in 1 g Gravity	Presence of Mechanical Moving Component	Inherent Power Consumption	Main Functions in TCS
Thermal switch	0.01–0.025 m	YES	NO	$\Delta T_{reg} = 5$–20 K $M = 0.1$–0.2 kg $GL_{max} = 0.73$ W/K $GL_{min} = 0.0075$ W/K $GL_{max}/GL_{min} > 80{:}1$ (Gilmore 2002)	HIGH	YES	NO	Device thermal connection with radiator, with spacecraft structure
Loop heat pipe	0.1–4 m	YES	YES	$\Delta T_{reg}^{\ active} = 1$–6 K $Q_{max}/Q_{min} = 15{:}2$ Mass $= 0.1$–0.5 kg (without radiator) (Maydanik et al., 2003, Goncharov et al., 2009)	HIGH	YES	NO [passive] YES [active]	Heat removal from device to structure or radiator and temperature regulation
Capillary pump loop	0.5–6 m	NO	YES	$Q = 200$ W on distance of 10 m at 20°C with Refrigerant-11 (Capillary pump loop heat pipe 1982)	NO	NO	YES [active]	Heat removal from device to spacecraft structure or radiator and temperature regulation

(Continued)

TABLE 10.1 (*Continued*)

Comparison of Different Space Thermal Control Means (Heat Transfer of 20–100 W)

Thermal Control Means	Ability to Transport Heat over Distances	Control Temperature in Passive Mode	Control Temperature in Active Mode	Technical Characteristics	Ability to Operate in 1 g Gravity	Presence of Mechanical Moving Component	Inherent Power Consumption	Main Functions in TCS
Louvre	0.01–0.06 m	YES	YES	$\Delta T_{reg} = 10$ K $m = 3\text{–}5\ \text{kg/m}^2$ $\varepsilon_{min} = 0.1$; $\varepsilon_{max} = 0.7$ $\varepsilon_{max}/\varepsilon_{min} < 7{:}1$ (Gilmore 2002)	HIGH	YES	NO	Device thermal coupling with the space; temperature regulation
Controllable heater, thermostat	0	NO	HIGH	$Q_{max}/Q_{min} = Q_{max}{:}0$ $\Delta T_{reg} = 0.01\text{–}10$ K Mass $= 0.05\text{–}0.1\ \text{kg}^a$ For operation, SC power is necessary	HIGH	NO	YES	Compensation of thermal energy deficit in device heat balance at required temperature level
Heat accumulator	0	YES	NO	$\Delta T_{reg} = 0.4$ K $E = 360$ KJ, $Q = 100$ W $M = 4.3$ kg $V = 0.3{\times}0.3{\times}0.02\ \text{m}^3$ (Gilmore 2002)	HIGH	NO	NO	Absorption of excess heat with its subsequent release

Single-phase pumped loop	NO	Tens of metres	YES	HIGH	$\Delta T_{reg} = 0.2\text{–}10$ K $G = 0.75$ l/min $\Delta P = 27$ kPa, $Q_{el} = 10$ W $M_{pump} = 8$ kg (Gilmore 2002)/for 30 kW system $m_{tr} = 0.138$ kg/W (Sementsov 2003, Delil 2004)	YES	YES $Q_{el}/Q = 0.07$ (Capillary pump loop heat pipe 1982)	Heat removal from device to structure or radiator, heat exchanger and temperature regulation
Two-phase pumped loop	NO	Tens of metres	YES	HIGH	It is preferable to compare with single phase at $Q > 10$ kW (Sementsov 2003, Delil 2004); $m_{tr} = 0.093$ kg/W (Sementsov 2003, Delil 2004)	YES	YES $Q_{el}/Q = 0.006$ (Capillary pump loop heat pipe 1982)	Heat removal from device to spacecraft structure or radiator, or heat exchanger and temperature regulation
Variable emittance coating	YES	0	YES	HIGH	$m = 0.4$ kg/ m2 (without substrate) $\varepsilon_{min} = 0.3$ $\varepsilon_{max} = 0.8$	NO	YES	Device thermal coupling with the space; temperature regulation

(Continued)

TABLE 10.1 (*Continued*)

Comparison of Different Space Thermal Control Means (Heat Transfer of 20–100 W)

Thermal Control Means	Ability to Transport Heat over Distances	Control Temperature in Passive Mode	Control Temperature in Active Mode	Technical Characteristics	Ability to Operate in 1 g Gravity	Presence of Mechanical Moving Component	Inherent Power Consumption	Main Functions in TCS
GLHP	0.1–0.7 m	YES	YES	$\Delta T_{reg}^{active} = 0.5\text{--}2$ K $\Delta T_{reg}^{passive} = 5\text{--}20$ K $M = 0.2\text{--}0.7$ kg $^{without\ radiator}$ $GL_{max}/GL_{min} > 15{:}1$	LOW	NO	$NO^{passive}$ YES^{active} (1–3 W)	Heat removal from device and its thermal coupling with the space; temperature regulation

ΔT_{reg} – temperature variation at regulation (K); m – specific mass, ratio of equipment mass and radiating surface (kg/m²); m_{tr} – specific mass, ratio of system mass and transferred power (kg/W); M – mass (kg); GL_{min} – minimal conductance (W/K); GL_{max} – maximal conductance (W/K); ε_{min} – minimal emittance of radiating surface; ε_{max} – maximal emittance of radiating surface; E – accumulated energy (J); Q – transferred power (W); G – heat-carrier mass rate (l/min); ΔP – available pressure drop in liquid line (Pa); active – active mode with subsidiary power consumption; passive – passive mode without subsidiary power consumption.

ª Without mass of solar panel, power supply unit and harness.

Fields of application of TCSs on variable conductance heat pipes

- For autonomous optoelectronic units (sensors, transmitters, amplifiers, measuring units, mounting places), which use the radiative heat sink
- As isothermal mounting places with regulating temperature for equipment layout
- As base plates for equipment layout with a narrower temperature diapason and isothermal volume canister with stabilised case temperature
- For equipment with adjustable temperature levels during exploitation
- As a variable thermal conductance element, connecting heat source and heat sink to compensate inaccuracy in the definition of transferred heat power and thermal properties of elements involving in heat transfer
- As a thermal insulation element to essentially reduce heat flux from device to space in certain exploitation modes (inner calibration, switch off, standby, safe mode), returning to high conductivity at full power transferred
- For replacement of thermal control by electrically controllable heaters, in cases when SC energy saving is important
- As replacement for convection heat exchangers in liquid cooling systems to enhance whole system operation at wide ambient temperature excursions and rejected power variation (that actually occur in long-life systems based on lunar and Martian surfaces)

Acknowledgements

Elaboration of TCS principles with the use of GLHP advantages was conducted because of tight cooperation with the Special Design Bureau of Space Research Institute of Russian Academy of Sciences, Bishkek, Kyrgyzstan and with the Space Research Institute of Russian Academy of Sciences. The author thanks Dr. A. Dudeev, N. Gudkova, M. Grechina, Dr. V. Kostenko, Dr. M. Semena, K. Shcoda, O. Savchenko and Dr. S. Zhuk for collaborating and assisting in all stages of realisation of ideas and implementing these ideas in practice.

The author also thanks the team from the Department of System Conditioning, Institute of Space Systems, Berlin/Bremen, DLR, Germany for their assistance in surveying and summarising references and chapter review, and for their systematic support in thermal modelling and consultations on thermal vacuum testing procedures.

Nomenclature

C	Total heat capacity (J/K)	**Subscripts and Abbreviations**	
c_p	Specific heat capacity (J/kg·K)	a	Active zone
F	Cross-sectional area (m²)	ab	Absorbed
K	Thermal conductance (W/K)	abs	Absorber
k	Coefficient of heat transfer (W/m²·K)	ad	Adiabatic zone
L	Length (m)	con	Condenser, heat output zone
M	Molecular mass (kg/kmol)	cond	Conductive
m	Mass (kg)	cont	Contact
P	Total pressure (Pa)	dev	Device
P_v	Partial vapour pressure (Pa)	dif	Diffusion
R	Thermal resistance (K/W)	dir	Direct
R_μ	Universal gas constant (8314 J/kmol·K)	eff	Effective
T	Temperature (K, °C)	el	Element
Q	Heat power (W)	ev	Heat input zone, evaporator
q	Heat flux (W/m²)	g	Gas
V	Volume (m³)	GLHP	Gas-loaded heat pipe
Z	Coordinate	HP	Heat pipe
		max	Maximal
Greek Letters		min	Minimal
α_s	Solar absorptance, non-dimensional	MLI	Multi-layer insulation
χ	Coefficient, non-dimensional	mp	Mounting place
ε	Emittance, non-dimensional	o	External environment
η	Efficiency, non-dimensional	rad	Radiator
φ	Angle of solar incidence	res	Reservoir
λ	Thermal conductivity (W/m·K)	rev	Reverse
Π	Perimeter of heat exchange (m)	S	Solar
ρ	Density (kg/m³)	so	Heat source
σ	Boltzmann constant (5.67×10^{-8} W/m²·K⁴)	sh	Shell
Ω	Axial thermal resistance (K/W)	sub	Substrate
Γ	Time (s, min, h)	t	Connecting tube
		v	Vapour

References

ACT. 2010. http://www.1-act.com/products/satellite-thermal-control/variable-conductance-heat-pipe.

Anderson, W., M. Ellis, and K. Walker. 2008. Variable Conductance Heat Pipe Radiators for Lunar and Martian Environments. http://www.1-act.com/pdf/lunandmar.pdf.

Anderson, W.T., D.K. Edwards, J.E. Eninger, and B.D. Marcus. 1974. Variable conductance heat pipe technology. Final Research Report. CR – 114750. TRW Systems Group.

Andreanov, V., V. Artamonov et al., 1973. *Autonomous orbital stations*. Moscow: Nauka (in Russian).

Antoniuk, D. 1987. Generalized modelling of steady state and transient behaviour of variable conductance heat pipes. Proc. of AIAA 22nd Thermophysics Conf., Honolulu, HI, USA.

Antoniuk, D., and D. Edwards. 1980. Depriming of arterial heat pipes: an investigation of CTS thermal excursions. Final Report. NASA CR – 165153. TRW Systems Group.

Barsukov, V.V. 1981. Research of steady and start modes of gas-loaded heat pipes. Abstract of PhD diss., Odessa Technological Institute of Refrigeration Industry (in Russian).

Baturkin, V.M. 1979. Research of thermoregulating properties of low-temperature gas-loaded heat pipes. Abstract of PhD diss., Kiev Polytechnic Institute (in Russian).

Baturkin, V.M. 2011. Thermal control systems on the base of heat pipes for space instrument-making. Abstract of Doctor of technical sciences dissertation, National Technical University of Ukraine 'Kyiv Polytechnic Institute' (in Russian).

Baturkin, V.M., N.K. Grechina, S.K. Zhuk et al., 1992. Component of thermocontrol system on the basis of heat pipes for scientific device compartment of small space laboratory 'REGATA'. Proc. of 8th Int. Heat Pipe Conf., Beijing, China, Report No E-1.

Bienert, W.B., P.J. Brennan, and J.P. Kirkpatrick. 1971. Feedback controlled variable conductance heat pipes. *AIAA Paper* 421:1–11.

Bienert. W.B. 1972. Application of heat pipes for temperature regulation. *Heat pipes: Collection of scientific proceedings*. Moscow: Mir Publishers (in Russian), 349–370.

Bobco, R.P. 1989. VCHP Performance prediction: comparison of first-order and flat front models. *Journal of Thermophysics* 3/4:401–405.

Brennan, P., and E. Kroliczek. 1979. *Heat Pipe Design Handbook*. Maryland: B & K Engineering, Inc. Prepared for NASA Goddard Space Flight Center under contract No NAS5-23406.

Capillary pump loop (CPL) heat pipe. 1982. Development status report. CR-175273. OAO Corporation.

CAST. 2012. http://www.cast.cn/CastEn/Show.asp?ArticleID=17504.

Cima, R.M., and J.H. Abrose. 1995. Elimination of natural convection heat transport in variable conductance heat pipes. Proc. of 9th Int. Heat Pipe Conf., Albuquerque, NM, USA.

Delil, A. 2004. Ongoing Developments of Mechanically Pumped Two-Phase Thermal Control Systems. Presentation on 15 Spacecraft Thermal Control Technology Workshop. El Segundo.

Delil, A.A.M., and J. van der Vooren. 1981. *Uniaxial model for gas–loaded variable conductance heat pipe performance in the inertial flow regime*. Advances in Heat Pipe Technology. Proc. of 4th Int. Heat Pipe Conf., London, United Kingdom, 359–372.

Doctor, F., and R.C. van Benthem. 2011. Prototype thermal design module for automated designs of space harness. Presentation on 25th European Workshop on Thermal & ECLS Software, ESTEC, Noordwijk, The Netherlands.

Donabedian, M. ed. 2004. *Satellite Thermal Control Handbook*, Vol. 2: Cryogenics. El Segundo, California: The Aerospace Corporation Press.

Dulnev, G.N., and A.P. Belyakov. 1985. *Heat pipes in electronic systems of temperature stabilization*. Moscow: Radio i svyaz (in Russian).

Dunn, P.D., and D.A. Reay. 1976. *Heat pipes*. Oxford: Pergamon Press.

Dvirnyi, V., O. Zagar, Yu. Golovanov, S. Ermilov, K. Smirnov-Vasiliev, V. Khalimanovich et al., 1999. Aluminium variable-conductive heat pipes of the communication satellites. AIP Conf. Proc., Int. Forum 'Space Technology and Application', Albuquerque, New Mexico, USA, 949–953.

Edelstein, F., and H. Flieger. 1978. Satellite battery temperature control. Collection of technical papers of 3rd Int. Heat Pipe Conf., Palo Alto, CA, USA, 361–365.

Edom, A. 2001. Development and dynamic study of a space radiator integrating multiple gas–loaded heat pipes. PhD dissertation, Santa Catarina Federal University.

Edwards, D.K., G.L. Fleischman, and B.D. Marcus. 1972. Theory and design of variable conductance heat pipes: steady state and transient performance, Research Report N3, CR – 114530. TRW Systems Group.

Edwards, D.K., G.L. Fleischman, and B.D. Marcus. 1973. User's manual for the TRW gaspipe 2 program. TRW document No 13111-6054-R0-00, CR – 114672. TRW Systems Group.

ESATAN-TMS Thermal User Manual. 2011. ITP Engines UK Ltd. Whetstone, Leicester: 3-65–3.66, 6-118–6-119.

Faghri, A. 1995. *Heat pipe science and technology*. Taylor & Francis. Publishing office: Taylor &Francis, 1900 Vermont Avenue, N.W., Suite 200, Washington, DC 20005–3665

Favorskiy, O.N., and Ya.S. Kadaner. 1972. *Problems of heat exchange in space*. Textbook for institutes of higher education, 2nd issue. Moscow: Vysshaya shkola (in Russian).

Fleischman, G.L., G.R. Pasley, and R.J. McGrath. 1978. A high reliability variable conductance heat pipe space radiator. Collection of technical papers of 3rd Int. Heat Pipe Conf., Palo Alto, CA, USA, 216–226.

Furukawa, M. 1983. Analytical studies for temperature predictions and control with emphasis on space application. TKU-80604. National Space Development Agency of Japan.

Gilmore, D. ed. 2002. *Satellite Thermal Control Handbook*. El Segundo, CA: The Aerospace Corporation Press.

Gnilichenko, V.I. 1982. Research of dynamic characteristics of low-temperature heat pipes at start-up from the state with frost heat carrier. Abstract of PhD dissertation, Odessa Technological Institute of Refrigeration Industry (in Russian).

Goncharov, K., Al. Kochetkov, V. Buz, and R. Schlitt. 2009. LHP in European satellite, Proc. of Int. Conf. 'Heat pipes for space application', Moscow, Russia.

Goncharov, K.A., O.A. Golovin, Y.V. Panin, K.N. Korzhov, and A.Yu. Kochetkov. 2007. Variable conductance heat pipe. Preprint, 14th Int. Heat Pipe Conf., Raia do Santinho – Florianopolis, Brazil.

Groll, M., and M. Hage. 1974. Development of an electrical feedback controlled variable conductance heat pipes for space applications, *AIAA Paper* 752:1–7.

Hinderman, J.D., E.D. Waters, and R.V. Kaser. 1971. Desing and performance of non-condensible gas-controlled heat pipes. *AIAA Paper* 420:1– 8.

International Astrophysical Observatory 'GRANAT'. 2010. http://sigma-2.cesr.fr/sigma/Sigma.Home.html.

Kaganer, M.G. 1979. Heat and mass exchange in low-temperature thermoinsulated constructions. Moscow: Energia (in Russian).

Karam, R.D. 1998. *Satellite thermal control for systems engineers*, Vol. 181, Progress in Astronautics and Aeronautics. VA: American Institute of Aeronautics and Astronautics.

Kirkpatrick, J.P., and B.D. Marcus. 1972. A variable conductance heat pipe/radiator for the lunar surface magnetometer. *Thermal control and radiation. Progress in Astronautics and Aeronautic* 31:83–102.

Konev, S.V. 1976. Research of heat and mass exchange in gas-regulated heat pipes. Abstract of PhD diss., Heat Mass Transfer Institute. Minsk (in Russian).

Kreb, H., M. Perdu, and C. Savage. 1978. Design of a gas-controlled radiator for a Marots–type TPA–radiator. Collection of technical papers of 3rd Int. Heat Pipe Conf., Palo Alto, CA, USA, 233–240.

Lanteri, A., and B. Henning. 1988. Tele-X antenna module thermal control. Proc. of 3rd European Symp. on Space Thermal Control and Life Support Systems, Noordwijk, The Netherlands, 3:319–326.

Marcus, B.D. 1972. Ames heat pipe experiment (AHPE), Experiment description document. TRW document No 13111-6033-R0-00. Report NASA CR – 114413. TRW Systems Group.

Marcus, B.D. 1972. Theory and design of variable conductance heat pipes. Report NASA CR – 2O18. TRW Systems Group.

Marcus, B.D., and G.L. Fleischman. 1970. Steady-state and transient performance of hot reservoir gas-controlled heat pipes. *ASME Paper* HT/SpT-11.

Marshburn, J., and R. McIntosh. 1978. Design & evaluation of a rotating variable conductance heat pipe system. Collection of technical papers of 3rd Int. Heat Pipe Conf., Palo Alto, CA, USA, 211–215.

Materials Properties (database). 2012. K&K Associates. http://www.tak2000.com/ThermalConnection.htm.

Matonak, B., and H. Peabody. 2006. An introductory guide to variable conductance heat pipe simulation. www.tfaws.nasa.gov/TFAWS06/Proceedings/ThermalControlTechnology//Papers/

Maydanik, Yu., S. Vershinin, and C. Gerhart. 2003. Miniature LHPs with a flat and a cylindrical evaporator – which one is better? Presentation on International Two-Phase Thermal Control Technology Workshop. Noordwijk.

McIntosh, R., and S. Ollendorf. 1978. A thermal canister experiment for the space shuttle. Collection of technical papers of 3rd Int. Heat Pipe Conf., Palo Alto, CA, USA, 402–407.

Mock, P.R., E.A. Marcus, and B.D. Edelman. 1974. Communications technology satellite – a variable conductance heat pipe application. Proc. of AIAA/ASME Thermophysics and Heat Transfer Conf., Boston, MA, USA, 743–749.

NX7.5 (software product). MAYA Heat Transfer Technologies Ltd. http://www.mayasim.com.

Palatnik, L.S., M.Ya. Fuks, and V.M. Kosevich. 1972. *Mechanisms of formation and structure of condensed films*. Moscow: Nauka (in Russian).

Prasher R., C.-P. Chiu, and R. Mahajan. 2004. Thermal interface materials: A brief review of design characteristics and materials. *Electronics Cooling*, February 2004. http://www.electronics-cooling.com/2004/02/thermal-interface-materials-a-brief-review-of-design-characteristics-and-materials/.

Red Book, The. 2013. http://www.sheldahl.com/documents/RedBook.pdf.

Savage, C.J., and G.M. Aalders. 1978. Development of a technological model variable conductance heat pipe radiator for Marots type communication spacecraft. Collection of technical papers of 3rd Int. Heat Pipe Conf., Palo Alto, CA, USA, 227–232.

Savage, C.J., B.G.M. Aalders, and H.A. Kreeb. 1979. Variable conductance heat pipe radiator for MAROTS–type communication spacecraft. *Journal of Spacecraft* 16/3:176–180.

Schlitt, R. 1973. Design and testing of a passive, feedback-controlled, variable conductance heat pipe. NASA TM X-62, 293.

Semena, M.G., A.N. Gershuni, and V.K. Zaripov. 1984. *Heat pipes with metal-fibrous capillary structures*. Kyiv: Vyshcha shkola (in Russian).

Semena, M.G., V.M. Baturkin, B.M. Rassamakin, and N.K. Grechina. 1990. Analytical and experimental study of operating characteristics of low temperature variable thermal resistance heat pipes. Proc. of 7th Int. Heat Pipe Conf., Vol.1: Fundamental and Experimental Studies, Minsk, Belarus, 525–532.

Sementsov, A.N. 2003. Modelling of two-phase heat transfer loop of centralized heat removal system of Russian segment of international space station in conditions of space flight. Abstract of PhD diss., Moscow Aviation Institute (in Russian).

Shimoji, S., and H. Kimura. 1978. Prediction of evaporator temperature of a gas loaded heat pipe by the diffuse front model. Collection of technical papers of 3rd Int. Heat Pipe Conf., Palo Alto, CA, USA, 155–161.

SINDA/G User's Guide. 2003. Version 2.3:288–317. Copyright by Network Analysis Inc., Arizona Sate Research Park, 7855 S.River Parkway #101, Tempe, AZ 85284, USA, www.sinda.com

Smirnov, H.F., and A.D. Tsoy. 1999. *Heat exchange at vapor generation in capillaries and capillary-porous structures*. Moscow: Publishing house of Moscow Energy Institute (in Russian).

Suntsov, S., V. Derevyanko, A. Makukha, D. Nesterov, A. Burov, O. Ivanov, et al., 2009. Terminally hyper-conductive porous structures in astrionics units. Proc. of Int. Conf. 'Heat pipes for space application', Moscow, Russia.

Systema 4.5.0 (software product). MAYA Heat Transfer Technologies Ltd. http://www.systema.astrium.eads.net.

User's Guide MSC Sinda 2008 r1, MSC. 2009. Software Corporation. Version 2.6.00:312–339.

Van Benthem, R.C., W. de Grave, F. Doctor, K. Nuyten, S. Taylor, and P.A. Jacques. 2011. Thermal analysis of wiring for weight reduction and improved safety. Report NLR-TP-2011-469, National Aerospace Laboratory NLR of the Netherlands.

Vasiliev, L.L., and S.V. Konev. 1977. Regulated heat pipes. *Engineering-Physical Journal* 32/5:920–938 (in Russian).

Vasiliev, L.L., L.P. Grakovich, and V.G. Kiselyov. 1973. Experimental study of heat and mass transfer in heat pipes with non-condensable gas. Int. Heat Pipe Conf., Stuttgart, Germany, 1–10.

11

Thermosyphon Technology for Industrial Applications

Marcia B. H. Mantelli

CONTENTS

11.1 Introduction 412
    11.1.1 Configurations and Geometries 413
11.2 Modelling of Physical Principles 415
    11.2.1 Nusselt Models for Condensation in Walls 415
    11.2.2 Thermal Modelling 418
        11.2.2.1 Thermal Resistance Models 421
    11.2.3 Heat Transfer Limits 427
        11.2.3.1 Entrainment or Flooding Limits 427
        11.2.3.2 Sonic Limit 430
        11.2.3.3 Viscous Limit 431
        11.2.3.4 Dry-Out Limit 431
        11.2.3.5 Oscillation Limit 432
        11.2.3.6 Boiling Limit 433
11.3 Thermosyphon Design and Fabrication 434
    11.3.1 Selection of Working Fluid 435
    11.3.2 Selection of Casing Materials 438
    11.3.3 Thermosyphon Fabrication 441
        11.3.3.1 Cleaning Process 442
        11.3.3.2 Out-Gassing of Tube Materials and
                  Working Fluids 443
        11.3.3.3 Charging Procedures 443
    11.3.4 Design Methodology 446
    11.3.5 Thermal Testing 448
11.4 Industrial Applications 450
    11.4.1 Heat Exchangers 450
    11.4.2 Ovens 452
    11.4.3 Oil Tank Heaters 457
    11.4.4 High Heat Power Cooling 458
11.5 Concluding Remarks 460

Acknowledgements 460
Nomenclature 461
References..462

11.1 Introduction

Thermosyphons, also known as gravity-assisted heat pipes, are very efficient heat transfer devices. They work in two-phase close cycles where latent heat of evaporation and condensation is used to transfer heat.

Basically, a thermosyphon consists of a metallic evacuated tube where a certain amount of working fluid is inserted. They are composed of three different regions: evaporator, condenser and adiabatic sections.

In the evaporator region, heat is supplied to the tube, vaporising the working fluid contained in the evaporator. The generated vapour, due to pressure differences, travels to the condenser region. In the condenser, heat is removed, condensing the vapour, and the resulting liquid returns, by gravity, to the evaporator, closing the cycle. To work properly, the thermosyphon evaporator must be located in a region below the condenser. Figure 11.1 illustrates the physical working principles of thermosyphons.

Between the evaporator and condenser, an adiabatic region can be found. For practical applications, this intermediate region may be small or even not existent, depending on the geometry of the device where the thermosyphon will be mounted.

The main difference between thermosyphons and heat pipes is that in heat pipes the liquid returns from the condenser to the evaporator by means of a porous medium, usually located inside the case, attached to the internal

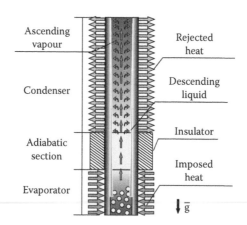

FIGURE 11.1
Schematic of the working principles of thermosyphon.

walls. Efficient porous media require a sophisticated fabrication process which tends to increase the cost of the device. One example of application of heat pipes is in space vehicles, where the lack of gravity and the high cost involved in space missions make these devices convenient. Another very typical application of heat pipes is for cooling of electronic devices in portable computers, such as laptops and notebooks. In this case, the computer industry has made large investments to produce low-cost, easy-to-fabricate devices.

Thermosyphons are more suitable than heat pipes for most of the industrial applications, especially those involving large equipment as, in general sense, they are more efficient and cheaper to fabricate. Gravity can be used to promote the working fluid displacement from the evaporator to the condenser, eliminating the need of a porous medium. They can be used in any application that requires efficient heat conduction or isothermalisation of walls and/or environments. In this chapter, some modelling of the working principles will be presented as well as some industrial applications of thermosyphon heat transfer devices.

11.1.1 Configurations and Geometries

Thermosyphons can have different configurations. The simplest and most common configuration is the straight geometry, where the thermosyphons are made of a single tube, as illustrated in Figure 11.1. Several equipments make use of straight thermosyphons, with heat exchangers being the most known industrial application, as discussed in Section 11.4. Non-condensable gases can be included inside the thermosyphon so that a controlled temperature distribution in the wall can be observed, as required for some applications.

Thermosyphons can also be built in circuits, where the evaporator and condenser are connected by tubes that conduct only liquid or vapour, avoiding working fluid droplets from being carried by the vapour in countercurrent flows, as observed in straight thermosyphons. This configuration, denominated as loop thermosyphons, is illustrated in Figure 11.2. Thermosyphons in circuits are also known as gravity-assisted loop heat pipes. One should note that, also for this configuration, the evaporator must be located in an inferior position relative to the condenser.

Thermosyphon-type vapour chambers can be described as 'inverse thermosyphons': the external surfaces are adiabatic, while all the heat exchanges happen inside the device. Figure 11.3 illustrates this device. In this case, the heat source (a hot stream flowing inside the inferior tube) supplies heat to the working fluid, which evaporates. The resulting vapour rejects heat to a heat sink (in this case, the stream inside the upper tube) and condenses over the tube external surface. The condensate returns to the liquid pool by means of gravity. Large-size vapour chamber applications include the heating of natural gas in gas distribution stations, where cold gas is heated before being delivered to the final users. Vapour chambers can also be found in small sizes, in flat shapes, the flat heat pipes. They are usually applied as

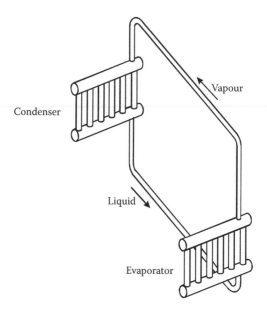

FIGURE 11.2
Schematic of a loop thermosyphon.

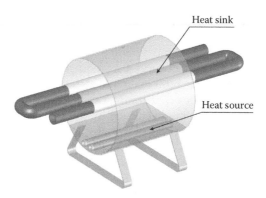

FIGURE 11.3
Vapour chamber thermosyphon configuration.

heat spreaders in electronic cooling and possess a porous medium inside the chamber to ensure that the liquid is delivered to the heat source.

In thermosyphons with tree configurations, a common evaporator (trunk) is connected to several condensers (branches). These branches can be vertical, for uniform heating of vertical surfaces, or inclined, for near horizontal or inclined surfaces. Figure 11.4 shows schematics of vertical and horizontal tree configuration thermosyphons. In these configurations, the amount of

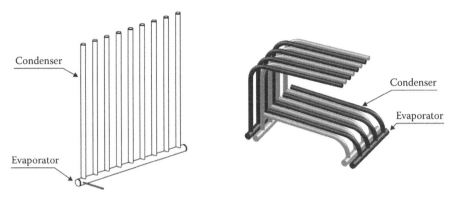

FIGURE 11.4
Tree configuration thermosyphons.

liquid inside the evaporator should be carefully determined to avoid blocking the vapour connection between evaporator and condenser. The main feature of this configuration is that any point of any condenser has basically the same temperature. This configuration is very useful when one desires to keep a surface in a uniform controlled temperature level. Therefore, tree thermosyphons are very useful for ovens for many industrial and domestic applications.

11.2 Modelling of Physical Principles

11.2.1 Nusselt Models for Condensation in Walls

The vapour condensation over cooled walls may be homogeneous or heterogeneous. The heterogeneous condensation happens over small particles (like dust): when close to these particles, the saturated vapour suffers a small pressure drop, creating condensation conditions. On the other hand, to homogeneous condensation to happen, the main condition to be fulfilled is that the vapour temperature must be over the saturation temperature.

The Nusselt model for condensation in vertical cooled walls can be used for describing the homogeneous condensation phenomenon that happens inside the condenser section of the thermosyphon. This model is especially valid when the casing tube has large dimensions, so that the casing curvature does not affect very much the condensation process. Most of the condensation models developed to predict the condensation in thermosyphon condensers are based on the Nusselt analysis, which will be briefly described in this chapter.

A liquid film in steady-state conditions over a vertical cooled surface is considered, as shown in the schematic of Figure 11.5. The equation of conservation of movement, in the vertical direction, where the inertia effects (advective terms) and the diffusion are neglected is from [1]:

$$\frac{\partial^2 v}{\partial y^2} = \frac{1}{\mu_l}\frac{dp}{dx} - \frac{X}{\mu_l}$$ (11.1)

where v is the vertical velocity, p is the pressure, X represents the body forces, μ_l is the liquid viscosity and x and y are the coordinates.

The body force (X) within the liquid film is given by the liquid density times the gravity ($\rho_l g$). According to the boundary layer hypothesis, the pressure gradient within the boundary layer is the same within the vapour, far from the boundary layer, and is given by vapour density times gravity, $\rho_v g$. Therefore, Equation 11.1 takes the following form:

$$\frac{\partial^2 v}{\partial y^2} = -\frac{g}{\mu_l}(\rho_l - \rho_v)$$ (11.2)

Integrating twice and applying the boundary conditions $v(0) = 0$ and $(\partial v/\partial y)\big|_{y=\delta} = 0$, the following velocity profile is found:

$$v(y) = \frac{g(\rho_l - \rho_v)\delta^2}{\mu_l}\left[\frac{y}{\delta} - \frac{1}{2}\left(\frac{y}{\delta}\right)^2\right]$$ (11.3)

where δ is the thickness of the liquid film.

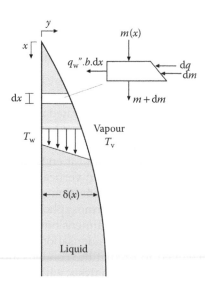

FIGURE 11.5
Physical model for the condensation over a vertical wall.

From this last equation, one observes that the liquid velocity depends directly on the distance y and implicitly on x, through the film thickness. One can determine the heat absorbed in the vapour condensation, considering the saturation temperature in the interface liquid–vapour, using the following expression:

$$dq = h_{lv}dm = q''_w b dx \tag{11.4}$$

where b is the width of the vertical wall and m is the mass flux. Considering linear temperature gradient within the liquid film, the convection heat transfer coefficient is $h = k_l/\delta$, where k_l is the thermal conductivity of the liquid. The Fourier law can be used to predict the heat flux as:

$$q''_w = \frac{k_1\left(T_{sat} - T_w\right)}{\delta} \tag{11.5}$$

The condensate mass divided by the cooling wall width is

$$\frac{m}{b} = \int_0^{\delta(x)} \rho_1 v(y) dy = \frac{g\rho_1\left(\rho_1 - \rho_v\right)\delta^3}{3\mu_1} \tag{11.6}$$

Substituting Equation 11.5 in Equation 11.4 and into Equation 11.6, the derivative of the last equation gives

$$\delta^3 d\delta = \frac{k_1\mu_1\left(T_{sat} - T_w\right)}{g\rho_1\left(\rho_L - \rho_v\right)h_{lv}} dx \tag{11.7}$$

If this last equation is integrated from $\delta(0) = 0$ to x, one gets the following equation for the liquid thickness:

$$\delta(x) = \left[\frac{4k_1\mu_1\left(T_{sat} - T_w\right)x}{g\rho_1\left(\rho_1 - \rho_v\right)h_{lv}}\right]^{1/4} \tag{11.8}$$

From the definition of the local convection heat transfer coefficient $\left(h = k_1/\delta\right)$, one can get the mean convection heat transfer coefficient over a length L, which is

$$\bar{h}_L = \frac{1}{L}\int_0^L h dx \tag{11.9}$$

According to [1], the result of this integral can be approximated by the following correlation:

$$\bar{h}_{Nu} = 0.943\left[\frac{\rho_L g\left(\rho_L - \rho_v\right)h'_{lv}k_L^3}{\mu_L\left(T_v - T_w\right)L}\right]^{1/4} \tag{11.10}$$

A modified latent heat of vaporisation parameter that considers linear temperature profile and that corrects the absorbed heat during the vapour condensation is employed. According to Incropera and DeWitt [1], Rohsenow suggests the following expression:

$$h'_{lv} = h_{lv} + 0.68 Cp_l (T_{sat} - T_w) \tag{11.11}$$

where Cp_l is the liquid specific heat at constant pressure.

The ascending vapour affects the descending liquid film, causing a stress in the liquid–vapour interface. The vapour is able to push some liquid upward, increasing the thickness of the liquid layer. To reach the vapour, the heat has to cross the liquid film. Therefore, thick layers induce lower heat transfer coefficients. The shear stress is neglected in the Nusselt model for film condensation just presented. Chen [2] shows that the heat transfer coefficient decreases with the Prandtl number of the liquid $Pr_l = Cp_L \mu_L / k_l$ and increases with the Jacob number $Ja = Cp(T_x - T_{sat})/h_{fg}$, and suggests the following expression, valid for $Ja < 2$ and $(Pr_l / Ja) < 0.05$:

$$h = h_{Nu} \left[\frac{1 + 0.68 Ja + 0.02 (Ja^2/Pr_L)}{1 + 0.85 (Ja/Pr_l) - 0.15 (Ja^2/Pr_l)} \right]^{1/4} \tag{11.12}$$

This decrease in the heat transfer coefficient is important for flows with low Reynolds number, such as those involving a liquid metal. Bejan [3], through parametric analysis, verified the importance of the inertia effects, which are directly proportional to the Prandtl number, concluding that Nusselt analysis is valid only when the shear stress effects are negligible. The following expression for the film Prandtl number is proposed:

$$Pr_f = \frac{Pr_l (1 + Ja)}{Ja} \approx \frac{Pr_L}{Ja} \tag{11.13}$$

This parameter, valid for $Pr_f < 0.3$, works as a correction factor for the heat transfer coefficient given by Equation 11.10. Hewitt [4] also proposed the following correction factor for the case when the average vapour temperature is above the saturation temperature:

$$h = h_{Nu} \left[1 - \frac{Cp_v (T_v - T_{sat})}{h_{lv}} \right]^{1/4} \tag{11.14}$$

In this case, the vapour has to cool down before condensate in the external surface, and therefore the heat transfer coefficient is decreased.

11.2.2 Thermal Modelling

Engineers must determine the heat transfer capacity of thermosyphons to be able to design equipment assisted by this technology. The temperature distribution in thermosyphons is also important for the design of equipment involving isothermal walls, such as ovens.

The literature presents efforts in modelling temperature fields and pressures in thermosyphons using numerical models [5], but the computational cost and time involved in such procedure make the simplified models the most used in engineering.

The analogy between electric and thermal circuits is a powerful tool for the design of strongly one-dimensional, steady-state systems. This modelling technique is simple and precise for many applications. Therefore, it can be employed very successfully for the determination of the heat transfer capacity and the temperature distribution of thermosyphons.

The global or effective thermal resistance (R) is defined as the ratio between ΔT, the evaporator and condenser mean temperature ($\overline{T}_e$ and $\overline{T}_c$, respectively) difference and the transferred power q:

$$R = \frac{\overline{T}_e - \overline{T}_c}{q} = \frac{\Delta T}{q} \tag{11.15}$$

As well known, the thermal resistance represents the difficulty of the device to transport the heat power. Larger thermal resistances represent lower heat transfer capacity and larger temperature differences between evaporator and condenser.

For a thermosyphon under steady-state conditions, heat follows through a path represented by 10 thermal resistances, as shown schematically in Figure 11.6 [6]. From this figure, one can see that the heat coming from the heat source must first flow through the evaporator external thermal resistance, which depends on the heat transfer mechanism between heat source and outer wall of the thermosyphon (i.e. conduction, convection or

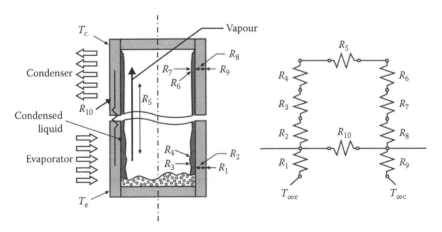

FIGURE 11.6
Thermal circuit of a thermosyphon.

radiation.) This thermal resistance is represented in the thermal circuit by R_1. For the conduction case, R_1 is a thermal contact resistance. In the same way, an external thermal resistance R_9 is associated between the condenser external wall and the heat sink.

Once the heat reaches the thermosyphon, it is conducted through the wall material in two major directions: radial (R_2) and axial (R_{10}). The heat conducted axially follows through a longer path compared with the axial direction way.

The fluid inside the thermosyphon works in its saturated condition. When heat reaches the fluid in the evaporator, vapour is generated. The thermal resistance associated to the evaporation is R_3. Due to the density variations caused by temperature raise, gravity moves the vapour from warmer to colder regions of the thermosyphon. Most of the heat is transferred by the vapour, which heat transfer path is associated with the thermal resistance R_5. This resistance also includes the adiabatic section (if there is one).

In the condenser region, heat is removed from the vapour and condensate is formed over the internal walls. The thermal resistance associated to the condensation process is R_7. The generated droplets of liquid flow back to the evaporator, and depending on the volume of the liquid formed in the condenser, rivulets or films can be created. As happens in the evaporator section, the heat removed in the condenser crosses the condenser wall through the conduction resistance R_8. In this region, heat eventually conducted by the axial wall reaches the evaporator.

In Figure 11.6, one can observe that, between the evaporation R_3 or the condensation R_7 resistances and the vapour resistance R_5, there are two resistances associated in series: R_4 and R_6. These resistances are associated with the pressure drops due to the curvature of the liquid–vapour interface and therefore are located in the interface between liquid and vapour in the evaporator and condenser sections. Although, in steady-state conditions, the liquid–vapour interface is considered in equilibrium, this region is quite active, with liquid and vapour molecules changing places due to vaporisation and condensation processes. This dynamic behaviour of the interface produces a pressure drop, which, in turn, creates a thermal resistance. A deeper discussion of this subject can be found in [7]. These resistances are usually small, but they can be important for very small thermosyphons, such as those developed for electronic applications [8].

From Figure 11.6, combining all the resistances of the thermal circuit, the following expression can be obtained:

$$R = R_1 + \left[\left(R_2 + R_3 + R_5 + R_7 + R_8 \right)^{-1} + R_{10}^{-1} \right]^{-1} + R_9 \qquad (11.16)$$

Table 11.1 shows the order of magnitude of the thermal resistances, as presented in Figure 11.6, for a typical straight thermosyphon (steel–water, 1 m long, with 50 mm of diameter, for instance). Usually, R_{10} is much larger than other thermal resistances of the circuit; as this resistance is

TABLE 11.1

Order of Magnitude of Thermosyphon Resistances of Thermal Circuit Shown in Figure 11.6

Thermal Resistance	(°C/W)
R_2, R_8	10^{-1}
R_4, R_6	10^{-5}
R_3, R_7	10
R_5	10^{-8}
R_{10}	10^5

Source: Reay, D., and Kew, P., *Heat Pipes Theory Design and Applications*, 5th Edition, Elsevier, 2006.

in parallel with R_2 to R_8 resistances, it can be removed from the circuit. The following criterion for the removal of the resistance R_{10} from the thermal circuit is recommended [9]:

$$\frac{R_{10}}{R_2 + R_3 + R_5 + R_7 + R_8} > 20 \qquad (11.17)$$

R_5 is also very small, and as it is in series with the others resistances, it can also be removed from the circuit. Furthermore, the resistances associated with the pressure drops due to the curvature of the liquid–vapour interface are also very small and usually neglected. Therefore, the overall thermal circuit can be reduced to the combination of the following resistances in series: evaporator external resistance (R_1), conduction in the evaporator wall (R_2), evaporation (R_3), condensation (R_7), conduction in the condenser wall (R_8) and condenser external resistance (R_9), resulting in the following expression:

$$R = R_1 + R_2 + R_3 + R_7 + R_8 + R_9 \qquad (11.18)$$

In most applications, the larger resistances of the circuit are the external resistances R_1 and R_9. They usually determine the temperature level of the equipment and, in some devices, can be considered as the only resistances of the thermal circuit.

In the next section, the models employed to predict these thermal resistances are presented.

11.2.2.1 Thermal Resistance Models

11.2.2.1.1 Conduction Resistances

The conduction thermal resistances R_2 and R_8 are obtained from well-known expressions [1,10,11]. For flat plates, the following expression is applied:

$$R_2 = R_8 = \frac{L}{kA} \qquad (11.19)$$

where L is the thickness of the plate and A is the heat transfer cross-sectional area, which depends on the geometry of the evaporator and condenser. For cylinders, the expression is

$$R_2 = R_8 = \frac{\ln\left(d_o/d_i\right)}{2\pi L k} \tag{11.20}$$

where L is the length of the evaporator L_e or condenser L_c and d_o and d_i are the external and internal diameters of the casing tube, respectively. Heat conduction resistances for other geometries can be obtained using appropriated expressions that can be easily found in the literature.

R_{10} is the material axial thermal resistance and is determined by the equation:

$$R_{10} = \frac{L_{ef}}{A_w k_w} \tag{11.21}$$

where A_w is the cross-sectional area of the casing wall, k_w is the thermal conductivity of the wall material and L_{ef} is the effective length given by

$$L_{ef} = L_a + \frac{1}{2}\left(L_e + L_c\right) \tag{11.22}$$

where L_a is adiabatic section length.

11.2.2.1.2 Vapour, Vapour–Liquid and Liquid–Vapour Resistances

The vapour–liquid and liquid–vapour interface thermal resistance can be estimated using expressions from the literature [12]. The following expression is proposed:

$$R_4 = R_6 = \frac{RT^2\left(2\pi RT\right)^{1/2}}{h_{lv}^2 p_v A} \tag{11.23}$$

where R is the universal gas constant and A is the heat exchange area between the heat source (or sink) and thermosyphon (A_e for evaporator and A_c for condenser).

The vapour thermal resistance is determined from expressions for the vapour pressure differences between evaporator and condenser. The Clausius–Clapeyron equation for the equilibrium between liquid and vapour is [13] as follows:

$$\frac{dp}{dT} = \frac{h_{lv}}{T_s\left(v_v - v_l\right)} \tag{11.24}$$

where v_v and v_l are the specific volumes of the vapour and liquid, respectively, and T_s is the saturation temperature, which, in the present case, is considered equal to the vapour temperature T_v. Usually, $v_l \ll v_v$ resulting in $dp_v/dT = h_{lv}/T_s v_v$, which, combined with the ideal gas law, gives

$$\frac{dp_v}{dT} = \frac{h_{lv} p}{RT_v^2} \tag{11.25}$$

where dp_v/dT can be approximated to $\Delta p_v/\Delta T$. Equation 11.21 can be used to estimate the vapour thermal resistance as

$$R_5 = \frac{\Delta T_v}{q} = \frac{RT_v^2}{h_{lv} p_v} \frac{\Delta p_v}{q} \tag{11.26}$$

Again, applying the ideal gas model, the following expression can be found [14]:

$$R_5 = \frac{T_v \left(p_{ve} - p_{vc} \right)}{h_{lv} \rho_v q} \tag{11.27}$$

where p_{ve} and p_{vc} are the evaporator and condenser vapour pressures, respectively, and $\Delta p_v = p_{ve} - p_{vc}$. Therefore, to estimate R_5, the Δp_v parameter still has to be determined.

The vapour pressure difference within a thermosyphon can be obtained from several models available in the literature. A simplified model [12] that treats the vapour as incompressible fluid considers that the vapour pressure decreases along the tube due to viscous effects. In addition, as vapour is generated along the evaporator, the vapour velocity decreases in the evaporator section, increasing again along the condenser, as vapour is converted into liquid. The most simplified model [12] considers that there is complete recovery of the vapour velocity drop in the evaporator along the condenser, whereas another more realistic model Cotter (see Peterson [4]) considers that the velocity recovery is only partial. These velocity profiles create pressure distribution effects, which are added to the pressure variation due to viscous forces. The following expression was obtained for fully developed vapour flow in the thermosyphon core and is suggested for the determination of the pressure drop along the tube:

$$\Delta p_v = -\left(1 - \frac{4}{\pi}\right) \frac{\dot{m}^2}{8\rho_v r_v^4} - \frac{8\mu_v \dot{m}}{\rho \pi r_v^4} L_a \tag{11.28}$$

where r_v is the vapour radius. Usually, the vapour pressure drop is very small, and therefore the vapour thermal resistance can be removed from the thermal circuit.

11.2.2.1.3 Condenser Resistance

R_7 is the thermal resistance associated with the condensation within the condenser region. Actually, condensate is moving downward, from the end of the condenser region to the evaporator working fluid pool's upper surface. In the condenser region, as heat is removed, the saturated vapour condensates and a film is formed over the internal wall surface. The thickness of the film increases as the liquid film moves down. When the condensate reaches the evaporator region, the film is in contact with the heated wall and evaporates. The thinner the liquid film, the more efficient is the evaporation. The remaining liquid accumulates in the evaporator region, forming a pool, and evaporation occurs in boiling regime. In most of the applications (such as those where water is the working fluid), the volume of condensate is small; when the liquid film reaches any kind of surface imperfection, small rivulets are formed. If one knows the heat transfer coefficient in the condenser h_c, the following expression can be used to determine the thermal resistance R_7:

$$R_7 = \frac{1}{h_c\, 2\pi\, r_i L_c} \tag{11.29}$$

If the diameter of the thermosyphon casing tube is large, the Nusselt model correlation (Equation 11.10), obtained for flat surfaces, can be used to predict the heat transfer coefficient for the condensation region. This parameter, in turn, is used to determine the amount of heat exchanged q, which is used to predict the condenser thermal resistance.

Faghri [5] and Mantelli et al. [15], among other researchers, made comparative studies based on several literature condensation heat transfer correlations and models for thermosyphon applications. They observed differences of up to two orders of magnitude between correlations. Mantelli et al. [15] compared correlations with experimental data for the thermosyphon configuration: stainless steel and water, of 1 m long, with 20 cm of evaporator and no adiabatic section, transporting up to 600 W and working in average temperatures from ambient to 350°C. They concluded that the correlation from Kaminaga et al. [16,17] was the one that compared better with the data. This correlation is given by

$$\mathrm{Nu} \equiv \frac{h_c d_i}{k_l} = 25\,\mathrm{Re}_l^{0.25}\,\mathrm{Pr}_l^{0.4} \tag{11.30}$$

where h_c is the condenser coefficient of heat transfer, Pr_l is the liquid Prandtl number, k_l is the liquid thermal conductivity, d_i is the internal tube internal diameter and Re_l is the liquid flow Reynolds number, determined by

$$\mathrm{Re}_l = \frac{4q}{\pi\, d_i h_{lv} \mu_l} \tag{11.31}$$

Groll and Rösler [6], based on the Nusselt analysis for condensation, suggest the following expression for the condensation thermal resistance:

$$R_7 = \frac{0.345\, q^{1/3}}{d_i^{4/3} g^{1/3} L_c \Psi^{4/3}} \tag{11.32}$$

where Ψ is given by

$$\Psi = \left(\frac{h_{lv} k_l^3 \rho_l^2}{\mu_l} \right)^{1/4} \tag{11.33}$$

11.2.2.1.4 Evaporator Resistance

As already observed, two different regions can be found in the evaporator, subjected to two different heat transfer regimes: boiling in the liquid pool volume, represented by R_{3p}, and liquid film evaporation, represented by R_{3f}, in the rest of the evaporator.

The filling ratio F is defined as the ratio between the volume of working fluid V_1 and volume of the evaporator V_e ($F = V_1/V_e$). Using this parameter, the evaporator resistance can be taken as a combination of the boiling and film resistances, as given by the expression

$$R_3 = R_{3p} F + R_{3f} (1-F) \tag{11.34}$$

The pool evaporation process was exhaustively studied in the last decades. Theoretical models and experimental correlations are available in the literature to estimate the heat transfer coefficient for liquid pools. Carey [18] presents a review of some literature correlations to determine the heat transfer between the wall and the pool. In most thermosyphon applications, the temperature difference between the wall and the liquid is of only a few degrees, meaning that the predominant regime is nucleate boiling. Groll and Rössler [6] suggest the following expression:

$$R_{3p} = \frac{1}{g^{0.2} \phi q^{0.4} (\pi d_i L_e)^{0.6}} \tag{11.35}$$

where ϕ is given by

$$\phi = \frac{\rho_l^{0.65} k_l^{0.3} C_{pl}^{0.7}}{\rho_v^{0.25} h_{lv}^{0.4} \mu_l^{0.1}} \left(\frac{p_v}{p_{atm}} \right)^{0.23} \tag{11.36}$$

where p_{atm} is the atmospheric pressure.

The evaporation process of the liquid film in a vertical heated wall is similar to the condensation, and the same Nusselt theory can be used to determine the heat exchange associated with liquid film evaporation. Groll and

Rösler [6] proposed the same expression for the determination of the film evaporation resistance, where the condenser length l_c is substituted for the evaporator length l_e:

$$R_{3f} = \frac{0.345\, q^{1/3}}{d_i^{4/3} g^{1/3} l_e \Psi^{4/3}} \tag{11.37}$$

Mantelli et al. [15] observed that the heat transfer coefficients determined from the Nusselt theory present results much larger than those obtained experimentally for thermosyphons. This happens because the condensed film is divided into small flow currents, the rivulets, leaving the rest of the tube dry. When the evaporator heat flux is high, nucleate boiling is observed in the descending liquid. On the other hand, the bubbles in the liquid pool grow until they collapse abruptly, spreading the fluid over the bubble, wetting the wall and increasing the heat transfer coefficient. These authors compared several correlations proposed in the literature with data obtained in their laboratory and concluded that the El-Genk and Saber [19] correlation, for small Reynolds number, and Gross [20] correlation, for higher Reynolds number, presented the best comparison. The Gross correlation is

$$\mathrm{Nu} \equiv \frac{h_e}{k_1} \left[\frac{\vartheta_l^2 \rho_l}{g(\rho_l - \rho_v)} \right]^{1/3} = f_d\, 0.925 \left(\frac{\mathrm{Re}_l}{4} \right)^{-1/3} \tag{11.38}$$

where h_e is the evaporation heat transfer coefficient and ϑ is the kinematic viscosity. The f_d parameter takes into account the influence of the tube geometry being estimated by

$$f_d = 1 - 0.67\frac{d}{20} \qquad \text{for} \qquad 6 \text{ mm} < d \leq 20 \text{ mm} \tag{11.39}$$

$$f_d = 0.33 \qquad \text{for} \qquad d > 20 \text{ mm} \tag{11.40}$$

The sub-indexes v and l refer to vapour and liquid, respectively. One may use Equation 11.29 (substituting h_c by h_e), which, together with the evaporation heat transfer obtained from Equation 11.38, determines the resistance R_{3f}.

The liquid pressure in the pool base is larger than the vapour pressure, due to the column of liquid in the pool. The boiling phenomena almost do not influence the pressure in the evaporator basis. This pressure is determined by the following expression:

$$p_p = p_v + \rho_{1g} FL_e sen\beta \tag{11.41}$$

where β is the thermosyphon inclination angle with the horizontal position.

Therefore, if the volume of the liquid pool is large, a temperature gradient can be observed within the evaporator. The temperature in the pool base T_p

is assumed as the saturation temperature corresponding to p_p. Assuming that the pool temperature increases linearly with the liquid depth, the average temperature in the evaporator is

$$T_m = T_v(1-F) + \frac{T_v + T_p}{2}F \qquad (11.42)$$

Therefore, the average temperature difference due to the hydrostatic pressure is

$$\Delta T_h = T_m - T_v = \frac{T_v - T_p}{2}F \qquad (11.43)$$

For most of the operation conditions, ΔT_h is very small and can be neglected. On the other hand, this difference may be important for fluid operating in low temperature levels and, therefore, low vapour pressures.

11.2.2.1.5 External Resistances

Finally, the thermal resistances R_1 and R_9 represent the heat exchanges between the evaporator and condenser walls with the heat sources or sinks, and are defined as, respectively,

$$R_1 = \frac{1}{h_e A_e} \qquad (11.44)$$

and

$$R_9 = \frac{1}{h_c A_c} \qquad (11.45)$$

where A_e and A_c are the areas of the evaporator and condenser. The expressions that can be used for the determination of the external coefficients of heat transfer depend on the heat transfer mechanisms and are well studied. Literature correlations [1,21,22] can be used to determine h_e and h_c for the evaporator and condenser, respectively.

11.2.3 Heat Transfer Limits

Although very efficient heat transfer devices, thermosyphons are subjected to limitations, which determine the maximum heat power they can transport under pre-determined operation conditions.

11.2.3.1 Entrainment or Flooding Limits

In thermosyphons, liquid and vapour flow in different directions. As the vapour has much higher velocity than the liquid, shearing forces happen between the vapour and the liquid. If these forces are large enough, the return of the liquid can be affected and the liquid is accumulated in the condenser which is flooded. In this case, the flooding limit is reached.

An increase in the heat input in the evaporator results in an increase of the vapour velocity, which can cause instabilities in the liquid flux. In the most severe cases, waves can be observed in the liquid film. The crests of these waves are subjected to shearing forces caused by the vapour in countercurrent flow. If these liquid–vapour interface shearing forces are larger than the liquid tension forces, liquid drops are pulled out from the film and are dragged by the vapour flow in the condenser section direction. This causes excess of liquid in the condenser and consequently lack of liquid in the evaporator. The conditions in which this occurs are known as entrainment limit.

The main difference between flooding and entrainment limits is the drag of liquid drops that happens in the entrainment limit. In this chapter, both limits are treated equally and are denominated entrainment limit. They are found especially in thermosyphons in which the ratio between length and diameter are large. This may be the most severe limitation for thermosyphons and is the object of several studies in the literature. Reay and Kew [23] list 13 literature models and correlations for the prediction of this limit. Peterson [14] compared the predictions of several correlations for a cooper–water thermosyphon. They observed differences among the correlations of up to four orders of magnitude, showing that still work has to be done in this field.

Fortunately, in most thermosyphon applications for the industry, this heat power transfer limit is much larger than the heat to be transferred. If, for a specific case, the designed heat power is close to the entrainment limit, the designer must go to the literature and find the most appropriate expression for the specific application. In this section, some models and correlations are presented.

One of the most used correlations for the entrainment limit prediction is proposed by Groll and Rösler [6]:

$$q_{ent} = f_1 f_2 f_3 h_{lv} \rho_v^{1/2} \left[g \left(\rho_l - \rho_v \right) \sigma \right]^{1/4} \tag{11.46}$$

where σ is the surface tension and f_1 is a parameter which is a function of the Bond number (Bo) defined as

$$Bo = d_i \left[\frac{g \left(\rho_l - \rho_v \right)}{\sigma} \right]^{1/2} \tag{11.47}$$

The parameter f_2 is a function of the non-dimensional parameter Kp, given by

$$Kp = \frac{p_v}{\left[g \left(\rho_l - \rho_v \right) \sigma \right]^{1/2}} \tag{11.48}$$

For $Kp \leq 4 \times 10^4$, $f_2 = Kp^{-0.17}$ and for $Kp > 4 \times 10^4$, $f_2 = 0.165$.

The parameter f_3 corrects the correlation expression for tilted thermosyphon position and its values are functions of Bo number (see Ref. [6]). For the vertical position, $f_3 = 1$.

Wallis (see Faghri [5]) proposed the following semi-empirical correlation for the entrainment limit for thermosyphons, considering the vapour as incompressible, of constant film thickness, two-phase flow in equilibrium, steady-state, one-dimensional and homogenous flows. From the equilibrium between inertial and hydrostatic forces, the following expression is suggested:

$$q_{ent} = A \frac{C_w^2 h_{lv} \sqrt{g d_i \left(\rho_l - \rho_v\right)\rho_v}}{\left[1+\left(\rho_v/\rho_l\right)^{1/4}\right]^2} \tag{11.49}$$

where C_w is a non-dimensional constant empirically determined and is a function of the fluid thermophysics properties. For most of the cases, C_w is between 0.7 and 1.

Another expression based on the flooding correlation of Kutatekadze [5,14] was developed by Tien and Chung [24], where the effect of the tube diameter was included:

$$q_{ent} = C^2 \left[\frac{1}{4}\left(\pi d_i^2\right)\right] h_{lv} \left(\rho_l^{-1/4} - \rho_v^{-1/4}\right)^{-2} \left[\sigma g \left(\rho_l - \rho_v\right)\right]^{1/4} \tag{11.50}$$

where

$$C = 3.2^{1/2} \tan h \left(\frac{1}{2} Bo^{1/4}\right) \tag{11.51}$$

Faghri et al. [25] extended the Tien and Chung [24] model, using the constant C, given by Equation 11.51 for water [14] and $C^2 = \left(\rho_l/\rho_v\right)^{0.14} \tan h^2 Bo^{1/4}$ for other fluids [5].

Nguyen-Chi and Groll [26] noticed that the thermosyphon tilt angle β affects the entrainment limit and proposed the following expression, to be applied to the selected correlation, dividing it by

$$f\left(\beta\right) = \left[\frac{\beta}{180°} + \sqrt{sen 2\beta}\right]^{0.65} \tag{11.52}$$

Experimental data showed that this correction is well succeeded when applied to the Wallis correlation (Equation 11.49).

As stated by Peterson [14] verified the influence of the working fluid volume by studying the behaviour of thermosyphons, which present different working fluid charges. For small working fluid charges, these authors used the Cohen and Bayley (1955) correlation

$$q_{ent} = \frac{\pi g d_c \rho_l^2 h_{lv}}{3\mu} \left(\frac{V_e \left(V_l/V_e - \rho_v/\rho_l \right)}{\left[\pi d_c \left(0,8L_c + L_a \right) + \left(d_e/d_c \right)^{2/3} \left(L_a + 0.75L_c \right) \right] \left[1 - \left(\rho_v/\rho_l \right) \right]} \right)^3 \tag{11.53}$$

This correlation followed the observed trends but overestimated Shiriashi et al. data by a factor of 5. For large amounts of fluid, these authors employed successfully Katto correlation:

$$q_{ent} = \frac{0.01\pi \, d_e L_e \rho_v^{0.5} h_{lv} \left[\sigma g \left(\rho_l - \rho_v \right) \right]^{1/4}}{\left[1 + 0.0491 \, L_e/d_e \mathrm{Bo}^{0.3} \right]} \tag{11.54}$$

Imura et al. [27], using a similar approach, obtained the following correlation, which compares with other authors experimental data within ±30%, for an optimal charge of working fluid:

$$q_{ent} = 0.64 \left(\frac{\rho_l}{\rho_v} \right)^{0.13} \left(\frac{d}{4L_e} \right) h_{lv} \left[\sigma g \rho_v^2 \left(\rho_l - \rho_v \right) \right]^{1/4} \tag{11.55}$$

11.2.3.2 Sonic Limit

For some thermosyphons, especially those that use liquid metal as working fluid, the vapour velocity can reach sonic levels in start-up or in steady-state conditions. After the sonic speed is reached, the vapour, usually located in thermosyphon core, experiences a shock wave and the vapour is considered 'blocked'. In this case, even if more vapour is generated in the evaporator, the vapour flow does not increase. Therefore, increasing the evaporator heat power increases only the evaporator region temperature. On the other side, if the sonic velocity is reached, condenser temperature changes are not sensed in the evaporator.

Models applied for the sonic heat transfer limit prediction are based on the converging–diverging nozzle theory [28]. Actually, compressible mass flow in a duct of constant cross-sectional area with mass addition and removal has approximately the same behaviour of a constant mass flux in a converging–diverging nozzle. The sonic limit represents the heat applied to the thermosyphon when vapour reaches the sonic speed. This limit is influenced by the size of the vapour core. The following equation, also applicable for heat pipes, was proposed by Busse [29] for the determination of the sonic limit:

$$q_{son} = 0.474 h_{lv} A_v \left(\rho_v p_v \right)^{1/2} \tag{11.56}$$

where A_v is the vapour core cross-sectional area.

11.2.3.3 Viscous Limit

Especially for thermosyphons working at low temperature levels, the vapour pressure differences between the evaporator and condenser can be very small. If this pressure difference is smaller than the viscous forces, the vapour is not able to move and the viscous limit is reached. This limit may happen during start-up, and the best way to avoid it is to increase the heat flux and so to increase the vapour temperature until the pressure gradient along the tube exceeds the viscous forces. The main concern about this procedure is that in vapour pressures close to the viscous limit, where small amounts of vapour are present, the sonic limit can be easily reached. Therefore, in some cases, there is a transition between viscous and sonic limits. The same Busse work [29], which resulted in the sonic limit expression, also developed the following equation for the viscous limit:

$$q_{vis} = d_v^2 h_{lv} A_v \frac{\rho_v p_v}{64 \mu_v l_{ef}}$$ (11.57)

where d_v is the diameter of the vapour core. An experimental work presented in the Engineering Science Data Unit (ESDU) [14] showed that the vapour flux may be limited by the pressure gradient between evaporator and the minimum pressure in the condenser. They propose a ratio between the pressure drop and the absolute vapour pressure, p_v, which establishes a criterion to avoid the viscous limit:

$$\frac{\Delta p_v}{p_v} < 0.1 .$$ (11.58)

11.2.3.4 Dry-Out Limit

Basically, there are two means to reach dry-out in a thermosyphons. First, when the radial heat flux is very small, so that the rate of vapour produced is not enough to guarantee a continuous circulation of vapour and liquid. In this case, the small amount of vapour condenses in regions near the evaporator, and the rest of the condenser does not participate in the heat exchange, generating cold spots. This limit is also commonly observed during thermosyphon start-ups. A similar effect is observed when, instead of the heat, the total volume of working fluid is very small. In this case, the small volume of condensate evaporates in the upper regions of the evaporator and liquid is not able to reach the bottom of the evaporator, forming hot spots, close to the evaporate base.

The second is associated with the entrainment limit. When the heat flux reaches the entrainment limit, the liquid film is not able to reach the pool, as it is dragged back to the condenser, forming dry regions in the internal thermosyphon wall. Extended dry regions in the evaporator are denominated dry-patches. Both dry-out and dry-patch increase the thermosyphon thermal resistance and so increase the wall temperature, close to the locations of the dry regions. Models for determining dry-out limits are basically the same as those for entrainment limit.

11.2.3.5 Oscillation Limit

The oscillation limit happens at heat fluxes higher than the entrainment limit. Figure 11.7 illustrates the phenomenon. It starts with the drag of small liquid drops to the condenser region (Figure 11.7a), with a formation of a dry-out region in the upper part of the evaporator (Figure 11.7b). Dry-out regions increase, forming dry-patches, at the same time that a thick liquid film is formed at the top of the condenser (Figure 11.7c). The temperature of the dry regions increases as the heat applied to the evaporator is not able to be employed in the vapour production. As less vapour is produced, the vapour pressure inside the thermosyphon decreases. When the pressure is not able to hold the liquid film at the top of the condenser, the liquid column collapses (Figure 11.7d) and the whole surface wall, including the dry regions, is re-inundated. As the wall temperature is very high, a very strong liquid boiling phenomenon is observed causing a quick pressure increase in the thermosyphon (Figure 11.7e). The large amount of vapour generated moves upward in high velocity, keeping, for a few moments, the liquid film in the condenser region. The vapour pressure stops increasing when the condensing rate is

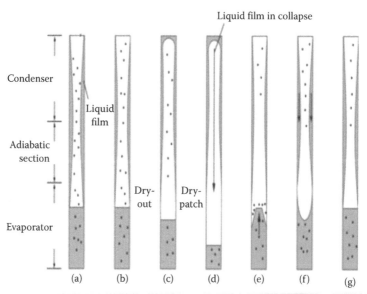

FIGURE 11.7
Oscillation phenomena in thermosyphons. (a) Start up. (b) Dry out. (c) Dry patch. (d) Accumulation of liquid in the condenser. (e) Collapse of the liquid film with high film evaporation rate. (f) System stabilisation. (g) Back to (a) condition.

larger than the evaporation rate, and this happens quickly because the liquid dragging decreases the liquid in the evaporator (decreasing the rate of vapour production) and causes an increase in the condenser pressure (increasing the rate of condensation) (Figure 11.7f). The liquid film is then formed again (Figure 11.7g) and the cycle re-starts. The oscillation limit is reached when the oscillation behaviour persists with time. In some applications, when there is some oscillation dumping, the thermosyphon oscillations can be temporary.

One way to prevent the oscillation behaviour in thermosyphons is to maintain the maximum transferred heat power in levels under the entrainment limit, so that no dragging of liquid drops happens, and no dry-out regions are formed, avoiding the onset of the oscillation. For the estimative of this maximum heat flux, the entrainment limit correlations and models can be used.

Usually, oscillations are observed during the thermosyphon start-up (geyser boiling [15]), but they disappear and do not represent any heat transfer limitation for the device. In these cases, the period of the oscillations can be shortened if more heat is applied to the thermosyphon, taking care that the sonic limit is not reached.

11.2.3.6 Boiling Limit

Boiling limit is observed for large amount of working fluid in thermosyphons that are subjected to large heat fluxes in the evaporator. This limit is observed in the transition between nucleate pool to vapour film boiling, when the so-called critical heat flux, q_{crit}, is applied. Under this heat flux, the generated bubbles coalesce into one vapour film, thermally insulating the tube wall. Due to the poor thermal conductivity of vapour, the wall temperature keeps increasing and can quickly reach the tube metal fusion point. This limit is also known as burn-out. Literature correlations can be used for pool boiling limit determination models, as presented by Carey [18] and by Collier and Thome [22]. Lienhard and Dihr propose the following expression, for maximum heat to be applied at the tube base, in the pool region:

$$q_{bol} = 0.12 h_{lv} \rho_v^{1/2} \left[g\sigma\left(\rho_l - \rho_v\right) \right]^{1/4} \tag{11.59}$$

This expression can be applied for the rest of the evaporator if it is subjected to heat fluxes similar to the heat applied to its base. Usually, this hypothesis is reasonable if the tube is designed to avoid the burn-out. One way to avoid this limit is, for the same heat flux, to increase evaporator heating area, by increasing the length or diameter of the casing tube. The designer must be careful that, by avoiding the boiling limit with long tubes, the entrainment limit can be reached.

As observed by Faghri [5], Gorbis and Savchenkov proposed the following correlation for the boiling limit, valid for thermosyphons working from vertical to 86° of tilt angle with vertical, filling ratio F between 0.029 and 0.6

and the ratio between non-condensable gas volume V_g and total volume of the thermosyphon V_t ranging from 0.0006 and 0.1:

$$q_{bol} = q_{crit} C^2 \left[0.4 + 0.012 \frac{d}{2} \sqrt{\frac{g(\rho_l - \rho)}{\sigma}} \right]^2 \tag{11.60}$$

where q_{crit} is the pool boiling critical heat flux, given by

$$q_{crit} = 0.142 \sqrt{\rho_v} \left[g\sigma(\rho_l - \rho_v) \right]^{1/4} \tag{11.61}$$

and

$$C = A' \left(\frac{d_i}{l_c} \right)^{-0.44} \left(\frac{d_i}{l_e} \right)^{0.55} \left(\frac{V_l}{V_e} \right)^{n'}$$

where

$$A' = 0.538 \quad \text{and} \quad n' = 0.13 \quad \text{for} \quad \frac{V_l}{V_e} \le 0.35$$

$$A' = 3.54 \quad \text{and} \quad n' = -0.37 \quad \text{for} \quad \frac{V_l}{V_e} \ge 0.35$$

11.3 Thermosyphon Design and Fabrication

The designer must know very well the thermal and geometrical conditions of the equipment where the thermosyphon will be employed, including the heat power to be transferred, the acceptable temperature differences, operation conditions, weight and possible volume of the device. With these data, it is possible to select the most appropriate technology. In industrial applications, economic aspects usually are dominant over the other constraints. Thermosyphons and loop thermosyphons are preferable over heat pipes and similar devices, which are usually more expensive than thermosyphons due to the need of porous media.

Once the operational temperature conditions are known, the designer is able to choose the materials and working fluids. This selection must take into consideration the affinity between the working fluid and the case material and between the case and the environment where the equipment will be installed. Then, the thermal and mechanical properties of the liquid and tube casing must the gathered. With this information, it is possible to design the thermosyphon. Before the thermosyphon is installed in the equipment,

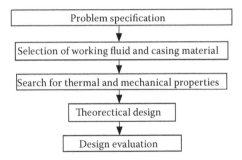

FIGURE 11.8
Thermosyphon design procedure.

the theoretical design must be checked, usually through an experimental study. Figure 11.8 shows a flow chart of this procedure.

Reliability and safety are also important aspects to be taken into consideration in thermosyphon design.

In this section, practical aspects of thermosyphon design methodology, involving selection of materials, working fluids, fabrication and tests of thermosyphons, are treated.

11.3.1 Selection of Working Fluid

Among all the parameters relevant for thermosyphon designs, the selection of the working fluid is one of the most important. The following aspects must be taken into consideration:

- Range of working temperature
- Chemical compatibility between the working fluid and casing material
- Vapour pressure over the operation temperature
- Latent heat
- Stability
- Toxicity
- Thermal conductivity
- Surface tension
- Viscosity
- Wettability and so on

The temperature range is essential for the selection of the working fluid. Theoretically, any working fluid can operate between its critical and triple-state temperatures. Beyond the critical point, any fluid is considered to be in plasma state, where a temperature increase does not result in a pressure increase. Below the triple state, the working fluid is in equilibrium between solid and vapour. Actually, these are extreme points and the working fluid should operate in a much narrower ranges. Figure 11.9 presents the diagram $p \times v$ of a pure substance showing the critical and triple-state line point.

Figure 11.9 also presents two processes at a constant volume (AB and CD). These processes happen inside a thermosyphon when heat is added to the working fluid. As thermosyphons are closed systems, the specific volume of the working fluid v does not change during these processes. If v is smaller than the specific volume at critical conditions (v_c), the heat addition, represented by AB process, causes a decrease in the quality (ratio between mass of vapour and total mass of the fluid). On the other hand, if v is larger than v_c, the heat addition process (CD) increases the working fluid quality, and more vapour is available inside the thermosyphon. In both cases, the maximum working temperature of the thermosyphon is smaller than that of the critical state.

The working fluids can be divided into four major groups according to the operation temperature levels:

- Cryogenic (1–200 K): helium, hydrogen, neon, nitrogen, oxygen and so on. These fluids are in gas state at ambient temperature and pressure conditions.

- Low temperatures (150–600 K): methane, ethane, acetone, ammonia, water and thermal fluids used in refrigeration systems (CFC, HCFC, HFC and other working fluids).

- Medium temperatures (500–800 K): mercury, naphthalene, sulphur and other organic fluids.

- High temperature (800–3000 K): caesium, potassium, sodium, lithium, lead, indium and silver. These working fluids are solids at ambient pressure and temperature.

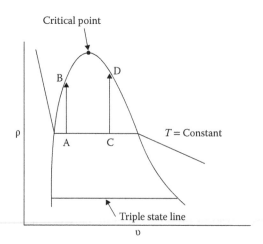

FIGURE 11.9
Typical $p \times v$ diagram of a pure substance.

For many industrial applications, where the operating temperature range is between 300 and 600 K, such as heat exchangers, heaters and electronic cooling devices, water is the most appropriate working fluid, since it has high latent heat, is stable and is non-toxic. Using small volumes of water, large amount of heat can be transported, as small volumes of vapour are generated by large amounts of heat. The main concern about water is the vapour pressure, which can increase significantly with the temperature. In these cases, the casing material must be selected so that it can bear the vapour pressure. Ammonia is a very interesting fluid for low-temperature ranges, but its application is very limited as it is also considered toxic. The Freons and other refrigeration fluids are considered harmful to the environment and must also be avoided.

From the thermal properties point of view, mercury is one of the best working fluids for medium-temperature ranges, but its toxicity makes this fluid condemned for most industrial applications. Many heat exchangers operate at this temperature range. In these cases, naphthalene is being used as working fluid [30]. Sodium is an interesting working fluid for higher operating temperature levels (above 800 K), but it requires care in its manipulation as it is highly reactive with water (as well as humidity), releasing highly inflammable gases. Potassium is also reactive with water and forms explosive mixtures with air at ambient temperature.

A parameter that relates working fluid properties with its maximum heat power capacity, denominated number of merit (N), is used to classify working fluid for thermosyphons and is expressed as follows [23]:

$$N = \left(\frac{h_{lv} k_l^3 \sigma_l}{\mu_l} \right)^{1/4} \tag{11.62}$$

The thermophysical properties that compose the merit number are dependent on the temperature level. The higher is N, the best is the working fluid. Water presents a very interesting number of merit as, at 450 K, $N \approx\sim 7500 \left(kg/K^{3/4}s^{5/2} \right)$, remaining above 4000 for temperatures ranging from 150 to 600 K. Ammonia is also a suitable working fluid: for 235 K, $N \approx 4800 \left(kg/K^{3/4}s^{5/2} \right)$. Toluene, when compared to other fluids, shows a worse performance as, for 325 K, $N \approx\sim 1055 \left(kg/K^{3/4}s^{5/2} \right)$.

Working fluid impurities may decrease the thermal performance of thermosyphons. Non-condensable gases are one of the most common impurities. These gases already can be mixed with the fluid when they are introduced in the thermosyphon. In addition, gases can be generated by chemical reaction after the working fluid and the casing metal are brought to contact during the liquid charging process and/or during thermosyphon operation. Actually, during the operation, the non-condensable gases are pushed to the upper end of the condenser region of the thermosyphon, blocking part of the condenser and causing a reducing in the thermal performance of the device. The impurities can also decrease the viscosity of the working fluid.

11.3.2 Selection of Casing Materials

The selection of the casing material is a very important parameter for the design of thermosyphons. The casing has to be compatible with the working fluid and environment to bear the mechanical strength needs of the thermosyphon.

As already observed, the working fluid and casing material should not react chemically to avoid the generation of non-condensable gases. Table 11.2, based on information presented by Reay and Kew [23] and Faghri [5], shows a list of compatibility between common casing materials and working fluids.

TABLE 11.2

Working Fluid and Casing Material Compatibility List

	Recommended	Not recommended
Ammonia	Aluminium Steel Nickel Stainless steel	Copper
Acetone	Copper Silica Aluminium Stainless steel[a]	
Methanol	Copper Stainless steel Carbon steel Silica	Aluminium
Mercury	Stainless steel	Nickel Inconel Titanium Niobium
Water	Copper Monel Silica[a,b] Nickel[a,b] Stainless steel[a,b] Carbon steel[a,b]	Stainless steel[a,b] Carbon steel[a,b] Aluminium Silica Inconel Nickel
Dowtherm A	Copper Silica Stainless steel[b]	
Naphthalene	Carbon steel Stainless steel	
Potassium	Stainless steel Lnconel	Titanium
Sodium	Stainless steel Lnconel	Titanium
Silver	Tungsten Tantalum	Rhenium

[a] Considered compatible for some authors and incompatible for others.

[b] Recommended with caution.

Most of the compatibility studies conducted in the literature are intended to supply data for space applications, where heat pipes are the most common technology applied for thermal management, and where durability and safety requirements are much stricter than those for industrial applications. Therefore, water and steel (carbon or stainless), for example, are considered incompatible for space applications and compatible for most industrial applications. This combination is the most common for thermosyphon heat exchangers, especially due to cost and good thermal performance characteristics. Although there will be generation of non-condensable gases, this volume is small, so the decrease of the thermal performance of the equipment is slow. In fact, the equipment will need to undergo regular maintenance due to other operational conditions much sooner than the thermosyphon's low performance is noticeable.

As already observed, non-condensable gases tend to accumulate in the rear condenser regions during the thermosyphon operation. These gases do not participate in the heat transfer processes as they do not change phase and are thermal insulators. Therefore, the most obvious symptom of the presence of non-condensable gases is the cold tip of thermosyphons in the rear condenser regions.

Mantelli et al. [30] studied the temperature distribution inside naphthalene carbon steel thermosyphons where a controlled amount of a non-condensable gas, argon in this study (represented by the charging vessel pressure at room temperature, during the charging procedure) is inserted in the tube. Figure 11.10 shows the main results obtained. It can be observed that the volume occupied by the non-condensable gases increases with the amount of non-condensable gases but decreases with the operation temperature. At higher temperatures, the vapour pressure increases, squeezing the non-condensable gases in the rear regions of the condenser, decreasing its volume.

The thermosyphon casing material must resist the mechanical forces caused by the presence of the pressurised vapour in its operation temperature. The working fluid and the operation temperature are prime input parameters for the selection of the tube material and thickness. Elevated vapour pressure demands thicker walls. In the case of water, the vapour pressure can quickly reach high levels for temperatures above 350°C. Therefore, appropriate material and casing geometry (tubes resist more pressure than parallelepipeds, for instance) are required.

In the selection of the tube material, the designer must consider materials that can be easily welded to avoid leakages, ensuring the integrity of the system during operational conditions. Stainless steel can be employed for thermosyphons operating at temperatures up to 800°C. Other carbon steels are used to temperatures up to 1000°C. At temperatures over 850°C, ceramics are considered, which present high mechanical strength at high temperature, have high resistance to corrosion and erosion and have low cost. The main concern is that ceramics can be fragile and have low thermal conductivity.

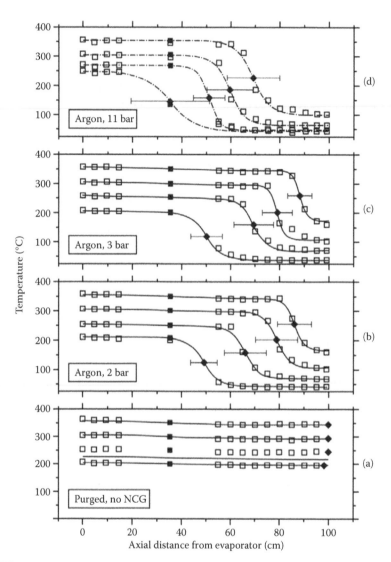

FIGURE 11.10
Temperature distribution for naphthalene thermosyphons with the presence of controlled amount of non-condensable gases. (a) Free of non-condensable gases. (b) Small amount of non-condensable gases. (c) Medium amount of non-codensable gases. (d) Large amount of non-condensable gases.

In industrial applications, where cost is one of the most important parameters to be considered, commercial tubes must be, as much as possible, employed in thermosyphons. As already mentioned, in some cases a moderate incompatibility between tube and working fluid must be tolerated as the degradation of the thermosyphons is usually much slower than the actual maintenance periods, due to other reasons.

11.3.3 Thermosyphon Fabrication

Basically, a thermosyphon casing is usually composed of a metal tube with two closing lids, composed of round flat plates welded in the tube tip borders. In one of the round lids, a small tube is welded, which is used for vacuum and filling procedures.

In Figure 11.11, they are presented four different types of closing lids designed to be welded to the tube using TIG (tungsten inert gas) soldering process. In Figure 11.11a, the closing lid has exactly the same diameter of the tube and the welding happens in the external surfaces, resulting in a small protuberance formed by the soldering material. The lid shown in Figure 11.11b has basically the same dimensions of the lid shown in Figure 11.11a; the major difference is in a small groove that is machined in the soldering area. This geometry avoids the formation of the soldering material protuberance, but is more subjected to vacuum leakage. In Figure 11.11c, the external diameter of the lid is almost of the same dimension of the internal tube diameter, so that the lid fits inside the tube. Figure 11.11d shows a similar geometry, but with a groove to fit the soldering material. In these last cases, the welding is performed inside the tube. This soldering execution can be difficult, especially for small-diameter tubes, as it is hard to keep the lid in the correct position during the welding, but the results in terms of leakage possibility can be rewarding.

The TIG welding process is recommended for this case, as this process is suitable for welding small metal pieces, such as the thermosyphon components. Also, this procedure is relatively clean and provides the necessary

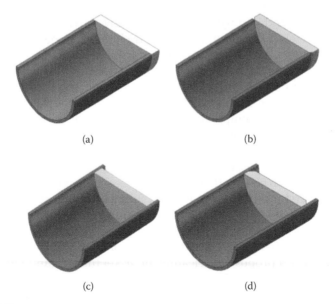

(a)   (b)

(c)   (d)

FIGURE 11.11
Geometry for the thermosyphon closing lids. (a) Closing lid of the same diameter of the tube. (b) As in (a) with small groove. (c) Lid fits inside the tube. (d) Same as in (c) with small groove.

strength to the thermosyphon. Other welding procedures may cause bubbles that retain gases and can be considered dirty processes, damaging the quality of the vacuum.

A small umbilical tube, used for charging and cleaning of the inner surfaces of the thermosyphon (typically ¼ external diameter), is welded in one of the lids.

11.3.3.1 Cleaning Process

Depending on the thermosyphon, the cleaning of the components is better performed before any welding procedure, as the access to the thermosyphon inner parts is easier. Therefore, two cleaning procedures are adopted: primary, before the major welding processes, and secondary, after most the thermosyphon is fabricated and before the working fluid is charged.

Based on a carbon steel–water thermosyphon, the following procedure is proposed:

  Primary cleaning: Its objective is to remove the protection waxes (used by the tube manufacturer to protect the tubes against rust during the storage and transportation) and other gross impurities which are found in commercial tubes. This cleaning must be employed before any soldering or machining processes. The cleaning procedure can be made by means of alumina microspheres jets inside the tube.

  Secondary cleaning: The secondary cleaning is performed after the machining and cap-welding processes are carried out. The tube is completely filled with acetone or trichloroethylene, which work as detergents, and left in an ultrasonic bath for about 15 minutes to remove the tube internal impurities. After this first cleaning, the acetone or trichloroethylene is removed from the thermosyphon, which is filled with isopropyl alcohol to remove the detergent residues. The tube is emptied again and left to dry. To accelerate the drying process, the tube is heated by means of a blowtorch.

For stainless steel tubes, the process is simplified, as the tubes bought from the market are much cleaner than the carbon steel ones. In this case, the thermosyphon is first machined and welded, followed by the secondary cleaning process, just before the vacuum and charging procedures are performed.

Copper tubes are cleaned by immersing the parts into a 10% sulphuric acid. The cleaned material is rinsed in water for 10 minutes and left to dry. Reay and Reay [23] propose the cleaning of nickel tubes by immersing them into an acid solution with 25% of nitric acid.

The cleaning process, for the fabrication of liquid metal working fluid, thermosyphons must be meticulous, as the presence of small amounts of non-condensable gases can affect very much the thermosyphon thermal behaviour, especially for low temperature levels.

11.3.3.2 Out-Gassing of Tube Materials and Working Fluids

For some applications, it is interesting to perform out-gassing of the tube materials and working fluids, after the cleaning and before the working fluid charging procedures. The objective is to remove gases absorbed in the metal that could be released to the working fluid during the operation of the thermosyphon. This is important for high-temperature thermosyphons once these working temperature levels may start up metal out-gassing. Out gassing is performed by subjecting the tube to vacuum after the cleaning is performed.

The working fluid must be bought as clean as possible from the market and, in some cases, must be purified before being inserted in the thermosyphon. Distillation is a process highly employed for fluids that operate at low temperature levels such as acetone, methanol and ammonia, as the presence of water can cause chemical incompatibilities between the casing metal and the working fluid.

The out-gassing of low temperature level working fluids (up to 200°C) can be realised by freezing the fluid, followed by the application of vacuum to remove the absorbed gases, which are not frozen. The liquid is then thawed. For high working fluid cleaning standards, the process can be repeated many times. Of course, more sophisticated working fluid cleaning procedures result in more expensive fabrication costs.

In most of the water–steel thermosyphons for industrial applications such as heat exchangers, the working fluid treatment is quite simple: first, the water is distilled and then out-gassed by means of a vacuum pump.

11.3.3.3 Charging Procedures

After the welding and cleaning procedures and before the working fluid charging procedures, the thermosyphon casing must be subjected to vacuum leakage tests. Small leakages can compromise the operational behaviour of thermosyphons, as non-condensable gases can intrude the device. The leak tests must be performed by means of very sensible leak detectors, which use a mass spectrometer to detect very small amounts of sprinkled helium from outside of the thermosyphon casing. The vacuum is considered of good quality if the leak detector indicates 10^{-9} mbar/s of helium.

The thermosyphon tubing case is also tested for the mechanical strength by subjecting the tube to elevated pressures, by means of pressurised water at pressures 150% above the operation pressures. After the pressurised water test, the tube is evacuated and dry.

The working fluid is then charged through the umbilical tube. Two processes can be applied for charging the thermosyphon: by heating and by vacuum pump.

11.3.3.3.1 Charging by Heating

The following equipments are necessary for this procedure: electrical power source, electrical resistance, beaker, hose, valve and connections. Figure 11.12 presents a schematic of this procedure.

After the tube is cleaned, it is completely filled with the working fluid, which is already treated, such as distilled water at ambient temperature. A valve connects the umbilical tube to the hose, which is introduced inside a beaker with a certain amount of working fluid. The power source feeds an electrical resistance installed in the thermosyphon evaporator. A continuous and slow heating must take place to allow enough time for the small air bubbles trapped in the thermosyphon's internal wall and to have enough energy to escape to the beaker and then to the environment. The heating

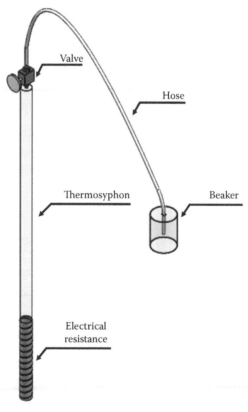

FIGURE 11.12
Working fluid filling rig by electrical heating.

process proceeds until boiling takes place. The vapour is expelled to the beaker, which is cold, so that the vapour condensates again. When the vapour pressure is not sufficient to expel the working fluid, the valve is closed until pressure grows again. The valve cannot be closed for a long time, avoiding the pressure increase, which could damage the fragile parts of the thermosyphon-filling rig.

After the working fluid is almost completely removed from the thermosyphon, the valve is closed and the electrical source is turned off. The tube cools down and the internal pressure decreases. The valve is then opened again, so that the tube sucks the working fluid from the beaker, making sure that air is not introduced in the tube again.

Heating and cooling down procedure is repeated several times (five or six) until air bubbles are not observed during the slow heating process. After the fluid is free of air bubbles and the appropriate amount of working fluid is inside the thermosyphon, the valve is closed and the hose removed.

The amount of working fluid is determined through the comparison of the thermosyphon casing volume and the water originally inserted on the beaker. For example, if the objective is to insert 48 mL of working fluid inside the thermosyphon, whose internal volume is of 300 mL, the beaker can be originally filled with 148 mL and, by the end of the process, the beaker would present with 100 mL.

The great advantage of using this procedure is the very simple apparatus needed and the low operational cost. On the other hand, the following disadvantages can be cited: complexity, time-consuming operation, difficulty in guaranteeing the working fluid cleaning quality and small precision in the fluid charging volume (some working fluid is lost by evaporation and is trapped in the hose and valves).

11.3.3.3.2 *Charging by Vacuum*

The following equipments are necessary for charging by vacuum: vacuum pump, vacuum gauge, beaker, valves, hoses and connections.

After the cleaning process is carried out, the thermosyphon casing is connected to a vacuum pump, as schematically shown in Figure 11.13. The vacuum system is turned on until the vacuum of the order of 10^{-3} Torr is reached. Then, valve V1 is closed and the tube is disconnected from the vacuum pump and connected to the hose that, in turn, is connected to the bottom side of the beaker, taking care that all the connections between the thermosyphon and the beaker are filled with the working fluid, without any air bubble. With the beaker filled with more working fluid than necessary to fill the thermosyphon, the valve V2 is opened slowly so that the empty thermosyphon, in vacuum, sucks the working fluid until the necessary volume of working fluid is inserted, which is measured through the scale of the beaker, for removal of the thermosyphon and later sealing. For example, if one desires to fill a thermosyphon with 30 mL of working fluid, the beaker may contain 100 mL of the fluid and the valve is opened until the 70 mL liquid level.

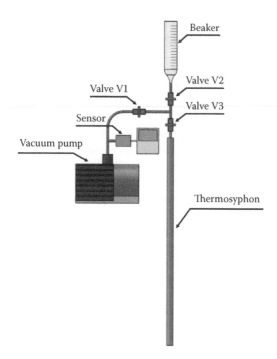

FIGURE 11.13
Filling rig using vacuum pump system.

The advantage of this procedure is its simplicity, quick operation and the precise control of the working fluid to be inserted inside the thermosyphon. The main disadvantage is the cost of the equipments and the necessity of mounting a filling set-up. The cost of the equipment and set-up construction is diluted along the time, as a large number of thermosyphons are fabricated.

Therefore, between these methods, the vacuum procedure is recommended due to the larger precision and smaller charging time. The heating procedure can be employed only when a vacuum system is not available.

After the filling procedure, the umbilical tube is sealed in two steps. First, the umbilical is pinched to provide a mechanical sealing, using a hydraulic press. Two metal bars are used together with the hydraulic press in the sealing process. Care must be taken so that the metal bars do not cut the filling tube. The metal bars must also be designed so that the contact area between the pressing bars and tube is the largest possible. After the pinching, the tube is cut and welded. Figure 11.14 shows the metal bars and the sealing process.

11.3.4 Design Methodology

The following design methodology based on the experience of the Labtucal/ UFSC [31] and on procedures proposed in the literature [6] is suggested for the thermosyphon operating in steady-state conditions.

FIGURE 11.14
Sealing process including pinching, cutting and welding.

Basically, it consists of the determination of all thermal resistances of the thermal circuit as presented in Figure 11.6. The operation limits must also be determined. Based on this information, the total heat power is calculated. The heat transportation capacity of the thermosyphon is then compared with the operational limits. If the heat to be transported is within the operational limits, optimisation procedures, aiming the reduction of cost, improvement of heat transport capacity and geometry, can be applied. Testing may be necessary for exquisite thermosyphon designs.

The following steps are proposed for a thermosyphon exchanging heat with a source and sink by convection:

1. Design parameter specification: lengths, tube diameters, inclination angle, evaporator and condenser external areas, external evaporator and condenser convection coefficients (obtained from literature correlations and models), source $T_{\infty e}$ and sink $T_{\infty c}$ temperatures, filling ratio and thermal conductivity of the casing metal.

2. Determine R_1, R_2, R_8 and R_9.

3. Estimate the vapour temperature by $T_v = T_{\infty c} + \dfrac{R_8 + R_9}{R_1 + R_2 + R_8 + R_9}(T_{\infty e} - T_{\infty c})$

4. Obtain the following working fluid properties, considering T_v as the saturated temperature: p_v, ρ_l, ρ_v, h_{lv}, μ_l, μ_v, σ, k_l and c_{pl}.

5. Determine the pressure in the pool basis using Equation 11.41 and obtain the saturated temperature for p_p.

6. Determine ΔT_h using Equation 11.43.

7. Determine the global temperature difference: $\Delta T = (T_{\infty e} - T_{\infty c}) - \Delta T_h$

8. Determine approximately the heat load: $q = \Delta T / (R_1 + R_2 + R_8 + R_9)$

9. Determine R_3 and R_7.

10. Determine the global thermal resistance (using Equation 11.16) and recalculate the heat load.

11. Compare the heat load obtained in step 10 with 8. If the difference is not acceptable, return to step 9 until convergence is reached.

12. Compare the heat load with the operational limits. If the limits are smaller, the thermosyphon must be redesigned and the whole process starts again.

11.3.5 Thermal Testing

Thermosyphons work under complex physical principles usually difficult to predict analytically. Experimental evaluations of thermosyphon thermal resistances and/or temperature distributions are recommended when they work close to their heat transfer limits, new geometries, working fluids, casing materials and different applications are envisioned; or if one desires to validate mathematical models.

A well-succeeded experiment must be designed carefully. The experimental set-up must reproduce the operation conditions as close as possible. Therefore, information such as heat power, operation temperature level and temperature distribution must be taken into account to plan the experiment. The operation conditions of small devices that work in moderate temperature ranges are usually easy to reproduce in the laboratory and the set-up may be simple and inexpensive. On the other hand, large equipments, such as heat exchangers, may have several meters of length, height and depth. The construction of an experimental rig for testing heat exchangers in their actual operation conditions may be expensive. In most cases, the external heat source and sink coefficients of heat transfer are the dominant resistances of a thermal circuit. Therefore, it is important to know very well the external heat transfer conditions and try to reproduce them in the laboratory.

A thermosyphon thermal performance testing apparatus must include a controlled heat power source that delivers heat to the evaporator and a controlled cooling device that removes heat from the condenser. Usually, the heat power source is responsible for the heat power input, while the cooler is responsible for the thermosyphon temperature level. The cooler usually consists of a small heat exchanger where cold water, controlled by a thermal bath, removes heat from the thermosyphon condenser. The amount of heat input is determined from the electrical voltage and current readings. The heat output is usually measured by an energy balance of the cooling water. These heat transfer measurements must be compared and the difference between these data can be attributed to heat losses. Although a high cooling water mass flow rate would provide a high heat cooling capacity in the condenser, the uncertainty in the measurement of the cooling heat power increases as the difference between the water input and output temperatures decreases, increasing the relative significance of thermocouple uncertainty level.

The measurement of temperatures along the thermosyphon is made by means of thermocouples installed over its external surface. In some cases, internal temperatures need to be accessed, but these measurements are difficult to perform as the thermosyphon usually operates with high internal pressures and any hole in the casing wall (necessary to introduce thermocouples) can cause leakages. Data acquisition systems are used to read the temperatures, which are stored in computers. Several software developed to control the experiment (heat power source input, readings etc.) can be used. Figure 11.15 presents a schematic of an experimental set-up mounted for testing the thermal performance of thermosyphons, including all its major components.

Figure 11.16 shows a schematic of a thermosyphon heat exchanger test rig available at Labtucal/UFSC in Brazil. It reproduces the forced convection heat transfer in the condenser and evaporator sections of actual heat exchangers. Two wind tunnels are available where air circulates by means of fans. In the lower loop (dark grey), hot air, with temperatures up to 500°C, delivers heat to the evaporator region of the heat exchanger to be tested, which is represented by the dark box in the drawing. Heat is supplied through a liquefied petroleum gas (LPG) or natural gas burner, or by electrical heaters. In the upper loop, ambient-temperature air circulates and removes heat from the evaporator regions of the heat exchanger. Chimneys are provided to release air form the tunnels if necessary. The gas mass flows and pressure in both wind tunnels are measured with Pitot tubes and pressure gauges. This set-up is quite large, occupying a volume of 15 × 8 × 3 m. This test rig, installed at Labtucal/UFSC [31], has been used for testing new heat exchanger technologies, to measure the external heat transfer coefficients in the evaporator and condenser, to study the behaviour of finned tubes and so on.

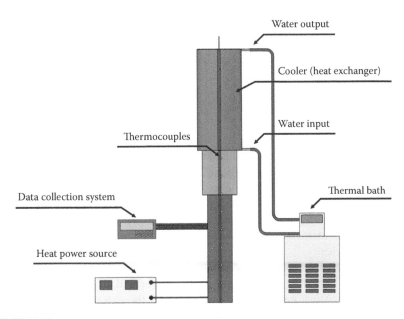

FIGURE 11.15
Schematic of an experimental set-up for straight thermosyphon testing.

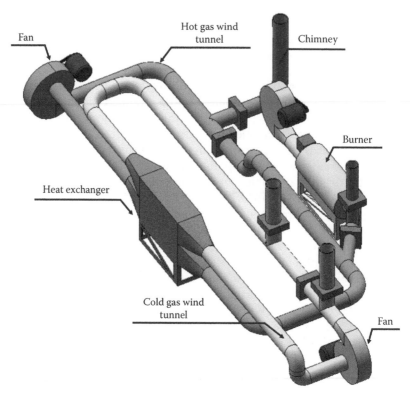

FIGURE 11.16
Experimental set-up for testing technologies for thermosyphons heat exchangers.

11.4 Industrial Applications

11.4.1 Heat Exchangers

Probably the most usual application of thermosyphon technology for industrial equipment is in heat exchangers. In many countries, the use of thermosyphon (or heat pipe) heat exchangers is quite new, but other countries, such as China, has applied this technology for decades. The geometric flexibility, low maintenance, low cost [32,33] and compactness [34] make thermosyphon heat exchangers attractive when compared to the traditional tube-shell and plate technologies.

A schematic of a typical thermosyphon heat exchanger is shown in Figure 11.17. It basically consists of a beam of vertical straight thermosyphons in parallel. A hot gas stream flows through the external surfaces of the evaporators, while a cold stream flows through the condensers. The thermosyphon set removes heat from the hot stream and delivers it to the cold

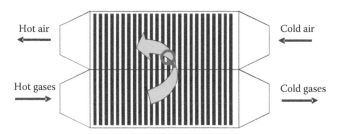

FIGURE 11.17
Schematic of a thermosyphon heat exchanger.

stream. Gas or liquid flow (or a combination of them) can be used as the heat source or sink in theses equipments. The adiabatic section is usually very small. The external surfaces of the equipment are thermally insulated to avoid heat losses to the environment.

The thermosyphon beams can work in vertical or inclined positions, where the minimum angles usually adopted is 7° with the horizontal position. If the application requires the tubes to be aligned in the horizontal position, heat pipes (with the porous medium for capillary pumping) are recommended.

The main thermal resistances of thermosyphon heat exchangers are associated with the heat transfer between the external flows and the tubes [35–37]. Usually, the convection coefficients of heat transfer between gas flows and tubes are not high, and one way to improve the heat exchange is increasing the tube external area by using finned tubes. Another way to improve this heat exchange is to provide chicane to the external flow. Thermosyphons are quite nice for this case as they can homogenise the tube surface temperature. Figure 11.18 shows a schematic of a gas (heat source) and liquid (heat sink) heat exchanger, where finned tubes are used in the evaporator and chicanes are used in the liquid cold stream, in the condenser region.

As already mentioned, the low-cost maintenance is one of the features of thermosyphon heat exchangers. This equipment can be constructed in modules, which fit in the equipment frame as a vertical drawer. Each module is composed of a group of thermosyphons. If several thermosyphons of a same module are not working, the module is removed and a new module inserted in the place. The whole equipment does not need to be removed, simplifying the maintenance and reducing the down time of the equipment. This is a very interesting feature especially for petrochemical applications. This configuration of thermosyphon heat exchangers also allows for the modification of the total conductance of the equipment. If the industry operational conditions change, modules can be removed or inserted, changing the total heat transfer between the hot and cold streams. Figure 11.19 shows the module configuration of thermosyphon heat exchangers.

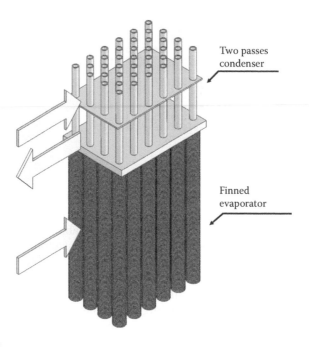

Two passes
condenser

Finned
evaporator

FIGURE 11.18
Gas–liquid thermosyphon heat exchanger, with finned tubes in the evaporator and chicane in
the cold stream.

11.4.2 Ovens

There are many different types of ovens in the market, for different applications, different temperature levels, different sizes and so on. Basically, an oven is a heated cavity, insulated from the external environment, where some material receives some kind of heat treatment. Usually, the heat is delivered to the oven by means of electricity, gas (GLP, natural etc.), burning wood and so on, although other more sophisticated technologies can also be employed for heating the material, such as microwave and infrared radiation. The cavity can be at atmospheric pressure (most applications), filled with air or other gas, pressurised, in vacuum and so on.

Usually, two features are very important for industrial ovens: to present a uniform temperature distribution in the cavity and low operational costs. The thermosyphon technology [38] is appropriate to be applied for many industrial ovens as they present uniform temperatures in the condenser and evaporator sections and allow for the heat to reach more directly the regions to be heated, saving energy.

Domestic and industrial food-cooking ovens usually work at temperatures ranging from 150°C to 250°C [39]. Much of the energy employed in cooking is used to transform the water of the food into vapour, so that a typical food temperature during the cooking process is around 100°C. In these

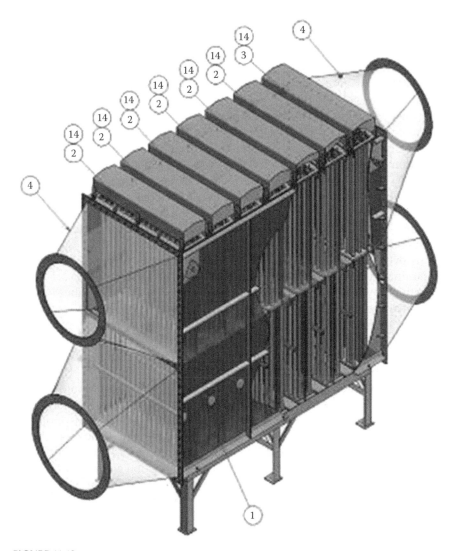

FIGURE 11.19
Module configuration thermosyphon heat exchangers.

applications, electricity or hot gases from burning of gas or wood are usually used as the heat source. The heat reaches the food mainly by hot air natural or forced convection and by radiation, mainly emitted by the heated walls.

Electricity is more efficient as the electrical heaters are usually installed inside the cavity. The problem with this design is that the temperature distribution in the cavity is not uniform, as the electrical resistances supply very concentrated heat. In these cases, fans are used to improve the temperature distribution, but many cooking processes (cake baking for instance) can be largely affected by strong airflow caused by fans. Even if the temperature is homogenised by air movement in the cavity, the heat distribution may not be uniform, as the electrical heater temperature can be very high and large amount of radiation heat is released to the food that 'sees' the electrical heater, so this food can experience surface burning.

Usually, in ovens heated by hot gases resulted from burning of gas or wood, heat reaches the food by convection and radiation. In most domestic gas-powered ovens, the burner chamber, located below the cavity bottom plate, is connected to the cooking chamber by small holes. Therefore, the gases resulting from the gas combustion mix with cavity air. As the cavity bottom plate is hotter than the rest of the cooking chamber, the food is heated by natural convection. Fans located in the rear region of the cooking chamber may also be applied, improving the convection. The radiation comes from the heated walls, especially from the bottom one, which is hotter than the others. If the food receives direct radiation from the bottom surface, burning in the lower surface of the food may occur.

This oven design is not suitable for the food industry, as the product is contaminated by the combustion gases mixed with air. Old ovens are being substituted by new designs, where the hot gases flow through tubes located in the rear region of the cooking cavity and delivered to the atmosphere by a chimney. A fan blows air against the tubes to remove and spread the heat by convection inside the cooking chamber. This configuration is not energy efficient, as typical convection heat transfer coefficients of gas is low and the air released by the chimney is very hot.

Silva and Mantelli [38], Milanez and Mantelli [40,41] and Mantelli et al. [42] propose the use of straight-tree configuration thermosyphons (see Figure 11.4), located close to the two vertical lateral walls, to transport with negligible thermal resistance the heat generated in the combustion chamber to the cooking environment. Flute-type natural or GLP gas burners are located in the bottom region of the oven. The combustion gases are in contact with the evaporator and the heat is uniformly released inside the cooking chamber by the several condensers. The combustion gases are not mixed with the hot cooking air, as the combustion is separated of the cooking by a plate. A fan is also employed inside the cooking cavity to promote convection. The fan is necessary because the bottom and the lateral vertical walls have very uniform temperatures, so natural convection is not promoted.

Figure 11.20 shows a drawing of the bread-baking oven assisted by thermosyphon technology, highlighting the location of the evaporator, condensers and gas burners. Figure 11.21 shows a comparison between the temperature distributions of electrical ovens (left side) and thermosyphon-assisted gas ovens. The quality of the bread cooked in the electrical ovens varies according to the position of the bread inside the cooking chamber: overcooked when close to the electrical resistances and other hot areas, and undercooked in other colder regions. On the other hand, the bread produced in thermosyphon-assisted ovens has uniform cooking quality, as shown in Figure 11.22.

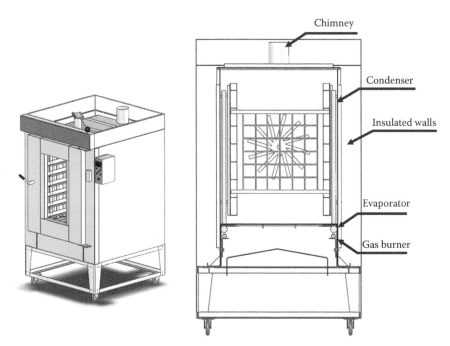

FIGURE 11.20
Bread-baking oven assisted by thermosyphon technology.

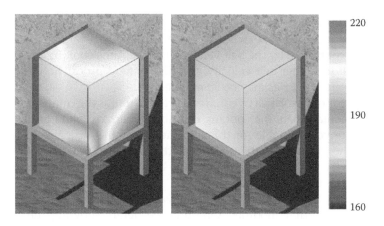

FIGURE 11.21
Temperature distribution of electrical oven (*left*) and thermosyphon-assisted oven (*right*).

Only a small share of energy applied to the oven is actually used for the cooking process. Most of the energy is used to heat up and keep the oven hot. Therefore, to increase the efficiency of the baking ovens, the heat must be directly applied to the air cooking cavity, not to the mechanical structure. In the thermosyphon-assisted ovens, heat is first applied to the cavity and then

FIGURE 11.22
Uniform bread cooking in thermosyphon-assisted ovens.

the hot air heats up the walls. Their heating up time is considerably smaller when compared to other technologies, and they consume around only 50% of the gas used in the traditional gas ovens [40].

In large food industries, ovens are designed for high production. Electrical-conveyor-belt ovens are usually employed. They basically consist of a long externally insulated electrical heated cavity, where the conveyor belt runs with a controlled velocity. Electrical resistances are spread along the belt way over and above the belt in an attempt to promote uniform temperature distribution. The resulting temperature distribution is not good in the cooking cavity as the colour of the food changes in the belt direction, especially due to the radiation influence of the electrical resistance over the food along the cooking cavity.

The baking time depends on the food. As an example, cookies may take around 20 minutes to bake, whereas cakes can take up to 1 hour. The cooking time is controlled by the belt velocity. For the same cooking time, longer ovens result in higher belt velocity and higher food production. Industrial ovens in biscuit factories may be 50 m in length.

Gas-driven conveyor-belt ovens are not very usual, as the combustion gases can easily contaminate the cooking chamber. Thermosyphon technologies, however, are very suitable for this application, as they are able to transfer heat from the combustion chamber, located in the lateral lower regions of the cooking cavity to the cooking cavity. The combustion and cooking cavities are isolated by means of plates.

Figure 11.23, left side, shows a thermosyphon-assisted conveyor belt oven made of four similar modules. One module detail, showing the thermosyphon sets, is shown in the right side of Figure 11.23. Two sets of water–stainless steel thermosyphons are employed for heating the cooking cavity: located below and above the belt (see Figure 11.4 for thermosyphon geometry details). Each set contains two independent tree configuration thermosyphon, with the condensers being almost horizontally positioned. Actually, the evaporator external area of just one thermosyphon is not enough to capture the necessary heat from the flute-type burners, which are located just below each pair of evaporators. The left set furnishes heat for the cooking

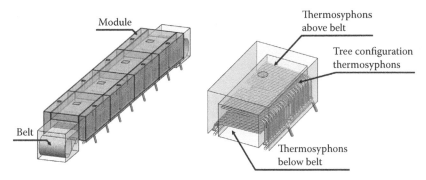

FIGURE 11.23
Schematic of the conveyor belt oven with upper and lower tree configuration thermosyphons.

chamber from below the belt in horizontal position. Hot air coming from these condensers circulates through the belt, usually made of metallic wire screens, delivering heat by convection to the food. Most of the baking energy comes from the left-side set of thermosyphons. The right-side thermosyphon set provides heat to the condensers located above the belt and furnishes radiation heat transfer—these are mostly employed to blush the food. In several cooking processes, these thermosyphons can be deactivated. These many thermosyphons make the system robust, so that the effect of the failure of one of the tubes is almost not sensed.

Another advantage of this configuration is that the oven is modular and the number of modules determines the total length. Each module's temperature can be controlled independently by means of the burning gas control, so the oven can accommodate different cooking process with different temperature levels. Fans can also be used to improve thermal convection. The temperature distribution of the thermosyphon conveyor belt oven is very uniform along one module. As for the baking oven, the gas consumption is lower than the usual gas-driven equipment available in the market. Actually, this configuration can be used whenever a conveyor belt oven is applied.

11.4.3 Oil Tank Heaters

In petrochemical industries, oils need to be stored in tanks. At ambient temperatures, some of these materials solidify, and therefore they require constant heating to be at the liquid state to be processed. This is the case of asphalt in refineries.

The usual way to heat asphalt in tanks is by using vapour, which circulates inside serpentines, located in the bottom region of the tanks. The main problem with that technology is that usually vapour is available only in a predetermined pressure and temperature, and thus the temperature level of the asphalt cannot be properly controlled. Besides, the stored asphalt does not need to be heated at high temperature levels all times. High temperatures are necessary when asphalt needs manipulation, such as when it is pumped to trucks, for example. Unnecessary energy is used to keep the asphalt at high temperature levels. Furthermore, high temperatures intensify the deterioration of the physical properties of the product.

Thermosyphon heaters have been employed to heat asphalt inside storage tanks [43,44]. It consists of a bundle of thermosyphons in parallel, which are partially inserted into the asphalt tank from its lateral vertical wall. An inclination of about 7° is provided. The evaporators of the thermosyphons are located outside the tank and receive heat from a gas burner, whereas the condensers are located inside the asphalt tank and deliver the heat to the asphalt. The asphalt temperature can be controlled by means of a gas-burning controller. Figure 11.24 shows the schematic of the asphalt tank with a bundle-type thermosyphon.

A study was performed [37] showing that the temperature distribution inside asphalt tanks can be quite uniform (within 5°C), even when the heat source is concentrated. Serpentines in a loop thermosyphon, located in the rear region of the tank (in substitution of traditional serpentines), were also considered, but the idea was discarded as the cost is much higher and the risk of employing this technology is high, as a small damage in the serpentine would ruin the entire loop thermosyphon heating system. On the other hand, if one of the tubes of the bundle is lost, the system would be negligibly affected.

11.4.4 High Heat Power Cooling

In many applications such as computers, radars and electric power transformers, electric or electronic devices dissipate high amount of heat energy in small volumes. To control their temperature and so guarantee the proper functioning of the devices, the generated heat must be dissipated. Conventional heat dissipation systems include cooled water circulation within serpentines in refrigeration plates, which are attached to the electronic systems. This solution may not be safe, as water circulates close to the

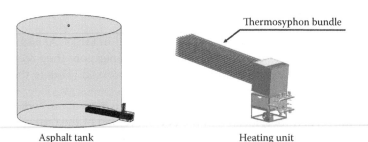

Asphalt tank   Heating unit

FIGURE 11.24
Schematic of the asphalt tank assisted by thermosyphon technology.

electric or electronic devices and any leakage can damage the equipment. Another solution is to blow cooled air over the heated electronic devices. In this case, the refrigeration unit may be large (some time larger than the electronic equipment itself), may vibrate and are usually expensive.

Thermosyphons may provide an efficient and passive way to remove high amount of concentrated heat from within electronic boxes. A typical device consists of a bunch of vertical thermosyphons in parallel, whose evaporators are in mechanical contact with the electronics to be cooled. Care must be taken to minimise the contact resistance between the evaporator and the electronic box. The condensers are located outside the box and the heat can be removed by cooled liquid (water, for instance) or ambient air flux. If air is used to remove the heat, the condensers of thermosyphon compact heat dissipaters must be connected to several thin narrow-spaced plates in parallel, which form efficient fins. Figure 11.25 shows a schematic of the proposed device, which can remove large amount of heat (e.g. more than 2 kW in a volume of 0.09 m^3). Actually, if convenient, the thermosyphons can have inclinations of up to 7° with the horizontal position or even present different evaporator and condenser inclinations.

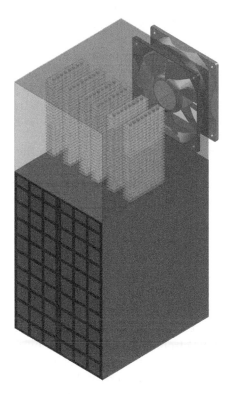

FIGURE 11.25
Schematic of a compact cooler for dissipation of high concentrated heat energy.

11.5 Concluding Remarks

Thermosyphons are devices that transfer high amounts of heat energy even when subjected to small temperature differences. In vapour state, the working fluid spreads out very easily along the tube and thus is able to homogenise its temperature. Thermosyphons also present high geometric flexibility and low cost. These characteristics make thermosyphons important devices to be applied for the solution of several industrial heat transfer and temperature control problems.

The first patents involving in heat pipes and thermosyphon equipment were registered in the eighteenth century [23], but these technologies were not employed widely for more than one century. Around 1960s, due to the space industry development in the major world countries, heat pipes were considered as a solution for the thermal management in spacecrafts and so government budgets for heat pipe research and development sharply increased. With the end of the cold war, since the last decade of the twentieth century, these financial supports have been slowly decreasing. Actually, the electronic industry brought a great revival in heat pipe technology research. As a result, many of the computers nowadays have heat pipes or vapour chambers for the thermal control of highly concentrated heat dissipation.

In the last three decades, thermosyphons (gravity-assisted heat pipes) have received increasing amounts of financial support around the world aiming to provide thermal solutions for several industrial problems, especially those concerning efficient use of energy. The industries and governments are investing in new or redesigning traditional equipments, with the major objective of using heat rationally and ecologically.

Therefore, considering the current energy crises the world faces, which tend to expand in the next years, it is expected that the budget for the thermosyphon technology development will continue to grow, with an increasing number of new and thermally efficient equipment being delivered to the market.

Acknowledgements

The author wishes to acknowledge the support from Federal University of Santa Catarina (UFSC), Petrobrás for financing many projects and the construction of the Labtucal building, FINEP for supporting financially the conveyor belt oven project, CNPq for research grants, CNPq and CAPES for supporting graduate students with scholarships, and other several industries that supported equipment development projects.

Nomenclature

A	Area (m²)
b	Wall width (m)
Bo	Bond number
C	Non-dimensional constant
Cp	Specific heat capacity (J/kg·K)
d	Tube diameter (m)
f	Entrainment limit parameter
F	Filling ratio
g	Gravity acceleration (kg·m/s²)
h	Coefficient of heat transfer (W/m²·K)
$\bar{h}_L$	Average latent heat (kJ/kg)
h_{lv}	Latent heat (kJ/kg)
Ja	Jacob's number
k	Thermal conductivity (W/m·K)
L	Length (m)
$\dot{m}$	Mass flux (kg/s)
N	Number of merit
Nu	Nusselt number
p	Pressure (Pa)
Pr	Prandtl number
q	Heat power (W)
q''	Heat flux in the wall (W/m²)
r	Radius (m)
R	Thermal resistance (K/W), universal gas constant
Re	Reynolds number
T	Temperature (K)
v	Vertical velocity (m/s)
V	Volume (m³)
x, y	Coordinates (m)
X	Body forces (kg/m²·s²)

Geek Letters

β	Inclination angle
δ	Liquid film thickness (m)
μ	Viscosity (kg/m·s)
ϑ	Kinematic viscosity (cm²/s)
ρ	Density (kg/m³)
σ	Surface tension (N/m)

Subscripts

a	Adiabatic

(*Continued*)

atm	Atmospheric
bol	Boiling
c	Condenser
crit	Critical
e	Evaporator
ef	Effective
ent	Entrainment
f	Film
h	Hydrostatic
i	Internal
p	Pool
l	Liquid
lv	Liquid–vapour
m	Medium
o	External
p	Pool
Nu	Nusselt
s	Saturação
son	Sonic
v	Vapour
vis	Viscous
sat	Saturated
w	Wall

References

1. Incropera, F.P., DeWitt, D., Bergman, T.L., and Lavine, A.S., *Fundamentals of Heat and Mass Transfer*, 6th Edition, John Wiley (2007).
2. Chen, M.M., An Analytical Study of Laminar Film Condensation Part I—Flat Plates, *Journal of Heat Transfer*, 83(Series C), 48–55 (1961).
3. Bejan, A., *Heat Transfer*, John Wiley, New York (1993).
4. Hewitt, G.F., Shires, G.J., and Bott, T.R., *Process Heat Transfer*, New York, Begell House (1993).
5. Faghri, A., *Heat Pipe Science and Technology*, Taylor & Francis, New York (1995).
6. Groll, M., and Rösler, S., Operation Principles and Performance of Heat Pipes and Closed Two-Phase Thermosyphons, *Journal of Non-Equilibrium Thermodynamics*, 17, 91–151 (1992).
7. Paiva, K., Desenvolvimento de Novas Tecnologias para Minitubos de Calor: Análise Teórica e [Development of New Technologies for Mini Heat Pipes: Theoretical and Experimental Analysis], Doctoral Thesis, EMC/UFSC, Brazil (2011).
8. Kaminaga, F., Matsumura, K., and Horie, R., A Study on Thermal Conductance in a Looped Parallel Thermosyphon, 16th IHPC, France, 048 (2012).
9. Brost, O., Closed Two Phase Thermosyphons, *Class Notes*, EMC/UFSC Florianópolis, Brazil (1996).

10. Özisik, N., *Heat Transfer: A Basic Approach*, McGraw-Hill, New York (1989).
11. Holmann, J.P., *Heat Transfer*, 9th Edition, McGraw-Hill, New York (2002).
12. Dunn, P.D., and Reay, D.A., *Heat Pipes*, 4th Edition, Pergamon, Oxford, England (1994).
13. Moran, M.J., and Shapiro, H.N., *Fundamentals of Engineering Thermodynamics*, 3rd Edition, John Wiley, New Jersey (1995).
14. Peterson, G.P., *Heat Pipes, Modeling, Testing, and Applications* John Wiley and Sons (1994).
15. Mantelli, M.B.H., Colle, S., Carvalho, R.D.M., and Moraes, D.U.C., Study of Closed Two-Phase Thermosyphons for Bakery Oven Applications, NHTC99-205, 33rd National Heat Transfer Conference, Albuquerque, USA (1999).
16. Kaminaga, F., Hashimoto, H., Feroz, M.D., Goto, K., and Matsumura, K., Heat Transfer Characteristics of Evaporation and Condensation in a Two-Phase Closed Thermosyphon, 8th IHPC, China (1992).
17. Kaminaga, F., Okamoto, Y., and Suzuki, T., Study in Boiling Heat Transfer Correlation in a Closed Two-Phase Thermosyphon, 8th IHPC, China (1992).
18. Carey, V.P., *Liquid-Vapor Phase Change Phenomena: An Introduction to the Thermophysics of Vaporization and Condensation Process in Heat Transfer Equipment*, Hemisphere, New York (1992).
19. El-Genk, M.S., and Saber, H.H., *Thermal Conductance of the Evaporator Section of Closed Two-Phase Thermosyphons (CTPTs)*, AIAA/ASME Joint Thermophysics and Heat Transfer Conference, 3, ASME, 99–106, Albuquerque, USA (1998).
20. Gross, U., Falling Film Evaporation Inside a Closed Thermosyphon, 8th IHPC, China (1992).
21. Bejan, A., *Convection Heat Transfer*, 3rd Edition, John Wiley, New Jersey (2004).
22. Collier, J.G., and Thome, J.R., *Convective Boiling and Condensation*, 3rd Edition, Oxford, New York (1994).
23. Reay, D., and Kew, P., *Heat Pipes Theory Design and Applications*, 5th Edition, Elsevier (2006).
24. Tien, C.L., and Chung, K.S., *Entrainment Limits in Heat Pipes*, Proceeding of the 3rd IHPC, USA, 36–40 (1978).
25. Faghri, A., Chen, M.M., and Morgan, M., Heat Transfer Characteristics in Two-Phase Closed Conventional and Concentric Annular Thermosyphons, *ASME Journal of Heat Transfer*, 111(3), 611–618 (1989).
26. Nguyen-Chi, H., and Groll, M., Entrainment or Flooding Limit in a Closed Two-Phase Thermosyphon, *Heat Recovery Systems*, 1, 274–286 (1981).
27. Imura, H., Sasaguchi, K., and Kozai, H., Critical Heat Flux in a Closed Two-Phase Thermosyphon, *International Journal of Heat and Mass Transfer*, 26 (8), 1181–1188, New York (1983).
28. Street, R.J., Watters, G.Z., and Vennard, J.K., *Elementary Fluid Mechanics*, 7th Edition, John Wiley, New York (1996).
29. Busse, C.A., Theory of the Ultimate Heat Transfer Limit of Cylindrical Heat Pipes, *International Journal of Heat and Mass Transfer*, 16, 169–186 (1973).
30. Mantelli, M.B.H., Ângelo, W.B., and Borges, T., Performance of Naphthalene Thermosyphons with Non-Condensable Gases Theoretical Study and Comparison with Data, *International Journal of Heat and Mass Transfer*, 53, 3414–3428 (2010).
31. Heat Pipe Laboratory (Labtucal), Mechanical Engineering Department, Federal University of Santa Catarina. http://www.lepten.ufsc.br/pesquisa/tucal/linha3/trocater.html

32. Borges, T.P.R., Mantelli, M.B.H., and Persson, L.G., Techno-Economic Optimization of Thermosyphon Heat Exchanger Design Using Mathematical Programming, 14th IHPC, Brazil (2007).
33. Borges, T.P.F., Mantelli, M.B.H., and Persson, L.G., Trade off Between Prices and Pressure Losses for Thermosyphon Heat Exchangers, 39th AIAA Thermophysics Conference, Miami (2007).
34. Costa, C., and Mantelli, M.B.H., ANSYS-CFX as a Tool in the Development of Heat Exchangers Assisted by Heat Pipe Technology, ANSYS Conference and 27o CADFEM User's Meeting, Germany (2009).
35. Isoppo, D., and Borges, T.P.F., Development of a Detailed Thermal Model for Designing Heat Pipe Heat Exchangers, ECOS 2009, Foz do Iguaçu (2009).
36. Molz, M., Mantelli, M.B.H., and Landa, H.G., Transient Modeling of a Closed Two-Phase Thermosyphon for Heat Exchanger Applications, 13th IHPC, Shanghai (2004).
37. Mantelli, M.B.H., Martins, G.J., Reis, F., Zimmerman, R., Rocha, G.K.L., and Landa, H.G., Vertical Thermosyphons for Industrial Heat Exchanger Applications, 13th IHPC, Shanghai (2004).
38. Silva, A.K., and Mantelli, M.B.H., Thermal Applicability of Two Phase Thermosyphon in Cooking Chambers—Experimental and Theoretical Analysis, *Applied Thermal Engineering*, Elsevier, 24, 717–733 (2003).
39. Santos, B.S., and Mantelli, M.B.H., Determination of Thermal Conditions of Pizza's Baking Process, COBEM, Gramado, Brazil (2009).
40. Milanez, F.H., and Mantelli, M.B.H., Analytical Model for Thermal Performance Analysis of Enclosure Heated by Aligned Thermosyphons, *Journal of Thermophysics and Heat Transfer*, 20(2), 267–275 (2006).
41. Milanez, F.H., and Mantelli, M.B.H., A New Methodology for Measuring Heat Transfer Coefficients—Application to Thermosyphon Heated Enclosures, 13th IHPC, Shanghai (2004).
42. Mantelli, M.B.H., Milanez, F.H., and Mielitz, G., Tree Configuration Thermosyphon Study, AIAA 38th Thermophysics Conference, Toronto (2005).
43. Costa, C.A.S., Estudo do Comportamento Térrmico de Tanques de de Armazenamento de Asfalto [Study of the Thermal Behavior of Storage Asphalt Tanks], M.Sc. Dissertation, POSMEC-EMC/UFSC (2008).
44. Mantelli, M.B.H., and Milanez, F.H., A Loop Thermosyphon for Asphalt Tank Heating, 8th IHPS, Kumamoto, Japan (2006).

12

Fluid Flow and Heat Transfer with Phase Change in Minichannels and Microchannels

V. V. Kuznetsov and S. A. Safonov

CONTENTS

12.1 Introduction 465
12.2 Surface Tension and Capillarity 469
12.3 Two-Phase Gas–Liquid Flow in Minichannels and Microchannels .. 470
    12.3.1 Capillarity-Controlled Adiabatic Flow Patterns 470
    12.3.2 Capillarity-Controlled Diabatic Flow Patterns......................... 474
    12.3.3 Modelling of Annular Flow in a Rectangular Channel........... 475
12.4 Heat Transfer with Evaporation and Condensation in
    Minichannels and Microchannels.. 478
    12.4.1 Evaporation and Condensation in Annular Flow..................... 478
    12.4.2 Heat Transfer with Evaporation in the Microregion 480
12.5 Flow Boiling Heat Transfer 484
    12.5.1 Heat Transfer at Predominant Nucleate Boiling 484
    12.5.2 Heat Transfer at Predominant Forced Convection and
         Evaporation.. 487
12.6 Conclusions.. 490
12.7 Nomenclature... 491
12.8 Acknowledgements 492
References.. 492

12.1 Introduction

A heat pipe is a high-performance heat transmission device with an extremely high thermal conductivity, which strongly depends on capillarity induced flow in the mini- and microchannels. There are many variants of heat pipes, but in all cases the working fluid must circulate, when temperature difference exists between the evaporator and the condenser, and capillary forces play a governing role in this circulation. In this chapter, we discuss the concept of fluid flow and heat transfer with the phase change in microscale, which is the novel cooling strategy for the devices requiring successful thermal management.

In the literature, the criterion of transition to microscale is not precisely defined. The geometric definition of microchannels by Mehendal et al. (2000) proposes that the channels with diameters of 1–100 μm can be considered as microchannels, 100 μm to 1 mm as mesochannels, 1–6 mm as compact channels and diameter greater than 6 mm as usual channels. Classification by Kandlikar and Grande (2003) suggests that the channels in the range of hydraulic diameters from 200 μm to 3 mm should be classified as minichannels and those less than 200 μm should be classified as microchannels. Taking into account that the effects of dilution for many gases occur in channels with the range of hydraulic diameters from 0.1 to 10 μm, such channels are considered as the transitional ones. Such a classification ignores the influence of physical properties of liquids and gases for two-phase flows, so it should be considered as conditional. Kew and Cornwell (1997) classify the transition to microscale using the 'confinement number', defined as the ratio of capillary constant $\delta_c = \left[2\sigma/\left(\rho_{liq} - \rho_{gas}\right)g \right]^{1/2}$ to diameter of the channel, and determine the transition to microscale behaviour in a two-phase flow. While comparing the experimental data on two-phase heat transfer coefficient with the results of calculations based on conventional correlations, it was noticed by Kew and Cornwell (1997) that the measured heat transfer coefficients differ from the calculation results when the confinement number is less than 0.5. The confinement number provides criterion determination for transition to the microchannel behaviour.

Existing definitions are insufficient to determine the exact criterion of transition from macro- to microscale behaviour and give only a rough estimate of the channel size at which this transition occurs. If one reduces the size of a channel, the wide variety of phenomena that are not typical for the large scale becomes apparent. The degree of their appearance depends not only on the geometrical scale but also on the shapes of the channels, heat fluxes, pressure and so on. Identification of these phenomena and determination of the conditions when their influence on heat and mass transfer with phase changes becomes determining is considered in this chapter. As the initial classification, all systems with transverse size less than the capillary constant are considered as systems with the possibility of microscale effects manifestation.

Heat transfer under conditions of phase change in mini- and microchannels is a fast-developing field of science. The reasons for this are a tendency towards miniaturisation in various fields, giving landmark technologies such as thermal management of electronics using heat pipes, green energy technologies based on the application of fuel cells and so on. The ratio of heat transfer area to volume of the channel is inversely proportional to the lateral size of the channel, and its decrease causes high-intensity heat transfers in microsystems. It is the reason why considerable emphasis was placed last time on the study of flow boiling and condensing in open and closed microchannels. However, this complex object is still poorly understood due to the large number of determinative parameters influencing the degree of boiling, evaporation and condensation enhancement in confined flows.

In this context, an important parameter that affects heat transfer with phase change is flow patterns in mini- and microchannels. Flow pattern provides important information to the analysis of data on the mechanism of boiling and condensing. Many studies on upward and horizontal gas–liquid flows in small-diameter tubes were conducted in recent years. These studies showed that the two-phase flow in a channel with a gap less than the capillary constant differs significantly from that in a conventional tube. Three basic flow patterns such as isolated bubbles, compressed bubbles and annular/slug flow were observed by Nakoryakov et al. (1992), Triplett et al. (1999), Kawahara et al. (2002) and Revellin and Thome (2006); the flow pattern map was created for both adiabatic and diabatic flows. A peculiarity of the annular gas–liquid flow in a microchannel is that capillary forces endeavour to aggregate the waves on opposite sides of the channel into the liquid bridge. The liquid bridges can be unstable and permeable for gas at high flow rates (Nakoryakov et al., 1992). At large values of Weber number, microdroplet clots instead of liquid bridges were observed by Wambsganss et al. (1991) and Kuznetsov et al. (1997). The analysis of flow boiling heat transfer data using flow mapping showed that the correlation is not so obvious. There are data that show the deterioration of heat transfer at transition to the annular flow (Karayiannis et al., 2010), and there are data that show an increase of heat transfer in this area (Kuznetsov and Shamirzaev, 2007a).

At low heat fluxes, evaporation of thin films and condensation at interfaces are the predominant mechanisms of heat transfer. Many reviews and books, for example, those by Mehendal et al. (2000), Stephan (2002) and Kandlikar et al. (2006), consider the theoretical and experimental study of heat transfer with phase change on the microscale. The model based on conduction and convection through the film flowing down the walls of channels is usually used to calculate heat transfer with evaporation and condensation. The pioneering works in the field of contact evaporation of menisci (DasGupta et al., 1993; Wayner, 1994) presented a model designed to describe heat transfer and fluid flow in extremely thin films of liquids based on disjoining pressure, capillary pressure and the Clapeyron–Clausius concept. When comparing the influence of dynamic contact angle and overheating in the prediction of local wall superheating at the beginning of heat transfer crises, it was shown by Wayner (1994) that heat flux, velocity of contact line and local overheating depend on the apparent contact angle. Stephan (2002) showed that a combination of the microscopic and macroscopic models of evaporation provides a tool for the development of microstructure heat exchange devices and processes, such as an improved capillary structure for a heat pipe with microgrooves.

The mechanisms of heat transfer for flow boiling in minichannels and microchannels are discussed in many books and reviews, for example, those by Kandlikar et al. (2006) and Thome (2006). It was shown that the relative contributions of nucleate boiling, thin film evaporation and convective boiling in flow boiling heat transfer remain unclear. In some studies, the heat transfer coefficients were unaffected by vapour quality or mass flux, but they strongly depended on heat flux, for example, those by Lazarek and Black (1982) and Owhaib et al. (2004). There are studies in which this trend was not observed and the heat transfer coefficients were dependent on vapour quality and much less on heat flux, as reported by Karayiannis et al. (2010). The correlation between flow pattern and heat transfer for flow boiling in microchannels was considered by Bar-Cohen and Rahim (2009) using the flow pattern mapping methodology of Taitel and Dukler. Characteristic M-shaped heat transfer coefficient variation versus quality for flow boiling of refrigerants and dielectric liquids in miniature channels was identified, and the inflection points in this M-shaped curve are seen to be approximately equal to flow regime transitions. This heat transfer coefficient behaviour was observed by Consolini and Thome (2009) for flow boiling of R-134a, R-236fa and R-245fa in two microtubes with 510 and 790 µm internal diameters. The growth of heat transfer at low heat fluxes was observed by Kuznetsov and Shamirzaev (2007a) when transition to annular flow occurs at flow boiling of R-21 in a vertical rectangular minichannel.

Many studies were focused on modelling flow boiling heat transfer in mini- and microchannels. Kandlikar and Balasubramanian (2004) and Kandlikar (2010) proposed a model for flow boiling heat transfer prediction based on the modification of models developed for conventional tubes. The model by Liu and Winterton (1991) was corrected by Kuznetsov and Shamirzaev (2007a) for the prediction of flow boiling heat transfer in rectangular and annular minichannels at low Reynolds numbers. Special-purpose models for the calculation of flow boiling heat transfer were designed by Tran et al. (1997) and Bertsch et al. (2009). The three-zone model for heat transfer prediction was proposed by Thome et al. (2004), who take into account the features of gas–liquid flow in microchannels. Initially, this model was designed for the elongated bubbles regime, where transient evaporation of a thin film surrounding an elongated bubble was essential, and then it was also developed for annular flow. Nevertheless, Bertsch et al. (2009) noted that the application of existing models to different modes of heat transfer in a microchannel cannot be considered successful.

The review shows that if one reduces the size of the channel, the wide variety of phenomena that are not typical for conventional tubes becomes apparent. The degree of their manifestation depends not only on the geometrical scale but also on the shapes of channels, heat fluxes, pressure and so on. Identification of these phenomena and determination of the conditions when their influence on heat transfer with phase change becomes essential in mini- and microchannels is considered in this chapter. The objective of this chapter is a discussion on the fundamentals of gas–liquid flow in mini- and microchannels and the approach to annular flow modelling when capillary forces are predominant. Later in Section 12.4, using the results of annular flow modelling in a rectangular channel the mechanism of heat transfer with evaporation and condensation in small-sized channels is discussed. Finally in Section 12.5, a new approach to predict flow boiling heat transfer in a channel with transverse size less than the capillary constant based on accounts of nucleate boiling enhancement and nucleate boiling suppression in thin films is proposed and discussed using available experimental data.

12.2 Surface Tension and Capillarity

To analyse fluid flow and heat transfer with phase change in mini- and microchannels, it is necessary to understand the phenomena that govern the behaviour of the liquid–vapour interface for this case. The boundary of liquid, vapour and solid phases typically exists through a narrow transition layer, which can be treated as the interface. For a curved interface, the surface tension, wettability and contact angle result in a pressure misbalance that provides the capillary pumping required for proper microchannel operation.

The pressure misbalance at a curved interface, which is called the capillary pressure, can be determined using the assumption of thermodynamic equilibrium of liquid and gas taking into account the interface properties. It gives us the Laplace equation for capillary pressure calculation:

$$p_{gas} - p_{liq} = \sigma \left(\frac{1}{R_1} + \frac{1}{R_2} \right) \tag{12.1}$$

The pressure for gas is higher than that for liquid because of the convex surface for the gas phase. For plane surfaces, $R_1 = R_2 = \infty$ and the pressures in both phases are equal.

To determine the equilibrium shape of the interface, it is necessary to solve the variation problem on definition of the minimum of free energy of a gas–liquid system. The internal free energy depends on the volume of the system and not on its shape. If a liquid–gas system is within the gravitational field with acceleration due to gravity g, the equilibrium condition can be written as follows:

$$\sigma \int dS + g\rho_{liq} \int z\, dV = \min, \int dV = \text{const} \tag{12.2}$$

where ρ is liquid density, gas density is assumed to be small, and z is the vertical coordinate. The second equation in Equation 12.2 expresses the invariance of liquid volume, whereas the first term in the first equation corresponds to free surface energy and the second one corresponds to energy in a gravity field. The physical properties under the equilibrium condition form the complex $\sigma/g\rho_{liq}$, which has the dimension of square of length. Then the interface shape will be determined by a square root from this complex named the capillary constant:

$$l_\sigma = \sqrt{\frac{2\sigma}{g\left(\rho_{liq} - \rho_{gas}\right)}} \qquad (12.3)$$

If the capillary constant is much higher than the system size, gravity effects can be neglected.

The important number that characterises the behaviour of gas–liquid flow is the Weber number, which measures the relative importance of a fluid's inertia compared to the surface tension:

$$We = \frac{\rho_{liq} U_{liq}^2 D}{\sigma} \qquad (12.4)$$

where D is the characteristic scale and U_{liq} is the liquid velocity. Using D as the gas bubble size, the condition $We \ll 1$ shows that the bubble will not be deformed in the flow.

12.3 Two-Phase Gas–Liquid Flow in Minichannels and Microchannels

It is known that the patterns of gas–liquid flow in channels with a small gap are significantly different from those in conventional tubes. Let us consider a channel with a transverse size compared with the capillary constant as a minichannel and another channel with a transverse size much less than the capillary constant as a microchannel. From the definition of capillary constant, it follows that two-phase gas–liquid flows in minichannels are characterised by considerable influence of capillary forces and gravity on the flow patterns. In microchannels, the capillary forces play the dominant role and gravity effects can be neglected.

12.3.1 Capillarity-Controlled Adiabatic Flow Patterns

In minichannels and microchannels, the flow pattern depends on the value of superficial velocities of gas and liquid, channel geometry and surface tension. The study on ascending and horizontal flow patterns in transparent rectangular minichannels showed that the basic flow regimes are flow with elongated bubbles, transition flow and annular flow (see Figure 12.1a).

The mode with elongated bubbles combines with the bubbly slug flow and slug flow, which are usually allocated in conventional channels.

These data were obtained for a glass minichannel with cross section 0.72 × 1.5 mm^2 and length 0.5 m described by Kozulin and Kuznetsov (2010). Nitrogen and water were supplied through the adjusting valve and flow meter to a T mixer placed before the test section. Using high-speed video and dual-laser flow scanning, the flow patterns were identified for different gas and liquid superficial velocities. Dual-laser flow scanning showed

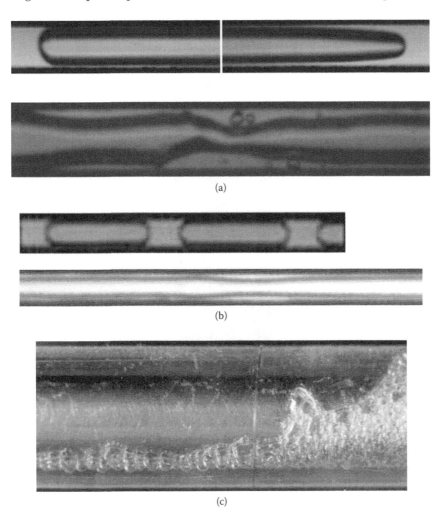

(a)

(b)

(c)

FIGURE 12.1
Adiabatic and diabatic flow patterns in minichannels and microchannels: (a) nitrogen–water flow in a rectangular minichannel with a cross section of 0.72 × 1.5 mm^2 at j_{liq} = 0.29 m/s, j_{gas} = 0.78 m/s (top) and j_{liq} = 0.12 m/s, j_{gas} = 3.35 m/s (bottom); (b) microchannel with a cross section of 217 × 370 μm^2 at j_{liq} = 0.083 m/s, j_{gas} = 0.035 m/s (top) and j_{liq} = 0.052 m/s, j_{gas} = 4.15 m/s (bottom); (c) diabatic flow patterns at RC-318 flow boiling in an annular channel with 0.95-mm gap at p_r = 0.12 and We = 9.2; and (d) R-11 flow boiling in a rectangular channel at p_r = 0.04, We = 0.2 and q_w = 1.4 kW/m^2.

(d)

FIGURE 12.1 (*Continued*)

that for an elongated bubble flow the signal from a photodiode is the series of alternating gas slugs and liquid plugs. The elongated bubble flow is characterised by periodicity and the lack of small bubbles in the flow. The length of elongated bubbles and liquid plugs is determined by gas–liquid interface instability in T mixer branches. In a range of liquid superficial velocities, j_{liq}, from 0.08 m/s and gas velocities j_{gas}, from 0.78 to 3.05 m/s, the flow loses its periodicity and the transition flow is detected as the flow characterised by destruction of the rear meniscus and chaotic behaviour of the optic signal immediately behind the bubble. The annular flow is observed in the range of superficial velocities j_{liq} from 0.29 m/s and j_{gas} from 4.5 m/s. This mode differs from the others by chaotic bursts of optic signal amplitude due to propagation of interface waves and absence of steady liquid plugs. Similar features of flow patterns were observed for the horizontal minichannel.

To predict the boundary between elongated bubble flow, transition flow and annular flow, let us extend the model of Mishima and Ishii (1984) for the case of minichannels. In the original model, it was postulated that the transition from slug to churn flow occurs when

$$\alpha_{s1} = \alpha_m \tag{12.5}$$

where α_m is the average void fraction of slug flow and α_{s1} is the void fraction of slug flow, based on the drift flux model. For the minichannel, the rise in velocity of a single bubble is much smaller than the mixture superficial velocity and expressions for α_m and α_{s1} have the following form:

$$\alpha_m = 1 - 0.813 \times \left[\frac{(C_0 - 1)\,j_m}{j_m + 0.75\sqrt{\Delta\rho g D_h/\rho_{liq}}\,(\Delta\rho g D_h^3/(\rho_{liq} v_{liq}^2))^{1/18}} \right]^{0.75} \tag{12.6}$$

$$\alpha_{s1} = \frac{j_{gas}}{(C_0 j_m)}$$

Here, C_0 is distribution parameter; j_m is mixture superficial velocity; and j_{gas} and j_{liq} are the gas and liquid superficial velocities, respectively. Using Equations 12.5 and 12.6, the transition line from elongated bubble flow to transition flow was calculated.

To predict the transition to annular flow in the original model of Mishima and Ishii (1984), liquid superficial velocity was set to zero in the void fraction for annular flow, α_{an}, based on the drift flux model:

$$\frac{j_{gas}}{\alpha_{an}} - j_m = \frac{1 - \alpha_{an}}{\alpha_{an} + \left[(1 + 75\{1 - \alpha_{an}\}) \dfrac{\rho_{gas}}{\rho_{liq}\sqrt{\alpha_{an}}} \right]^{1/2}} \times \left\{ j_m + \sqrt{\frac{\Delta\rho g D_h (1 - \alpha_{an})}{0.015 \rho_{liq}}} \right\} \tag{12.7}$$

Taking into account the fact that transition flow in a microchannel is the elongated bubble flow with destruction of the rear meniscus, one can assume that the transition to annular flow occurs when the void fraction of the annular flow based on Equation 12.7 exceeds the average void fraction for the slug flow according to Equation 12.6. In the extended model of Mishima and Ishii (1984), we assume that the transition from churn to annular flow occurs when the void fraction of annular flow equals the average void fraction of the slug flow:

$$\alpha_{an} = \alpha_m \tag{12.8}$$

Calculation of the lines of flow patterns transition for upward flow according to Equations 12.5 and 12.8 shows that they predict the experimental data Kozulin and Kuznetsov (2010) well and can be used for estimation of the flow patterns transition lines.

Typical flow patterns for horizontal microchannels are shown in Figure 12.1b. These data were obtained for a microchannel with a cross section of 217 × 370 μm² and length of 0.2 m using the methodology described by Kozulin and Kuznetsov (2010). For a channel with transverse size much less than the capillary constant, the basic flow regimes are flow with periodic elongated bubbles, non-periodic elongated bubble flow and annular flow. Similar to minichannels, the length of an elongated bubble and a liquid plug is determined by the frequency of gas–liquid interface motion in the T mixer governed by capillarity action. Dual-laser flow scanning showed that relative length of the elongated bubble increases continuously with an increase in gas flow rate. Non-periodic elongated bubble flow was observed for j_{gas} from 0.4 m/s to 1 m/s, and annular flow was observed in the range of superficial velocities, j_{liq}, over 1 m/s. These values do not correspond to the calculations in Equations 12.4 and 12.6. It shows that gravity does not play an important role in the mechanism of flow pattern transition in microchannels.

12.3.2 Capillarity-Controlled Diabatic Flow Patterns

In many studies, the flow patterns at flow boiling in microtubes were obtained using a transparent section after heating the tube, where boiling was stopped (e.g. Revellin and Thome, 2006). Under constrained conditions, the mechanism of flow boiling depends on the ratio of departure bubble diameter to transverse size of the channel, because the flow patterns at high and low Weber numbers based on the transverse size of the channel will be essentially different (Kuznetsov, 2010). If the Weber number based on superficial liquid velocity is much higher than one, then intensive nucleate boiling can exist in the liquid plugs at high reduced pressures. The flow visualisation of R-318C flow boiling in an annular gap of 950 μm with transparent outer wall and central heating tube is presented in Figure 12.1c. The flow boiling picture was obtained using the experimental set-up described by Kuznetsov and Shamirzaev (2007b). At high mass fluxes and high Weber numbers, intensive nucleate boiling was observed in the liquid plugs even if the gap size was less than the capillary constant. For this case, the detachable bubble diameter was smaller than the gap size for both subcooled and saturated boiling. Boiling was observed in the crest of waves on the film surface and in liquid plugs. The length of a liquid plug decreases with increase in mass vapour quality. At transition to annular flow nucleate boiling can occur in the waves in the same regimes, and the transition to annular flow does not immediately cause heat transfer deterioration.

Flow boiling of refrigerants at low flow rates and low reduced pressures has a different character. Figure 12.1d shows the picture of R-11 flow boiling in a vertical 2×7 mm^2 channel at a reduced pressure, $p_r = 0.038$. This picture was obtained using the set-up described by Kuznetsov et al. (2004). For a heat flux density of about 1.4 kW/m^2 and a mass flow rate of $G = 20$ kg/m^2·s, nucleate boiling does not exist and evaporation occurs in the thin liquid film and at the gas–liquid interface. Flow with elongated bubbles and annular flow with liquid concentrated in the corners of the channel are the typical flow patterns for both diabatic and adiabatic flows at low Weber numbers.

12.3.3 Modelling of Annular Flow in a Rectangular Channel

Prediction of flow patterns in non-circular channels and grooves is very important for the design of internals of heat pipes. Annular flow is found to be the dominant regime in microgap channels. Kuznetsov et al. (1997) showed that the important future of annular flow in a rectangular channel is essentially liquid accumulation in the channel corners. It was observed that almost all liquid flows in the corners even at short distances from the channel inlet at downward flow of kerosene in a channel with cross section 2×7 mm^2. This shows that capillary forces are crucial for liquid distribution in non-circular minichannels and microchannels. They move the liquid to the corners due to the gradient of capillary pressure on curved interface forming a thin film on the walls, which can be ruptured in the case of poor wall wettability.

The numerical modelling of annular flow in a rectangular minichannel based on commercial CFD codes cannot be applied due to progressive thinning of the film near the channel corners. According to Kuznetsov et al. (1997), to avoid this singularity, a model of capillarity-controlled annular gas–liquid flow can be proposed; this model is based on the allocation of liquid flow in two zones: (1) flow in a corner of the channel bounded by the meniscus with radius R_m and (2) thin film flow. When solutions for these zones are obtained, they should be matched with one another using the conjugation condition. In this model, there is a small parameter $\varepsilon = \delta_0 / a \ll 1$, where δ_0 is initial thickness and a is half the width of the long side of the channel. Further, half the width of the short side of the channel is denoted as b.

To model the descending film flow in a co-current gas flow, we introduce Cartesian coordinates as follows: x-axis is along the liquid flow direction, y-axis is crosswise to the channel's surface, and z-axis is perpendicular to the channel wall. We assume that the gravity vector follows the x-axis direction. Skipping the convective terms, the Navier–Stokes equations can be written for the steady film as follows:

$$-\frac{1}{\rho_{liq}} \nabla p_{liq} + \nu_{liq} \nabla^2 \mathbf{U} + \mathbf{g}_e = 0, \quad \mathbf{U} = (u,v,w), \quad \mathbf{g}_e = \left(g - \frac{1}{\rho_{liq}} \frac{dp_{gas}}{dx}, 0, 0 \right) \quad (12.9)$$

where p_{gas} is pressure in the gas flow. The mass balance equation for the flow is as follows:

$$\int_0^\delta \left(\frac{\partial u}{\partial x} + \frac{\partial v}{\partial y} \right) dz + w\big|_{z=\delta} = 0 \tag{12.10}$$

In the presence of gas flow, the thinning of the liquid film by interfacial shear stress should be taken into account. The following conditions are valid on the interface:

$$w = u\frac{\partial \delta}{\partial x} + v\frac{\partial \delta}{\partial y}$$

$$p_{liq} - p_{gas} + \sigma k = e_{ij}n_i n_j \tag{12.11}$$

$$\mu e_{ij}\tau_i n_j = \chi, \quad e_{ij}t_i n_j = 0$$

Here, k is the curvature of liquid surface; eij are the components of the tensor of deformation rates; n_i, τ_i and t_i are the components of the normal, binormal and tangential vectors, respectively, on the liquid surface; μ is the liquid viscosity; and χ is the shear stress produced by the gas flow. The intermolecular forces are accounted as a disjoining pressure component: $p_d = A_0/(6\pi\delta^3)$, where A_0 is the Hamaker constant.

Scaling Equations 12.9 through 12.12 by typical velocity δ_0^2/ν_{liq} for u and v, $\varepsilon\delta_0^2/\nu_{liq}$ for w and $p_{liq}ga$ for pressure and typical size in direction of the x axis – a/ε, direction of the y axis – a and direction of the z axis – δ_0 and skipping the terms that are $O(\varepsilon^2)$ or smaller, one obtains the following equations:

$$u = (\gamma - \varepsilon p_x)(mz - z^2/2) + \kappa z/\varepsilon$$
$$v = -p_y(mz - z^2/2) \tag{12.12}$$

Here, $\gamma = 1 - \left(dp_{gas}/dx\right)/p_{liq}g = 1 - 4a\kappa/d_{hgas}$, $\kappa = \chi/p_{liq}ga$ and d_{hgas} is the hydraulic diameter of the part of channel cross section occupied by the gas phase, and the same symbols for dimensionless variables are used again. Equations 12.9 through 12.12 can be applied also for the upward gas–liquid flow, when the sign of χ should be changed from + to – in all equations.

Taking into account the mass balance equation (Equation 12.10) and the kinematic conditions on the interface (Equation 12.11), Equation 12.12 takes the following form:

$$\left(\gamma m^3 + 1.5\frac{\kappa}{\varepsilon}m^2 \right)_\xi + (m^3 m_{\eta\eta\eta})_\eta = \frac{3}{\varepsilon^4}Ga\left(\frac{m_\eta}{m} \right)_\eta. \tag{12.13}$$

Here, $Bo = p_{liq}ga^2/\sigma$; $Ga = A_0/(6\pi a^2\sigma)$; the dimensionless film thickness is scaled via initial film thickness $m = \delta/\delta_0$; longitudinal, ξ, and transverse, η, dimensionless coordinates are scaled via aBo/ε and a, respectively; and A_0 is the Hamaker constant. A similar equation for curved film evolution without the influence of interfacial friction and disjoining pressure was considered by Wong et al. (1995) for the motion of long bubbles in polygonal capillaries. The disjoining pressure should be taken into account for walls with very low roughness, which provides the terms on the right side of Equation 12.13.

This equation together with Poisson's equation for liquid flow in the corners of a channel allows us to calculate the isothermal annular flow in a rectangular channel. Poisson's equation for liquid flow in a channel corner bounded by interface meniscus can be presented as follows:

$$\nabla^2 U = -\frac{g_e}{\nu_{liq}} \tag{12.14}$$

This equation is supplied by the boundary condition that takes into account the shear stress at the interface:

$$\frac{\partial u}{\partial n} = \frac{\chi}{\mu_{liq}} \tag{12.15}$$

On solving Poisson's equation for liquid flow in a corner of a channel bounded by a meniscus of constant curvature, the dependence of flow rate on the radius of curvature of meniscus, R_m, and its configuration, contact angle and shear stress at the interface was obtained in the form of polynomials. If the liquid flow is bounded by the meniscus and the contact angle is zero, the liquid flow rate can be presented as the sum of two terms:

$$Q_m = 0.00234 \frac{g_{ef}}{\nu_{liq}} R_m^4 + 0.0135 \frac{\chi}{\mu_{liq}} R_m^3 \tag{12.16}$$

The relationship between flow rate, Q_m, in meniscus and radius of the meniscus becomes more complex if the meniscus connects the film with the thicknesses at the conjunction points of δ_1 and δ_2. For negligibly small shear stresses, this correlation can be presented as follows:

$$q_m = \frac{Q_m \nu_{liq}}{ga\delta_0^3} = \varepsilon^{-3} f(R_m, \delta_1, \delta_2, \Theta), \quad \Theta = \arctan\left[\frac{R_m + \min(\delta_1, \delta_2)}{R_m + \max(\delta_1, \delta_2)}\right], \quad \delta_2 > 0$$

$$q_m = \frac{Q_m \nu_{liq}}{ga\delta_0^3} = \varepsilon^{-3} f(R_m, \delta_1, \phi, \Theta), \quad \Theta = \arctan\left[\frac{R_m \cos\phi}{R_m + \delta_1}\right], \quad \delta_2 = 0 \tag{12.17}$$

where ϕ is the wetting (contact) angle. These polynomials were used as solutions to the flow problem in the meniscus.

The boundary conditions are the symmetry conditions at the middle of the sides of the channel: $m_\eta = 0$ at $\eta = 1$ and $\eta = b/a$; and $m_\eta = 0$ and $m_{\eta\eta} = 1/(r_m \varepsilon)$

at the point of conjunction of the meniscus and the film solution, where $r_m = R_m/a$. The value $\eta = 0$ corresponds to a channel corner. This condition is followed from the continuity of capillary pressure in the conjugation. Even for complete wetting, the minimal thickness of the film cannot be lower than the height of roughness.

The flow in the meniscus becomes unstable at high gas velocities. This development of instability moves the liquid back from the meniscus to the film. The flow stability in a corner area bounded by the meniscus was calculated in long-wave approximation, taking into account the variation of shear stress and pressure on the slightly wavy surface (Abrams and Hanratty, 1985). These calculations give the increments of perturbation growth for the radius of the meniscus and define the critical radius, which is stable for a given gas velocity (Kuznetsov, 2010). For smaller radii the perturbations decay, and these radii were named the 'limit' ones. On reaching the radius limit, accumulation of liquid in the corners of a channel is terminated and, later, the flow rate in the meniscus is determined by the velocity of gas in the channel.

12.4 Heat Transfer with Evaporation and Condensation in Minichannels and Microchannels

Evaporation always happens in mini- and microchannels when the temperature of the wall exceeds the saturation temperature. Film rupture and rivulet formation are the typical processes of liquid film evaporation. In these cases, heat transfer near the contact line will be determined by microscopic phenomena in a very thin layer called the 'microregion', where the liquid–vapour interface reaches the wall.

12.4.1 Evaporation and Condensation in Annular Flow

As discussed before in Section 12.3.3, capillarity pulls a liquid to the corners of a small-sized rectangular channel and the film thickness on walls decreases with time. Thin film rupture and formation of a contact line are the important phenomena that can occur when liquid accumulates in a corner. Taking into account accumulation of liquid in the corners and rupture of the film, the model of phase change in a small-sized channel can be constructed using the model of liquid flow allocation in two zones: (1) flow in the channel corner bounded by a meniscus with radius R_m, and (2) thin film flow on channel walls, which is described in Section 12.3.3. For the case of phase change, the Navier–Stokes equations for steady film flow should be completed by the mass balance equation

$$\int_0^\delta \left(\frac{\partial u}{\partial x} + \frac{\partial v}{\partial y} \right) dz + w\big|_{z=\delta} = \frac{-G_{fg}}{\rho_{liq}} \tag{12.18}$$

and the total flow conservation equation

$$\frac{d(Q_m + Q_f)}{dx} = -\frac{1}{\rho_{liq} h_{fg}} \oint q_h \, dl, \quad Q_m + Q_f \big|_{x=0} = Q_0, \tag{12.19}$$

where G_{fg} is the mass flux due to evaporation or condensation. Then, Equations 12.12, 12.18 and 12.19 can be reduced to the equation for dimensionless variables:

$$(\gamma m^3 + 1.5 \frac{\kappa}{\varepsilon} m^2)_{\xi} + (m^3 m_{\eta\eta\eta})_{\eta} = \frac{3}{\varepsilon^4} \left[Ga \left(\frac{m_{\eta}}{m}\right)_{\eta} - \frac{G_0 \Theta_{w,i}}{m \varepsilon} \right] \tag{12.20}$$

where the last term in the right part takes into account the mass flux due to phase change. Here, $G_0 = \lambda_{liq} T^* v_{liq}/(h_{fg} \sigma a)$ and $\Theta_{w,i} = (T_{w,i} - T_{sat})/T^*$ are determined by the inner wall temperature, $T_{w,i}$; saturation temperature, T_{sat}; and latent heat of vaporisation, h_{fg}. The temperature is reduced to a dimensionless form via the characteristic temperature T^* defined as $T^* = T_{w,e} - T_{sat}$ for a specified temperature of external wall, $T_{w,e}$, and $T^* = H \times a^2/\lambda_f$ for a specified volume thermal-power production density, H, in the channel walls that could be conditioned by electric heating. The mass flux at the interface was determined by the heat transfer model based on conduction through the film, $G_{film} = \lambda_{liq}(T_{w,i} - T_{sat})/(\delta h_{fg})$. A similar equation was proposed by Hirasawa et al. (1980) to calculate condensation in a rectangular semi-channel without vapour flow. The singularity due to film's thickness tending to zero in the vicinity of the contact line is eliminated by introducing wall roughness, which limits the minimum thickness of the film. Equation 12.20 is solved together with the equation for liquid flow in the meniscus and the total fluid flow conservation equation (Equation 12.19). The inspection of the rivulet half-width corresponding to the current liquid flow is required to calculate the shape of the interface, if the contact angle differs from zero. If rivulet half-width is less than the distance from the centre of the channel to the edge of the meniscus, the film is broken and a new configuration of the flow is established in the channel, consisting of rivulet, menisci and dry spots between them. Accumulation of liquid in the corners stops when the curvature radius of the meniscus reaches its limit value discussed in Section 12.3.3, and further liquid flow rate in the meniscus will be determined by the evaporation rate and the current velocity of vapour in the channel. When the interface shape is determined, the heat conduction equations for liquid area and wall area should be solved together considering that heat conduction through the film is subjected to enhancement by the waves (Gimbutis, 1988).

For the case of uniform volume of heat production in the wall, it is necessary to solve the two-dimensional Fourier's heat conduction equation in the symmetry element of the channel wall,

$$\lambda_w \nabla^2 T + H = 0, \tag{12.21}$$

and for the liquid area,

$$\nabla^2 T = 0 \qquad (12.22)$$

The wave effect on heat conduction enhancement was applied only for the film area.

Boundary conditions are $T = T_{sat}$ at the gas–liquid interface, $\partial T/\partial n = h_{gas}\left(T_w - T_{sat}\right)/\lambda_w$ (n is the normal to the interface) at the gas–solid interface and $\partial T/\partial n = 0$ at symmetry lines as far as the external wall surface of the channel for uniform heat production in the wall and continuity of both heat flux and temperature at the solid–liquid interface.

The resulting equations were solved numerically using the procedure described by Kuznetsov et al. (1997). Simultaneously with the calculation of liquid surface configuration, the heat transfer problem has been solved in the symmetry element of the channel. The Gauss–Zeidel iterative procedure has been used to solve the heat problem. We used a non-uniform grid pattern near the vapour–liquid interface for higher computational accuracy.

Figure 12.2a shows an example of the calculation of liquid surface shape for the evaporation of refrigerant R-21 in a rectangular minichannel of 6.3 × 1.6 mm² at $G = 50$ kg/m²· s with an initial vapour quality of 0.2 and volume heat production density of 60 MW/m². Initially, liquid was distributed uniformly along the channel perimeter. During evaporation, liquid is concentrated in the corner and at a distance of 0.4 m from the inlet film rupture occurs with the formation of contact line. In calculations, a contact angle equal to 3° was used. Formation of dry spots and increasing heat transfer coefficient in the vicinity of the contact line are typical for evaporation in microchannels. Figure 12.2b represents the distribution of the local heat flux along the channel perimeter for the surface configuration shown in Figure 12.2a. Here, the coordinate ς is measured from the middle of the long side of the channel. The local heat flux in the vicinity of the contact line is limited by wall roughness or suppression of evaporation for ultra-thin films. The mechanism of heat transfer for the data presented in Figure 12.2 is evaporation of a curved liquid film, including predominant evaporation near the contact line.

At condensation, the establishment of the limit radius of meniscus in a corner provokes the alignment of the film and lowers the coefficient of heat transfer in comparison with the curved film, which is typical for evaporation. The example of interface shape for R-21 downward condensation in a rectangular channel is shown in Figure 12.2a.

12.4.2 Heat Transfer with Evaporation in the Microregion

When thin film rupture occurs, heat transfer near the contact line is determined by microscopic phenomena in a very thin layer called the microregion. In this area the film's thickness goes down to that of non-evaporating

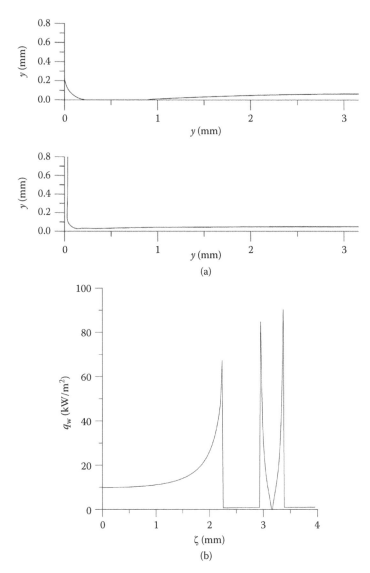

FIGURE 12.2

Refrigerant R-21 phase change in a rectangular minichannel of 6.3×1.6 mm^2 at $G = 50$ kg/m^2·s and $T_{sat} = 303$ K: (a) shape of the interface for upward evaporation at vapor quality $X = 0.855$, $H = 60$ MW/m^3 (top), and downward condensation at vapor quality $X = 0.823$, $\Delta T = 1$ K (bottom) in a quarter of the channel; (b) local heat flux distribution along the channel perimeter for R-21 upward evaporation.

adsorbed films, where the attractive forces between the atoms of the wall and fluid molecules cause a disjoining pressure and prevent evaporation. Rapid evaporation near the contact line can lead to an increase in apparent contact angle. It occurs when the flow of evaporating liquid is not compensated by capillary inflow, and microscale heat transfer has an important impact on film flow through a change of the flow boundary conditions.

The problem of heat transfer in a microregion will be considered for rivulet evaporation in the frame of the creeping flow model described in Section 12.4.1. Let us consider rivulet downward evaporation, when the vapour shear stress is negligible over a vertical plate with a constant temperature, T_w. For this case, the size of flow region in the direction of gravity is much larger than that in the perpendicular direction. We suppose that the film thickness is much smaller than the rivulet width, a. Let us introduce Cartesian coordinates as they were done in Section 12.3.3. The governing equations are the Navier–Stokes equations for liquid flow area (Equation 12.9), the conditions on a free liquid surface (Equation 12.18) and the mass balance equation (Equation 12.19) taking into account mass flux due to evaporation. For the constant wall temperature, we can specify the mass flux as $G_{fg} = \lambda_{liq} \Delta T / (h_{fg} \delta)$, where $\Delta T = T_w - T_{sat}$.

Let us scale the initial equations using initial rivulet width a; typical velocity $U = q_0 / \delta_0$; pressure $\rho g b$; time b/U; typical size b in directions x and y and δ_0 in z direction, where $\delta_0 = (3 q_0 \nu_{liq} / g)^{1/3}$ is the initial liquid layer thickness at the inlet section; $q_0 = Q/a$; Q is the volumetric flow rate; and $b = a/2$ is the initial half-width of the rivulet. The dimensional numbers for rivulet evaporation are $\varepsilon = \delta_0 / b$, Reynolds number $Re = q_0 / \nu_{liq}$ and Bond number $Bo = \rho_{liq} g b^2 / \sigma$. The typical size of the rivulet in the x direction is much greater than that in the y direction, and for the creeping flow case ($\varepsilon Re \ll 1$) we can reduce the Navier–Stokes equations to a parabolic equation (Equation 12.20) for dimensionless film thickness, m. The boundary conditions for Equation 12.20 are as follows: the condition of symmetry at line $\eta = 0$ is $m(0,\eta) = 1$ for $\eta \le 1$ and $m(0,\eta) = 0$ for $\eta > 1$; $m = 0$ far from the rivulet.

Equation 12.20 was solved using the implicit difference scheme with respect to x. To solve a system of non-linear equations appearing after the discretisation of governing equations, iteration procedures were implemented. To avoid the singularity caused by the condition $m = 0$ far from the rivulet, this condition was replaced by the condition $m = m^* > 0$. The value of m^* was selected as 10^{-5} such that the solution in the rivulet area was not dependent on this parameter. The mesh size on coordinate ξ was equal to 0.005 (it corresponds to 1 μm), and along the coordinate η it was equal to 0.01.

Figure 12.3a shows the calculated shape of the interface and local heat flux density along the width of the rivulet at 5 cm from the inlet for $Re_{ini} = 2.5$ and $\Delta T = 0.5$ K. Here, the coordinate ς is measured from the middle of the rivulet. The simulations show that there is an abnormally high heat flux in the vicinity of the liquid–vapour–solid contact line, where the liquid film has minimal thickness. The rapid evaporation of liquid in the microregion could not be compensated by the liquid flow induced by capillarity, so the interface is deformed in the vicinity of the contact line and an apparent contact angle is established. Its magnitude is greater than that of the equilibrium contact angle and increases with the rise in wall superheating. The apparent contact angle, φ, was determined as the tangential angle at the inflection point of the

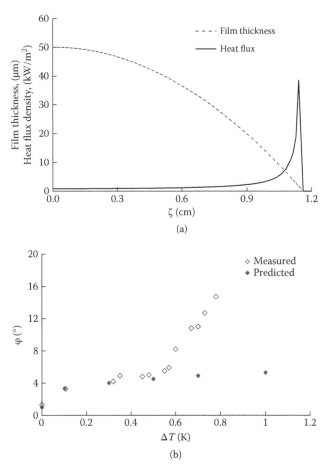

FIGURE 12.3
Film thickness and local heat flux distribution in the rivulet cross section (a) at a distance of 5 cm from the inlet for $Re_{ini} = 2.5$ and $\Delta T = 0.5$ K; (b) the dependence of apparent contact angle on wall superheating: open rhombs are experimental data (Kuznetsov et al., 2007) and black points are the calculations according to Equation 12.20.

rivulet profile. The dependence of the calculated apparent contact angle on wall superheat is shown in Figure 12.3b as black points. The data on apparent contact angle measurement from the study by Kuznetsov et al. (2007) are shown in Figure 12.3b as open rhombs. These data were obtained for refrigerant R-11 rivulet evaporation along the heating polished copper surface. In the experiments, rivulet flow was generated after film rupture in a semi-rectangular channel or by liquid jet spreading. The apparent contact angle was determined by Kuznetsov et al. (2007) through the measured rivulet height using the assumption of a semi-parabolic profile of velocity in the rivulet. Good comparison of calculation results and experimental data occurs up to a wall superheat of 0.4 K. In the experiments, rivulet spreading was replaced by its compression if the wall superheat exceeded 0.4 K. For these superheats swelling of the interface near the contact line was observed, and it is responsible for a sharp increase of the apparent contact angle (see Figure 12.3b).

12.5 Flow Boiling Heat Transfer

Flow boiling in minichannels and microchannels is characterised by a significant influence of capillary forces on the pattern of flow and heat transfer. In conventional tubes, the diameter is much higher than the detachable bubble size and it has no effect on the condition of bubble separation from the wall. Another situation occurs in microchannels where the bubble cannot break away from the wall and its compression occurs during the bubble motion at low Weber numbers (see Figure 12.1d). It is obvious that bubble compression has a decisive influence on flow boiling heat transfer.

12.5.1 Heat Transfer at Predominant Nucleate Boiling

To determine the regularity of microscale heat transfer during flow boiling, the experimental data on heat transfer coefficients were compared with the models of heat transfer at boiling in conventional tubes. The Liu–Winterton (1991) model combines heat transfer at nucleate boiling and forced convection and takes into account their mutual influence. The heat transfer coefficient in the Liu–Winterton model is calculated as follows:

$$h^2 = \left(Fh_{cf}\right)^2 + \left(S_{bo}h_{pb}\right)^2 \tag{12.23}$$

$$F = \left[1 + X \times Pr_{liq}\left(\frac{\rho_{liq}}{\rho_{gas}} - 1\right)\right]^{0.35}, \; S_{bo} = \left(1 + 0.55F^{0.1}Re_{liq0}^{0.16}\right)^{-1} \tag{12.24}$$

Here h_{cf} and h_{pb} are the heat transfer coefficients for forced convection and pool boiling, respectively; F is the factor of forced convection enhancement; and S_{bo} is the factor of nucleate boiling suppression. Liu and Winterton (1991) recommend the use of the Cooper (1984) pool boiling model to calculate h_{pb} and the Dittus and Boelter (1930) equation to calculate h_{cf} for all cases. The Dittus and Boelter equation is valid only for the developed turbulent flow, whereas the Cooper model does not predict correctly heat transfer at pool boiling for several refrigerants. It was suggested by Kuznetsov and Shamirzaev (2007a, 2007b) to use Equations 12.23 and 12.24 in the extended Liu–Winterton model for the calculation of heat transfer coefficient and to choose h_{cf} and h_{pb} in such a way as to make the models of pool boiling and forced convection correspond to the conditions of the experiments. For example, for refrigerant RC-318 the pool boiling model of Cooper (1984) was recommended, for R-21 the model of Danilova (1970) was recommended, and for R-134a the model of Gorenflo (1993) was recommended, accordingly. Equation 12.25 shows the Danilova (1970) pool boiling model, which takes into account wall roughness:

$$\alpha_{pb} = 550 \frac{P_{cr}^{0.25}}{T_{cr}^{0.875} M^{0.125}} q_w^{0.75} R_z^{0.2} \left(0.14 + 2.2 \frac{P}{P_{cr}} \right) \qquad (12.25)$$

The influence of thermophysical properties of liquid and wall material on heat transfer coefficient was also taken into account. The Dittus and Boelter equation has been replaced by the equation of Gnielinski (1976) for forced convection calculation in turbulent flow and the Stephan and Preußer equation (1979) for laminar flow.

The calculations of total heat transfer coefficient for different refrigerants discussed by Kuznetsov (2010) according to the extended Liu–Winterton model showed that the share of nucleate boiling and forced convection depends on the refrigerant flow rate, reduced pressure and heat flux density considered in the test. For example, for flow boiling of RC-318 in an annular channel (Kuznetsov and Shamirzaev, 2007b) and of R-134a in minitubes with diameters from 1.1 to 4.3 mm (Karayiannis et al., 2010), the effect of nucleate boiling is dominant. For R-21 flow boiling in a rectangular minichannel (Kuznetsov and Shamirzaev, 2007a), the share of nucleate boiling is comparable with the share of forced convection only for heat fluxes exceeding 40 kW/m². The discussion of the shares of nucleate boiling and forced convection for these three refrigerants can be useful in explaining the peculiarities of the heat transfer behaviour.

Let us consider microscale heat transfer in channels where the nucleate boiling impact is dominant. One can see two opposite behaviours of flow boiling heat transfer in a minichannel on comparing the data of Karayiannis et al. (2010) with the predictions made according to the Liu–Winterton model. At low vapour quality, small heat fluxes and small Weber numbers, the heat transfer coefficient is higher than the calculated ones. This is also observed for the data of Kuznetsov and Shamirzaev (2007a) obtained for refrigerant R-21 flow boiling in a rectangular channel of 1.6 × 6.3 mm² at a low mass flow rate of 50 kg/m²·s when the measured heat transfer coefficients exceed by two times the calculated ones. A reason for this trend can be the change in heat transfer mechanism, when there is variation in the ratio of capillary pressure at the scale of channel size to the dynamic pressure in the flow. As a result, the flow patterns were changed and they affected flow boiling heat transfer. The enhancement of heat transfer at small Weber numbers can be taken into account by considering the factor of boiling and forced convection development, which was obtained using the data of Karayiannis et al. (2010) and of Kuznetsov and Shamirzaev (2007a):

$$\Psi_{dev} = 1 + \frac{1.2}{\exp\left(0.8 \times We^{0.35}\right)} \tag{12.26}$$

The use of the development factor (Equation 12.26) in the modification of the Liu–Winterton model to magnify nucleate boiling suppression factor and forced convection enhancement factor can improve the accuracy of heat transfer prediction, especially at low mass flow rates.

Another opposite behaviour of flow boiling heat transfer in minichannels is heat transfer deterioration occurring at high vapour qualities, as discussed by Karayiannis et al. (2010) and Kuznetsov and Shamirzaev (2007b). It was associated by Kuznetsov (2010) with boiling suppression in annular flow. However, the transition to annular flow does not cause immediate deterioration of heat transfer. Another necessary condition is the suppression of nucleate boiling in a thin film. For the wavy, free-falling turbulent liquid film, it was shown by Marsh and Mudawar (1989) that the assumption of a linear temperature profile is applicable only within a viscous sub-layer. Turbulence keeps the liquid temperature close to the surface temperature outside the sub-layer, and vapour nuclei with diameters larger than the sub-layer thickness cannot be activated. It means that for intensive boiling the diameters of vapour nuclei should be much less than the thickness of the viscous sub-layer. The diameter of an active nucleus based on the Hsu's tangential criteria for nucleation can be calculated by assuming the linear temperature profile in liquid and using Clapeyron—Clausius equation for vapour pressure in the nucleus (Davis and Anderson, 1966):

$$d_{tan} = \sqrt{8\sigma T_{sat}\left(\rho_{liq} - \rho_{gas}\right)\lambda_{liq}/\rho_{gas}\rho_{liq}h_{fg}q_w}. \tag{12.27}$$

As a rule, in annular flow the vapour stream is turbulent and the film surface is wavy. The thickness of viscous sub-layer for the wavy film can be estimated as that for the turbulent film:

$$y_{vis} = \frac{5\nu_{liq}}{\sqrt{\chi/\rho_{liq}}} \tag{12.28}$$

Here, χ is the wall shear stress in the film flow. To calculate shear stress, the model of wavy annular flow by Asali et al. (1985) was used. Comparing the diameter of active nucleus with viscous sub-layer thickness, one can see that for the annular flow in experiments Karayiannis et al. (2010) they were comparable and will suppress nucleate boiling in the film.

The degree of boiling suppression depends not only on the ratio y_{vis}/d_{tan}, although the latter is decisive, but also on the magnitude of heat flux (Kuznetsov, 2010). Boiling is more intense at low heat fluxes due to the lower ratio of evaporated liquid to flux flowing in a channel. The experimental data of Karayiannis et al. (2010) shows that the ratio of measured heat transfer coefficients to calculated ones depends on a suppression parameter:

$$\theta_{sup} = \frac{\left(y_{vis} / d_{tan} \right)^{0.6}}{\left(Bo_x^{0.4} We^{-0.08} Pr_{liq}^{1/3} \right)} \tag{12.29}$$

The suppression parameter, θ_{sup}, includes the ratio y_{vis}/d_{tan}, the number Bo_x based on the local liquid flow rate, Weber number and Prandtl number. The dependence of boiling suppression factor on suppression parameter can be presented as follows:

$$\Psi_{sup} = \tanh^2 \left(2.5 \times 10^{-3} \theta_{sup}^2 \right) \tag{12.30}$$

which can be used for predicting the degree of nucleate boiling suppression in a thin film.

The use of both boiling development factor (Equation 12.26) and boiling suppression factor (Equation 12.30) as multipliers to factors S_{bo} and F in the Liu–Winterton model can improve the prediction of experimental data. The developed flow boiling model includes Equation 12.23, new forced convection enhancement factor, and new boiling suppression factor, as follows:

$$F = \left[1 + X \times Pr_{liq} \left(\frac{\rho_{liq}}{\rho_{gas}} - 1 \right) \right]^{0.35} \left(1 + \frac{1.2}{\exp(0.8\,We^{0.35})} \right) \tag{12.31}$$

$$S_{bo,new} = \left(1 + 0.55 F^{0.1} Re_{liq0}^{0.16} \right)^{-1} \left(1 + \frac{1.2}{\exp(0.8 \times We^{0.35})} \right) \times \tanh^2 \left(2.5 \times 10^{-3} \theta_{sup}^2 \right) \tag{12.32}$$

Figure 12.4a shows the comparison of heat transfer coefficients, measured by Karayiannis et al. (2010), for R-134a flow boiling in a tube of inner diameter (ID) 1.1 mm at $G = 400$ kg/m²·s with calculations according to Equations 12.23, 12.31 and 12.32. One can see the good prediction of heat transfer coefficient within the whole range of vapour qualities, including the area of heat transfer enhancement at low qualities and the area of heat transfer deterioration at high qualities. This approach takes into account the microscale effects on heat transfer at upward flow boiling and describes well other data for R-134a flow boiling.

12.5.2 Heat Transfer at Predominant Forced Convection and Evaporation

Nucleate boiling is not the dominant mechanism of heat transfer at low heat fluxes and low reduced pressures (see Figure 12.1d). It was found by Kuznetsov and Shamirzaev (2007a) that heat transfer deterioration does not occur for the

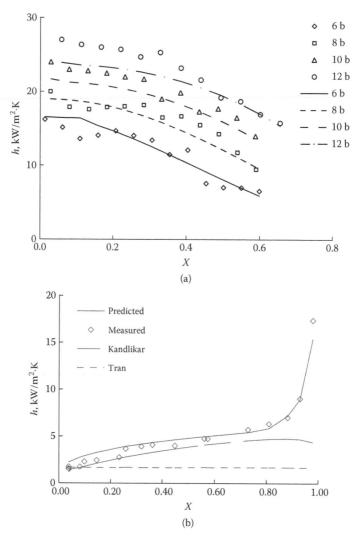

FIGURE 12.4
Local heat transfer coefficient versus vapour quality for R-134a flow boiling in a minitube with internal diameter equal to 1.1 mm for different pressures: (a), (b) R-21 flow boiling data for 1.6×6.3-mm^2 channel at $p_r = 0.041$. Points on (a) are data from the study by Karayiannis et al. (2010), and lines show the calculations according to Equations 12.23, 12.31 and 12.32. Points on (b) are data from the study by Kuznetsov and Shamirzaev (2007a), and lines show the calculations according to Equations 12.33 through 12.35.

flow boiling of refrigerant R-21 in a rectangular minichannel of 1.6×6.3 mm^2 for a mass flow rate of 215 kg/m$^2 \cdot$s and a heat flux of up to 30 kW/m^2. Heat transfer coefficient is weakly dependent on heat flux, and two-phase forced convection is the dominant mechanism of heat transfer. Figure 12.4b shows the experimental data of Kuznetsov and Shamirzaev (2007a) on heat transfer coefficient versus vapour quality for R-21 flow boiling at $G = 215$ kg/m^2·s and $q_w = 6$ kW/m^2. The fine dotted line shows calculations according to the model by Tran et al. (1997), and the dotted line shows calculations according to the Kandlikar and Balasubramanian (2004) model. The modified Liu–Winterton model (Kuznetsov and Shamirzaev, 2007a) has almost the same results as the model of Kandlikar and Balasubramanian (2004). The models of Kandlikar and Balasubramanian (2004) and Kuznetsov and Shamirzaev (2007a) predict the data well for a vapour quality less than 0.6.

When the vapour quality is higher than 0.6, the data in Figure 12.4b go up contrary to calculations according to the models of Kandlikar and Balasubramanian (2004) and Kuznetsov and Shamirzaev (2007a). Similar results were obtained for flow boiling of refrigerants in a 0.52-mm-ID microchannel (Consolini and Thome, 2009; Karayiannis et al., 2010). An increase in heat transfer coefficient at high vapour qualities could occur due to the contribution of film evaporation to heat transfer. In a microchannel, the liquid film is very thin due to the high surface area and the contribution of film evaporation can exceed the contribution of forced convection. The weak dependence of temporal dispersion and standard deviations of wall temperature distribution on vapour quality obtained by Kuznetsov and Shamirzaev (2007a) confirms the high share of evaporation at high vapour qualities, when heat transfer is almost uniform along the channel perimeter. These data were obtained using microthermocouples, glued around the perimeter of the 1.6 × 6.3-mm^2 rectangular channel. Non-uniformity of temperature distribution along the perimeter increases significantly with the increase in heat flux and standard deviation grows almost linearly up to $q_w = 30$ kW/m^2, when heat transfer deterioration occurs. It shows the liquid film rupture at high heat fluxes.

For annular flow with a high interface shear stress and an extremely thin liquid film, the film evaporation term should replace the forced convection term in Equation 12.23 and the flow boiling heat transfer model for the microchannel becomes the following:

$$h_{fb}^2 = \left(h_{nonb}\right)^2 + \left(S_{bo,\,new} h_{pb}\right)^2. \tag{12.33}$$

$$h_{nonb} = \max\left(h_{evap}, F_{new} h_{cf}\right). \tag{12.34}$$

For calculating the forced convection enhancement factor, F_{new}, Equation 12.31 should be used, and for calculating the boiling suppression factor, $S_{bo,new}$, Equation 12.32 should be used. In calculations of heat transfer coefficients for pool boiling, h_{pb}, and forced convection, h_{cf}, they are selected such that they correspond to the conditions of the experiments. Also, surface roughness and influence of thermophysical properties of liquid and wall material should be taken into account. In vapour shear induced flow, the conduction through the thin film could be the possible mechanism of heat transfer. In this case, it is proposed to calculate the film thickness using the wavy annular flow model of Asali et al. (1985) and apply the wave enhancement factor of Gimbutis (1988) for calculating the heat transfer coefficient:

$$h_{evap} = \lambda_{liq}/\delta \times \left(1 + 0.02\mathrm{Re}_{liq}^{0.2} + 0.0009\mathrm{Re}_{liq}^{0.85}\mathrm{Pr}^{0.65}\right) \qquad (12.35)$$

The solid line in Figure 12.4b is calculated according to Equations 12.33 through 12.35. The results of calculation of flow boiling heat transfer coefficient are in good agreement with the experimental data of Kuznetsov and Shamirzaev (2007a). This model predicts the growth of heat transfer coefficients at high vapour qualities as it was observed for the 0.52-mm channel (Consolini and Thome, 2009; Karayiannis et al., 2010) for which the evaporative mechanism also becomes dominant. The stability of the thin film should be analysed for high vapour quality to predict the threshold, where the forced convection heat transfer mechanism should be used again because of film rupture.

12.6 Conclusions

The success of heat pipe operation is determined by the intensity of capillarity-induced flow in minichannels and microchannels. In this chapter, the concept of fluid flow and heat transfer with phase change in small-sized channels was discussed; this is the novel cooling strategy in devices requiring successful thermal management. The experimental observation of capillarity-controlled gas–liquid flow in rectangular minichannels and microchannels shows that elongated bubble flow, transition flow and annular flow are the dominant flow patterns. Gas–liquid flow in minichannels is characterised by the considerable influence of capillary forces and gravity on flow patterns, and the Mishima–Ishii model can be extended to predict the transition between flow patterns. Flow boiling in minichannels is characterised by the same flow patterns, and nucleate boiling intensity depends on the value of Weber number and reduced pressure.

The influence of capillary forces, disjoining pressure, wall roughness and vapour shear stress on the configuration of the gas–liquid interface and local heat transfer were analysed using the proposed model of capillarity-controlled annular flow in a small-sized rectangular channel. The microscale evaporation and condensation model shows the significant effect of capillary forces on film surface curvature and local heat transfer. When the film ruptures at evaporation, a new interface configuration is formed by menisci in the channel corners and rivulets on the channel walls. In this case, the enhancement of heat transfer occurs due to the curved surface of the film and the high heat flux in the microregion. The stability of menisci in the channel corners and the establishment of a limit radius of meniscus curvature at high vapour velocities seem to be the limiting factors for the accumulation of liquid in the corners especially for vapour condensation. Rapid evaporation of liquid in a microregion cannot be compensated by the capillarity-induced liquid flow. As a result, an apparent contact angle is established at the rivulet boundary. Its magnitude is much higher than that of the equilibrium contact angle, and it increases with the rise in wall superheating.

To determine the regularity of microscale heat transfer during flow boiling, the experimental data on heat transfer coefficients were compared with the Liu–Wintertion model of heat transfer at flow boiling in conventional tubes. The degree of boiling development and the degree of boiling suppression in thin liquid films were determined for minichannels based on the experimental data for refrigerant R-134a obtained at high heat fluxes and reduced pressures, when nucleate boiling is the determining factor. The Liu–Winterton model was extended to incorporate microscale behaviour, and the modified model can predict the experimental data well for both low and high vapour qualities. For microchannels, evaporation of a thin liquid film becomes the important mechanism of flow boiling heat transfer. Consideration of film evaporation in flow boiling models allows us to predict heat transfer enhancement, which is observed in the experiments at high vapour qualities and low reduced pressures, when two-phase evaporation is dominant. The proposed models of flow boiling heat transfer were tested for a microchannel gap larger than 300 micron. This represents the range of dimensions where the results can be applied.

12.7 Nomenclature

A_0	Hamaker constant (J)	X	Vapour quality
a	Half-width (m)	x	Cartesian axis direction along the flow (m)
Bo	Bond number	y	Cartesian axis direction crosswise the flow (m)
D_h	Hydraulic diameter (m)	**Greek letters**	
F	Forced convection enhancement factor	α	Void fraction
G	Mass flow rate (kg/m²·s)	ΔT	Wall superheat (K)
Ga	Hamaker's parameter	θ_{sup}	Suppression parameter
g	Acceleration due to gravity (m/s²)	φ	Wettability angle (°)
h_{fg}	Latent heat (J/kg)	ξ	Distance along the channel perimeter (m)
j	Superficial velocity (m/s)	λ	Thermal conductivity (W/m·K)
h	Heat transfer coefficient (W/m²·K)	μ	Dynamic viscosity (kg/m·s)
M	Molecular mass (g/mol)	ρ	Density (kg/m³)

(Continued)

m	Dimensionless film thickness	δ	Film thickness (m)
Pr	Prandtl number	σ	Surface tension (N/m^2)
p	Pressure (Pa)	ν	Kinematic viscosity (m^2/s)
p_r	Reduced pressure	ε	Small parameter
q_w	Heat flux density (W/m^2)	χ	Shear stress (N/m)
Re	Reynolds number	Ψ	Factor
R_z	Roughness (μm)	**Subscripts and Abbreviations**	
R_m	Radius of meniscus (m)	cf	Forced convection
S_{bo}	Boiling suppression factor	dev	Development
t	Time (s)	evap	Evaporation
U	Velocity vector (m/s)	pb	Pool boiling
u	Dimensionless velocity in x direction	sat	Saturation
v	Dimensionless velocity in y direction	sup	Suppression
We	Weber number	w	Wall
T	Temperature (K)	0	Initial

12.8 Acknowledgements

This work was supported in part by the Russian Foundation for Basic Research grants 08-08-00953-a and 11-08-01140a.

References

Abrams, J., and Hanratty, T.J. 1985. Relaxation Effects Observed for Turbulent Flow over a Wavy Surface. *Journal of Fluid Mechanics* 151:443–55.

Asali, J.C., Hanratty, T.J., and Andreussi, P. 1985. Interfacial Drag and Film Height for Vertical Annular Flow. *AIChE Journal* 31:886–902.

Bar-Cohen, A., and Rahim, E. 2009. Modeling and Prediction of Two-Phase Microgap Channel Heat Transfer Characteristics. *Heat Transfer Engineering* 30:601–25.

Bertsch, S.S., Groll, E.A., and Garimella, S.V. 2009. Composite Heat Transfer Correlation for Saturated Flow Boiling in Small Channels. *International Journal of Heat and Mass Transfer* 52:2110–18.

Consolini, L., and Thome, J.R. 2009. Micro-Channel Flow Boiling Heat Transfer of R-134a, R-236fa, and R-245fa. *Microfluid and Nanofluid* 6:731–46.

Cooper, M.G. 1984. Heat Flow Rates in Saturated Nucleate Pool Boiling-A Wide Ranging Examination Using Reduced Properties. *Advances in Heat Transfer* 16:157–239.

Danilova, G.N. 1970. Correlation of Boiling Heat Transfer Data for Freons. *Heat Transfer-Soviet Research* 2(2):73–8.

DasGupta, S., Schonberg, J.A., Kim, I.Y., and Wayner, P.C. 1993. Use of the Augmented Young-Laplace Equation to Model Equilibrium and Evaporating Extended Menisci. *Journal of Colloid Interface Science* 157:332–42.

Davis, E.J., and Anderson, G.H. 1966.The Incipience of Nucleate Boiling in Forced Convection Flow. *AIChE Journal* 12:774–80.

Dittus, F.W., and Boelter, L.M.K. 1930. Heat Transfer in Automobile Radiators of the Tubular Type. *University of California Publications on Engineering* 2:443–61.

Gimbutis, G. 1988. *Heat Transfer of a Falling Fluid Film.* Vilnius: Mokslas Publishers.

Gnielinski, V. 1976. New Equations for Heat and Mass Transfer in Turbulent Pipe and Channel Flow. *International Chemical Engineering* 16:359–67.

Gorenflo D., 1993. *Pool boiling.* In VDI-Heat Atlas (Chapter Ha). Dusseldorf: VDI-Verlag GmbH.

Hirasawa, S., Huikata, K., Mori, Y., and Nakayama, W. 1980. Effect of Surface Tension on Condensate Motion in Laminar Film Condensation (Study of Liquid Film in a Small Through). *International Journal of Heat and Mass Transfer* 23:1471–78.

Kandlikar, S. G. 2010. Similarities and Differences Between Flow Boiling in Microchannels and Pool Boiling. *Heat Transfer Engineering* 31(3):159–67.

Kandlikar, S.G., and Balasubramanian, P. 2004. An Extension of the Flow Boiling Correlation to Transition, Laminar, and Deep Laminar Flows in Mini-Channels and Micro-Channels. *Heat Transfer Engineering* 25(3):86–93.

Kandlikar, S.G., and Grande, W.J. 2003. Evolution of Microchannel Flow Passages-Thermohydraulic Performance and Fabrication Technology. *Heat Transfer Engineering* 24:3–17.

Kandlikar, S. G., Li, D., Colin, S., and King, M.R. 2006. *Heat transfer and fluid flow in minichannels and microchannels.* UK: Elsevier Ltd.

Karayiannis, T.G., Shiferaw, D., Kenning, D.B.R., and Wadekar, V.V. 2010. Flow Patterns and Heat Transfer for Flow Boiling in Small to Micro Diameter Tubes. *Heat Transfer Engineering* 31(4):257–75.

Kawahara, A., Chung, P., and Kawaji, M. 2002. Investigation of Two Phase Flow Pattern, Void Fraction and Pressure Drop in a Microchannel. *International Journal of Multiphase Flow* 28:1411–35.

Kew, P.A., and Cornwell, K. 1997. Correlations for the Prediction of Boiling Heat Transfer in Small-Diameter Channels. *Applied Thermal Engineering* 17:705–15.

Kozulin, I.A., and Kuznetsov, V.V. 2010. Statistic Characteristics of the Gas-Liquid Flow in a Vertical Minichannel. *Thermophysics and Aeromechanics* 17(1):101–08.

Kuznetsov, V.V. 2010. Heat and Mass Transfer with Phase Change and Chemical Reactions in Microscale. *Proceedings of the International Heat Transfer Conference IHTC14*, Washington, DC, IHTC14–22570.

Kuznetsov, V.V., and Shamirzaev, A.S. 2007a. Boiling Heat Transfer for Freon R21 in Rectangular Minichannel. *Heat Transfer Engineering* 28:738–45.

Kuznetsov, V.V., and Shamirzaev, A.S. 2007b. Flow Patterns and Flow Boiling Heat Transfer of Freon R318C in an Annular Minichannel. *Thermophysics and Aeromechanics* 14(1):53–61.

Kuznetsov, V.V., Shamirzaev, A.S., and Ershov, I.N. 2004. Flow Boiling Heat Transfer and Regimes of Upward Flow in Minichannels, *Proceedings of the 3rd International Symposium on Two-Phase Flow Modeling and Experimentation*, Pisa, ven 03.

Kuznetsov, V.V., Vitovskii, O.V., and Krasovskii, V.A. 2007. An Experimental Investigation of Modes of Flow under Conditions of Evaporation of Liquid on a Vertical Heating Surface. *High Temperature* 45(1):68–75.

Kuznetsov, V.V., Safonov, S.A., Sunder, S., and Vitovsky, O.V. Capillary Controlled Two-Phase Flow in Rectangular Channel. 1997. *Proceedings of the International Conference on Compact Heat Exchangers for the Process Industries*, 291–303. Snowbird: Begell House.

Lazarek, G.M., and Black, S.H. 1982. Evaporative Heat Transfer, Pressure Drop and Critical Heat Flux in a Small Vertical Tube with R-113. *International Journal of Heat and Mass Transfer* 25:945–60.

Liu Z., and Winterton, R.H.S. 1991. A General Correlation for Saturated and Subcooled Flow Boiling in Tubes and Annuli, Based on a Nucleate Pool Boiling Equation. *International Journal of Heat and Mass Transfer* 34:2759–66.

Marsh, W.J., and Mudawar, I. 1989. Predicting the Onset of Nucleate Boiling in Wavy Free-Falling Turbulent Liquid Films. *International Journal of Heat and Mass Transfer* 32:361–78.

Mehendal, S.S., Jacobi, A.M., and Shah, R.K. 2000. Fluid Flow and Heat Transfer at Micro- and Meso-Scales with Application to Heat Exchanger Design. *Applied Mechanics Reviews* 53(7):175–93.

Mishima, K., and Ishii, M. 1984. Flow Regime Transition Criteria for Upward Two-Phase Flow in Vertical Tubes. *International Journal of Heat and Mass Transfer* 27:723–37.

Nakoryakov, V.E., Kuznetsov, V.V., and Vitovsky, O.V. 1992. Experimental Investigation of Upward Gas-Liquid Flow in a Vertical Narrow Annuls. *International Journal of Multiphase Flow* 18:313–26.

Owhaib, W., Martin-Callizo, C., and Palm, B. 2004. Evaporative Heat Transfer in Vertical Circular Microchannels. *Applied Thermal Engineering* 24:1241–53.

Revellin R., and Thome, J.R. 2006. New Diabatic Flow Pattern Map for Evaporating Flows in Microchannels. *Annals of the Assembly for International Heat Transfer Conference 13*, Sydney, Australia, MIC-07.

Stephan, P. 2002. Microscale Evaporation Heat Transfer: Modelling and Experimental Validation. *Proceedings of the 12th International Heat Transfer Conference*, Volume 1:315–27. Grenoble, France.

Stephan, K., and Preusser, P. 1979. Heat Transfer and Critical Heat Flux in Pool Boiling of Binary and Ternary Mixtures. *German Chemical Engineering* 2:161–169.

Thome, J.R. 2006. State-of-the-Art Overview of Boiling and Two-Phase Flows in Microchannels. *Heat Transfer Engineering* 27(9):4–19.

Thome, J.R., Dupont, V., and Jacobi, A.M. 2004. Heat Transfer Model for Evaporation in Microchannels. Part I: Presentation of the Model. *International Journal of Heat and Mass Transfer* 47:3375–85.

Tran, T. N., Wambsganss, M.W., Chyu, M.C., and France, D.M. 1997. A Correlation for Nucleate Flow Boiling in a Small Channel. *Proceedings of the International Conference on Compact Heat Exchangers for the Process Industries*, 291–304. Snowbird: Begell House.

Triplett, K.A., Ghiaasiaan, S.M., Abdel-Khalik, S.I., and Sadowski, D.L. 1999. Gas-Liquid Two-Phase Flow in Microchannels. Part I: Two-Phase Flow Patterns. *International Journal of Multiphase Flow* 25:377–94.

Wambsganss, M.W., Jendrzejczyk, J.A., and France, D.M. 1991.Two-Phase Flow Patterns and Transition in a Small, Horizontal, Rectangular Channel. *International Journal of Multiphase Flow* 17:327–42.

Wayner, P.C. 1994. Thermal and Mechanical Effects in the Spreading of a Liquid Film Due to Change in the Apparent Finite Contact Angle. *Journal of Heat Transfer* 116:938–45.

Wong, H., Radke, C.J., and Morris, S. 1995. The Motion of Long Bubbles in Polygonal Capillaries. Part 1. Thin films. *Journal of Fluid Mechanics* 292:71–94.

Index

A

A ACD cycle, *see* Advanced adsorption cooling cum desalination cycle
Absorbed flux imitation, 383
Absorption energy, 285
AC, *see* Activated carbon
ACF, *see* Activated carbon fibre
Activated carbon (AC), 31–33, 48
Activated carbon fibre (ACF)
  adsorption kinetics of, 38
  Busofit-M8, 287, 288, 291
  characteristics of, 33
  mesopores of, 273, 277
Active cooling methods, 189
Active regulation, 391
Ad-HEx unit, *see* Adsorbent-Heat Exchanger unit
Adsorbed gas storage cylindrical vessel, 296
Adsorbed vessels
  with heat pipe, mathematical model of, 293–296
    calculated cells and boundary conditions, 296–299
    solution algorithm, 299–300
Adsorbent density, 79
Adsorbent grain, 84–86, 95
  size, 70–71
Adsorbent-Heat Exchanger (Ad-HEx) unit, 64, 65, 78, 81, 95
Adsorbent-insensitive regime, 75
Adsorbent/refrigerant pairs
  for adsorption cooling applications, 30–33
Adsorbent-sensitive regime, 75
Adsorbers
  sorbate mass content in, 276
  unit, sketch of, 266–267
Adsorption
  characteristics of ACF, 33
  chiller, silica gel/water, 42, 47
  cooling applications, 30–33
  cycle, 48, 291, 292

  for AC/ammonia, 32
  direct contact, 43–45
  parameters of, 52
  silica gel/water, 42
  energy balance, 38–39, 51
  isobar, 65, 67–69, 92
  isotherms, mathematical modelling, 36–38
  kinetics of ACF, 33, 38
  process, 286
  rate, 75, 80, 87, 92, 100
  time, 73, 102
  vs. desorption time, 69–70
Adsorption dynamics
  in AHT
    adsorbent grain size, effect of, 70–71
    adsorption *vs.* desorption time, 69–70
    boundary conditions, effect of, 70
    driving force for ad-/desorption process, 75
    exponential kinetics, 67–68
    grain-size-sensitive and grain-size-insensitive modes, 74–75
    number of layers, effect of, 71–73
    shape of adsorption isobar, 68–69
    S/m ratio, effect of, 73
  numerical modelling of, 81–82
    monolayer configuration, *see* Monolayer configuration
    multilayer configuration, *see* Multilayer configuration
Adsorption-evaporation process, 34, 36
Adsorption heat, 91
  converters, 259
    computer tool for, 260
    integrated models of, 268–269
    numerical modelling of, 271

Adsorption heat transformers (AHTs)
    adsorption dynamics in,
        see Adsorption dynamics
    dynamic behaviour of, 64
    dynamic optimisation of, 101–103
Adsorption refrigeration system, 32
    silica gel, 43
Advanced adsorption cooling cum
        desalination (AACD) cycle, 30
    mathematical modelling of, 50–52
    performance results of, 52–55
    working principle of, 48–50
Advanced adsorption cooling cycles,
        thermally powered
    mass recovery technique, 44–47
    multi-bed dual-mode adsorption
        system, 41–43
    recent trends in, 47–48
Agricultural products, cold energy
        storage for, 175–177, 209
AHTs, *see* Adsorption heat transformers
Air conditioner system adsorption, 31
Air-conditioning system with SHP,
        252–253
Air cooling, 146
    thermal resistances for, 151
Air syphons, 320
Alcohol pairs for adsorption cooling
        applications, 33
Aluminium flat condenser, external view
        and cross section of, 224, 225
Aluminium loop heat pipe, 223, 224
Aluminium oxide porous coating, 246,
        247
Ambient temperature, sky temperature
        vs., 178–179
Ammonia, 437
    capillary pressure for, 265
    in LT and HT adsorber, 277
    pair for adsorption cooling
        applications, 32, 33
Analytical models, 82, 174
Ancillary thermal energy source, 398
Annular flow, 472–473
    evaporation and condensation in,
        478–480
    in rectangular channel, 475–478
Antifreezes, 333
Asphalt temperature, 458

Automotive fuel cells with hybrid
        power systems, 215
Autonomous devices
    with ratio of heat generation, 399
    with TCS, 392
Average grain temperature, 84–85, 88,
        90–92
Axial DC fan for thermal
        comparison, 159
Axially grooved heat pipes, 148
Axial one-dimensional temperature
        distribution, 369
Axial symmetry of cooling
        process, 138
Azeotropic solutions, 334

B

Bakery waste heat recovery, case
        study of, 180–182
Balance equations, 369–370, 373
Biot number, 84, 98
Boiling, 109
    convective heat flux, 118
    film, *see* Film boiling
    limits, heat transfer, 433–434
    nucleate, *see* Nucleate boiling
    transition, *see* Transition boiling
    zone of heat medium, 339–342
Boiling water reactor (BWR), 196
Bond number (Bo), 428
Boundary conditions, 375
    effect of, 70
Boundary layer hypothesis, 416
Boyko–Krushilin formula, 122
Brinkman–Darcy–Forchheimer
        model, 22
Brinkman equations application
        mode, 300
Bubble generation, 239–242
Bubble growth rate, 112, 118
Bubble regimen, 328–332, 344, 345
Burn-out, 433–434
Busofit-M8, 286, 308, 310, 311, 314
    activated carbon fibre, 287, 288, 291
    hydrogen sorption for, 292
    physical sorption of, 287
Buttercup heat exchanger, 181
BWR, *see* Boiling water reactor

C

Cables
ohmic power generation in, 379
thermal resistance, 390
Capillary condensation effects on
characteristics of heat
converters, 273
Capillary cylindrical evaporator, cross
section of, 7–8
Capillary forces, 243
Capillary–porous medium, 263
Capillary pressure, 16
Capillary pumped loops (CPLs), 1
Capillary structure (CS)
configurations, 382
in gas-loaded heat pipe zones,
configurations of, 395
non-linear processes on boiling in,
374
Capillary surface, 469–470
Carbon materials, hydrogen
sorption on, 300
Carbon nanotubes, 286
Carbon sorbents
and hydrogenous gas, 292–293
advantages of combined system,
313–314
mathematical model of adsorbed
vessels with heat pipes,
293–296
numerical modelling, 300–313
Carbon steel–water
thermosyphon, 442
Cascading adsorption cycle, 32
Casing materials, selection of, 438–440
Chemical sorption models, 261–262
Chemisorption heat converters, 259, 260
Chronological survey of GLHP
models, 363
Clapeyron–Clausius equation, 119, 295
for vapour pressure, 486
Clausius–Clapeyron equation, 262, 422
Clausius–Clapeyron relationship, 10
Cleaning process, thermosyphons
fabrication, 442–443
Closed thermosyphons, 320, 325
Closed two-phase thermosyphons
(CTTs), 320, 322, 341–343

concrete constructions of, 335, 349
crisis phenomena in, 352
implementation of, 323, 324
operation of, 353
thermodynamic conditions of effective
work of, 324–328
'Closed type' SHP, 248
Coefficient of performance (COP), 66, 71
of AC/methanol pair, 33
and SCHP, 102
of silica gel/water-based
adsorption, 30
Cold conditions, heat generation,
380–381
Cold energy storage for agricultural
products, 175–177
Cold unwicked reservoir, scheme
with, 364
Cold water storage, 171
Commercial thermal analysers, 379
Compensation chamber, 7
Complex geometry, 4
Composite sorbents, adsorbers with, 275
Computers
heat pipe application in
desktop thermal solutions,
155–156
formulation, 149–151
notebook thermal solutions,
151–154
for nuclear reactor, 196–197
thermal solution for graphics
processors, 156–158
thermal solution using
piezo fan, 158–161
modelling, 260
COMSOL Multiphysics® simulation
environment, 96, 101
Concave adsorption isobar, 67
Concentrated photovoltaic (CPV)
technology, 177–179
Concentrated thermoelectric generator
(CTEG) with PCM thermal
storage, 189–191
Concrete exploitation conditions, 376
Condensation
in adsorber, models for, 262–266
variation of, 359
zone, heat exchange in, 343–346

Condenser
  design, 231–232
  energy balance, 39–40, 50
  and heat sink, variable radiation
      thermal coupling
      between, 374
  LHP, 224–225
  resistance, 424–425
  VDT, 235
Conductance
  heat transfer by, 362
  variation of, 358–359
Conduction
  and radiation heat transfer
      problems, 379
  resistances, 421–422
Conductive heat exchange, parameters
      of, 381
Confinement number, 466
Consistent model, 95
Contact resistance improvement,
      methods of, 367
Continual approaches, adsorption/
      chemisorption heat converter,
      260–261
Continuity equation, 225–226
Continuum approach models
  capillary pressure, 16
  evaporators application, 21–23
  heat and mass transfer governing
      equations, 15–16
  local thermal equilibrium models,
      18–19
  multi-phase mixture model, 19–21
  non-equilibrium models, 16–18
  porous material, 14
Convectional thermosyphons, 321
Convection heat exchangers in liquid
      cooling systems, replacement
      for, 405
Convection heat resistance, 200
Conventional heat dissipation systems,
      458–459
Conventional heat pipes, 358
Conventional resorption systems,
      216–217
Converging–diverging nozzle
      theory, 430
Convex exponential isobar, 87

Cooling capacity
  of AACD, 54–55
  of silica gel/water refrigeration, 30
Cooling data centres
  advantages, 147
  cold energy storage for, 169
  heat pipe cold storage system for,
      167–169
Cooling desktop PCs, thermal design
      trends for, 155, 156
Cooling rate, 134–135
COP, *see* Coefficient of performance
Copper–constantan
      thermocouples, 383
Copper tubes, 442
Cost-effective system of gas
      storage, 284
CPLs, *see* Capillary pumped loops
CPV technology, *see* Concentrated
      photovoltaic technology
Critical heat flux, 433
Cryogenic fluids, 436
Crystallisation temperatures of water
      solution, 333
CS, *see* Capillary structure
CTEG, *see* Concentrated thermoelectric
      generator
CTEG-PCM system, thermal
      performance of, 191, 192
CTT, *see* Closed two-phase
      thermosyphons
CTTs, *see* Closed two-phase
      thermosyphons
Cylindrical CPL evaporator, behaviour
      of, 22
Cylindrical fuel cell, heat pipe-based
      thermal management
      of, 220
Cylindrical reactor, 267
Cylindrical vessel, hydrogen filling and
      extraction for, 301–304

D

D-A equation, *see* Dubinin–Astakhov
      equation
Darcy's law, 264, 266
Darcy's scale, 2–3
Data acquisition systems, 449

Data centres
    cooling systems for, 147
    power consumption in, 168
    thermal management of, 167, 208
DC blower for thermal
        comparison, 159, 161
Decay heat by nuclear reactor, 197–198
Desalination applications, 30–33
Desktop thermal solutions, 155–156
Desorption
    cycle, 292
    energy balance, 38–39, 51
    process, 49
    reaction rates, 285
    time, adsorption *vs.*, 69–70
Desorption-condensation process, 34, 36
Device case temperature, 385
Device compartment, design of
        isothermal panel for, 400
Device–device–space environment'
        system, mounting place of, 357
Device power generation, function
        of, 374
Device temperature
    control, thermal conception of, 396
    device and radiator heat
        balances, 394
    regulation range, 397
Device thermal balance, components
        of, 378
Device thermal control, isothermal
        panel for, 398
Device thermostating, passive concept
        of, 394
Digital porous media, 3–4
Digital random packing of spherical
        particles, 3–4
Direct simulations, 23–24
    of two-phase flow, 4–6
Discharging process, 312
Discrete approaches, adsorption/
        chemisorption heat converter,
        260–261
Dispersion of fluid, 322
Dittus and Boelter equation, 484, 485
Domestic gas-powered ovens, 454
Draining film of fluid, 329
D-R equation, *see* Dubinin–
        Radushkevich equation

Drift flux model, 473
Driving force
    for ad-/desorption process, 75–81
    for heat and mass transfer, 88
Dry-out limits, heat transfer, 431–432
Dual-fuel vehicles, application in, 284
Dual-laser flow scanning, 471–472, 474
Dubinin–Astakhov (D-A) equation,
        50, 261
Dubinin–Radushkevich (D-R) equation,
        32, 37, 294, 300, 301
Dynamic thermal behaviour of storage
        system, 286

E

Earth, ultra-large heat pipes for cooling,
        191–195
ECCS, *see* Emergency core cooling
        system
Effective thermal conductivity of
        porous systems, 244–246
Effective thermal resistance, 419
Effective vapour condensation, 360
8800 kW data centre, pre-cooler
        design for, 173, 174
Electrical-conveyor-belt ovens, 456
Electrical ovens, temperature
        distribution of, 454, 455
Electrical resistances, 456
Electric power consumption, 147
Electrohydrodynamical
        thermosyphon, 322
Electronic elements, temperatures
        equalisations of, 393
Electronics, *see also* Computers, heat
        pipe application in
    PCB, 387
Element temperature,
        smoothing of, 390
Elongated bubble flow, 472–475
Emergency core cooling system
        (ECCS), 148
    BWR with, 197
    electric generators for, 196
    failure condition, 203
Energy balance
    adsorption and desorption, 38–39, 51
    condenser, 39–40

equation
in grain, 88
for single adsorbent grain, 82–83
evaporator, 40, 50
Energy conservation equation, 226,
293, 295
Energy consumption in Japan, 194
Energy equations, 16, 20–21
Engineering Science Data Unit
(ESDU), 431
Entrainment limits, heat transfer,
427–430
Equilibrium state equation, 294
Equivalent liquid film, 112
ESDU, *see* Engineering Science
Data Unit
Evaporation/condensation process, 214
Evaporation model, adsorber, 262–266
Evaporation zone, heat exchange in,
339–342
Evaporative heat transfer, 242
Evaporators
applications, continuum approach
models, 21–23
concept of ice removal on, 175
energy balance, 40, 50
heat transfer coefficient in, 201
resistance, 425–427
vapour from, 322
Exponential kinetics, 67–68
External environment, heat exchange
between device case and, 377
External thermal boundary
conditions, 378
External thermal resistances, 427

F

Fickian diffusion model, 82
Film boiling, 110–111
of subcooled liquids
characteristics of, 132–135
experimental result, 135–139
Film condensation, 343–344
Filonenko formula, 122
Fin efficiency, 202
Finned tube heat exchanger, 79
First law of efficiency, 214
Flat heat pipe, 399

Flat minichannel condenser, 224
Flat sectional vessel
gas extraction, 309–313
gas filling, 305–309
Flat vapour–gas front theory, 362
Flexible transport lines, LHPs
with, 224
Flooding limits, heat transfer, 427–430
Flow boiling
heat transfer, 121–123
analysis of, 467
predominant forced convection
and evaporation, 487–490
predominant nucleate boiling,
484–487
in minichannels and
microchannels, 484
of refrigerants, 474–475
Flow visualisation of R-318C flow
boiling, 474
Fluids
dispersion of, 322
draining film of, 328–329
flow (propane) visualisation in
annular minichannel,
231, 232
method, volume of, 226
Fourier law, heat flux prediction, 417
Freezing index, defined, 175, 176
Fresnel lens concentrator, 190, 191
Freundlich equation, 36
Friction factor, 122
Fuel cell
efficiency, 214–215
stack, 220
thermal management, heat pipes
for, *see* Heat pipe, for fuel cell
thermal management
Fuji silica RD, grains of, 70, 71, 75, 95, 97

G

Gas adsorption, reaction rate of,
302–303
Gas desorption, isothermal conditions
of, 303
Gas-driven conveyor-belt ovens, 456
Gas extraction, flat sectional vessel,
309–313

Gas filling, flat sectional vessel, 305–309
Gas–liquid flow
  behaviour of, 470
  fundamentals of, 468
  in minichannels and microchannels
    adiabatic flow patterns, 470–474
    annular flow in rectangular
      channel, 475–478
    diabatic flow patterns, 474–475
Gas–liquid thermosyphon heat
      exchanger, 451, 452
Gas-loaded heat pipe (GLHP), 373
  appearance of thermal control
    system with, 395
  for autonomous devices, principles of
    TCSs operation on
    design features of TCS, 380,
      381–382
    gas-loaded heat pipe, thermal
      control system on, 383
    heat exchange, joint analysis
      of, 378
    heat leak via cable, calculation
      of, 379
    imitation of satellite
      movement, 385
    structure of heat fluxes in
      TCS, 377
    temperature distribution in
      gas-loaded heat pipe, 384
    temperature stabilisation of
      device case, 376
    thermal control system operation
      in unsteady regime, 386
  configurations, 363
  design of, 368
  feature of, 359
  heat output by, 390
  modelling, distributed codes for,
    368–369
  models, 369
  operation, unsteady regime of,
    370–371
  principle of, 360
  regulation, accuracy of, 362
  scheme of, 382
  start-up performance of, 372
  temperature distribution in, 384, 392
  thermal control system on, 383

Gas–mass balance, partial pressure of
      gas in, 378
Gas partial pressure, equation for
      distribution of, 369
Gas sorption processes, mathematical
      model of, 293
Gas sorption storage
  heat pipes for, 297
  system, 294
Gas storage
  cost-effective system of, 284
  hydrogenous, advantages of
    combined system, 313–314
  vessel, thermodynamic efficiency
    of, 293
Gauss–Zeidel iterative procedure, 480
Gear's BDF method, 52
Geometric scheme of adsorber,
      266–268
Geothermal heat extraction, large-scale
      loop heat pipe for, 179–180
GHP, *see* Grooved heat pipe
GHPPL, *see* Grooved heat pipe with
      porous layer
GLHP, *see* Gas-loaded heat pipe
Global thermal resistance, 419
Global warming potential (GWP),
      47–48
Gluckauf equation, 294
GPU, *see* Graphics processor unit
Grain-size-insensitive regime, 67,
      74–75
Grain-size-sensitive modes, 74–75
Grain temperature, 85, 89
Graphics processors, thermal solution
      for, 156–158
Graphics processor unit (GPU), 156
Gravimetric sorption density, 313
Gravimetric storage density, 313
Gravity-assisted heat pipes,
      see Thermosyphons
Gravity-assisted loop heat pipes,
      413, 414
Gravity-assisted water charge,
      198–199
Grooved heat pipe (GHP), 242–243
  evaporator, 244–247
Grooved heat pipe with porous layer
      (GHPPL), 243

H

Hamaker constant, 476, 477
Heat
  balance
    in cold conditions, quantitative
      content of, 382
    components for TCS operation
      under cold conditions, 381
    equations of, 357, 371
    for radiator and reservoir,
      components of, 377–378
    structure of, 396
  exchange, joint analysis of, 378
  generation
    autonomous devices with
      ratio of, 399
    cold and hot condition, 380–381
    thermal control of devices
      without, 393–397
  leak, calculation of, 379
  and mass transfer, mechanisms
    of, 359
  power budget for assembly device
    and TCS, 397
  rejection from radiator into
    environment, 390
  source temperature, 365, 366
  transfer by conductance and vapour
    diffusion, 362
  transport function, 357
Heat carrier switching, 275–276
Heat conduction path length, 80
Heat conversion
  cycle, 273
  performance characteristics of,
    277, 278
Heat converters
  capillary condensation on
    characteristics of, 273–280
  design concepts of, 269
  operation cycle of, 270
  single-unit adsorber, *see* Single-unit
    adsorber, adsorption and
    phase transitions in
Heat exchange
  in condensation, 343–346
  in evaporation, 339–342
Heat exchange coefficients, 267, 341–345

Heat-exchange intensification,
    electrohydrodynamical
    methods of, 322
Heat-exchange recuperators, 324
Heat exchangers
  thermosyphons, 450–453
  of two-bed adsorption cooling
    cycles, 33
Heat-exchange surface, 323
Heat flux, 113, 371, 383
  components of, 379
  convective constituent of, 118
  in CTT, 335
  degree of influence, 340–341
  density of, 110, 115, 339, 475
  directly based on approximate
    mechanistic mode, 115
  dry-out limit, 431–432
  GLHP, 378
  limiting, 346–349
  prediction, Fourier law, 417
  in transition boiling, 130
Heat-generating object, decreasing
    temperature of, 324
Heating devices, based on
    vapour-dynamic
    thermosyphons, 233–234
Heating factor, 312
Heat medium for thermosyphons,
    332–335
  compatibility of, 337
  evaporation zone of, 339–342
Heat pipe, 217, 465
  application, 358, 413
  for cooling and temperature
    regulation, 375
  for fuel cell thermal management
    LHPs, 223–229
    MHP, 218–219
    micro/miniature, 220–221
    mini loop heat pipes, 222–223
    SHPs, 219
  for hydrogen storage tanks, 293
  improvement, 164–165
  mathematical model of adsorbed
    vessels with
    calculated cells and boundary
      conditions, 296–299
    solution algorithm, 299–300

movement of vapour from, 360
new generation of, 238–242
into panels, integration method of,
  398
proposed conceptions of systems
  with, 357–358
shell temperature distribution
  in, 361
thermal analysis, 200–203
Heat pipe–based spreader, 220–221
Heat pipe–based thermal management
  of cylindrical fuel cell, 220
Heat pipe ECCS, 198, 203–205
  with initial water charge, 205–207
Heat pipe evaporators, nanoporous
  coatings of, 242
Heat pipe heat exchanger (HPHE), 171,
  180, 182
Heat pipe revolution, 148–149
Heat pipe sizes, heat transfer capability
  vs., 193
Heat pipe turbine, 182–183
Heat sink
  thermal conductance between source
    of energy and, 359
  thermal resistance between device
    case and, 384
  variable radiation thermal coupling
    between condenser and, 374
Heat transfer, 323
  crisis of, 344, 347, 350–351
  devices, *see* Thermosyphons
  distance, 78–79
  driving force for, 88
  with evaporation and condensation
    in annular flow, 478–480
    in microregion, 480–484
  in film boiling, 110, 111
  intensity of
    GHP evaporator, 244–247
    SHP, 252
  limits
    boiling, 433–434
    dry-out, 431–432
    entrainment/flooding, 427–430
    oscillation, 432–433
    sonic, 430
    viscous, 431
  mathematical model of, 293

mechanisms of, 467–468
model of, 125
one-phase thermosyphons, 321
predominant forced convection and
  evaporation, 487–490
predominant nucleate boiling,
  484–487
processes of, 322
ratio of, 466
in vapour-dynamic thermosyphon
  annular channel, 232
Heat-transfer agent, intermediate,
  see Intermediate heat-transfer
  agent
Heat transfer capability *vs.* heat pipe
  sizes, 193
Heat transfer coefficient (HTC), 228–229,
  242, 488
  based on typical boiling curve, 110
  cooling process of, 138
  in film boiling, 139
  in nucleate boiling, 111–113
  at transition boiling, 131
Heat-transmitting capacity
  of thermosyphons, 348–352
  of two-phase thermosyphons, 347
Heat-transmitting devices, 325, 332,
  334, 349
  air syphons as, 320
  classification of, 321, 322
Heat-transmitting processes, 338
Heat-transmitting properties of
  thermosyphon, 325
Heat upgrading, modelling of, 272
Hertz–Knudsen formula, 264
Heterogeneous condensation, 415
High-aspect extrusion heat sink,
  155–156
High axial conductive radiator, designs
  of, 368
High heat power cooling, 458–459
High Speed Gas Sorption Analyzer
  NOVA 1200, 287, 289
High temperature (HT) adsorber, 269,
  275–276
  detailed distribution of ammonia
    in, 277
  heat conversion, performance
    characteristics of, 278, 279

High temperature fluids, 436
Hot condition heat generation,
        380, 381
HPHE, *see* Heat pipe heat exchanger
HT adsorber, *see* High temperature
        adsorber
HTC, *see* Heat transfer coefficient
Hybrid cooling system, 151–152
Hydraulic diameter
    of annular channel, 232
    of condenser annular channel, 230
    of MHPs, 220
Hydride alloys, 290
Hydrodynamic processes in
        thermosyphons, 328–332
Hydrogen
    adsorption
        capacity, 285
        cycle of, 291
        isotherms, 301
        molecules, 286
    charging, thermodynamic process
        of, 300
    compression, 313
    filling and extraction for cylindrical
        vessel, 301–304
    gravimetric storage density
        of, 286
    intake, 306
    isotherms of, 287, 289
    physisorption, 285
    storage systems, 308
        metal hydride based, 284–285
    volumetric storage density of,
        304, 308
Hydrogenous gas
    carbon sorbent and, 292–293
        advantages of combined system,
            313–314
        mathematical model of adsorbed
            vessels with heat pipes,
            293–296
        numerical modelling, 300–301
Hydrogenous gas sorption storage,
        thermally controlled sectional
        vessel for, 297
Hydrogenous gas storage, advantages
        of combined system,
        313–314

Hydrogen sorption, 291
    Dubinin–Radushkevich Equation
        for, 300
    storage, 305
Hyperbolic tangent, 138

I

Ice formation, test results of, 175
Ice storage system to support cooling
        system failure, 169–171
IDS method, *see* Isothermal differential
        step method
IHCP, *see* Inverse heat conduction
        problem
IHS, *see* Integrated heat spreader
Impulse laser beam, 239
Imura's correlation, 201
Inherent power generation, passive
        thermal control system for
        device without, 396
Inlet flow rates of gas, 307
Integrated heat spreader (IHS),
        150, 164
Integrated models of adsorption heat
        converters, 268–269
Integration
    of GLHP into passive TCS, 375
    of heat pipes into panels, method of,
        398
Intensive evaporation heat
        transfer, 242
Intermediate heat medium, 335, 341
    circulation of, 322
    state change of, 325
Intermediate heat-transfer agent, 332
    thermophysical properties of, 335
Intermediate heat-transmitting
        devices, 323
International Mathematics and Statistics
        Library (IMSL), 52
Intra-diffusion adsorption process, 294
Inverse heat conduction problem
        (IHCP), 138
Inverse thermosyphons, 413–414
Investigated activated carbon
        fibres, 288
Isobaric adsorption stage, 67, 70
Isobaric desorption stage, 67, 70

Isobaric stages of AHT cycle, 66, 67, 69, 77
Isosteric heat of adsorption, 51
Isothermal adsorption dynamics, 93
Isothermal adsorption, equation of, 294
Isothermal device case, heat balance of, 393
Isothermal differential step (IDS) method, 97, 101
Isothermal panel for device thermal control, 398
Isotherms of hydrogen, 287, 289
Istituto di Tecnologie Avanzate per l'Energia (ITAE-CNR), 66

J

Jacob number, 418

K

Katto correlation, 430
Kelvin equation, 264
Kinetic equation, approximate, 294–295
Knudsen diffusion, 93

L

Laplace equation, 116
    for capillary pressure, 469
Laptops, thermal solution trends for, 154
Large-scale loop heat pipe for geothermal heat extraction, 179–180
Large-size vapour chambers, applications of, 413–414
Large temperature jump method (LTJM), 64–66, 77, 80, 101
Latent heat of vaporisation, 39
Lattice Boltzmann method (LBM), 4
LBM, *see* Lattice Boltzmann method
LCZ, *see* Lower convective zone
LDF equation, *see* Linear driving force equation
Leidenfrost effect, 199
Leverett function, 265
LHPs, *see* Loop heat pipes
Lid operation, 397

Linear driving force (LDF) equation, 38, 50
Linear isobar, 87
Liquid cooling systems, 146
    heat transfer capability of, 151
    replacement for convection heat exchangers in, 405
Liquid film, 113, 116–118
Liquid–gas system, 469
Liquid microlayer evaporation, 113
Liquid–vapour thermal resistances, 422–423
Literature correlations, 433
Liu–Winterton model, 484–489
LMTD, *see* Log mean temperature difference
Local thermal equilibrium models, 18–19
Log mean temperature difference (LMTD), 39, 40, 51
London–van der Waals forces, 114
Long-term alternative refrigerant, 242
Loop heat pipes (LHPs), 1, 217, 223–225
    evaporator cooling, 252
    evaporator porous wick, 2
    mathematical model of heat and mass transfer in, 225–227
    mini, 222–223
    modeling results of, 227–229
    temperature field evolution in, 250, 251
Loop thermosyphons, 229, 413, 414, 434
Loop-type heat pipe system, 199
    overall thermal resistance of, 202, 208
    thermal performance of, 200
    water temperature with, 204, 205
Loose grains configuration
    adsorption dynamics for, 65, 66
    schematics of, 71
Lower convective zone (LCZ), 184
Low temperature (LT)
    adsorber, 275–276
        capillary condensation, 281
        detailed distribution of ammonia in, 277
        heat conversion, performance characteristics of, 277–279
    fluids, 436
    power systems, 214

LT, *see* Low temperature
LTJM, *see* Large temperature jump
    method
Luikov's approach, 263
Lumped approach, 82
Lumped regime, 75, 77

M

Macropores, 221, 222, 265
Mass balance, 127
    of AACD Cycle, 50
    equation, 83, 476, 478
    of refrigerant, 40
    in vector form, 296
Mass balance equations, 15
Mass filling degree, 326–328
Mass flux, 479, 482
Mass gas balance equation, 371
Mass recovery adsorption refrigeration
    cycle, 44–47
Mass transfer
    driving force for, 88
    mechanisms of heat and, 359
    processes of, 322
MATLAB®, 227
Maxsorb, 286
Maxsorb III activated charcoal, 32, 48
Mean sorbent pressure, 312
Mean sorbent temperature, 306, 310
Medium temperature fluids, 436
MEMS process,
        see Microelectromechanical
        systems process
Mendeleev–Clapeyron equation, 268
Mercury, working fluid, 437
Mesopores, 265, 266
    of ACF, 273, 277
Mesoscale, 2
Metal hydrides
    advantages of, 286
    applications, 289
    micro- and nanoparticles of,
        289–290
    microparticles, 289
Metal plate, 75
Methane
    desorption of, 301
    storage system, 285

Methanol pair for adsorption cooling,
    31–33
MHP, *see* Micro heat pipes; Miniature
    heat pipes
Microbubble boiling, 110
Microchannels, 466
    in MHP, 220
Micro-channel two-phase pump loop,
    166–167
Micro-channel vapour chamber
    (MVC), 162
Micro combined heat and power
    (micro-CHP), 215
Microelectromechanical systems
    (MEMS) process, 223
Micro heat pipes (MHP), 217
    microchannels in, 220
    porous media, effect in, 221–222
Micro/miniature heat pipes for fuel
    cell thermal management,
    220–221
Micropores, 221, 286
    in Maxsorb, 286
    size distribution, 287
    surface of, 295
Microregion, 117, 478
    heat transfer with evaporation in,
    480–484
Microscale effect, 248
Microscopic approaches, 2
Micro x-ray computerised
    tomography, 3
Miniature heat pipes (MHP), 217
Mini loop heat pipes, 222–223
Minimal heat flux, 362
Minipores, menisci of evaporation in,
    244
Miniscale effect, 248
Mixed PNM, 10
MLI, *see* Multi-layer insulation
Mobile air-conditioning (MAC) systems,
    47, 48
Momentum balance, 296
Momentum equations, 15, 16
Monolayer configuration, 82–84
    adsorbent grain size, effect
        of, 70–71
    boundary conditions, effect
        of, 70

driving force for heat and mass transfer, 88–95
number of layers, effect of, 71–73
temperature and pressure, 84–86
uptake/release curves, 86–88
Monolithic carbon for adsorption cooling applications, 32
Morphological pore networks, 7
Motor cooling, liquid system of, 309
Mounting panel, 398
Mounting place
heat exchange between device case and, 377
temperature, impact of, 384, 385
M-shaped heat transfer coefficient for flow boiling, 468
Multi-bed dual-mode adsorption system, 41–43
Multi-component mixtures, use of, 333
Multilayer configuration, 95–97
of loose grains, 65
simulation of, 97–101
Multi-layer insulation (MLI), 374
blankets, 377
design, improvement of, 384
Multi-phase filtration models, 263, 264
Multi-phase mixture model, 19–21
Multi-phase transport, adsorber, 262–266
Mutual heat exchange, 378
MVC, *see* Micro-channel vapour chamber

N

Nanocoatings (NCs), 218
of heat pipe evaporators, 242–248
thermosyphons with, 238–242
Nanofluids (NFs), 217–218
thermosyphons with, 238–242
Nanomaterials, generation of engineered, 217
Nanoporous technology, 246
Nanotubes, cooling of, 286
Naphthalene, 334
Naphthalene carbon steel thermosyphons, temperature distribution, 439, 440
Natural gas, 284

Navier–Stokes equations, 4, 225–227, 475, 478, 482
NCs, *see* Nanocoatings
NCZ, *see* Non-convective zone
Newtonian law, 268
NFs, *see* Nanofluids
Nominal regulation diapason, 361
Non-condensable gases, 336, 413, 437, 439
Non-convective zone (NCZ), 184
Non-equilibrium models, 16–18
Non-equilibrium thermodynamics, 263
Normal extrusion heat sink, 155
Notebook thermal solutions, examples of, 151–154
Nuclear fission, 196
Nuclear reactor, 196–198
Nucleate boiling, 425, 426, 487
cooling process in, 135
definition of, 109–110
flow boiling heat transfer, 121–123
HTC in, 111–113
mechanism of, 113–115
of nanoliquids, 239
of NFs, 218
share of, 485
Nucleate boiling heat transfer
model of, 115–120
theory of, 110
Nucleation site, 113–114, 118
Nucleation site density, 116, 119
Numerical microstructure, 4
Numerical modelling, 273, 275
of adsorption dynamics, 81–82
monolayer configuration, *see* Monolayer configuration
Nusselt analysis, 415, 418, 425
Nusselt equation, 345
Nusselt models for condensation, in walls, 415–418
Nusselt's correlation, 201
Nusselt theory of film condensation, 343

O

Ohmic power generation in cables, 379
Oil tank heaters, 457–458

One-dimensional conductivity
equation, 370
One-phase thermosyphons, 320, 321
Opened thermosyphons, 320
Open-type MHP, 218
Open type SHP, 248
Operation diapason of temperatures,
332, 333, 336
Operation temperature, 436, 439, 448
Optical properties of coating,
deviation in, 374
Optoelectronic equipment, hierarchy of
VCHP integration into, 375
Oscillation limits, heat transfer,
432–433
Ovens
bread-baking oven, 454–455
domestic and industrial
food-cooking, 452–453
electricity, 453–454
industrial, 452
straight-tree configuration
thermosyphons, 454
thermosyphon-assisted conveyor belt
oven, 456–457
Ozone depletion potential (ODP), 47

P

Paraffin wax, thermophysical properties
of, 190, 191
Parasitic heat flux, 12, 14
Partial gas pressure in reservoir, 363, 364
Passive cooling, 146, 189, 190, 197
Passive heating element–absorber,
concept of integration of, 394
Passive regulation, 391
Passive TCS design, principle of, 397
Passive thermal control system for
device, 396
PCBs, *see* Printed circuit boards
PCMs, *see* Phase change materials
PEG, *see* Polyethylene glycol
Perkins pipes, 323
Permafrost storage system, 175–176, 208
Permeability, 223
of network, 9
of porous medium, 10
Phase change materials (PCMs), 177

CPV cooling by use of, 178
TEC generator module with
thermosyphon and, 190
thermal storage, CTEG with, 189–191
Photothermal conversion of energy, 240
Photothermal signals, 241
Photovoltaic (PV) cells, 177, 178, 192
Physical sorption models, 261–262
Physisorption process, 48
Piezo fan, thermal solution using, 158–161
Plate heat exchanger, 79–80
PNMs, *see* Pore network models
PN simulations, *see* Pore network
simulations
Poisson's equation for liquid flow, 477
Polyethylene glycol (PEG), 178
Pool evaporation process, 425–427
Pore
geometry and size distribution, 285
in sorbent, types, 265
Pore-network approach
boundary conditions, 11
capillary cylindrical evaporator,
cross section of, 7–8
morphological pore networks, 7
PNMs, 8
pore scale models, 6
vapour–liquid wick, 8–9
Pore network models (PNMs), 6–9, 261
mixed, 10
parasitic heat flux, 12
Pore network (PN) simulations, 2, 23–24
Pore scale models, 6
Porous adsorbent bed models, 82
Porous matrix
metal hydride in, 290
numerically generated, 4
thermal conductivity of, 13
Porous media, 14
boiling problem in, 18
digital, 3–4
mass transport through, 223
MHP effect in, 221–222
numerical simulation of transport
in, 2
transport phenomena in, 23
Porous medium, 262–263, 414
heat transfer in, 16
permeability of, 264

Porous systems, effective thermal
        conductivity of, 244–246
Porous texture, 287
Porous wick, 248, 251
    heat transfer in, 10, 23
    of LHPs, 6
    numerical simulation tools, 2
Potassium, 437
Power consumption, 248
    in data centres, 167, 168
    of piezo disc fan, 160
Prandtl number, 121, 418, 424
Pre-cooler design for chiller inlet water,
        171–174
Pressure, distributions of, 84–86
Pressure gradient, 82, 89–90, 117
Pressure jump, 119, 127
Pressure regulator, 207, 309
Pressurised water reactor (PWR), 196
Principle of temperature regulation, 379
Principle of temperature
        stabilisation, 400
Printed circuit boards (PCBs), 382
    design, integration of heat pipe TCSs
        in, 389, 393
        cable thermal resistance, 390
        isothermal substrate and GLHP,
            design of autonomous
            electronic device with, 388
        temperature distribution along
            GLHP, 392
        vapour temperature, 391
    heat generation, 383
    principle of electronic elements
        cooling in, 387
    with TCS, design of, 386
    thermal design, proposed, 399
Processor die, heat spreading from,
        161–164
Proposed model, 125, 126, 131
    of flow boiling heat transfer, 491
    of nucleate boiling, 140
PV cells, *see* Photovoltaic cells
PWR, *see* Pressurised water reactor

Q

Quasi-equilibrium conditions, 281
Quasi-stationary mode, 90, 91

R

Radiation heat transfer problems, 379
Radiator
    density of flux by, 395
    into environment, heat rejection
        from, 390
    and reservoir, components of heat
        balance for, 377–378
Raked piezo fan, thermal performance
        of, 160
Rapid evaporation of liquid, 482, 491
Reactor thermal analysis, 203–207
Refrigerants
    mass balance of, 40
    pairs for adsorption cooling
        applications, 30–33
    valves operation in, 35–36
Regulating mechanism of GLHP, 360
Relative permeability of porous
        medium, 263, 264
Remote heat exchanger, 152, 153
Representative elementary volume
        (REV), 5–6
Reservoir
    components of heat balance for
        radiator and, 377–378
    CS with and without, 363–364
    external heat inputs to radiator and,
        390
Reservoir–evaporator liquid path,
        hydraulic resistance of, 385
Resorption coolers, 215–216
Resorption heat pumps, 215–216
Resorption systems, 216
Resorption technology, advantage of,
        216
REV, *see* Representative elementary
        volume
Reynolds analogy, 121
Reynolds number, 418, 424, 426
Rivulet evaporation, 482–483
Robust numerical tool, adsorption heat
        converters, 261

S

SCHP, *see* Specific cooling/heating
        power

SCP, *see* Specific cold production;
  Specific cooling power
SDWP, *see* Specific daily water
  production
Sealing process, 446, 447
Serpentines, 458
Shell/CS–working fluid–gas,
  combinations of, 368
Shell temperature distribution in heat
  pipe, 361
SHP, *see* Specific heat production
SHPs, *see* Sorption heat pipes
Silica gel
  adsorption chiller, 42, 47
  for adsorption cooling applications,
    30–31, 55
  adsorption refrigeration system, 43
Simulations, direct, 23–24
Single adsorbent grain, 81–82
Single-phase convective heat transfer,
  111–112
Single-unit adsorber, adsorption and
  phase transitions in
  discrete and continual approaches,
    260–261
  geometric scheme of, 266–268
  integrated models of, 268–269
  models for condensation,
    evaporation and multi-phase
    transport, 262–266
  physical and chemical sorption
    models, 261–262
  two-channel chemical-adsorption
    system, 271–272
  validation, 269–271
Sintered powder heat pipes, 148
Sintered powder wick, 221
Skive micro-fin structure, 163
Sky temperature *vs.* ambient
  temperature, 178–179
Slug-flow regime, two-phase layer for,
  330–332, 344, 350
Small satellite compartments, thermal
  control of, 398–399
Small-size heat transfer devices, 218
Snow melting VDT designs, 237–238
Sodium, 437
SOFC, *see* Solid oxide fuel cell

Software for GLHP design and model
  integration, 368
Solar pond, thermosyphon-based
  thermoelectric generation
  system for, 183–188
Solar solid sorption coolers, VDT
  thermal control system for,
  234–237
Solid oxide fuel cell (SOFC), 215
Solid sorbents, 264, 284
Solid sorption finned heat pipes, 252
Solid sorption heat pumps, 33, 44
  with VDT thermal control, 234, 235
Solid sorption transformer
  development, 215
Sonic limits, heat transfer, 430
Sorbate, adsorption and thermophysical
  parameters of, 273, 274
Sorbent bed, 273, 296, 310
  gravimetric storage density of gas
    in, 307
  heating conditions, variants of, 303
  heating/cooling of, 297
  mean temperature of, 302
  pressure and temperature in, 305
  regeneration, 290
  SHP, 248, 249
  temperature, 312
  total sorption capacity of, 289, 290
Sorbents
  adsorption and thermophysical
    parameters of, 273, 274
  convertor designs and combinations
    of, 260
  cooling of, 309
  development and testing of, 287–292
  elements, 293
  mean temperature of, 302
  multi-phase transport in, 263
  temperature in discharge process,
    310
Sorption
  capacity registration, 290
  elements in two-bed adsorption
    chiller, 35–36
  finite rate of, 294
  of hydrogen, 285
  hydrogen isosteres of, 291

Sorption cooler with composite sorbent, optimisation of
  definition of system, 272–273
  heat converters, capillary condensation on, 273
Sorption heat convertor, cooling efficiency for, 281
Sorption heat pipes (SHPs), 217, 248–252
  advantages of conventional heat pipes, 219
  air-conditioning systems based on, 252–253
Sorption heat pump, 216, 217
Sorption models, physical and chemical, 261–262
Spacecraft design, VCHP implementation in, 398
Space thermal control means, comparison of, 401–404
Space thermal control techniques, development of, 358
Specific cold production (SCP), 31, 45, 52, 277, 279
Specific cooling/heating power (SCHP), 64–66, 102
Specific cooling power (SCP), 31, 45, 52, 102, 277
Specific daily water production (SDWP), 52, 54
Specific heat production (SHP), 269
Stainless steel tubes, cleaning process, 442
Standard copper–constantan thermocouples, 233
Steady-state heat balance for system, 393
Steady-state regimes, 8, 377
Storage density of hydrogen, 313–314
Storage temperature, 312
Stored hydrogen, gravimetric and volumetric density of, 307–308
Straight thermosyphons
  geometry of, 413
  testing, experimental set-up, 449
Sub-atmospheric pressures, 32
Subcooled liquids, film boiling of, *see* Film boiling, of subcooled liquids
Substrate–electronic PCB, advantages of, 393

Suppression parameter, 487
Supra-atmospheric pressures, 32
Surface plots, two-dimensional, 305
Surface tension, 469–470

T

TCSs, *see* Thermal control systems
TDs, *see* Thermal diodes
TECs, *see* Thermoelectric cells
TEG, *see* Thermoelectric generator
Temperature distribution, 84–86
  of electrical oven, 454, 455
  in GLHP, 384, 392
  inside asphalt tanks, 458
  inside naphthalene carbon steel thermosyphons, 439, 440
  of thermosyphon conveyor belt oven, 457
Temperature disturbances, analysing, 364
Temperature field prediction, modelling approaches for GLHP, 362–363
Temperature regulation
  accuracy of, 400
  principles, 373, 379
Temperatures measurement, 448–449
Temperature stabilisation, principle of, 400
Thermal balance components for device, quantitative estimation of, 379
Thermal circuits of thermosyphon, 419
Thermal conductance, 9, 10, 258, 359
Thermal conductivity of sorbent bed, 289
Thermal control means, comparison of different space, 401–404
Thermal control, potential ways of, 394
Thermal control systems (TCSs), 307, 357
  assembly of autonomous device with, 391
  behaviour, 385
  concept of two-level, 399
  designs, 380, 381–382, 400
  elements, 380, 384
  exploitation in earth orbit, 385
  with GLHP, 374, 383, 395

heat power budget for assembly
device and, 397
manifold, 359–360
operation in unsteady regime, 386
operation under cold conditions,
heat balance components
for, 381
in PCB design, integration of heat
pipe, 386, 389, 393
cable thermal resistance, 390
isothermal substrate and GLHP,
design of autonomous
electronic device with, 388
principle of electronic elements
cooling in, 387
temperature distribution along
GLHP, 392
vapour temperature, 391
structure of heat fluxes in, 377
thermal and mechanical design,
387–388
in thermovacuum chamber, 382, 383
use of classic passive, 393
on variable conductance heat pipes,
fields of application of, 405
VDT, 234–238
Thermal design for cooling desktop
PCs, 155, 156
Thermal diodes (TDs), 358, 359
Thermal disturbance, 382–383, 400
Thermal energy sources, 260
Thermal fluids management system, 220
Thermal line vapour temperature–heat
sink, 364
Thermally activated adsorption heat
pumping technology, 214
Thermally controlled sectional vessel
for hydrogenous gas sorption
storage, 297
Thermally powered advanced
adsorption cooling cycles,
see Advanced adsorption
cooling cycles, thermally
powered
Thermally regulated vessel with carbon
sorbent and hydrogenous gas,
292–293
advantages of combined system,
313–314

mathematical model of adsorbed
vessels with heat pipes,
293–296
numerical modelling, 300–301
Thermal modelling, thermosyphons
under steady-state conditions,
419–421
temperature distribution, 418–419
thermal resistance models,
see Thermal resistance
models
Thermal models of cables, 379
Thermal network in panels, 398
Thermal parameters, variation
of, 359
Thermal resistance models
condenser resistance, 424–425
conduction resistances, 421–422
evaporator resistance, 425–427
external resistances, 427
vapour–liquid and liquid–vapour
resistances, 422–423
Thermal resistances, 419
for air cooling, 151
calculation of, 371
conduction, 421–422
between device case and heat sink,
384
external, 427
formulation, 149–151
models, *see* Thermal resistance
models
order of magnitude of, 420–421
of thermal circuit, 447
of thermosyphon heat exchangers,
451
VDT, 233, 234
Thermal solution
for graphics processors, 156–158
improvement, 161–167
trends for laptops, 154
using piezo fan, 158–161
Thermal spreading resistance,
157, 163
Thermal–technological processes, 323,
324
Thermal testing thermosyphons,
448–450
Thermal vacuum tests (TVTs), 391, 392

Thermochemical resorption machines, 217
Thermocouple, 133, 250, 290
    junctions of, 135
    location of, 136
Thermodynamic process
    of gas adsorption and desorption, 309
    of hydrogen charging, 300
    modelling of, 287
Thermoelectric cells (TECs), 184–186
Thermoelectric cooling, 146
Thermoelectric generator (TEG), 185
Thermophysical properties of paraffin wax, 191
Thermosiphons, *see* Thermosyphons
Thermostating autonomous devices, passive TCSs for, 373–376
Thermostating objects, 366, 374
Thermostating properties of GLHPs, 383
Thermosyphon-assisted conveyor belt oven, 456–457
Thermosyphon-assisted ovens, uniform bread cooking in, 454–456
Thermosyphon-based thermoelectric generation system for solar pond, 183–188
Thermosyphon Rankine engine (TSR), 182
Thermosyphons
    characteristics, 460
    configurations and geometries, 413–415
    constructive material, 336
    cooling, 321–322
    correlative characteristics of, 327
    description, 412
    design
        casing materials selection, 438–440
        methodology, 446–448
        procedure, 434–435
        working fluid selection, 435–437
    diode characteristics, 169
    divided by circulation method, 322
    evaporating, 324
    fabrication
        charging procedures, 443–447
        cleaning process, 442–443
        closing lids, geometry for, 441
        out-gassing, 443
        TIG welding process, 441–442
    film condensation, 343–344
    heaters, 458
    heat exchanger test rig, 449
    heat medium for, *see* Heat medium for thermosyphons
    heat transfer from, 174
    heat-transmitting capacity of, 348–352
    heat-transmitting properties of, 325
    hydrodynamic processes in, 328–332
    industrial applications
        heat exchangers, 450–453
        high heat power cooling, 458–459
        oil tank heaters, 457–458
        ovens, 452–457
    mass filling of, 326
    one-phase, 320, 321
    operation of vacuum, 346
    oscillation phenomena in, 432–433
    physical principles, modelling of
        heat transfer limits, *see* Heat transfer, limits
        Nusselt models for condensation in walls, 415–418
        thermal modelling, *see* Thermal modelling, thermosyphons
    recuperators, use of, 324
    TEC generator module with, 190
    thermal performance testing apparatus, 448
    thermal testing, 448–450
    two-phase, *see* Two-phase thermosyphons
    vs. heat pipes, 412–413
    working principles of, 167–168, 412
Thermosyphon thermoelectric module (TTM), 188
Thermosyphon tubing case, 444
Thermosyphon-type vapour chambers, 413–414
Three-dimensional computational domain, 22, 23
Three-dimensional images of liquid- and vapour- phase distributions, 227, 228
Three-dimensional numerical models, 22

Three-dimensional numerical porous
    structures, 3
Three-dimensional thermal
    management device, 156–157
    use of, 147
Three-temperature model, 16
TIG welding process, *see* Tungsten inert
    gas welding process
Time to reach temperature, equation of,
    372
TM, *see* Transient mode
Toluene, 348, 352, 437
Total heat generation, quantity of, 390
Traditional passive system, 384
Transfer processes, two-phase
    thermosyphons, 338–339
Transient mode (TM), 90
Transition boiling, 110
    experiments in, 123–125
    model of heat transfer, 125
Transition flow, 472, 473
Transparent evaporators, 240
Trapezoidal grooves, 242, 243
Tree configuration thermosyphons,
    414–415, 457
Tri-generation system, 214, 215
TSR engine, *see* Thermosyphon Rankine
    engine
Tsurumi activated charcoal, 32
TTM, *see* Thermosyphon thermoelectric
    module
Tube materials, out-gassing of, 443
Tungsten inert gas (TIG) welding
    process, 441–442
Turbulent film, 486
Turbulent flow regimes, 121
TVTs, *see* Thermal vacuum tests
Two-adsorber heat pump with heat
    pipe thermal control, 215, 216
Two-adsorber heat upgrading
    experiment, 270–271
Two-adsorber system, temperature
    evolution in, 270, 271
Two-bed adsorption cooling cycles, 30
    adsorption isotherms, 36–38
    energy balance, 38–40
    system performance, 41
    working principles, 33–36

Two-channel chemical-adsorption
    system, 271–272
Two-dimensional geometries, 298
Two-dimensional numerical model, 22
Two-dimensional transient model, 293
Two-phase conventional
    thermosyphons, 229
Two-phase gas–liquid flow,
    minichannels and
    microchannels
    adiabatic flow patterns, 470–474
    annular flow in rectangular channel,
        475–478
    diabatic flow patterns, 474–475
Two-phase heat transfer devices, 146
Two-phase loop thermosyphons, 229
Two-phase loop with nanocomposite,
    239
Two-phase mixture model, 19, 21, 22
Two-phase thermosyphons
    circuit designs modifications and
        classification of, 320–322
    conditions of effective work of, 324
    determination and operation
        principle of, 320
    heat-transmitting capacity of, 347
    hydrodynamic conditions of effective
        operating modes of, 328–332
    operation of, 327
    practical tasks solved by, 323–324
    transfer processes in, 338–339
Two-stage adsorption cooling cycle, 41
Two-temperature model, thermal
    energy equations
    for, 17–18

U

Ultra-large heat pipes for cooling earth,
    191–195
Upper convective zone (UCZ), 184
Uptake/release curves, 86–88

V

Vacuum leakage tests, 443
Vacuum pump system, filling rig using,
    445–446

Vaporisation, mass rate of, 17–18
Vapour bubble, growth rate of, 126–128
Vapour chamber, 152–154, 156
  improvement, 165–166
  isothermal surface characteristics
    of, 157
  thermal performance of, 158
Vapour chamber thermosyphon,
    configuration of, 413–414
Vapour condensation process,
    characteristic of, 343–345
Vapour diffusion, heat transfer by, 362
Vapour-dynamic thermosyphons
    (VDTs), 217, 229–233
  heating devices based on, 233–234
  thermal control system for
    adsorbers of solid sorption heat
      pumps, 234
    snow melting, 237–238
    solar solid sorption coolers,
      234–237
Vapour flow
  patterns of, 330
  structure of, 341
Vapour–gas front, 360, 384
Vapour–liquid thermal resistance,
    422–423
Vapour–liquid wick, 8–9
Vapour pipe, 230
Vapour space, saturated vapour
    pressure in, 370
Vapour temperature, 378, 391
Vapour temperature–heat sink, thermal
    line, 364
Vapour thermal resistance, 422–423
Variable conductance heat pipes
    (VCHPs), 399
  balance equations, 369–370
  chronological survey of GLHP
    models, 363
  cold unwicked reservoir, 364
  contact resistance improvement,
    methods of, 367
  fields of application of TCSs on, 405
  flat vapour-gas front theory, 362
  GLHP, 358, 360, 372
  heat balance equations, 371
  heat pipe thermal conductance, 358

  implementation in spacecraft design,
    398
  integration into optoelectronic
    equipment, hierarchy of, 375
  one-dimensional conductivity
    equation, 370
  shell/CS-working fluid-gas,
    combinations of, 368
  shell temperature distribution in
    heat pipe, 361
  TDs, 359
Variable radiation thermal coupling,
    374
Variation of conductance, 358–359
VCHPs, *see* Variable conductance heat
    pipes
VDTs, *see* Vapour-dynamic
    thermosyphons
Vessel behaviour, 312
Vessel filling, 300, 303
Vessel gas charging, 307
Viscous limits, heat transfer, 431
Visualisation of pool boiling
    process, 134
Volume filling degree, calculating
    relations of, 326–328
Volume of fluid (VOF), 4, 226
Volumetric weight method, 287

W

Wallis correlation, 429
Waste heat recovery system,
    180–182, 208
Water
  adsorption chiller, 42, 47
  adsorption isobar, shape of, 87
  evaporation, heat of, 74
  heat pipe, 248
  pair for adsorption cooling
    applications, 30–31, 55
  solution, crystallisation temperatures
    of, 333
Water–steel thermosyphons for
    industrial applications, 443
Weber numbers, 474–475, 484–485
Welding process, TIG, 441–442
Wetted surface, fraction of, 125, 129, 132

Wicked reservoir
    GLHP with, 381
    scheme with, 364–365
Working fluids, 330, 335, 336, 352
    and casing material compatibility
        list, 438
    filling rig by electrical heating,
        444–445
    gravity and, 413
    groups of, 436

and operation temperature, 439
out-gassing of, 443
properties and impurities, 437
selection of, 435–437
in vapour state, 460
water as, 179, 186, 188, 199

Z

Zeolites, type of, 31